GREEN POWER

PERSPECTIVES ON SUSTAINABLE
ELECTRICITY GENERATION

GREEN POWER

PERSPECTIVES ON SUSTAINABLE ELECTRICITY GENERATION

EDITED BY

João Neiva de Figueiredo • Mauro F. Guillén

CRC Press
Taylor & Francis Group
Boca Raton London New York

CRC Press is an imprint of the
Taylor & Francis Group, an **informa** business

A PRODUCTIVITY PRESS BOOK

CRC Press
Taylor & Francis Group
6000 Broken Sound Parkway NW, Suite 300
Boca Raton, FL 33487-2742

© 2014 by Taylor & Francis Group, LLC
CRC Press is an imprint of Taylor & Francis Group, an Informa business

No claim to original U.S. Government works

Printed on acid-free paper
Version Date: 20131227

International Standard Book Number-13: 978-1-4665-9048-9 (Hardback)

Visit the Taylor & Francis Web site at
http://www.taylorandfrancis.com

and the CRC Press Web site at
http://www.crcpress.com

Contents

List of Figures

List of Tables

Preface

Green power, or the generation of electricity in ways that do not pose tradeoffs for future generations, has become a central piece of the debate about humanity's long-term survival. This book describes different countries' perspectives on this important topic given their natural endowments, geography, history, and culture, with the objective of identifying best practices and contributing to their dissemination. It is the result of a joint effort by faculty, researchers, and both graduate and undergraduate students affiliated with the Joseph H. Lauder Institute of Management & International Studies and the Wharton School at the University of Pennsylvania. We are indebted to Selma Pastor, who provided invaluable copy-editing assistance; to Sean Michalson, whose efficiency and initiative helped smooth the laborious process of preparation for publishing; and to Kimberly Norton, who assisted the research team throughout the various stages of the project.

João Neiva de Figueiredo
Mauro F. Guillén
Philadelphia, May 2013

Editors

João Neiva de Figueiredo is associate professor in the Department of Management at Saint Joseph's University's Haub School of Business and Senior Fellow at the University of Pennsylvania's Wharton School. With over 20 years experience in international business, Dr. Neiva was a consultant at McKinsey & Company, a vice president at Goldman Sachs, and a partner at JPMorgan Partners. His research and teaching focus on the role and effects of sustainability practices in the areas of strategy and international management. Dr. Neiva holds electrical and systems engineering degrees from Rio de Janeiro's PUC, an MBA with high distinction (*Baker Scholar*) from the Harvard Business School, and a PhD in business economics from Harvard University.

Mauro F. Guillén is the director of the University of Pennsylvania's Joseph H. Lauder Institute of Management and International Studies, and Dr. Felix Zandman Professor of International Management at the Wharton School. He has written extensively about multinational firms, economic development, and the diffusion of innovations. He received his PhD in sociology from Yale and a doctorate in political economy from the University of Oviedo in his native Spain.

Contributors

José Normando Bezerra, Jr. graduated from the University of Pennsylvania's Lauder Institute joint MBA/MA program in 2012 and is with a top global business consulting firm in São Paulo.

Hannah Tucker González graduated from the University of Pennsylvania's Lauder Institute joint MBA/MA program in 2012 and now works at PIMCO Europe Ltd. in London.

Akhil Jariwala is a member of the University of North Carolina undergraduate class of 2014.

Saumil Jariwala is a member of the University of Pennsylvania undergraduate class of 2014.

Diana Townsend-Butterworth Mears graduated from the University of Pennsylvania's Lauder Institute joint MBA/MA program in 2011 and is with a top global business consulting firm in Washington, DC.

Sean Michalson is a member of the University of Pennsylvania undergraduate class of 2014.

José Luis González Pastor graduated from the University of Pennsylvania's Lauder Institute joint MBA/MA program in 2012 and is with Neuberger Berman in London.

Luca Ratto graduated from the University of Pennsylvania's Lauder Institute joint MBA/MA program in 2011 and is with Houlihan Lokey in London.

Andre Luiz Soresini graduated from the University of Pennsylvania's Lauder Institute joint MBA/MA program in 2011 and is with Bain & Co. in São Paulo.

José Carlos Thomaz, Jr. graduated from the University of Pennsylvania's Lauder Institute joint MBA/MA program in 2012 and is with Booz & Company in São Paulo.

Natalie Volpe is a member of the University of Pennsylvania's undergraduate class of 2013.

Julia Zheng was a member of the University of Pennsylvania undergraduate class of 2013 and is currently with the MIT Media Lab in Cambridge, Massachusetts.

Xiaoting Zheng was a member of the University of Pennsylvania undergraduate class of 2012 and the master's in engineering class of 2012 and is currently with Aberdeen Asset Management in Philadelphia, Pennsylvania.

1

Green Power: Perspectives on Sustainable Electricity Generation

João Neiva de Figueiredo and Mauro F. Guillén

Much has been written in recent decades about the need for sustainable practices if we are to offer future generations the same magnitude of benefit from natural resource utilization enjoyed by past generations. In 1987, the United Nations (UN) Bruntland Commission defined sustainable development as "meeting the needs of the present without compromising the ability of future generations to meet their own needs" (United Nations General Assembly 1987). Natural resources must be consumed by mankind in such a way that water, food, heat, clean air, energy, and other basic goods are provided to all, with no human being, alive or as yet unborn, being deprived. Clearly, we have not yet achieved this goal. Although there is a wide diversity of points of view and many opinions regarding the various components of this debate, including the means to achieve that overarching goal, there is general agreement on several basic facts. First, the population of the world continues to grow. Second, people in developing countries wish to increase their standard of living (if possible, to levels now enjoyed in the developed world). Third, natural resources cannot sustain an unlimited "carrying" population load. Finally, technology has helped to increase nature's population/load-carrying capacity.

Experts and scholars believe that our actions in the next several decades will help determine whether or not the planet will be able to sustain our species in the very long term (Rees 2012). In their opinion, current generations may, in their lifetimes, witness an inflection point in the history of mankind: either we collectively learn to live within the long-term load-carrying capacity of the planet or we leave unsolvable problems for our children's children's children. If we accept Edmund Burke's words that "Society is a contract… a partnership not only between those who are living, but between those who are living, those who are dead, and those who are to be born," it follows that each of us owes it to the future generations to leave environmental conditions at least as fertile as those we encountered at birth (cited in Stanlis 2003). Although technology has undoubtedly moved us in this direction by providing ever more efficient and advanced means to address our own needs, the reality is that many fundamental components of our lives continue to drive resource depletion, waste accumulation, and pollution-related degradation. Our fundamentally unsustainable lifestyles, therefore, pose a major difficulty, even in the face of increasing corporate, political, and consumer awareness of this issue. The solution is to adjust our day-to-day decision-making to incorporate fairness, even when the hidden costs of goods and services are transferred anonymously to other parties, including the environment. Each of these day-to-day decisions will depend on how we value our own and others' well-being, both now and in the future.

Can we collectively learn how to adjust incentive systems to lead us to value others' current and future well-being appropriately? This is a fundamental question but, as conceptual questions tend to be, one that does not offer easy answers. There are so many possible ways to address the issues at hand that we may find ourselves drowning in rhetoric. This difficulty in reaching consensus on actionable measures seems to surface whenever countries meet to decide how to share responsibility for reducing environmental effects. The recent United Nations Conference for Sustainable Development of June 2012, Rio+20, stands as a prominent example of this. Many officials labeled the conference "a failure of leadership" despite UN High Commissioner for Human Rights Mary Robinson's assertion that Rio+20 *needed* to be a "'once in a generation' moment when the world needs vision, commitment and, above all, leadership" (quoted in Black 2012). This comes as no surprise, given that most countries come to the table at such conferences with zero-sum game mindsets. As a first step for moving beyond declarations of intentions toward practical commitments, it would be useful for us all to realize that we are dealing with a non–zero-sum game on a global scale.

This book pragmatically approaches the issue of sustainable development by examining specific cases in which countries have attempted or are attempting to address electricity-generation sustainability issues. It emphasizes that countries must—and indeed do—tailor their approaches to their natural endowments, their cultural contexts, their historical legacies, and their institutional environments. The cases herein focus on a subset of the energy domain; namely, the realm of electricity generation, transmission, distribution, and storage.

Within this domain, we observe that although, historically, increases in population and economic output have usually led to higher per capita electrical energy consumption, select countries, such as Germany and Denmark, have been able to buck this trend. When faced with the right set of incentives, economic agents have, in fact, been able to reduce their energy usage, even as real output increases. Spreading this *decoupling of energy consumption from economic growth* will prove essential if we are to provide our children with an increasingly sustainable way of life.

Alone, however, such a decoupling will not make electricity sustainable. The other side of the equation is the need for the clean generation, transmission, and distribution of electric power, that is, the search for *sustainable electricity production*. Using the Bruntland Commission's definition, this means producing electricity in a way that does not compromise the ability of future generations to produce electricity or to fulfill their other needs. In particular, we feel this entails producing electricity in ways that do not unduly and anonymously transfer costs to others through pollution, depletion, relocation, or any other means.

The main premise of this book is that valuable lessons can be learned from other countries' experiences in harnessing the potential of green power, whether these experiences were positive or negative. The term *green power* refers herein to the subset of those renewable energy resources and technologies that provide the highest environmental benefit. The Environmental Protection Agency (EPA) defines green power as "electricity produced from solar, wind, geothermal, biogas, biomass, and low-impact small hydroelectric sources" (U.S. EPA 2013). "Renewable" energy sources, on the other hand, according to the EPA, restore themselves over short periods of time and do not decrease in size. For example, according to these definitions, large-scale hydroelectric power projects would be considered "renewable," but not "green" due to their potentially significant environmental effects resulting from land-use changes and ecosystem transformations. Although the words "green" and "renewable" may have slightly different technical definitions in different countries, this book uses the EPA definitions of these two terms.

As has been documented extensively in the literature, different countries have diverse power resource generation endowments. Therefore, each country's optimal mix of electricity generation sources is unique. Despite these unique characteristics, valuable lessons can be learned from countries that have harnessed or are attempting to harness renewable energy sources for electricity generation. Thus, the objective of this book is to describe selected experiences from other countries related to sustainable electricity. Through this, we hope to convey lessons drawn from other countries' incentive systems that will help deepen understanding of this challenging field. The ultimate goal of this work is to provoke new ideas about how to establish and expand sustainable electricity systems in different environments.

Chapter 2 describes the renewable and nonrenewable electric power sources that are expected to play a role in the twenty-first century electricity matrix. It provides a summary description of each electricity generation technology, an overview of geopolitical considerations, and observations on the potential of each source. Examples of countries that have developed each respective energy source—be it because of endowment, institutional, or technological factors—are also included. In addition, the chapter provides information on the worldwide growth in the share of natural gas and green sources of electricity and on the decline of oil's share over the past 40 years. For nonrenewable sources, it addresses technologies that can mitigate negative environmental effects, such as clean coal and carbon capture and sequestration. This overview has two objectives: first, to set the stage for the remainder of the book by providing contextual information and, second, to illustrate the differences among electricity generation technologies, thereby reinforcing the fact that each country's optimal electricity generation profile is highly dependent on its natural endowment.

Chapter 3 examines the cultural foundations and institutional environment that gave birth to Germany's energy revolution. The country has expanded its share of renewable energy generation from close to 1% in 1990 to 10% in 2010, with an expected expansion to 25% by 2025, a trajectory that has allowed for a significant decrease in its dependence on oil and coal. Even more impressive, considering that the country is the fourth-largest economy in the world, Germany has succeeded in increasing its gross domestic product/energy consumption ratio significantly over the past 20 years and is now at the forefront of the developed world on this metric. The chapter describes the history of this transformation, with emphases on the roles of the country's institutional environment, technological expertise, and cultural fabric. An examination of this historical narrative supports the view that a convergence of broad factors must precede the democratic adoption of a feed-in tariff incentive system. Germany's experience demonstrates, however, that once established, such an incentive system can lead to a virtuous cycle of domestic green power sourcing.

Chapter 4 focuses on China's current energy profile and the country's reliance on the "dirtiest" energy source: coal. The country faces daunting pollution issues, largely because its economy has grown at an annual rate of nearly 10% over the past 35 years and has not yet succeeded at decoupling output increases from energy production, with coal now supplying more than 70% of China's electricity. The chapter begins with an overview of China's energy consumption, including forecasts, and then proceeds to describe the country's electricity grid and the government's energy policy. The chapter then offers an in-depth analysis of the Chinese coal-based electricity sector, touching on coal consumption, external costs, and a detailed description of select clean coal technologies being developed to address the country's significant pollution problems.

Chapter 5 picks up where Chapter 4 leaves off and describes China's immense efforts to find electricity generation alternatives cleaner than coal that can be implemented at the necessary scale. The chapter begins by addressing natural gas, describing demand trends and supply aspects, including the opportunities posed by liquefied natural gas and unconventional gas sources. Next, it profiles the country's hydroelectric power generation installed capacity, which has increased significantly over the past 30 years. At the end of 2010, China was the nation with the highest installed wind generation capacity, with more than one-fifth of the world's total. The country plans to continue the sector's fast-paced growth, which will entail the creation of large-scale wind farms in the northern deserts. These efforts, as well as the ongoing investments in solar energy generation, are chronicled. The chapter concludes with a description of China's nuclear sector before discussing the country's role going forward as a leader in the worldwide green power industry, a position contradictory to China's addiction to coal.

Chapter 6 focuses on the renewable energy sector in Spain. During the decades before the recent economic crises in the United States and in the European Union, rapid economic growth and dwindling fossil fuel reserves led Spain to invest heavily in renewable energy. Although the country still relies on foreign suppliers for three-quarters of all the energy it consumes, Spain is at the forefront of wind and solar energy technologies. This is due, in part, to a successful feed-in tariff system, which has provided financial returns sufficient to attract private investments in those technologies. The chapter provides an overview of the renewable energy sector in Spain and offers an analysis of its future prospects in light of recent economic difficulties. It also illustrates the desirability of a close match between a country's institutional regulatory frameworks and its natural energy–producing endowment profile.

Chapter 7 tells the cautionary tale of French Polynesia, a country with ample renewable energy resources (solar, wind, marine) that remained, until recently, untapped. Paradoxically, although this country is geographically isolated and very distant from international oil supplies, it is heavily dependent on oil and is therefore subject to the commodity's significant price fluctuations. For example, in 2008, it needed imported oil to satisfy 89% of its energy requirements. That year, the country set a goal of generating 50% of its electricity from renewable sources by 2020. In addition, it set an even more ambitious goal to achieve complete renewable sourced electricity generation by 2030. This chapter analyzes how French Polynesia's transition to renewables has progressed since 2008, cautions that the targets set may not be realistic, and examines the future of renewable energy in this remote island nation. It demonstrates the importance of incentive alignment and solidarity among all the stakeholders.

Chapter 8 examines the feasibility of biomass as a large-scale electricity generation option. It begins with a description of biomass as a renewable energy source and examines its usage in different parts of the world. Through an uncomplicated exercise, the chapter estimates the feasibility of biomass and then explores its promise in Brazil, a country with vast biopower potential. The chapter's main conclusion is that the feasibility of biopower depends heavily on proximity to biomass sources (because of the high logistics and transportation costs/generated electricity revenues ratio) and also on whether the biopower can be coproduced within another process. This chapter emphasizes the importance of reflecting on the need to balance energy needs with other needs, such as food, land usage, etc. It concludes by pointing out that local conditions will likely determine the feasibility of this promising renewable energy source.

Chapter 9 explores the business case for power generation in space for terrestrial applications. Conceptually very promising, this system is based on massive satellite structures orbiting the earth. Their orbital locations would allow such satellites to convert solar energy perpetually into electricity, which would then be transmitted to reception stations on the surface of the planet. The proposal seems far-fetched at first glance because of inherent technological difficulties and the need to obtain broad international approval for such a large-scale project. However, the concept has the significant advantage of being the only source of energy that is completely exogenous to the biosphere. This means that it would neither deplete any of the planet's resources nor result in significant land usage. This chapter outlines the current state of this technology, describes the different options expected to be available in the short term, and introduces a business model to identify the financial conditions necessary for its successful deployment. Because of the significant capital requirements and the need for cooperation among different countries, one prerequisite for adopting this alternative is that political and corporate entities find common ground. Nevertheless, it is a visionary concept that, if conceived and implemented correctly, could solve mankind's energy needs on a permanent basis.

Chapter 10 extends beyond the topic of electricity generation to offer a perspective on electricity-powered transportation. In particular, this chapter details the roles of electric vehicles as alternatives to fossil fuel–based transportation and also as storage devices for electricity, examining the four main issues that must be resolved if electric vehicles are to be adopted on a large scale: battery technology, grid-support capability, battery-charging infrastructure, and the electric vehicle business model. The second issue, grid-support capability, is related intrinsically to the topic of this book, not only because of the projected increase in worldwide demand for electricity but also because in the presence of smart grids, electric car owners will be able to sell energy stored in their cars' batteries if they so choose. In short, the evolution of electric vehicle technology is linked closely to renewable electric power generation, storage, and distribution and therefore needs to be well understood.

This book presents real-life cases in each chapter to help find practical solutions to the difficult day-to-day issues related to renewable electric power generation, transmission, and distribution. Some countries, as described herein, are focusing intensely on the sustainability of their electric power systems. Lessons learned from their experiences can be very useful as other nations make their own energy-related policy decisions.

REFERENCES

Black, R. BBC News: Science & Environment. *Rio Summit: Little Progress, 20 Years On.* Last modified June 22, 2012. Accessed May 17, 2013. http://www.bbc.co.uk/news/science-environment-18546583.

Rees, M. From Here to Infinity: A Vision for the Future of Science. New York: W.W. Norton, 2012.

Stanlis, P.J. *Edmund Burke and the Natural Law* (originally printed in 1958 by The University of Michigan Press). New Brunswick, NJ: Transactions Publishers, 2003.

United Nations General Assembly. Report of the World Commission on Environment and Development: Our Common Future, by the World Commission on Environment and Development (WCED). New York: United Nations General Assembly, 1987.

U.S. Environmental Protection Agency (U.S. EPA). *Green Power Defined*. Last modified March 25, 2013. Accessed April 18, 2013. http://www.epa.gov/greenpower/gpmarket/index.htm.

2

An Overview of Electricity Generation Sources

Akhil Jariwala and Saumil Jariwala

CONTENTS

2.1 INTRODUCTION

I shall make electricity so cheap that only the rich can afford to burn candles.

—Thomas Edison

When Thomas Edison made pioneering advancements in electric power in the late 1800s, electricity was still a luxury good. Since then, it has become an integral part of industry, communication, transportation, food, and recreation. The evolution of national electric grids has provided the infra-structural backbone for national economies by connecting places where power is cheap to produce to places where power is needed. Continued innovation in the twenty-first century grid is essential for providing countries with new opportunities for growth and improved productivity. However, the dawn of this century has posed a number of grid-related challenges. In 2009, 80% of the world's population had access to electricity, the highest level ever (International Energy Agency [IEA] 2011). In addition, economic development in industry and commerce continues to provide and create more energy-intensive products and services, increasing electricity consumption per person. As a result, demand for electricity increased steadily in the past decade and will continue to increase. Finding a dependable supply to keep up with increasing demand is one of the most daunting energy challenges every country faces.

Global sustainability is the second hurdle that must be factored into an energy solution. An array of scientific research over the past half century has shown that CO_2 emitted by fossil fuel consumption is very likely contributing to global climate change (Intergovernmental Panel on Climate Change [IPCC] 2007). Rising sea levels, warmer temperatures, and changing precipitation patterns threaten to affect millions of people in this century. Currently, carbon-intensive coal dominates the electricity market and is also the single largest source (28%) of anthropogenic CO_2 emissions (IEA Coal Industry Advisory Board [CIAB] 2010). Transitioning from carbon-intensive energy sources, therefore, presents an opportunity to build a foundation for long-term, sustainable growth.

This chapter provides a review of the renewable and nonrenewable sources of energy that will contribute to electricity generation on the twenty-first century grid. Each section describes an energy source, the predominant methods of acquiring the energy, where the energy resource is most abundant, and the advantages and disadvantages of using the source in the energy supply. Broadly stated, renewable sources are those whose current production does not threaten the energy supply for future generations. Nonrenewable sources are limited and exhaustible in the absence of controlled exploitation. The renewable energies described in this chapter include wind, solar, hydroelectric, hydrokinetic, geothermal, and biomass; the nonrenewable sources are coal, natural gas, and nuclear power. The energy sources discussed here relate to electricity-generating technologies. This facilitates a specific focus on the issues that will be present in the electric grid.

Although oil is used today mostly for transportation and is quickly becoming a smaller and smaller segment of total electricity generation, it is included in this chapter. Figure 2.1 shows that the amount of oil used to generate electricity as of 2009 is used mostly to meet peak power demand

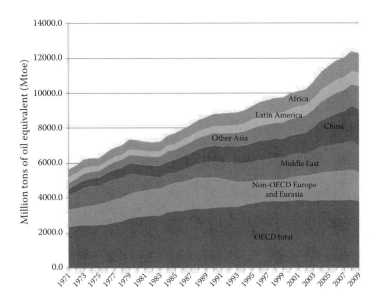

FIGURE 2.1
History of regional global electricity productions, 1971–2009. (Adapted from OECD. OECD Factbook 2011: Economic, Environmental and Social Statistics. http://www.oecd-ilibrary.org/sites/factbook-2011-en/06/01/04/06-01-04-g1.html.)

and accounts for just under 5% of the global power mix. Figure 2.2 and Table 2.1 illustrate that electricity generation from oil has remained stagnant over the past 40 years, despite significant increases in absolute total electricity generation (OECD 2011).

This chapter also concentrates on technologies that are implementable on a utility scale. Smaller-generation machines for individual businesses

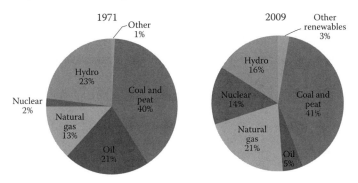

FIGURE 2.2
Comparing the global electricity mixes of 1971 and 2009. (Adapted from OECD Factbook 2011: Economic, Environmental and Social Statistics. http://www.oecd-ilibrary.org/sites/factbook-2011-en/06/01/04/06-01-04-g1.html.)

TABLE 2.1

Total World Electricity Generation by Energy Source in 1971 and 2009

1971		2009	
Power Source	**Electricity Generation (TWh)**	**Power Source**	**Electricity Generation (TWh)**
Coal and Peat	2102	Coal and Peat	8119
Oil	1097	Oil	1027
Natural Gas	695	Natural Gas	4230
Nuclear	111	Nuclear	2697
Hydro	1203	Hydro	3252
Other/renewables	36	Other/renewables	660

Source: Adapted from OECD Factbook 2011: Economic, Environmental and Social Statistics. http://www.oecd-ilibrary.org/sites/factbook-2011-en/06/01/04/06-01-04-g1.html.

and residences such as micro-hydro and home solar have become increasingly important in the twenty-first century, but their reach has been limited when compared with larger utility scale projects. Commercial electricity generation is an essential consideration in guaranteeing that the energy supply remains efficient, secure, and reliable.

At the end of each section, we highlight a country that has espoused a particular source to meet its energy demand. Finding the appropriate approach to handling energy needs is a challenge that individual countries must tackle within the context of their own energy resources. Countries that successfully surmount these hurdles in efficiently tapping their "energy endowment" will be those that carefully craft and follow a comprehensive energy plan.

Solving energy needs will not be easy, but national efforts to revamp the energy supply with new technologies, regulations, and investments will be imperative to address challenges in creating an electricity supply appropriate for this century. Developing an efficient, equitable, comprehensive, and modern energy economy is a must if we wish to improve the lives of the seven billion people on our planet.

Figures 2.1 and 2.2 delineate the changes in the size and composition of global electricity production from 1971 to 2009. Figure 2.1 shows rapid growth in the global electricity supply every year until the global financial crisis. Figure 2.2 demonstrates that this global growth was met by a change in how the nations of the world sourced their electricity production.

Four predominant trends have guided the global energy situation over the past 40 years and have allowed the world's electricity suppliers to keep up with the accelerating demand for energy among today's developed countries.

1. *Growth in nuclear power.* The majority of nuclear power capacity today is in the United States, France, Japan, Russia, and South Korea. Most of its growth occurred between the 1960s and the 1980s, when nuclear power supplied 16% of the global electricity mix. It has held fairly steady ever since (International Atomic Energy Administration [IAEA] 2008).
2. *Growth of natural gas.* Conventional gas production increased quickly in the 1990s. Total gas production increased 60% from 1990 to 2010. In the past 5 years, much of the expansion in North America has stemmed from shale gas. Today, natural gas has risen to more than one fifth of the global electricity supply (Tverberg 2012).
3. *Growth of renewables other than hydro.* The growth of renewables other than hydroelectricity has skyrocketed in the past 5 years as European countries have led the way in installing solar power, whereas China and the United States have led the way in installing wind power. Growth in these resources has mostly been driven by advancing technology and supportive state policies, allowing renewables to capture 3% of the market in 2009 (OECD 2011).
4. *Decline in oil.* A glut in global oil production kept oil prices between US$10/bbl and US$20/bbl for most of the twentieth century. Producing power is one of the least efficient uses of oil, however, and as crude oil prices increased during the oil crises of the 1970s, utilities around the world rapidly switched to natural gas and nuclear power sources. Transferring crude oil resources from power generation to liquid fuel uses pushed oil's share of the global generation mix down from 21% to 5% in 2009 (OECD 2011).

The outlook is very different looking ahead to the next 40 years. Demand for new energy resources will come from a different set of nations: the developing world. This demand for power resources will be structured in the context of global climate change, declining supplies of fossil fuels, and rapid advances in new technologies for renewable energy. The energy mix in 2050 will be informed by three major trends.

1. *Acceleration of unconventional natural gas.* Advances in horizontal drilling and hydraulic fracturing have suddenly opened up the harvesting of tight gas resources, or shale gas. In fact, unconventional

gas resources such as shale gas should account for nearly half the increase in global gas production by 2035 (OECD 2011).

2. *Growth in renewables.* Renewables, including hydroelectricity, will rival coal by 2035 as the primary sources of electricity generation. Investments in wind and solar technologies have risen every year for a decade, and that growth is projected to accelerate over the next 40 years. In addition, biomass for power generation will increase four-fold over the next 25 years. The growth in renewables will be driven by decreasing technology costs and state subsidies (OECD 2011).

3. *Decline in nuclear power.* In the wake of the catastrophic incidents that have eclipsed the nuclear industry over the past 40 years, policymakers and energy producers will be much more likely to turn to power sources other than nuclear power during the next 40 years. Three Mile Island, Chernobyl, and, most recently, Fukushima Daiichi have all substantially informed national nuclear energy politics in every developed country. Germany led the phase-out process after the Japanese tsunami in 2011 by committing to phase out all nuclear power plants by 2022 (Evans 2011). Other countries such as Japan and France have already committed to reducing their use of nuclear power production and will likely follow Germany's footsteps as nuclear power plant licenses come up for review in the years ahead (IEA 2012).

So, ultimately, what might the energy mix look like in 2050? One plausible scenario is that hydroelectricity, other renewables, natural gas, and coal will split the electricity supply almost evenly, with renewables accounting for the majority of the growth beginning in 2010. However, this future depends on the technologies, governments, and natural and anthropogenic events that occur in the next 40 years.

2.2 COAL

Coal is the world's most abundant nonrenewable energy source (Brus and Golob 1993, p. 4). The estimated level of total recoverable coal reserves is just over 900 billion metric tons worldwide, enough to meet current demand for almost 200 years (Massachusetts Institute of Technology [MIT] 2007, p. 125). Figure 2.3 shows that the majority of global coal reserves are concentrated in the United States, which controls approximately 27.6% of the total tonnage. Figure 2.4 illustrates that China leads the world in electricity production from coal as of 2011, accounting for approximately 49.5% of global production (BP 2012). In addition to being burned for electricity generation, coal has historically been used for heating purposes. Even in modern times, industry relies on coal to some degree to generate the heat needed for various industrial applications (U.S. Energy Information Administration [U.S. EIA] 2006a). Table 2.2 summarizes some of this resource's advantages and disadvantages as an energy source.

Coal is also the lowest-cost fossil source for base-load electricity generation. In many parts of the world, coal use costs US$1 to US$2 per million

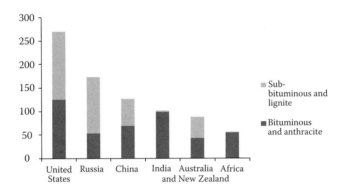

FIGURE 2.3
Proven reserves of coal by type around the world in billion short tons. (Adapted from U.S. EIA. *International Energy Outlook 2006* by the Office of Integrated Analysis and Forecasting. Washington, D.C.: U.S. EIA, 2006b.)

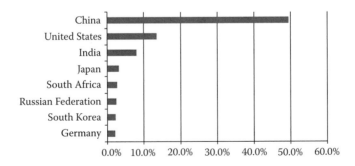

FIGURE 2.4
Proportion of total world energy consumption of coal by country in 2011. (Adapted from BP, P.L.C. *Statistical Review of World Energy June 2012*. Uckfield, United Kingdom: Pureprint Group, 2012. http://www.bp.com/sectionbodycopy.do?categoryId=7500&contentId=706848.)

TABLE 2.2

Advantages and Disadvantages of Coal as a Power Source

Advantages	Disadvantages
Currently cheaper than other fossil fuels per unit of energy	Older coal plants must be retrofitted for "clean coal" technologies
Most abundant nonrenewable energy source	"Clean coal" technologies are generally unproven
Widely distributed across the world	Arguably the dirtiest fuel in terms of emissions (i.e., CO_2, SO_2, NO_x, mercury, particulates) and in terms of CO_2 per unit of energy produced. This is both bad for the environment and makes it susceptible to government policies to reduce emissions such as a carbon tax
Can be synthetically converted into liquid fuels (i.e., for transport) by technology like the Fischer–Tropsch method, thereby mitigating some of the need for fuels to satisfy mobile energy needs (i.e., motor vehicle transportation)	Exhaustible, but there is enough to satisfy current global demand for approximately 200 years
Many coal plants exist and are operating today	Significant environmental challenges in mining
	Capital cost of a coal plant is greater than that of other plants (e.g., twice the cost of a natural gas plant)

Source: Adapted from MIT. *The Future of Coal: Options in a Carbon-Constrained World.* Boston: Massachusetts Institute of Technology. 2007; Brus, E. and R. Golob, *The Almanac of Renewable Energy.* New York: Henry Holt, 1993.

Btu, compared with US\$6 to US\$12 per million Btu for natural gas and oil. Along with its wide availability all over the world, this has made coal the energy source of choice for electricity generation in many places, particularly in developing countries (MIT 2007, pp. 17–21). Because of technologies such as the Fischer–Tropsch process, coal can also be transformed into synthetic liquids for conversion into petroleum products, including diesel, naphtha, and gasoline. Thus, extensive coal use can address both a country's stationary electricity needs and its transportation fuel needs.

Coal is created when dead plant matter transforms itself—first into peat, then into lignite coal, and eventually into anthracite coal. Although there are many different varieties and qualities of coals, they can be classified into four general types: anthracite, bituminous, sub-bituminous, and lignite (MIT 2007, pp. 21–22). The energy qualities of each type can vary widely. Anthracite has the highest energy density, with a heating value (HHV) upwards of 30,000 kJ/kg—about two-thirds the energy density of oil (Brus and Golob 1993, p. 5). Lignite has an HHV of approximately 14,000 kJ/kg—about a quarter of oil's energy density. Coal's sulfur content—responsible for the significant SO_2 that coal power produces—varies widely due to various geologic conditions; differences in sulfur and ash content do not pertain to the type of coal, but rather to the location from which it was harvested (MIT 2007, p. 126).

Compared with other energy sources, coal is the dirtiest, in terms of both general pollutant creation and CO_2-specific emissions. Burning coal produces a number of pollutants, including CO_2, SO_2, NO_x (nitric oxides such as NO and NO_2), mercury, and particulate matter. These pollutants contribute to acid rain, smog, mercury poisoning, and many other health issues. Moreover, coal generates the most CO_2 per unit of energy produced, contributing to an increase in greenhouse gas emissions. Interestingly, this also makes the fuel very susceptible to government policy initiatives such as the carbon tax, which discourages the consumption of coal and encourages the use of natural gas or other alternatives (MIT 2011, p. 66). There are three major methods for dealing with these harmful emissions: the use of coal with lower sulfur and ash content, the use of "scrubbers" such as limestone to precipitate out the pollutants, and precombustion cleaning to reduce the amount of pollutants beforehand.

Moving forward, carbon capture and sequestration (CCS) technologies—methods that aim to trap CO_2 in underground reservoirs such as saline aquifers, depleted oil and gas fields, and deep coal seams—could help reduce the environmental effect of CO_2 emissions from coal, but the technology will require a concerted effort from both policymakers and scientists to be effective. CCS could potentially prevent thousands of gigatons of CO_2 from entering the atmosphere. Although this technology is often discussed in conjunction with coal—probably because the fuel is both very dirty and inexpensive—theoretically, it can be used to mitigate carbon emissions from any source, albeit at significant added cost. The use of CCS has a number of benefits in its favor. For example, sequestering CO_2 after electricity generation could be easier to implement than changing existing energy infrastructure, and the implementation of CCS could utilize preexisting structures that are currently unused—such as depleted gas and oil fields—thereby reducing overall cost (MIT 2007, p. 60). At the same time, however, CCS technology does present certain liability issues that could hinder widespread adoption. After injection, the

CO_2 sequestered deep within the earth has the potential to leak out and contaminate groundwater through acidification, pollute the atmosphere, present human health risks, contaminate mineral reserves, and much more. These risks need to be better understood, and a framework must be established to ascertain the liability that operators of sequestration will be exposed to before implementation is feasible (ibid.).

Worldwide, the most commonly used method to generate electricity from coal is pulverized coal (PC) combustion, which is split into subcritical, supercritical, and ultrasupercritical. They require greater steam temperature and pressure at each stage. Another, less-commonly used method for generating electricity is fluid-based combustion on a circulating fluid bed (CFB). This method has lower costs than PC but also has lower thermal efficiency, resulting in less energy produced per unit of coal. The advantages of CFB are that it can be used with many types of coal—including low-quality coal lignite—and that it can capture the major pollutant SO_2 in the bed (MIT 2007, pp. 33–38). Note that water stress issues are pertinent to coal power. Coal plants are often located near rivers or other bodies of water so the water can be utilized to cool the plants and, more importantly, to act as steam to power the turbines. Water access may be just as important for the success of coal plants as inexpensive access to coal (Griffiths-Sattenspiel et al. 2012).

We now examine coal's specific role in the world. As one of the fastest-growing economies, India has been forced to choose between increasing affluence for its population and energy sustainability. The country has turned to coal consistently to fuel its own economic growth, and today it is the world's third-largest producer of coal after China and the United States. At projected rates, India will reach the current level of U.S. consumption by 2020, and it will reach the current level of Chinese consumption by 2030 (MIT 2007, pp. 74–75). However, economic progress has come at the cost of serious environmental harm. Consumption of coal at these levels could result in significant air and water pollution, which are detrimental to the population. Furthermore, studies of Indian power plants reveal that, although renovations to improve load capacity are commonplace, plants have failed to increase their operating efficiency significantly. A nationwide increase in plant efficiency could result in fewer emissions and decrease the country's dependence on foreign coal, while making coal production less expensive per kilowatt-hour in the long run. The use of only primitive "subcritical" coal electricity generation technology in India has similar economic and environmental ramifications (ibid., pp. 169–170).

2.3 OIL

Oil's valuable characteristics of being easily transportable and durable in storage have made it too valuable in recent times to be used for stationary electricity generation except under very special circumstances. However, during other periods, such as in the early 1970s, an oil surplus allowed oil to comprise 21% of the total load. In the United States, for example, low oil prices in the 1960s converged with concerns about coal smog to stimulate the construction of additional petroleum power plants that pushed petroleum to replace coal-fired power plants and capture nearly 20% of the domestic electricity market (U.S. EIA 2012). Today, oil is used in developed countries primarily as a valuable peak-hour energy source. Regionally, however, electricity generation from petroleum is more common in areas of the world with large supplies of this resource. In fact, the Middle East today uses oil exuberantly for electricity generation. Oil is plentiful, and electricity shortages occur in the summer due to the heavy use of air conditioning (Wigglesworth 2011). Although the United States uses petroleum products for less than 1% of its electricity load, the Middle East depends on petroleum for 35% of its generating load (Styles 2011). Because oil is used for both fuel consumption and electricity generation, it is difficult to isolate the exact amount of oil used for the latter purpose and

TABLE 2.3

Advantages and Disadvantages of Oil as a Power Source

Advantages	Disadvantages
Highly transportable, particularly for isolated economies like Hawaii	US$ per kWh costs of oil generation are significantly higher than other generation source costs
Turbine technology allows oil to be operable at smaller generation volumes than other sources	As the price of crude oil increases with heightened demand in the transportation sector, electricity generation uses become even less competitive
Low cost and time lag to build new plants because oil plants are at a size of 1 MW instead of coal power plants that operate at hundreds of megawatts (Ecoleaf 2010)	Several different types of emissions, including particulate matter, sulfur, and nitrous oxides
Easily storable for peak hour uses	Generation creates significant greenhouse gas emissions

Source: Adapted from Ecoleaf. Understanding Petroleum Power Plants. 2010. http://www.ecoleaf.com/green_energy/petroleumgaspowerplants.html.

to identify the largest consumers worldwide. Table 2.3 summarizes some of the advantages and disadvantages of oil as an energy source.

Although oil has a high energy density, its commercial electricity generation uses have been phased out recently in countries in the Organisation for Economic Co-operation and Development (OECD) countries due to its high price point (Downey 2009, p. 7). Over the past 40 years, petroleum use has decreased from 21% of the global electricity mix to just 5% as total global generation has remained stable at an annual amount of 1000 TWh. This is attributable largely to the price instability of oil over the past decades, as shown in Figure 2.5. The percentage will continue to decrease as the price of oil remains above US$100/bbl and countries such as Saudi Arabia resort to solar, natural gas, and nuclear power (Carlisle 2010). Remote, isolated economies such as Hawaii, which currently produces three-quarters of its electricity from imported oil, will likely be the last to switch from using oil as a generating source [State of Hawaii

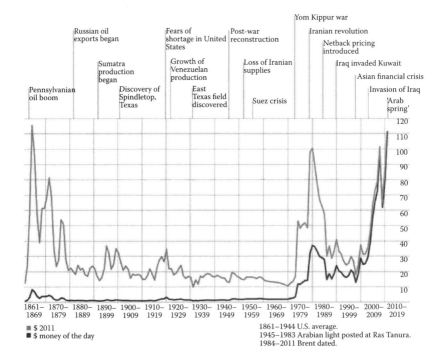

FIGURE 2.5

The effect of geopolitical events on the price of a barrel of crude oil from 1861 to 2011. (Reprinted with permission from BP, P.L.C. *Statistical Review of World Energy June 2012.* Uckfield, U.K.: Pureprint Group, 2012. http://www.bp.com/sectionbodycopy.do?categoryId=7500&contentId=706848.)

Department of Business, Economic Development & Tourism (HDBEDT) 2011]. In these locales, the price of oil at US$0.10/kWh to US$0.15/kWh (assuming costs of US$100/bbl) is competitive because of transportation costs and low economies of scale.

Producing electricity from raw crude has been practiced for nearly a hundred years. Conventionally, light oil is pumped into a boiler, where it is burned, converting chemical energy into thermal energy. Piped coils containing cooled water surround the boiler, so the burned oil heats up the water directly. This heated water is then pumped through a steam turbine, which turns a generator to produce electricity. Technological advances in generation technology have introduced (1) the combustion turbine process, which burns oil under pressure and uses the hot exhaust gases to spin a turbine, and (2) the combined cycle process, which uses the heated petroleum exhaust to spin one turbine and boiled water to spin another (EDF Energy 2013).

The location and concentration of oil reserves have important political ramifications. Approximately 60% of daily global oil production comes from elephant fields, that is, oilfields that produce at least 500 million barrels per day. Arguably, the largest of these is the Ghawar Oilfield, located in Saudi Arabia, which has the world's largest crude oil reserves, estimated at approximately 260 billion barrels. The nations assumed to control the next four largest reserves—Iran, Iraq, Kuwait, and the United Arab Emirates— are also in the Middle East. Because of their similar chemical composition of oil and natural gas, oilfields often include natural gas reservoirs, which can contain pockets of oil (Downey 2009, pp. 31, 92–95, 303).

Important political factors are associated with oil as an energy source. Because it can be used for heating, electricity generation, industrial applications, and transportation, politics are highly relevant to its use. Similar to natural gas, control over oilfields and oil pipelines can be a powerful asset for nations looking to exert influence over other countries. Before the United States hit peak oil production in 1970, the U.S. government strongly influenced global oil prices through a regulatory body known as the Texas Railroad Commission. Since the early 1970s—when most of the production shifted to other countries, including Saudi Arabia, Iraq, and Venezuela—the Organization of Petroleum Exporting Countries (OPEC) has been an important player in determining global oil prices. OPEC collusion has had significant political influence, most saliently during the Yom Kippur War of 1973, when the Arab nations instigated an oil embargo that cut world supply by 5% to 10% overnight. Although OPEC still has significant influence today, some argue that its control over the global oil markets has waned since 2005

(Downey 2009, pp. 8–13, 23–29). Finally, as shown in Figure 2.5, conflict, particularly in the Middle East, as in the cases of the Yom Kippur War and the Arab Spring, has influenced global oil prices significantly (BP 2012).

Although the use of oil for electricity generation has negative environmental effects, it is less harmful than coal combustion. Oil combustion results in a number of pollutants, including ozone (O_3), nitrogen oxides (NO_x), carbon monoxide (CO), sulfur oxides (SO_x), lead, and particulate matter, among others. These pollutants stem from oil's chemical nature. Similar to natural gas, oil is composed of hydrocarbons. However, these hydrocarbons are longer than the single-carbon methane in natural gas; oil, on average, is composed of 8 to 10 carbons. Its hydrocarbon composition depends heavily on the oilfield from which it was extracted (Downey 2009, pp. 51–55, 305).

Because of its glut of oil and natural gas resources, Saudi Arabia has become reliant on using petroleum for electricity generation. Just over 50% of the country's power comes from burning petroleum, and petroleum power plants produced approximately 110 TWh of power in 2010 (Trading Economics 2012). It burns one million barrels of oil per day for power generation purposes during the summer, which costs the country an extraordinary amount of money now that oil prices run more than US$100/bbl. Saudi power demand is growing at an annual rate of 6%, which is so rapid that the country's need is expected to double within the decade. As a result, several analysts project that Saudi Arabia could become a net importer of crude oil by 2013 without a major change in energy policy (Hinckley 2012). To meet incremental power consumption without bankrupting the country, the Electricity and Co-Generation Regulatory Authority has built new natural gas power plants, which now account for nearly 50% of the country's electricity generation capacity. To reduce costs and align the sector with the realities of new oil prices, the government is planning a US$100 billion investment over the next 20 years to use nuclear and solar power to satisfy the growing demand for electricity. The country is developing a 100 MW solar farm near Mecca as part of its renewable energy-generating plan (Taha 2012).

2.4 NATURAL GAS

Natural gas is an energy source that comprises primarily methane— a hydrocarbon whose chemical formula is CH_4—but also has trace proportions of heavier hydrocarbons such as ethane, propane, and butane. The energy from natural gas comes from combusting (burning) the CH_4, which yields CO_2 (carbon dioxide) and H_2O (water). Because methane is such a simple hydrocarbon (unlike crude oil, which consists of hydrocarbons 8 and 10 carbons in length), it burns cleanly and requires very little processing before it can be used as fuel. However, this characteristic also results in a low energy density and explains its gaseous state. These drawbacks have resulted in the development of liquid natural gas (LNG), which has a high energy density (approximately 600 times that of traditional natural gas) and is maintained in a liquid form. The major drawbacks of LNG include a more expensive production process and a highly explosive nature. It is chemically equivalent to natural gas but is stored simply in a liquid state of matter. Before combustion occurs, LNG is converted into natural gas in its gaseous state, a process that requires only a change in pressure, volume, or temperature and no chemical reactions. Natural gas that is stored in a compressed state, but not a fully liquid state, is known as compressed natural gas (CNG). CNG is a rarely used alternative to LNG. Further development of LNG technology will be important to reduce the transportation constraints currently imposed by the natural properties of natural gas. Once these barriers are removed, LNG transportation by boat across oceans and seas will help solve the energy demand problems of many countries (MIT 2011, pp. 12–29). Table 2.4 summarizes some of the advantages and disadvantages of natural gas as an energy source.

TABLE 2.4

Advantages and Disadvantages of Natural Gas as a Power Source

Advantages	Disadvantages
Among the fossil fuels, it has the lowest CO_2 emissions per unit of energy	Political and economic risk for countries that are downstream of the pipeline
Very few noncarbon emissions	Difficult to transport unless converted into LNG or CNG
Highly abundant in many countries around the world	For both traditional natural gas, CNG, and LNG, transportation is a significant component of cost
Inexpensive to recover from conventional sources	Exhaustible, but there is enough to satisfy current global demand for approximately 150 years

Source: Adapted from MIT. *The Future of Natural Gas: An Interdisciplinary MIT Study.* Boston: Massachusetts Institute of Technology, 2011.

Figure 2.6 shows that, as of 2011, most of the global conventional gas reserves are concentrated in the Middle East and Russia. At the same time, Figure 2.7 illustrates that the United States leads the world in natural gas production, accounting for approximately 21% globally. Worldwide, natural gas today is used in industrial applications, for hot water and space heating, for electricity generation, and, to a very small degree, as fuel for vehicles. The estimated level of recoverable gas reserves is approximately 16,200 trillion cubic feet worldwide, enough to meet current demand for almost 150 years (MIT 2011, pp. 11–36).

Although conventional natural gas extraction has provided an abundant supply for decades, technological breakthroughs have now opened up the extraction of unconventional gas resources, particularly those available in shale rock formations. Production levels have increased dramatically over the past decade as horizontal drilling and hydraulic fracturing have become commonplace in North America. The IEA projects that by 2035, the production of unconventional gas will more than triple to 1.6 trillion cubic meters, accounting for 32% of total gas extraction, as opposed to just 14% today. The largest unconventional gas producers and beneficiaries of shale gas extraction will be the United States and China, followed by Australia, India, Canada, and Indonesia. The recent shale gas boom in North America has been named the "Golden Age of Gas" because the

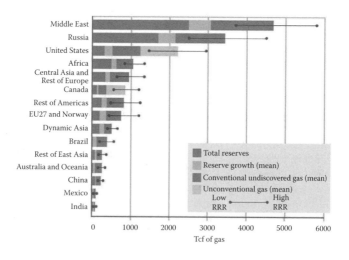

FIGURE 2.6

Total reserves of natural gas by country, with uncertainty. (Reprinted with permission from MIT. *The Future of Natural Gas: An Interdisciplinary MIT Study.* Boston: Massachusetts Institute of Technology, 2011.)

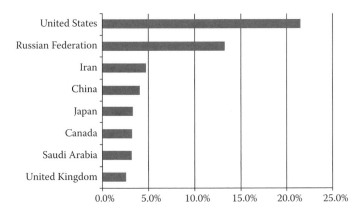

FIGURE 2.7
Proportion of total world energy consumption of natural gas by country in 2011. (Adapted from BP, P.L.C. *Statistical Review of World Energy June 2012*. Uckfield, United Kingdom: Pureprint Group, 2012. http://www.bp.com/sectionbodycopy.do?categoryId=7500&contentId=706848.)

sudden availability of natural gas, perhaps as much as 100 years' worth, has proponents arguing that this resource should replace oil for transportation and coal for electricity production (Nelder 2011). However, the extraction of shale gas has initiated a tense discussion in North America about the environmental consequences of the extraction process, particularly with regard to groundwater quality and water use. Green advocacy groups challenge that, without regulation and oversight, fracking could cause groundwater contamination through leaked toxic fracking chemicals or seeped shale gas (IEA 2010b).

Another drawback of natural gas is that, although it is inexpensive to recover, it is expensive to transport. Resources can be developed economically for US\$1 to US\$2 per million Btu, but it may cost an additional US\$3 to US\$5 per million Btu to transport. These costs are a function of the distance travelled and thus vary substantially. However, the status of natural gas as an affordable fuel option, greater development of LNG and CNG technology, and the fuel's low carbon emissions have resulted in increasing production over time. From 1990 to 2009, the production of natural gas increased 42%, compared with the 22% increase in oil production (MIT 2011, pp. 29–33).

Let us look at the sources of these vast reserves. Natural gas is formed through two different processes: biogenic and thermogenic. In the former, microbes produce methane in an oxygen-free environment. Recovering natural gas for commercial use, however, focuses on thermogenic gas, which is formed by applying enormous heat and pressure over time to organic matter that is buried in porous rock formations thousands of feet below the earth's surface. Thermogenic gas is sometimes formed in conjunction with oil, giving such gas reserves the title of "associated gas"; but most of the gas is formed independently and is known as "nonassociated gas" (MIT 2011, pp. 25–60).

Natural gas is present in different types of reservoirs. Conventional sources include high-quality and low-quality reservoirs, which are well-defined subsurface accumulations with permeability values greater than a specified lower limit. Conventional resources tend to be easier to develop because simple vertical wells can be utilized. These reserves can also be extracted with a high recovery rate (up to 80% or 90%), meaning very little of the gas initially in place is wasted. Unconventional sources for natural gas include "tight" sandstone formations, coal beds, shale sandstone formations, and methane hydrates. These sources are considered unconventional because reservoir permeability is low. Unconventional natural gas sources are defined as reservoirs that are distributed over a larger area than conventional reservoirs. Extraction from these deposits usually requires advanced technology, such as horizontal wells or artificial stimulation. One example of artificial stimulation is hydraulic fracturing, or "fracking," in which a pressured liquid is used to create fractures in porous rock that contains natural gas—in particular, unconventional reservoirs. Although fracking is a very effective tool for increasing the yield of unconventional natural gas sources, it is highly controversial because of its negative environmental effect (MIT 2011, p. 26).

Natural gas is recovered through an expansion process. In the ground, it is compressed due to the high pressure present thousands of feet below the surface. By reducing the pressure, extracting the gas in a controlled fashion, and treating it at the surface, it can be recovered at higher efficiencies when compared with other fuels such as coal or oil. At the end of the recovery process, the methane must be treated. Natural gas is known as "wet gas" before processing; it becomes "dry gas" once all the hydrocarbons other than methane are removed through a steam-based distillation process. The hydrocarbons that are removed are known as natural gas liquids (MIT 2011, p. 25).

Although energy usage is influenced primarily by economic, environmental, and geographic factors—how much it costs, how it affects the environment, and where it is located—there are other relevant influences. Natural gas is highly location-sensitive because it is difficult to store and transport. Thus, control of natural gas pipelines can lead to significant influence over the energy supply and even the heating and industrial capacity of other regions. In recent history, Russia has used its control of natural gas pipelines as a political tool to gain power and influence over the countries of the former Soviet Union. In 2006 alone, the government cut off or threatened to cut off the gas supply to many nations, including Belarus, Georgia, Ukraine, and Moldova (Gelb 2007). More recently, it cut off Georgia's access to natural gas during the 2008 Russo–Georgian War (Martin 2008). However, exploiting energy resources to gain political influence has had serious consequences. Russia's actions, in particular, have damaged its reputation as a reliable energy supplier, encouraged importing countries to seek other sources, and provoked international criticism (Gelb 2007).

2.5 NUCLEAR POWER

Although high-temperature fusion and fission both fall under the umbrella of nuclear power (note that the former is decades from development), today the term refers to nuclear fission, a process in which a free-moving atomic particle (a neutron) collides with the nucleus of an atom, causing it to split apart (Brus and Golob 1993, p. 7). The end products of this process include additional free neutrons, which then split other atoms in a chain reaction. In the fission process, fuel made up of uranium or plutonium is split through the fission reaction, producing immense amounts of energy from a tiny amount of matter. One ton of uranium can produce as much energy as 10,000 tons of oil. Fission generates power by splitting the uranium and plutonium isotopes. Isotopes of an element have varying masses, meaning that their chemical properties are almost identical, but their physical behaviors and how they react with other atomic particles vary widely. The fission process is started by neutrons in the reactor core in a process that both liberates considerable energy and produces more neutrons, which jumpstart new fission reactions through chain reaction (MIT 2003, p. 112). As a result of the vast energy yield of this process, nuclear power generates a significant amount of electricity for many countries. Figure 2.8 lists the four largest producers of nuclear fission power in 2011, namely, the United

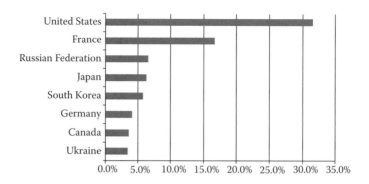

FIGURE 2.8

Proportion of total world energy consumption of nuclear by country in 2011. (Adapted from BP, P.L.C. *Statistical Review of World Energy June 2012*. Uckfield, United Kingdom: Pureprint Group, 2012. http://www.bp.com/sectionbodycopy.do?categoryId=7500&contentId=706848.)

States, France, Russia, and Japan. Table 2.5 summarizes some of the advantages and disadvantages of nuclear power as an energy source.

Two isotopes that have a high probability of undergoing fission are uranium 235 (U-235) and plutonium 239 (Pu-239), which does not occur naturally but is produced through nuclear reactions. Naturally occurring uranium contains only 0.7% of the isotope U-235. The rest is uranium 238 (U-238), which does not undergo nuclear fission. The total recoverable reserves for uranium are either three million or four million tons, based on reasonable cost-recovery assumptions. As shown in Figure 2.9, the four

TABLE 2.5

Advantages and Disadvantages of Nuclear Power as a Power Source

Advantages	Disadvantages
No emissions of any kind	There are important issues regarding safety and waste disposal
Broadly distributed across the world	Limited fuel supply prevents nuclear from being the dominant energy solution for any large nation
	Exhaustible, but there is enough to satisfy current global demand for approximately 80 years

Source: Adapted from MIT. *The Future of Nuclear Power: An Interdisciplinary MIT Study.* Boston: Massachusetts Institute of Technology, 2003; Abbott, D., *Proceedings of the IEEE* 99(10): 1611–1617, 2011.

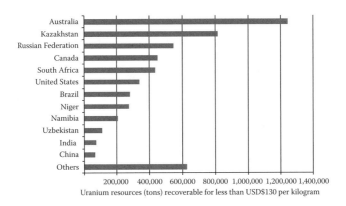

FIGURE 2.9

Global remaining recoverable uranium resources by country in 2007. (Adapted from IAEA. *World Distribution of Uranium Deposits (UDEPO) with Uranium Deposit Classification*. Vienna, Austria: IAEA, 2009. http://www.pub.iaea.org/MTCD/publications/PDF/te_1629_web.pdf.)

countries with the largest recoverable uranium reserves, as of 2007, are Australia, Kazakhstan, Russia, and Canada. However, if the price of uranium increases significantly, then approximately 30 million tons could be recovered because a significant amount of uranium is trapped in low concentrations in terrestrial deposits and seawater. After mining, uranium must be processed into fuel rods, an expensive and complicated process that is not addressed in this chapter (MIT 2003, pp. 44, 113).

Among the many different types of reactors, the two important designs currently are the light-water reactor (LWR) and the high-temperature gas-cooled reactor (HTGR). In a LWR, the moderation is accomplished by rapid collision of the neutrons with hydrogen nuclei (protons) in the water molecule. An HTGR, on the other hand, runs at a very high operating temperature (900°C) and uses helium instead of water as the reactor core coolant and driver for the power conversion cycle and compressors. There are many HTGR subtypes that have their own unique advantages and disadvantages. Although the LWR model is well-vetted and cost-effectively generates large quantities of energy, the HTGR model cost-effectively generates small quantities of energy, has a higher conversion efficiency, and may be safer in the event of a reactor failure (MIT 2003, pp. 59, 115). In lay terms, HTGR reactors allow for nuclear plants in areas with limited energy demand. As a result, these reactors theoretically allow nuclear energy to be used in smaller-scale projects.

Selection of the fuel cycle is just as important as, or even more important than, selecting a reactor type. There are three major fuel cycles. A conventional thermal reactor with "once-through" processing is the simplest of the three. After the reaction, discharged spent fuel is sent directly to disposal with no reprocessing. The majority of the world's reactors nowadays are LWRs that utilize this once-through pass. A conventional thermal reactor with "closed" reprocessing involves a closed fuel cycle and reprocessing of spent fuel to extract more energy. Waste products are separated from unused fissionable material and then recycled as fuel into the reactors. In some countries today, waste plutonium is extracted, fabricated into a mixed plutonium and uranium oxide fuel, and sent back through the reactor for one additional pass (MIT 2003, pp. 111–118).

A new, developing nuclear technology—known as a fast-breeder reactor or fast reactor—could potentially disrupt nuclear power generation. Fast reactors combine aspects of once-through and closed fuel cycles. This third—and least developed—cycle utilizes balanced closed reprocessing. Unused fission fuel is sent through the reactor in a once-through pass. However, instead of being sent out for disposal after this initial pass, the waste products are sent through a fast reactor that extracts any remaining energy. This reactor more readily breeds fissionable isotopes (potential fuel) because it utilizes higher energy neutrons that, in turn, generate more neutrons when absorbed. Thus, a fast-breeder reactor can provide a larger energy output per unit of initially invested fissile fuel. However, this technology is more capital-intensive and, unfortunately, promotes nuclear proliferation. The complexity of this process requires large, secure nuclear energy "parks" to deal with the fast reactors, reprocessing, and fuel fabrication. There is a significant risk of insurgency groups gaining access to weapons-grade nuclear material if fast-breeder reactor technology becomes widespread (MIT 2003, pp. 117–118, 127). Because these reactors utilize closed fuel cycles, implementation could significantly extend the life of global uranium fuel supplies. In contrast, once-through processing could quickly deplete reserves. By some estimates, the use of fast reactors could allow for the extraction of more than 60 times the energy of once-through processing in an LWR (Brus and Golob 1993, p. 8).

Regardless of the reactor type and fuel cycle, after power generation, spent fuel must be disposed of in a secure manner because this fuel is radioactive and heat-producing. Immediately after use, spent fuel rods are stored at individual reactor sites for at least 10 years in large pools of water. They are then stored in large concrete casks that provide air cooling, shielding, and physical protection. Eventually, all the spent fuel will be moved from the reactor sites to underground geologic storage, such as at Yucca Mountain in Nevada (MIT 2003, pp. 115–116).

Unlike other forms of energy, nuclear technological progress has been relatively homogenous across nations; most countries using nuclear technology have kept up-to-date in terms of implementing safer and more efficient forms of nuclear energy. However, the tension between public responsibility and the private operation of nuclear power continues to vary considerably from place to place.

In certain countries, such as the United States, Germany, and Japan, private operators have run nuclear plants under significant government oversight. In contrast, nuclear power has always been state-run in China and Russia. Although the role of nuclear energy in worldwide energy mixes tends to be quite similar from country to country, France stands out because of its considerable dependence on nuclear energy. It is the world's second-largest producer of nuclear power after the United States, with 41.8% of its energy needs satisfied by nuclear energy as of 2011. It has also implemented a unique approach to dealing with the ownership issue. All French nuclear plants are owned by a single state-operated utility: Électricité de France. This implementation strategy has not only allowed for expertise-sharing across plants and the rapid deployment of new power plants but is also affected somewhat by market forces and thus is more efficient than a purely state-operated function. However, researchers believe that state-operated utilities suffer from a lack of competitive pressures and, as a result, misallocate resources to potentially cost-inefficient sources of electricity, particularly when considering the development of new technologies. Consequently, in the European Union, there is a push to deregulate this single French operator and other utilities. Although Électricité de France is today a privately owned corporation under French law, a supermajority of the company's stock shares is still owned by the French government (MIT 2003, pp. 27, 43).

2.6 BIOMASS

Biomass energy involves converting organic feedstock—such as wood or peat—into useful forms of energy, such as heat, electricity, or liquid fuels. The energy contained within biomass ultimately comes from the sun, as biomass is created when plants and algae convert sunlight into organic material through photosynthesis. Exploiting this power releases this captive sunlight energy and converts it into something that is easier for people to use. Biomass energy is arguably the oldest method of energy generation used by humanity (Brus and Golob 1993, p. 42), and it still plays a significant role in satisfying the energy needs of the almost two billion people who still use solid fuels for their daily domestic energy needs (Goldemberg et al. 2000, p. 368). Although it is used largely in the developing world to generate thermal energy, there are a few places where it is burned to actually create electricity (World Energy Council [WEC] 2007). As shown in Figure 2.10, as of 2005, the largest user of biomass for electricity generation is the United States, followed by Germany, Brazil, and Japan. Table 2.6 summarizes some of the advantages and disadvantages of biomass as an energy source.

The simplicity of the biomass–energy creation process means that there are many different kinds of usable feedstock, including peat, fuel wood, domestic farm waste, and aquatic biomass, among others. Peat—the most prolifically used fuel source—consists of partially decomposed plant matter that has accumulated in a water-saturated environment, such as a bog, marsh, or swamp. Before it can be burned, its moisture content must be reduced from as high as 90% down to 35% or less (Brus and Golob 1993, p. 46). For other fuel sources—such as fuel wood—the feedstock must be

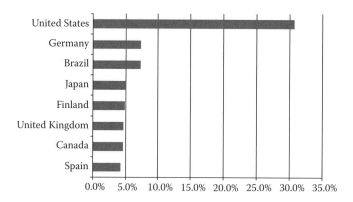

FIGURE 2.10

Proportion of total world energy consumption of biomass power by country in 2005. (Adapted from WEC. *Survey of Energy Resources 2007.* London, U.K.: WEC, 2007. http://www.worldenergy.org/documents/ser2007_final_online_version_1.pdf.)

TABLE 2.6

Advantages and Disadvantages of Biomass as a Power Source

Advantages	Disadvantages
Renewable	Creates significant particulate pollution and a moderate amount of other forms
Widely available across the world	Creates water stress and ecological harm due to the farming practices typically used to create it
Carbon-neutral energy source (combustion produces as much CO_2 as was absorbed during the creation of the biomass), ignoring carbon emissions associated with transportation, processing, drying, fertilizing, etc.	Low energy density means that a lot must be burned to create sufficient amounts of energy

processed into chips with a large surface area, resulting in secondary costs that affect the overall attractiveness of an otherwise inexpensive form of electricity generation (Brus and Golob 1993, pp. 42–60).

The two major methods for energy generation from biomass are direct combustion and biogas production. Combustion is by far the more common and simply involves burning feedstock either for heat directly or to drive electricity generation. It requires the existence of a very cheap feedstock, as raw biomass tends to have a very low energy density when compared with fossil fuels such as coal. Furthermore, this low energy density means that large quantities must be transported for energy generation, making transportation costs a significant component of the power generation cost. To address the energy density issue, biomass is often physically compressed, for example, by creating very dense pellets from raw feedstock (Brus and Golob 1993, pp. 43–44).

Biogas production is the other major method of energy generation for biomass. Biogas—a mixture similar to natural gas that contains mostly methane and carbon dioxide (CO_2)—is produced through the decomposition of organic matter through anaerobic digestion by bacteria. This digestion occurs in a low oxygen environment; consequently, biogas production occurs in containers known as digesters that limit the flow of oxygen to the organic matter. A side benefit of the contained nature of the process is that many parasites and disease-causing organisms are killed by the warm temperatures of the digesters, and leftover material is often used as fertilizer or fish food in developing countries (Brus and Golob 1993, pp. 43–44).

Although biomass combustion is highly versatile, the various fuel sources' chemically complex natures mean that biomass does not burn cleanly. Although combustion produces significantly fewer nitrogen oxides (NO_x) than natural gas and less sulfur dioxide (SO_2) than oil, it also releases a significant number of particulates, even though improvements in stove design have reduced emission levels. In addition, significant biomass generation through biomass plantations results in both large CO_2 emissions and ecological harm due to unsustainable farming practices on the plantations. Furthermore, widespread expansion of these plantations must overcome the scarcity of arable land and limited water supplies. In many countries, this has led to a widespread "plate versus tank" debate, as biomass production (in this case for biofuels, hence, "tank") often competes for land with conventional agricultural food crop production (Hasselbach and Jeppesen 2012).

An unconventional energy source used in recent years has been waste-to-energy conversion, which not only deals with the elimination of the billions of pounds of municipal solid waste (MSW) produced each year, but also generates a useful end product from an otherwise unwanted resource. Disposal of MSW in landfills typically comes at a considerable cost. As a result, some utilities, private companies, and local governments have built plants that generate electricity or industrial steam from the direct combustion of waste or from gaseous or liquid fuels derived from the waste. The U.S. Department of Energy estimates that approximately 60% to 70% of the MSW generated each year consists of organic materials. Consequently, if all the waste in the United States was used for energy combustion, waste-to-energy conversion would satisfy roughly 1% to 2% of the United States' total energy needs (Brus and Golob 1993, pp. 42–60). Currently, the United States has approximately 87 operational MSW-fired power plants, which generate approximately 2500 MW every year. This accounts for approximately 0.3% of the total national power generation (U.S. Environmental Protection Agency [U.S. EPA] 2012).

Although MSW-based electricity generation does conserve resources by reuse, a major drawback of this approach is that it produces highly toxic byproducts. MSW combustion creates nitrogen oxides (NO_x), sulfur dioxide (SO_2), carbon dioxide (CO_2), mercury compounds, and dioxins. Thus, it is one of the dirtiest fuels. Furthermore, careful monitoring is necessary to ensure that these toxins do not leach into groundwater tables (U.S. EPA 2012). Waste-to-energy conversion represents a paradigm shift in understanding energy and its costs. Costs that would normally be attributed to the collection of the fuel (or waste in this case) are sunk costs that cannot be attributed to the energy process. Thus, even the relatively low energy density harvested from this fuel source is still highly attractive (Brus and Golob 1993, pp. 42–60).

2.7 WIND

One of the most prominent renewable energy technologies is wind power, a source of power generation that uses the wind to spin a turbine and ultimately generate electricity. There are two major types of wind turbines: horizontal axis and vertical axis. The former, which is what people generally imagine when they think of windmills, involves three blades attached to an axis that is parallel to the ground. As the wind blows past the wind turbine, a pocket of lower air pressure forms behind the blade, which sucks the blade down. The resulting rotational motion spins a generator and forms AC current. Most horizontal wind turbines are suspended 100 feet or more in the air to take advantage of higher wind speeds at higher elevations (Union of Concerned Scientists 2009b). In contrast, a vertical axis turbine has a blade shaft perpendicular to the ground. Although it tends to be less efficient and less common at high wind speeds, it has the benefit of needing a lower initiation energy to get started and is, therefore, appropriate for use in areas with lower average wind speeds (Brus and Golob 1993, pp. 130–131). Table 2.7 summarizes some of the advantages and disadvantages of wind power as an energy source.

TABLE 2.7

Advantages and Disadvantages of Wind as a Power Source

Advantages	Disadvantages
Low maintenance and labor costs per power output after initial capital expenditure	Highly dependent on location—often transmission cables must connect supply with demand centers
No emissions of any kind	Variable, unpredictable power output—must couple with storage in significant wind economies
Very few safety risks	Noise pollution may be a nuisance
Distributed energy generation source that is easily scalable	Displeasing visual impact on landscape for some

Wind turbines must be installed in areas with a good wind resource level, which is determined by looking at historical wind speed data, topographical maps, and in-field survey results. These methods examine wind speed and the wind capacity factor. Wind speed determines the theoretical maximum of potential power generation (Boccard 2009). Most turbines are installed in places with average wind speeds of 7 to 10 m/s. It is important to note that the power output of wind turbines changes with the cube of the wind speed—that is, a twofold increase in wind speed results in an eightfold increase in power output. The wind capacity factor, or how consistently the wind blows, is the fraction of the peak power supply that is actually delivered on average. This factor, therefore, takes into account the following variables: daily variations in wind speed, seasonal variations in wind speed, and fluctuations in wind direction. For both horizontal and vertical axis wind turbines, the best available designs are able to convert approximately 35% of the kinetic energy in wind into electricity (Brus and Golob 1993, p. 131). A wind turbine's annual energy production is the product of the wind speed and the wind capacity factor.

Wind turbines come in many different sizes to meet various needs. Sometimes, small 50 kW turbines are constructed by businesses and small communities to save money on electric bills. These machines may bring electricity to off-grid locations. Larger 5 MW wind turbines are occasionally placed together in groups of 100 or more to create massive utility-scale wind farms capable of providing power to thousands of residences (Union of Concerned Scientists 2009b). Increasingly, there is interest in constructing these wind farms offshore, where the wind resources are greater. To this end, investments have been made around the world, but such farms are still uncommon. Nevertheless, offshore wind turbines today constitute the fastest growth in the wind industry. Offshore wind capacity has increased 600% since 2006 and now represents more than 4500 MW of generating capacity worldwide, as demonstrated in Figure 2.11 (Roney 2012). More than 90% of all offshore wind-generating facilities erected since 1990 are located in Europe. China also has installed hundreds of megawatts of capacity. The United States, on the other hand, has yet to install offshore facilities. Ultimately, the growth in the electricity capacity of offshore wind is expected to continue, with capacity reaching 26,000 MW by 2017 (Roney 2012). Figure 2.12 shows that China is currently the largest producer of wind power, followed by the United States, Germany, and Spain.

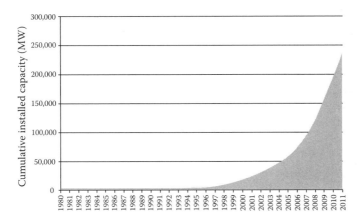

FIGURE 2.11
World cumulative installed offshore wind power capacity since 1991. (Adapted from J.M. Roney. Earth Policy Institute. Offshore Wind Development Picking Up Pace. 2012. http://www.earth-policy.org/plan_b_updates/2012/update106.)

Although the global financial crisis has had a strong effect on the wind industry, wind was still the fastest-growing source of electricity in the world in 2010. During that year, worldwide wind capacity increased by nearly 25% to reach 200 GW and thus satisfied 2.3% of the world's electricity needs (IEA Wind 2011). Wind was also the fastest-growing energy resource at the end of the last decade. The most efficient wind farms produce

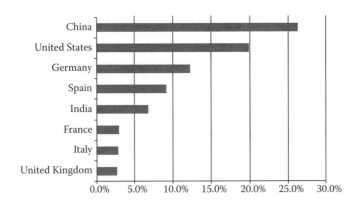

FIGURE 2.12
Proportion of total world energy consumption of wind power by country in 2011. (Adapted from BP, P.L.C. *Statistical Review of World Energy June 2012*. Uckfield, United Kingdom: Pureprint Group, 2012. http://www.bp.com/sectionbodycopy.do?categoryId=7500&contentId=706848.)

power at US$0.04/kWh, a price that is competitive with energy produced from coal in many countries. Wind power is a significant contributor to national electricity in countries all over the world, including Denmark, Portugal, Spain, Germany, the United States, and especially China, which took the lead in domestic wind production in 2010 (OECD 2011).

2.8 SOLAR

Solar energy takes advantage of the radiant light and heat emissions that come directly from the sun. It is the original source of many other energy resources, including wind, biomass, oil, coal, and natural gas. It is an exciting resource because it is infinite and completely clean with regard to environmental effects in its collection. In general, solar technologies can be labeled as passive or active. Passive solar entails designing buildings to maximize solar insolation, planning agricultural plots to optimize solar energy capture, and planning buildings around solar lighting. Active solar energy can be utilized for cooking, heating water, and generating electricity. Solar thermal concentrating systems and solar photovoltaic (PV) systems are the most common technologies used to generate electricity (Union of Concerned Scientists 2009a). Table 2.8 summarizes some of the advantages and disadvantages of solar power as an energy source.

Solar thermal concentrating systems utilize mirrors to focus sunlight onto a particular area, which heats a working fluid that spins a turbine to

TABLE 2.8

Advantages and Disadvantages of Solar as a Power Source

Advantages	Disadvantages
Low maintenance and labor costs per power output after initial capital expenditure	Dependent on location—utility scale construction requires vast open spaces
No emissions of any kind after production of the solar panels	Variable, unpredictable power output that depends on weather and cloud coverage
Highly scalable, since commercial solar PV facilities tend to be modular	Efficiency is usually low at approximately 10%, although efficiency is increasing with improving technology
Global solar resource is vast and inexhaustible as long as the sun continues to shine	Solar must be coupled with storage because peak solar generation occurs just after noon, which is before peak demand

generate electricity. The most common solar concentrator is a parabolic trough—a long, curved mirror with a thin tube of liquid at the focus. This liquid can be heated to temperatures as high as 300°C, at which point it is converted into steam, which is then directed to drive a turbine. Other common designs include parabolic dishes, which heat liquid at a single point, and central receivers, which feature a power tower that is super-heated by an entire field of mirrors and reflectors (Union of Concerned Scientists 2009a).

A solar PV cell converts sunlight directly into electricity by using two silicon semiconductors with slightly different charges. As photons reach these semiconductors, they excite electrons to jump from the *p*-type semiconductor to the *n*-type semiconductor. Wiring and circuitry aggregate the current from individual cells and produce DC voltage. Traditional solar cells are flat-plate silicon sheets that can be lifted and placed in grid patterns on flat surfaces. New thin-film solar cells are highly flexible layers of semiconductors that are just micrometers thick and can be unfurled on rooftops and building walls. Solar panels are best installed in areas that receive direct sunlight for consistently long durations, making climate the first characteristic installers consider. Many ground-mounted solar arrays are in deserts, where the long, direct sunlight and minimal ground obstructions and clouds provide optimal conditions for the absorption of insolation. Ground mounters place solar panels in large open areas away from trees and other obstructions that can cause shade (DIY-energy.org 2012). Like wind, solar power is considered a distributed energy generation source, meaning it can be installed modularly for individuals or in vast utility scale projects, such as the 166 MW Senftenberg solar plant in Germany (Meehan 2011).

In 2011, global solar PV reached 67 GW of installation, approximately 0.5% of global electricity generation capacity (REN21 2011). Today, the outlook for further expansion of this energy source looks increasingly positive, as prices have decreased dramatically in the past few years. The cost per watt of solar PV has followed a logarithmic curve, declining to US$0.15/kWh today compared with US$1.50/kWh 20 years ago, as shown in Figure 2.13. In fact, solar panel prices measured at the gates of manufacturers decreased by 50% in 2011 alone (Wesoff 2012). Figure 2.14 shows that, as of 2011, the four largest PV energy producers were Germany, Italy, Japan, and the United States. Increasing efficiencies, intense competition, and decreasing manufacturing costs led to this broad adoption in these countries. In most countries, including the United States, solar power is

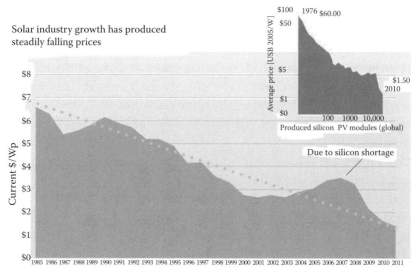

Solar industry growth has produced
steadily falling prices

Module pricing trends 1985–2011

FIGURE 2.13

Development of solar module prices and production volumes over time. (Reprinted with
permission from S. Lacey. Climate Progress. Must-See Photovoltaic Industry Graphs on the
Changing Economics of Solar. 2011. http://thinkprogress.org/romm/2011/06/09/241120/
solar-is-ready-now-%E2%80%9Cferocious-cost-reductions-make-solar-pv-competitive/.)

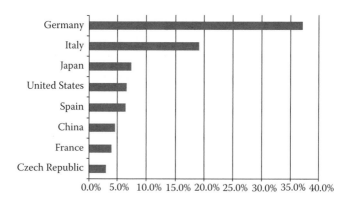

FIGURE 2.14

Proportion of total world energy consumption of PV by country in 2011. (Adapted from
BP, P.L.C. *Statistical Review of World Energy June 2012*. Uckfield, United Kingdom:
Pureprint Group, 2012. http://www.bp.com/sectionbodycopy.do?categoryId=7500&content
Id=706848.)

expected to reach grid parity with traditional fossil fuel sources within the decade. In fact, Germany, Italy, Spain, and China already have mature solar markets because they have very supportive policies and infrastructure (Chu 2012).

Italy is an example of a country with a robust solar development plan. It experiences one of the highest average solar irradiances in Europe at 4.5 kWh/m², roughly equivalent to the solar irradiance received by North Carolina (Enel 2012). At the end of 2011, it had installed 12.75 GW of solar-generating capacity. Moreover, virtually all of its solar capacity is in commercial PV arrays smaller than 200 kW, rather than the large megawatt installations that are more common in the United States. These arrays are established on small plots of land, often situated by highways and fields. In total, solar accounted for 3% of Italian energy consumption in 2011 (Clarke 2012).

The Italian solar industry has evolved around a robust feed-in tariff that reimburses building-mounted PV and green field PV owners for the electricity they export to the grid. In 2013, Italy's reimbursement rate is approximately €0.30/kWh. Its feed-in tariff has bolstered its ascendance as the second-largest market for solar PV by guaranteeing solar providers grid access, establishing price certainty in the market, and properly accounting for the long-term benefits of a healthy solar industry. The country is planning a total of 23,000 MW of solar PV installations by 2017 (Gipe 2012).

2.9 GEOTHERMAL ENERGY

Geothermal power plants capture the heat stored in the earth to generate steam and produce electricity. In 2011, 24 countries—including the United States, Iceland, the Philippines, and Costa Rica—were producing geothermal energy. This energy accounted for 11,014 MW of electricity generation, enough to meet the energy needs of approximately 60 million people (BP 2012). As shown in Figure 2.15, the four largest producers of geothermal energy that year were the United States, the Philippines, Indonesia, and Mexico. The two sources of this energy are heat flow from the core and the mantle of the planet and heat flow generated by continuing radioactive decay in the crust (IEA Geothermal 2011). Table 2.9 summarizes some of the advantages and disadvantages of geothermal power as an energy source.

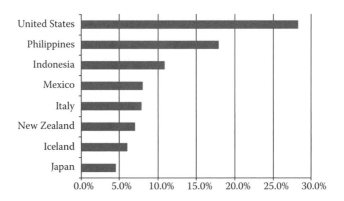

FIGURE 2.15

Proportion of total world energy consumption of geothermal power by country in 2011. (Adapted from BP, P.L.C. *Statistical Review of World Energy June 2012*. Uckfield, United Kingdom: Pureprint Group, 2012. http://www.bp.com/sectionbodycopy.do?categoryId=7500&contentId=706848.)

The geothermal resource best suited for commercial energy production is a hydrothermal system. This is a tectonically active hydrothermal reservoir, or an underground system of interconnected rock fractures that collects groundwater. When the water seeps down through these fractures, it comes into contact with rocks that have been heated by magma. This contact, combined with high pressure, heats the water past its normal boiling point. Usually, the water is heated to between 250°F and 400°F, but sometimes it can reach 600°F. When the hot water or steam migrates upward,

TABLE 2.9

Advantages and Disadvantages of Geothermal Energy as a Power Source

Advantages	Disadvantages
Cheap to exploit relative to other renewables at US$0.06–0.08 cents/kWh	Release of sulfurous gases into the atmosphere
Minor environmental effect	Projects have high upfront installation costs
Continuous generation source that can provide base power at no fuel cost	Only commercially viable in tectonically active areas
Renewable energy source	May be expensive to transmit energy to demand centers

Source: Adapted from Geothermal Energy Association. Geothermal 101: Basics of Geothermal Energy Production and Use. 2009. http://smu.edu/smunews/geothermal/Geo101_Final_Feb_15.pdf; Laszlo, E., *Energy: The Alternatives* 10(5): 248–249, 1981. http://www.jstor.org/stable/4312703.

in a process known as hydrothermal convection, it can be used to power a generator (Brus and Golob 1993, p. 63). The water is then pumped back underground as warm water to replenish the hydrothermal system.

There are three ways to capture the energy from these hydrothermal systems: (1) dry steam, (2) flash steam, and (3) binary cycle. If the water rises through the hydrothermal system as pure steam rather than as a liquid, a dry steam plant funnels the steam directly past a generator, through a condenser, and then back into the ground. A flash steam plant is used when the hydrothermal system is liquid-dominated, and the water comes out at more than 150°C (Geothermal Energy Association 2009). It depressurizes hot water at the surface, where it is separated into brine and steam. It then harnesses the steam to spin a turbine. A binary cycle system is a closed-loop process in which hot water is passed through a heat exchanger to boil a working fluid with a lower boiling point, such as isobutane. This type of plant can operate, even with water at a temperature of less than 150°C (National Renewable Energy Laboratory [NREL] 2012a).

Tectonically active areas have the greatest geothermal resources. Commercially useful resources are found near seismically active plate boundaries and near hotspots, which are areas of the earth's crust where magma rises toward the surface. Good hydrothermal systems are very hot and permeable and contain a lot of water. Often, these systems manifest themselves as hot springs or geysers, but they are not always marked by such visible signs (Geothermal Energy Association 2009). Worldwide, geothermal power production has increased rapidly. Since 2005, global production has grown by 20%, exceeding 11 GW in 2012. Much of the new generation is found in the United States, Indonesia, Iceland, and New Zealand. The cheapest sources have generated geothermal electricity at US$0.05/kWh. Although most geothermal development projects have been hydrothermal systems along plate boundaries, newer enhanced geothermal power systems, which pump cold water down an injection well into hot, dry rock, are growing in popularity (Geothermal Energy Association 2009).

Geothermal resources are limited throughout the world. However, their utilization suggests that this energy form could supply a significant portion of some countries' energy mixes. In particular, Iceland is often hailed as the model of geothermal development. The country's utilities are able to offer geothermal electricity to retail customers at US$0.043 cents/kWh, which has spurred economic development around cheap energy (LXRICHTER 2011). In 2010, Iceland met 575 MW of the demand, approximately 25% of its electricity load, using geothermal power (Geothermal

Energy Association 2010). This amount was nearly a threefold increase in supply over 2005, when the country met just 202 MW of its power supply with geothermal power (Bertani 2010). Overall, geothermal power has played a critical role in allowing Iceland to meet 100% of its electricity needs with renewable energy. Not only did the country produce 4.0 TWh of electricity in 2011, it also satisfied 87% of its need for hot water and heat through an extensive district heating system powered by geothermal energy. Reykjavik is a prime example of the benefits of geothermal energy. In the early 1930s, when nearly everyone used coal to heat their homes, coal soot caused a major air pollution problem in the capital. Today, the two largest Icelandic geothermal power plants, Hellisheioi and Nesjavellir, are just 30 km southeast of the capital and generate a combined 2.33 TWh of electricity per year while simultaneously providing heat and hot water for Reykjavik's buildings. In fact, Iceland's development has been so successful that supply has already outstripped domestic demand, and the country is now aggressively targeting the British Isles as a future market for electricity exports. The country has proposed a 700 mile subsea cable to Scotland and a 5 TWh geothermal development project that would cost approximately US$2 billion (Gipe 2012).

2.10 HYDROELECTRIC ENERGY

Hydroelectric power plants capture the potential energy of water through structures that take advantage of elevation differences along rivers. Hydropower is the largest source of renewable energy in the world and supplies approximately 16% of global electricity (IEA 2010a). Figure 2.16 shows that China, Brazil, Canada, and the United States are leaders in hydroelectricity production. Most of the remaining growth in the sector is expected in China, India, and Latin America (NRG Expert 2011). Table 2.10 summarizes some of the advantages and disadvantages of hydroelectric power as an energy source.

Hydropower takes advantage of the gravitational potential energy that comes from the natural gradient of river slopes. The predominant method of harnessing hydropower is damming or impoundment. A dam is a concrete structure that seals off a river basin and floods the valley upstream. In the process, water is stored upstream in a deep reservoir from which it is slowly released downhill, through a pipe called a penstock, straight

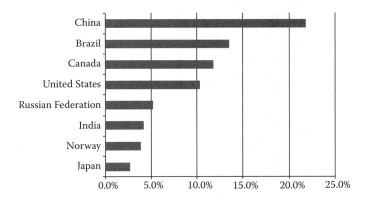

FIGURE 2.16

Proportion of total world energy consumption of hydroelectric power by country in 2011. (Adapted from BP, P.L.C. *Statistical Review of World Energy June 2012*. Uckfield, United Kingdom: Pureprint Group, 2012. http://www.bp.com/sectionbodycopy.do?categoryId= 7500&contentId=706848.)

into a turbine (Union of Concerned Scientists 2006). The dam controls the volume of the water flow. Impoundment structures are often constructed to control flooding, create economic development, and establish a reliable water supply before they are adapted to also generate hydroelectricity.

There is another method, called diversion or run-of-river, for capturing hydropower. Sections of fast-flowing rivers are channeled into an intake

TABLE 2.10

Advantages and Disadvantages of Hydropower as a Power Source

Advantages	Disadvantages
Nonpower benefits are the water reservoir, flood control, and recreation	Interrupt routes for migratory marine wildlife, such as salmon
Pump water storage systems	Modifies downstream aquatic habitats
Completely renewable resource with no man-made emissions	Flooding of upstream valley often forces families to evacuate
Often located near population centers, so lower transmission costs than other renewables	Lowers water flow, which can affect downstream farmers
Electricity can be generated as soon as it is needed	Alters water quality and temperature

Source: Adapted from U.S. DOE, Water Power Program. Benefits of Hydropower. http://www.eere. energy.gov/water/hydro_benefits.html; Union of Concerned Scientists. How Hydroelectric Energy Works, 2008. http://www.ucsusa.org/clean_energy/our-energy-choices/renewable-energy/how-hydroelectric-energy.html.

pipe that leads the water through a canal or penstock. The water that passes through this alternate route spins a turbine that generates electricity. All the water diverted through this alternate route is then returned to the mainstream through a tailrace (Kwagis Power 2012). Run-of-river is much less efficient than damming because less water passes through the alternate channel. On the other hand, run-of-river hydropower facilities do not create many of the ecological issues that are unavoidable in impoundment (Watershed Watch Salmon Society [WWSS] 2007).

Hydropower plants come in many different sizes. Large facilities generate 30 MW or more. Small facilities generate 100 kW to 30 MW. Microfacilities generate less than 100 kW and are commonly used by farms and villages. The amount of energy a hydropower facility can generate is the product of the flow—the volume of the water flowing through the penstock—and the head—the difference in elevation between the upstream and downstream reservoirs (Taylor 2010). The greater the head, the faster the water moves. Most hydroelectricity comes from large hydropower facilities on the largest rivers in the world, such as the Three Gorges Dam in China on the Yangtze River, the Itaipu Dam in Brazil on the Parana River, and the Guri Dam in Venezuela on the Caroni River. Each of these dams generates more than 10 GW of power (Lucky 2012).

One of the most important auxiliary benefits of dams is the opportunity to create pump-water systems. In these hydropower plants, water is pumped from the lower reservoir to the upper reservoir with a reversible turbine. This process is usually undertaken during periods of low energy demand, when electricity prices are cheaper, and allows the dams to serve as giant batteries because moving this water to a higher elevation increases its gravitational potential energy. Pumping water upstream, therefore, enables an energy supplier to send the water downstream during a later period of high demand, when energy prices are higher. Pumped-storage facilities are some of the most important tools that utilities have for storing unused electricity from coal and nuclear facilities during off-peak nighttime hours (Bonsor 2001).

Over the past 20 years, exploited hydropower resources have grown by more than 50% and constitute 16% of global electricity generation, producing more than 3500 TWh annually. This generation is supplied by the world's 45,000 dams. As shown in Figure 2.16, China, Brazil, Canada, and the United States constitute more than half the total global hydroelectric production. The greatest recent growth has been in China, which is expected to remain the global leader as electricity demand continues to increase (Moller 2012). It seems that this country will, indeed, have opportunities to expand its hydropower output as, globally, approximately 19% of hydropower resources have already been developed. About two-thirds of the remaining potential for development is located in only 10 countries. That this residual capacity will largely come to be exploited is almost assured, as the IEA states that hydropower will be an essential resource in the global energy mix because of its unique energy storage capability and its ability to respond quickly to fluctuating electricity demand. Both of these characteristics will prove highly valuable in supplementing the less predictable supplies of electricity from other renewables (IEA 2010a).

Canada is the world's second-largest producer of hydroelectricity after China. Overall, hydroelectricity satisfies 60% of Canadian electricity demand. In 2011, Canada was home to 475 hydropower plants that produced approximately 70,000 MW of electricity domestically. Almost all of Canada's major rivers, including the St. Lawrence River, the Mackenzie River, and the Columbia River, are dammed. Northern Quebec's Le Grande Complex is the second-largest hydropower facility on the planet, second only to the Three Gorges project in China (Ray 2011).

Canada's success is due, in large part, to its abundance of large rivers. The hydropower industry has helped the country's economy create tens of thousands of jobs in operations and upkeep. Hydropower will most likely continue to dominate the Canadian grid in the future, as there still exists the technical potential to more than double hydroelectric output (Canadian Hydropower Association 2009). Nevertheless, damming rivers has caused a number of problems for the Canadian boreal forests. Migratory fish, such as the Pacific pink salmon, have been blocked from reaching upstream spawning grounds and aboriginal communities have suffered community displacement and the loss of traditional hunting grounds due to the creation of new reservoirs (Herrmann 2011). As Canada seeks to expand its share of renewable fuels through increasing its number of hydropower facilities, it will have to reach socially acceptable consensual resolutions to address these problems.

2.11 HYDROKINETIC ENERGY

Hydrokinetic energy sources capture the kinetic energy that is available in moving bodies of water. Unlike hydropower, which deals strictly with elevation differences along rivers, hydrokinetic resources extend to (1) ocean waves, (2) tides, (3) ocean currents, and (4) fast-flowing rivers (Union of Concerned Scientists 2008). Hydrokinetic energy is largely an untapped, developing field. Table 2.11 summarizes some of the advantages and disadvantages of hydrokinetic power as an energy source.

Wave power is thought to be the greatest potential source of hydrokinetic energy. Wave power devices capture the energy of natural oscillations inherent in ocean waves both near shore and in the deep ocean. Waves are created by the drag exerted by the wind as it blows over the ocean's surface. The four types of wave power devices are (1) oscillating water columns, (2) point absorbers, (3) attenuators, and (4) overtopping devices (Minerals Management Service [MMS] 2006b). In an oscillating water column, waves move underwater to raise the water level inside a partly submerged concrete structure onshore, which compresses air trapped inside through a turbine. Point absorbers are floating buoys that are pushed up and down by the dynamic water level relative to a fixed mount. Wave attenuators are multisegment point absorbers that are aligned parallel to the direction of the ocean waves. Finally, overtopping devices take advantage of incoming waves that fill up a small reservoir by slowly releasing the extra water through a pump. Some of the best resources for wave power are off the coast of northwestern Europe, along the northwestern United States,

TABLE 2.11

Advantages and Disadvantages of Hydrokinetic Power as an Energy Source

Advantages	Disadvantages
Vast, untapped resource	Uncertain environmental consequences on benthic animals and from installation noise
Continuous energy supply	Few existing commercial models
No emissions of any kind	Transmission infrastructure from offshore supply to population centers is a big hurdle
Very predictable source—forecasts for waves are known days before and tides centuries before	High maintenance costs from biofouling, corrosion, and low system maintenance
	May compete with fisherman, shipping companies, and recreation

Source: Adapted from MMS, *Technology White Paper on Ocean Current Energy Potential on the U.S. Outer Continental Shelf.* Washington, D.C.: U.S. Department of the Interior. 2006. http://www.doi.gov/whatwedo/energy/index.cfm; Union of Concerned Scientists. How Hydrokinetic Energy Works. 2008. http://www.ucsusa.org/clean_energy/our-energy-choices/renewable-energy/how-hydrokinetic-energy-works.html.

off the southern tips of South America and Africa, and around Australia (Sea Power Ltd. 2011). Commercial devices are common throughout the United Kingdom, Australia, and the United States (Pike Research 2012).

Ocean tides are daily variations in sea level along the coast that are driven by the moon's gravity. Most commonly, a tidal barrage that operates similarly to a dam is built across an inlet with a high tidal resource potential. Barrage gates allow water to pass as the tide rises and falls, causing a height differential between the reservoirs on each side of the barrage. As of 2011, the only major tidal barrage was a 240-MW plant in Le Rance, France, although much smaller facilities can be found in locations across the globe (Altprofits 2011).

Ocean currents, as opposed to ocean tides, are continuous streams of water that are affected by wind, temperature, salinity, and the Coriolis effect, instead of the moon's gravitational pull. The energy in ocean currents is substantial because of the water's relatively high density. Most technologies exploit this energy by using submerged water turbines that look similar to wind turbines but are powered by moving water, which is more than 800 times as dense as air (MMS 2006a). These turbines are then anchored to the sea floor with a post or tethered by a cable. A 30 MW array of water turbines proposed in the Philippines would be installed in clusters analogous to a wind farm (MMS 2006a). Although China, Japan, the United States, and the European Union are pursuing ocean current technologies, as of 2010, there have been no commercial grid-connected current devices (ibid.). As a result, there are very limited data for the production of hydrokinetic electricity on a country-by-country basis.

River hydrokinetic turbines resemble the turbines detailed above, but are anchored to the bottoms of fast-flowing rivers and capture power without any of the harmful effects of dams. The ideal stream for most hydrokinetic devices would have a constant flow stream of 2.5 to 3.5 m/s. Hydro+, the first commercial in-river hydrokinetic project in the United States, was launched in 2008; today, it can generate up to 250 kW of electricity for the city of Hastings, Minnesota (Hydro Green Energy 2011).

A number of other experimental hydrokinetic energy technologies are currently in development. Ocean thermal energy conversion (OTEC) would exploit the natural temperature differences of ocean waters at varying depths. In closed-cycle OTEC systems, energy from warm ocean water boils a working fluid, such as ammonia, which then turns a generator. Then, cold deep seawater is pumped to the surface to condense the working fluid and complete the cycle. Areas with good OTEC resources are located in tropical and subtropical waters, where the temperature differential between warm surface water and cold deep water is at least 20°C (NREL 2012b). In the next 5 to 10 years, OTEC may become a viable alternative energy source for island nations that import oil and other petroleum products to meet all of their energy needs. However, there are currently no commercially viable plants.

Another largely experimental hydrokinetic energy technology involves creating electricity by utilizing the salinity gradient naturally present at river mouths. In addition to the natural gravitational potential energy that can be exploited by damming, rivers provide another source of potential energy because river mouths create a large amount of chemical potential energy that stems from a salinity gradient between the fresh river water and the salty ocean water. In fact, the chemical potential energy available per cubic meter of freshwater when it meets a typical salinity gradient is equivalent to the gravitational potential energy of a 225-m-high waterfall. A selective membrane that controls the rate of mixing the fresh water and salt water can exploit this chemical potential energy through pressure-retarded osmosis or reverse electrodialysis. This is actually the reverse of the electrodialysis process that desalinization plants use—current efficiencies allow for the recovery of 2 W/m^2 of membrane. Currently, no commercially viable salinity gradient power plants exist (Post 2009).

Hydrokinetic energy is currently a very small, developing form of power (Post 2009). In 2012, installed capacity worldwide was just 760 MW, but that number is expected to grow to 5.5 GW by 2017. Supporting this expected growth is the fact that tidal energy from certain sources can generate electricity at US$0.02/kWh in very limited locations (Pure Energy Systems 2011). The leaders in this field in the coming years are expected to be South Korea, the United Kingdom, and Canada (Pike Research 2012). Across the board, hydrokinetic technologies are incipient, and much research still needs to be conducted, not only to increase efficiency but also to understand the environmental and economic effects.

Although hydrokinetic energy is a fledgling, largely unproven, technology, certain developed nations have encouraged its development. The United Kingdom, in particular, has taken the lead in the field. As of 2011, the country had 1.3 MW of installed wave capacity and 2.1 MW of installed tidal capacity. Because of its physical and intellectual resources, the United Kingdom is well-positioned to serve as the center of hydrokinetic power technologies. Research suggests that marine hydrokinetic energy could meet as much as 20% of the country's electricity needs. Scotland, in particular, has one of the highest wave resource potentials in the world. Many of the United Kingdom's best tidal devices will be deployed around the Scottish islands of Orkney, Pentland Firth, and Shetland. Orkney, in fact, through a combination of wave and wind power, already meets 85% of its energy needs by using renewable power. Around the rest of the Scottish islands, which are home to some of the best wave and tidal resources within the United Kingdom, tidal power has an annual potential of 38,500 GWh (Carrell 2012). The United Kingdom also harbors testing grounds for innovative wave energy projects at sites such as WaveHub, an 8 km² government-supported facility in southwest England. WaveHub is currently the only multiberth, grid-connected, wave and tide generation open-sea test facility in the world, which is why the United Kingdom is considered the leader in the field (WaveHub 2010). In addition, a number of the most famous wave and tidal energy devices are located in England, such as the Pelamis 0.75 MW wave attenuator, the Oyster 1 nearshore 0.32 MW wave converter, the OpenHydro 2.5 MW tidal stream turbine, and the SeaGen 1.2 MW tidal turbine (RenewableUK 2011).

2.12 CONCLUSION

As this chapter has shown, there is no universal "best" source of energy. Countries' unique contexts make a variety of energy sources attractive in different areas of the world. Thus, whereas the global energy future may center on just a few sources, multiple forms of energy will be needed to satisfy total world demand. In this sense, all energy sources are viable candidates, and each is more or less appealing based on characteristics such as cost and environmental effects. Although economic and environmental factors are the most relevant, we have attempted to show that other considerations—including political factors, questions of storage and transportation, and technological progress, to name a few—also play a role in determining the attractiveness of different energy sources. Thus, understanding why countries choose to foster and implement different sources is critical because these examples reveal just how multifaceted and complex the process of choosing the optimal energy source configuration really can be.

Keeping this complexity and the idiosyncratic natures of each energy demand situation in mind, a number of individual forms of energy presently seem to be generally more advantageous than others, leading us to believe that they will flourish in the coming years. Among the nonrenewable fuels, for example, natural gas stands out as an effective alternative. It is very affordable, present in high quantities across the globe, and the cleanest of the fossil fuels. However, there are serious concerns regarding its transportation and storage.

At the other end of the spectrum, certain energy sources seem to be clear losers. Although coal is easily accessible around the world, it is by far one of the worst polluters and not as energy dense as other alternatives. Table 2.12 illustrates this by comparing the various green energy sources and fossil fuels according to a number of criteria.

The characteristics of the energy sources detailed in this table are essential for predicting the future of the world's energy mix, which is ultimately strongly tied to geopolitics. Although this chapter abstains from a long discussion of international relations, ultimately, the boundaries and flows shared among countries determine which energy sources are available, developed, and used. As the OPEC crisis and the more recent Russian natural gas cutoffs have shown, tenuous relationships among nations represent an unpredictable variable that can impair or facilitate the ability to exploit various energy sources worldwide.

It is partially attributable to this geopolitical uncertainty that, whereas the issue of global energy implementation progressing into the future has been studied extensively, forecasts of worldwide energy generation are highly divergent, depending on the source consulted. For example, some believe that nuclear energy will serve as the stopgap needed to reduce our fossil fuel consumption (MIT 2003). Others believe that nuclear power will be phased out in the next half century (Worldwatch Institute 2011). Some believe that the composition of our energy consumption will change dramatically to address the issue of climate change (MIT 2011), whereas others believe it will remain largely the same (BP 2013). Recent incidents, such as the Fukushima Daiichi nuclear disaster and the bottoming-out of natural gas commodity prices in the United States, have changed the future of global energy source composition in powerful ways in just the past 5 years; and emerging technologies, such as hydrokinetic power, have the potential to further disrupt the energy space. Thus, it is difficult to make predictions about worldwide energy generation. Who knows what global energy composition will look like in 2050?

TABLE 2.12

Comparison of Fossil and Renewable Energy Sources

	Coal	Oil	Natural Gas	Nuclear	Biomass	Wind	Solar	Geothermal	Hydroelectric	Hydrokinetic
Renewable	+	+	+	+	++	+++	+++	+++	+++	+++
No CO_2 emissions	+	+	+	+++	++	+++	+++	+++	+++	+++
No gaseous or particulate emissions	+	+	+	+++	+	+++	+++	+++	+++	+++
Could constitute a large fraction of total energy usage	+++	+++	+++	++	+	++	++	+	++	+
Technology could be implemented within the next 10 years	+++	+++	+++	++	+	++	++	++	++	+
Necessary infrastructure to add it to the grid is inexpensive	+++	+++	++	+	+	++	++	+	++	+

REFERENCES

Altprofits. Wave and Tidal Energy Report. Last modified May 2011 (accessed April 11, 2013). http://www.altprofits.com/ref/report/ocean/ocean.html.

Bertani, R. Geothermal power generation in the world: 2005–2010 update report. *Proceedings World Geothermal Congress 2010.* http://geotermia.org.mx/geotermia/pdf/WorldUpdate2010-Ruggero.pdf (accessed March 11, 2012).

Boccard, N. Capacity factor of wind power realized values vs. estimates. *Energy Policy* 37(7): 2679–2688, 2009.

Bonsor, K. How stuff works. How Hydropower Plants Work. Last modified September 2001 (accessed June 25, 2012). http://science.howstuffworks.com/environmental/energy/hydropower-plant2.htm.

BP, P.L.C. *Statistical Review of World Energy June 2012.* Uckfield, United Kingdom: Pureprint Group, 2012. http://www.bp.com/sectionbodycopy.do?categoryId=7500&contentId=706848 (accessed June 29, 2012).

BP, P.L.C. Energy Outlook 2030. Last modified January 2013 (accessed January 22, 2013). http://www.bp.com/extendedsectiongenericarticle.do?categoryId=9048887&contentId=7082549.

Brus, E., and R. Golob. *The Almanac of Renewable Energy.* 1st ed. New York: Henry Holt, 1993.

Canadian Hydropower Association. Hydroworld. Resource Overview: Hydropower in Canada: Past, Present, and Future. Last modified 2009 (accessed March 8, 2012). http://www.hydroworld.com/index/display/article-display/9103130396/articles/hydro-review/volume-28/issue-7/articles/resource-overview.html.

Carlisle, T. Saudi looks to nuclear future. *The National,* April 19, 2010. http://www.thenational.ae/business/energy/saudi-looks-to-nuclear-future (accessed January 10, 2013).

Carrell, S. Orkney, leader in green energy, launches wave power competition. *The Guardian,* August 28, 2012. http://www.guardian.co.uk/environment/2012/aug/28/orkney-green-energy-wave-power (accessed January 10, 2013).

Chu, T. Planetsave. Top 6 Countries using Solar Energy. Last modified March 6, 2012 (accessed April 11, 2013). http://planetsave.com/2012/03/06/top-6-countries-using-solar-energy/.

Clarke, C. Italy beats California in solar development. *KCET,* August 31, 2012. http://www.kcet.org/news/rewire/solar/italy-beats-california-in-solar-development.html (accessed January 10, 2013).

DIY-energy.org. Choosing the Best Location for Solar Panels. Last modified 2012 (accessed March 7, 2012). http://www.diy-energy.org/choosing-the-optimum-location-for-solar-panels.

Downey, M.P. *Oil 101.* 1st ed. New York: Wooden Table Press LLC, 2009.

Ecoleaf. Understanding Petroleum Power Plants. Last modified 2010 (accessed January 10, 2013). http://www.ecoleaf.com/green_energy/petroleumgaspowerplants.html.

EDF Energy. How Electricity is Generated through Oil. Last modified 2013 (accessed January 10, 2013). http://www.edfenergy.com/energyfuture/oil-generation.

Enel. Solar Power—Technology. Last modified 2012 (accessed January 10, 2013). http://www.enel.com/en-GB/innovation/project_technology/renewables_development/solar_power/.

Evans, S. Germany: Nuclear power plants to close by 2022. *BBC News,* May 30, 2011. http://www.bbc.co.uk/news/world-europe-13592208 (accessed January 10, 2013).

Gelb, B. CRS Report for Congress. Russian Natural Gas: Regional Dependence. Last modified January 5, 2007 (accessed March 11, 2012). http://www.usembassy.it/pdf/other/RS22562.pdf.

Geothermal Energy Association. Geothermal 101: Basics of Geothermal Energy Production and Use. Last modified February 15, 2009 (accessed March 11, 2012). http://smu.edu/smunews/geothermal/Geo101_Final_Feb_15.pdf.

Geothermal Energy Association. *Geothermal Energy: International Market Update* by A. Holm, L. Blodgett, D. Jennejohn, and K. Gawell. Washington, D.C.: Geothermal Energy Association, 2010. http://www.geo-energy.org/pdf/reports/gea_international_market_report_final_may_2010.pdf (accessed March 11, 2012).

Gipe, P. Iceland: A 100% renewables example in the modern era. *Reneweconomy*, November 7, 2012. http://reneweconomy.com.au/2012/iceland-a-100-renewables-example-in-the-modern-era-56428 (accessed January 10, 2013).

Goldemberg, J., A.K.N. Reddy, K. Smith, and R.H. Williams. Rural energy for developing countries. In *World Energy Assessment: Energy and the Challenge of Sustainability.* New York: UN Development Programme, 2000.

Griffiths-Sattenspiel, B., T. Leipzig, and W. Wilson. The River Network. Burning our Rivers: The Water Footprint of Electricity. Last modified April, 2012 (accessed June 29, 2012). http://www.rivernetwork.org/resource-library/burning-our-rivers-water-footprint-electricity.

Hasselbach, C. and H. Jeppesen. DW Akademie. Food or Fuel Debate Leads to EU Biofuel Changes. Last modified October, 2012 (accessed April 11, 2013). http://www.dw.de/food-or-fuel-debate-leads-to-eu-biofuel-changes/a-16313695.

Herrmann, L. Study: Call to protect Canada's boreal, the 'world's waterkeeper'. *Digital Journal*, March 18, 2011. http://digitaljournal.com/article/304805 (accessed January 10, 2013).

Hinckley, E. Why Saudi Arabia is taking a shine to solar. *Christian Science Monitor*, October 28, 2012. http://www.csmonitor.com/Environment/2012/1028/Why-Saudi-Arabia-is-taking-a-shine-to-solar (accessed January 10, 2013).

Hydro Green Energy. Hydro Green Energy's 1st Project. Last modified 2011 (accessed March 7, 2012). www.hgenergy.com/hastings.html.

Intergovernmental Panel on Climate Change (IPCC). *Climate Change 2007: Synthesis Report* by L. Bernstein, P. Bosch, O. Canziani, Z. Chen, R. Christ, O. Davidson, W. Hare, S. Huq, D. Karoly, V. Kattsov, Z. Kundzewicz, J. Liu, U. Lohmann, M. Manning, T. Matsuno, B. Menne, B. Metz, M. Mirza, N. Nicholls, L. Nurse, R. Pachauri, J. Palutikof, M. Parry, D. Qin, N. Ravindranath, A. Reisinger, J. Ren, K. Riahi, C. Rosenzweig, M. Rusticucci, S. Schneider, Y. Sokona, S. Solomon, P. Stott, R. Stouffer, T. Sugiyama, R. Swart, D. Tirpak, C. Vogel, G. Yohe, and T. Barker. Geneva, Switzerland: IPCC, 2007. http://www.ipcc.ch/pdf/assessment-report/ar4/syr/ar4_syr.pdf (accessed December 28, 2012).

International Atomic Energy Administration (IAEA). *Nuclear Power Global Status* by Alan McDonald. Vienna, Austria: IAEA, 2008. http://www.iaea.org/Publications/Magazines/Bulletin/Bull492/49204734548.pdf (accessed January 10, 2013).

International Atomic Energy Agency (IAEA). *World Distribution of Uranium Deposits (UDEPO) with Uranium Deposit Classification.* Vienna, Austria: IAEA, 2009. http://www.pub.iaea.org/MTCD/publications/PDF/te_1629_web.pdf (accessed January 22, 2013).

International Energy Agency (IEA). *Renewable Energy Essentials: Hydropower.* Paris, France: OECD/IEA, 2010a. http://www.iea.org/papers/2010/hydropower_essentials.pdf (accessed March 11, 2012).

International Energy Agency (IEA). *World Energy Outlook: Golden Rules for a Golden Age of Gas* by Marco Baroni, Laura Cozzi, Ian Cronshaw, Capella Feta, Matthew Frank, Timur Gül, Pawel Olejarnik, David Wilkinson, and Peter Wood. Paris, France: OECD/IEA, 2010b. http://www.worldenergyoutlook.org/media/weowebsite/2012/goldenrules/weo2012_goldenrulesreport.pdf (accessed January 10, 2013).

International Energy Agency (IEA). *Energy for All: Financing Access for the Poor*. Paris, France: OECD/IEA, 2011. http://www.iea.org/papers/2011/weo2011_energy_for_all.pdf (accessed March 11, 2012).

International Energy Agency (IEA). *World Energy Outlook 2012: Executive Summary*. Paris, France: OECD/IEA, 2012. http://www.iea.org/publications/freepublications/publication/English.pdf (accessed January 10, 2013).

International Energy Agency Coal Industry Advisory Board (CIAB). *Power Generation from Coal* by Mike Garwood, Allan Jones, Brian Heath, Colin Henderson, and Sankar Bhattacharya. Paris, France: OECD/IEA, 2010 http://www.iea.org/ciab/papers/power_generation_from_coal.pdf (accessed March 11, 2012).

International Energy Agency Geothermal (IEA Geothermal). *13th Annual Report—2009* by M. Mongillo, C. Bromley, R. Baria, S.J. Bauer, E. Gunnlaugsson, B. Goldstein, E. Naegele, F. Boissier, A. Desplan, L. Wissing, J. Ketilsson, P. Romagnoli, H. Muraoka, A. Takaki, Y. Song, R. Maya, L. Gutierrez-Negrin, D. Nieva, C. Bromley, C. M. Roa Tortosa, G. Siddiqi, R. Minder, J. Nathwani, A. Pressman, E. Thodal, D. Gowland, D. Wyborn, Geothermal Department of the Spanish Renewable Energy Association (APPA), Adrian Larkin, and Lucien Y. Bronicki. Paris, France: IEA, 2011. http://www.iea-gia.org/documents/2009GIAAnnReptVer2PDF17Apr11_001.pdf (accessed March 11, 2012).

International Energy Agency Wind (IEA Wind). *2010 Annual Report*. Boulder, CO: PWT Communications, LLC, 2011. http://www.ieawind.org/index_page_postings/IEA%20Wind%202010%20AR_cover.pdf (accessed March 11, 2012).

Kwagis Power. Kokish River Hydroelectric Project. Last modified 2012 (accessed March 5, 2012). http://kokishriver.com/content/hydropower-609.html.

Lacey, S. Climate Progress. Must-see Photovoltaic Industry Graphs on the Changing Economics of Solar. Last modified June 9, 2011 (accessed March 9, 2012). http://thinkprogress.org/romm/2011/06/09/241120/solar-is-ready-now-%E2%80%9Cferocious-cost-reductions-make-solar-pv-competitive/.

Laszlo, E. Geothermal energy: An old ally. *Energy: The Alternatives*. 10(5): 248–249, 1981. http://www.jstor.org/stable/4312703 (accessed March 11, 2012).

Lucky, M. All eyes on hydro. *Water Power Magazine*. Last modified June 22, 2012 (accessed April 11, 2013). http://www.waterpowermagazine.com/features/featurealleyes-on-hydro.

LXRICHTER. ThinkGeoenergy. Iceland Offers Geothermal Power at 4.3 cents per kWh. Last modified November 24, 2011 (accessed March 11, 2012). http://thinkgeoenergy.com/archives/9240.

Martin, R. Russia's Georgia invasion may be about oil. *ABC News*, August 16, 2008. http://abcnews.go.com/Business/story?id=5595811&page=1 (accessed March 11, 2012).

Massachusetts Institute of Technology (MIT). *The Future of Nuclear Power: An Interdisciplinary MIT Study*. Boston: Massachusetts Institute of Technology, 2003.

Massachusetts Institute of Technology (MIT). *The Future of Geothermal Energy: An Interdisciplinary MIT Study*. Boston: Massachusetts Institute of Technology, 2006.

Massachusetts Institute of Technology (MIT). *The Future of Coal: Options in a Carbon-Constrained World*. Boston: Massachusetts Institute of Technology, 2007.

Massachusetts Institute of Technology (MIT). *Update of the MIT 2003 Future of Nuclear Power: An Interdisciplinary MIT Study*. Boston: Massachusetts Institute of Technology, 2009.

Massachusetts Institute of Technology (MIT). *The Future of Natural Gas: An Interdisciplinary MIT Study*. Boston: Massachusetts Institute of Technology, 2011.

Meehan, C. Clean Energy Authority. German Senftenberg Solar Plant Now Largest PV Array in the World. Last modified October 05, 2011 (accessed April 11, 2013). http://www.cleanenergyauthority.com/solar-energy-news/senftenberg-solar-plant-largest-in-the-world-100511/.

Minerals Management Service (MMS). *Technology White Paper on Ocean Current Energy Potential on the U.S. Outer Continental Shelf*. Washington, D.C.: U.S. Department of the Interior, 2006a. http://www.doi.gov/whatwedo/energy/index.cfm (accessed March 11, 2012).

Minerals Management Service (MMS). *Technology White Paper on Wave Energy Potential on the U.S. Outer Continental Shelf*. Washington, D.C.: U.S. Department of the Interior, 2006b. http://ocsenergy.anl.gov/documents/docs/OCS_EIS_WhitePaper_Wave.pdf (accessed March 11, 2012).

Moller, H. Earth Policy Institute. Hydropower Continues Steady Growth. Last modified June 14, 2012 (accessed January 10, 2013). http://www.earth-policy.org/data_highlights/2012/highlights29.

National Renewable Energy Laboratory (NREL). Geothermal Electricity Production. Last modified May 18, 2012a (accessed June 24, 2012). http://www.nrel.gov/learning/re_geo_elec_production.html.

National Renewable Energy Laboratory (NREL). What is Ocean Thermal Energy Conversion? Last modified 2012b (accessed March 8, 2012). http://www.nrel.gov/otec/what.html.

Nelder, C. What the frack? *Slate*, December 29, 2011. http://www.slate.com/articles/health_and_science/future_tense/2011/12/is_there_really_100_years_worth_of_natural_gas_beneath_the_united_states_.html (accessed January 10, 2013).

NRG Expert. 2011. *Global Hydropower Report*. 1st ed. www.nrgexpert.com/wp-content/uploads/2011/09/nrg-expert-global-hydro-report-edition-1-2011.pdf (accessed March 11, 2012). OECD Publishing.

OECD. OECD Factbook 2011: Economic, Environmental and Social Statistics. Last modified 2011 (accessed March 11, 2012). http://www.oecd-ilibrary.org/sites/factbook-2011-en/06/01/04/06-01-04-g1.html.

Pike Research. Marine and Hydrokinetic Power Generation Installed Capacity to Increase Sevenfold by 2017. Last modified February 8, 2012 (accessed January 10, 2013). http://www.pikeresearch.com/newsroom/marine-and-hydrokinetic-power-generation-installed-capacity-to-increase-sevenfold-by-2017.

Post, J.W. Blue Energy: Electricity Production from Salinity Gradients by Reverse Electro-dialysis. Ph.D. Thesis. Wageningen University, 2009. http://www.waddenacademie.nl/fileadmin/inhoud/pdf/06-wadweten/Proefschriften/thesis_jan_Post.pdf (accessed March 11, 2012).

Pure Energy Systems. PESWiki. Cents per kilowatt-hour. Last modified December 19, 2011 (accessed March 7, 2012). http://peswiki.com/energy/Directory:Cents_Per_Kilowatt-Hour.

Ray, R. Canada hydropower: Liquid cornerstone. *Renewable Energy World*, April 21, 2011. http://www.renewableenergyworld.com/rea/news/article/2011/04/canada-hydropower-liquid-cornerstone (accessed March 8, 2012).

REN21. *Renewables 2011 Global Status Report.* Paris: REN21 Secretariat, 2011. http://www.ren21.net/Portals/97/documents/GSR/REN21_GSR2011.pdf (accessed March 11, 2012).

RenewableUK. *Wave and Tidal Energy in the U.K.: State of the Industry Report.* London, U.K.: RenewableUK, 2011. http://www.bwea.com/pdf/marine/wave_tidal_energy_uk.pdf (accessed March 11, 2012).

Roney, J.M. Earth Policy Institute. Offshore Wind Development Picking Up Pace. Last modified August, 2012 (accessed January 10, 2013). http://www.earth-policy.org/plan_b_updates/2012/update106.

Sea Power Ltd. Wave Energy. Last modified November 2011 (accessed March 9, 2012). http://www.seapower.ie/wave-energy/.

State of Hawaii Department of Business, Economic Development & Tourism (HDBEDT). *State of Hawaii Energy Data and Trends.* Hawaii, HI: State of Hawaii, 2011. http://hawaii.gov/dbedt/info/economic/data_reports/reports-studies/energy-data-trend-2011.pdf (accessed January 10, 2013).

Styles, G. Energy Outlook. Displacing More Oil for Power Generation. Last modified January 19, 2011 (accessed January 10, 2013). http://energyoutlook.blogspot.com/2011/01/displacing-more-oil-from-power.html.

Taha, S.M. Report: Half of Saudi Arabia's power generation comes from natural gas. *Arab News,* August 26, 2012. http://www.arabnews.com/report-half-saudi-arabia%E2%80%99s-power-generation-comes-natural-gas (accessed January 10, 2013).

Taylor, R. NREL. Tribal Energy Program Webinar Hydropower 101. Last modified September 17, 2010 (accessed March 11, 2012). http://apps1.eere.energy.gov/tribal-energy/pdfs/webinar_hydropower101.pdf.

Trading Economics. Electricity Production (kWh) in Saudi Arabia. Last modified 2012 (accessed January 10, 2013). http://www.tradingeconomics.com/saudi-arabia/electricity-production-kwh-wb-data.html.

Tverberg, G. Our Finite World. Why Natural Gas Isn't Likely to be the World's Energy Savior. Last modified 2012 (accessed January 10, 2013). http://ourfiniteworld.com/2012/10/17/why-natural-gas-isnt-likely-to-be-the-worlds-energy-savior/.

Union of Concerned Scientists. How Hydrokinetic Energy Works. Last modified April 28, 2008 (accessed March 7, 2012). http://www.ucsusa.org/clean_energy/our-energy-choices/renewable-energy/how-hydrokinetic-energy-works.html.

Union of Concerned Scientists. How Solar Energy Works. Last modified December 16, 2009a (accessed March 5, 2012). http://www.ucsusa.org/clean_energy/our-energy-choices/renewable-energy/how-solar-energy-works.html.

Union of Concerned Scientists. How Wind Energy Works. Last modified December 15, 2009b (accessed March 5, 2012). http://www.ucsusa.org/clean_energy/our-energy-choices/renewable-energy/how-wind-energy-works.html.

U.S. Department of Energy (U.S. DOE 2006a). Water Power Program. Benefits of Hydropower. Last modified October 11, 2011 (accessed March 5, 2012). http://www.eere.energy.gov/water/hydro_benefits.html.

U.S. Department of Energy (U.S. DOE 2006b). Water Power Program. Marine and Hydrokinetic Technology Database. Last modified October 11, 2011 (accessed March 7, 2012). http://www.eere.energy.gov/water/hydrokinetic/default.aspx.

U.S. Energy Information Administration (U.S. EIA 2006a). *Coal Production in the United States* by R. Bonskowski, W. Watson, and F. Freme. Washington, D.C.: U.S. EIA Office of Coal, Nuclear, Electric and Alternative Fuels, 2006. ftp://ftp.eia.doe.gov/coal/coal_production_review.pdf (accessed June 29, 2012).

U.S. Energy Information Administration (U.S. EIA 2006b). *International Energy Outlook 2006* by the Office of Integrated Analysis and Forecasting. Washington, D.C.: U.S. EIA, 2006.

U.S. Energy Information Administration (U.S. EIA 2012). Competition among Fuels for Power Generation Driven by Changes in Fuel Prices. Last modified July 13, 2012 (accessed January 10, 2013). http://www.eia.gov/todayinenergy/detail.cfm?id=7090.

U.S. Environmental Protection Agency (U.S. EPA 2012). Municipal Solid Waste. Last modified October 17, 2012 (accessed June 29, 2012). http://www.epa.gov/cleanenergy/energy-and-you/affect/municipal-sw.html.

Watershed Watch Salmon Society (WWSS). Green Hydro Power: Understanding Impacts, Approvals, and Sustainability of Run-of-River Independent Power Projects in British Columbia by Tanis Douglas. Coquitlam, British Columbia, Canada: WWSS, 2007. http://www.watershed-watch.org/publications/files/Run-of-River-long.pdf (accessed April 11, 2013).

WaveHub. About WaveHub. Last modified 2010 (accessed March 8, 2012). http://www.wavehub.co.uk/about/.

Wesoff, E. Greentech Media. Update: Solar Firms Setting New Records in Efficiency and Performance. Last modified February 28, 2012 (accessed April 11, 2013). http://www.greentechmedia.com/articles/read/Update-Solar-Firms-Setting-New-Records-in-Efficiency-and-Performance.

Wigglesworth, R. Gulf shifts focus to domestic energy. *Financial Times*, January 17, 2011. http://www.ft.com/intl/cms/s/0/be10db70-225e-11e0-b91a-00144feab49a.html#axzz1zxFc5jha (accessed June 29, 2012).

World Energy Council (WEC). *Survey of Energy Resources 2007*. London, U.K.: WEC, 2007. http://www.worldenergy.org/documents/ser2007_final_online_version_1.pdf (accessed January 10, 2013).

Worldwatch Institute. *The End of Nuclear* by Mycle Schneider, Antony Froggatt, and Steve Thomas. Washington, D.C.: Worldwatch Institute, 2011. http://www.worldwatch.org/system/files/pdf/WorldNuclearIndustryStatusReport2011_%20FINAL.pdf (accessed January 22, 2013).

3

Germany's Energy Revolution

José Carlos Thomaz, Jr. and Sean Michalson

CONTENTS

3.1 INTRODUCTION, DEFINITIONS, AND SETUP

Although Germany may have fallen out of the world's spotlight after reunification and the end of the Cold War more than two decades ago, the country continues to exert a major influence on the world. Its *Energiewende* (energy revolution) will most certainly increase this influence in the years to come. This term refers to the recent politically driven initiative to replace Germany's current energy mix, which is based on fossil and nuclear energy sources, with a matrix based almost exclusively on renewable energy sources. Although this effort has been underway since 2000, the 2011 nuclear disaster at Fukushima, by making the energy revolution more relevant to the public discourse, accelerated its pace. With this "revolution," Germany hopes to free itself from finite, environmentally harmful energy sources. The program has the mindset of the German population and the focus of German politicians to thank for its present success, which has, in addition to averting harmful atmospheric emissions, contributed to the development of cutting-edge renewable energy technology. In addition, the German people's willingness to take on this unprecedented program and the economic incentives created to fuel it has inspired many other countries.

This chapter explores the most important factors that influenced Germany's transformation from a country completely dependent on fossil fuels into a world leader in clean energy. Our first goal, therefore, is to discover what led the country to adopt the *Energiewende*. Our ultimate goal, however, is to seek lessons from the German experience that may benefit other countries developing their own renewable energy programs.

This first section will cover some basic information about renewable energy, while focusing specifically on the German context. Green energy has gradually increased its share of global energy production, especially in developed countries where environmental responsibility has become a contested topic. In fact, 9 of the top 10 countries that import and export renewable energy technology are considered "developed." Combined, they represent between 70% and 80% of the world's wind and solar energy device markets (Groba and Kemfert 2011). This demonstrates that the market for renewable energy, at least on a national level, is highly concentrated.

This chapter investigates the primary factors that led Germany to become one of these countries driving the push for renewables. Its influential role in the clean energy industry is especially interesting because it is somewhat counterintuitive, given that, despite its poor geographic location with respect to solar energy generation, the country has invested heavily in the expansion of clean energy, especially solar. This chapter, therefore, strives to understand what has prompted this massive investment and whether this multifaceted cause might be transferable to other countries as they follow their own paths to sustainable energy independence. To achieve this, we will analyze the factors (such as historical events and public policy) that compelled Germany to increase its renewable energy production capacity and to improve its energy efficiency.

Continuing along the lines of basic, introductory information, it is widely acknowledged that renewable energy sources are the most logical way to ensure a sustainable energy supply. The most common renewable energy sources are wind, solar, hydro, biomass, waste, geothermal, and tidal. However, not all of these sources are equal. "Green energy," in this chapter, is defined as a subset of renewable energy. Green energy's extraction, distribution, and consumption do not cause significant harm to the planet. For example, the production of heat by burning biomass is viewed as a renewable process, but might not be classified as green because burning this fuel will release substantial carbon emissions. Likewise, nuclear energy cannot be classified as green because it is not generated from a renewable source, as radioactive fuels (e.g., uranium, plutonium) are finite. Furthermore, nuclear power plants produce environmentally hazardous, long-lived nuclear waste as a by-product. Interestingly, although there is a conceptual difference between renewable and green sources, the energy sector does not seem to distinguish between the two.

There are also conceptual differences between the production and usage of energy resources. Table 3.1 illustrates the flow of energy during 2011 in Germany. It is clear from this diagram that most of the energy used by the country was imported (amounting to 11,288 PJ) in the forms of oil, gas, and nuclear fuel. These imports, along with the amount of primary energy produced domestically, yielded the total "energy available" (which amounted to 15,261 PJ). Apart from a small percentage that is used

TABLE 3.1

Breakdown of Germany's Energy Production, Transmission, Distribution, and Usage in 2011 (in petajoules)

		Amount of Energy (PJ)
Source		
Domestic production		4241
Import	+	11,090
Stock depletion	+	77
Total energy available inland		15,409
Consumption		
Total energy available inland		15,409
Export and bunkering	–	1887
Total energy available for primary energy usage[a] (coal, oil, gas, nuclear, renewable)		13,521
Non-energy usage	–	1004
Transformation losses	–	3292
Usage within the energy sector	–	501
Energy available for end energy usage[b] (electric/fuel thermal)		8744
Industry	–	2624
Transportation	–	2572
Household	–	2194
Services	–	1355

Source: Adapted from BMU. *Erneuerbare Energien in Zahlen: Internet-Update ausgewählter Daten* by F. Musiol, P. Bickel, T. Nieder, T. Rüther, M. Walker, M. Memmler, S. Rother, S. Schneider, and K. Merkel. Paderborn, Germany: Bonifatius GmbH, 2012.
Note: Statistical discrepancy: –19 PJ.
[a] From renewables: 1465 PJ.
[b] From renewables: 1054 PJ.

for bunkering (offshore fuel operations) and export, this is also the "primary energy used" (which amounted to 13,398 PJ; Bundesministerium für Umwelt, Naturschutz und Reaktorsicherheit [BMU] 2011a, p. 96).

Primary energy, according to the United Nations (UN), is the energy that can be extracted directly from natural (primary) sources (such energy is contained, for example, within crude oil, natural gas, coal, wind, the sun, etc.). Hence, it is a measure of the usable energy contained in the virgin forms of natural resources. The UN states that "Secondary energy should be used to designate all sources of energy that results [sic] from transformation of primary sources" (Overgaard 2008, p. 3). Secondary energy includes the electricity, heat, and fuel products (such as those produced by power plants and refineries) that are derived from primary energy sources.

Consequently, secondary energy comprises most of the energy used personally and economically by industrial customers, vehicles, households, and service providers. In the process of transforming primary energy into various forms of secondary energy, nearly a quarter of the energy is lost. As transformation losses increase, efficiency decreases, resulting in higher demand for primary energy sources. Increasing the efficiency of its energy transformation and usage has been, and continues to be, an essential component of Germany's strategy for achieving the *Energiewende's* goals.

3.2 THE GERMAN ENERGY SECTOR

Examining Germany's current energy matrix is a natural starting point for our analysis. This section will provide a brief overview of (1) the current composition of Germany's energy sector, (2) its historical development, and (3) the energy matrix's projected future development. We will concentrate not only on energy production and consumption but also on energy efficiency. It should also be noted that our analysis will, at times, concentrate specifically on renewable energy sources.

First, to provide a contemporary context, the German Federal Office for the Environment (Umweltbundesamt, UBA) has stated that significant changes have occurred in the energy sector since 1990 (UBA 2011). These changes have resulted in coal usage decreasing 50%, natural gas usage increasing nearly 25%, and a meaningful expansion of renewable energy generation capacity. As of 2010, renewables covered 10% of Germany's energy demand. The UBA expects this figure to increase to 18% by 2020 (ibid.). Figure 3.1 supports this assertion, showing the development of the main energy sources' shares of Germany's energy consumption matrix since 1950, as well as the forecast for primary energy consumption trends through 2050. Data from 1950 to 1987 were normalized to account for the absence of the former East German states by estimating how energy consumption by the former East Germany would have increased the total energy consumption in the current united Germany (Arbeitsgemeinschaft Energiebilanzen e.V. [AE] 2011; BMU 2010b; Burckhardt 1990). This estimation was based on a comparison of the gross domestic product (GDP) of West Germany in 1987 (US$1.26 billion) and the GDP of the united Germany in 1990 (US$1.71 billion; The World Bank Group [WBG] 2011c).

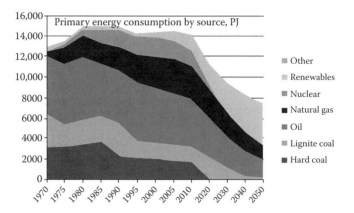

FIGURE 3.1

Share of each type of energy source. (Adapted from AE. Auswertungstabellen 1990–2011. http://www.ag-energiebilanzen.de/viewpage.php?idpage=139; BMU. Struktur und Höhe des Primärenergieverbrauchs im Leitszenario 2010. http://www.umweltbundesamt-daten-zur-umwelt.de/umweltdaten/public/document/downloadPrintImage.do?date=&ident= 22569; K. Borchardt. Zäsuren in der wirtschaftlichen Entwicklung. *Vierteljahrshefte für Zeitgeschichte* 61: 21–34, 1990.)

Figure 3.1 also highlights consequential moments in the energy sector's history. For example, in 1973, the first major oil shock of the 1970s, caused by the Organization of Arab Petroleum Exporting Countries (OAPEC) oil embargo, greatly reduced oil consumption and led Germany to promote nuclear energy. The oil shock of 1979 was followed by a further strengthening of the transition to nuclear energy. As a result, nuclear energy increased its share of the energy mix to more than 10%. The Chernobyl nuclear accident in 1986 interrupted this expansion, however, and spurred a search for alternative energy sources. Thereafter, nuclear power's share of the energy mix stopped expanding (as can be seen in Figure 3.1). However, this did not translate into other energy sources expanding their percentages of the energy mix (aside from the gains made by renewable energy sources), as energy production actually decreased between 1986 and the financial crisis in 2008 because of increased efficiency.

The global financial crisis in 2008 was the next event that had a major effect on Germany's energy trends. In the wake of the crisis, the demand for all energy sources declined due to the economic contraction. Nuclear power and coal power have not yet recovered, as their fractions of the energy mix remain near their depressed post-2008 levels. Conversely, renewable energy has maintained its pre–financial crisis growth. This trend is, in fact, expected to continue, as forecasts released in 2010 by the German Federal Ministry of the Environment, Nature Protection, and Nuclear Reactor Safety (BMU) state that renewable energy is expected to displace conventional energy sources over the next 40 years (BMU 2010b). In particular, by 2050, renewables are expected to satisfy 60% of Germany's total energy requirements and 80% of the country's electricity demand (BMU 2011a). This forecast represents an explicit goal of the *Energiewende* and is being supported by the German government, as will be detailed later in this chapter. Figure 3.2 displays this growth in renewables, tracing the share of renewable energy in the German matrix over time. Data after 2010 stems from the BMU's predictions. In accordance with another of the *Energiewende*'s goals, the BMU expects nuclear energy to disappear during the third decade of the twenty-first century (AE 2011; BMU 2010b).

The projected future growth of renewable energy's market share in Germany can be attributed to two distinct facts. First, as shown in Figure 3.2, green energy's installed capacity is expected to grow in absolute terms, whereas other sources' generation capacities are expected to decline. This is a positive shift from an environmental standpoint because coal currently supplies 23% of the country's primary energy (AE 2011). Shifting away from coal is necessary for Germany to reach its emission reduction goals, as coal is the most polluting fossil source, emitting 98 g of CO_2 per megajoule (compared with 50 g of CO_2 per megajoule for natural gas using modern technology, the cleanest fossil fuel [assuming no carbon capture and sequestration]; U.S. Energy Information Administration [U.S.

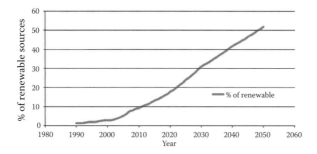

FIGURE 3.2
Percentage of renewable energy share in Germany. (Adapted from AE. Auswertungstabellen 1990–2011. http://www.ag-energiebilanzen.de/viewpage.php?idpage=139; and BMU. Struktur und Höhe des Primärenergieverbrauchs im Leitszenario 2010. http://www. umweltbundesamt-daten-zur-umwelt.de/umweltdaten/public/document/download-PrintImage.do?date=&ident=22569.)

EIA] 2011). Second, overall energy consumption in Germany is expected to decrease gradually. This is extraordinary for a country that has experienced an average (uncompounded) annual GDP growth rate of more than 5% since the 1970s and is expected to continue to grow (Statistisches Bundesamt [Destatis] 2012, p. 14). Since the reunification of the country in 1990, there has been no significant growth in energy consumption. Figure 3.3 shows how primary energy and GDP have trended over the past four decades (energy consumption data were normalized to account for the former East Germany; WBG 2011c).

According to Dr. G. Stadermann, of the Association for Research of Renewable Energy, the overall reduction in energy consumption per unit of GDP will continue due to increasing energy consumption efficiency (Stadermann 2011). He predicts that the decoupling of economic growth from energy consumption, as displayed in Figure 3.3, will continue, but will require broad efforts. Of the possible alternatives to foster more efficient consumption, changing consumer mindsets will prove the least expensive and most effective. Raising consumer awareness tends to cause individuals to (1) consume energy in a smarter way, (2) implement energy-saving measures such as home heating improvements, and (3) tilt

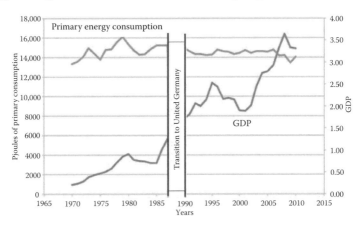

FIGURE 3.3

Energy consumption and GDP in Germany. (Adapted from AE. Auswertungstabellen 1990–2011. http://www.ag-energiebilanzen.de/viewpage.php?idpage=139; WBG. World Databank: World Development Indicators [WDI] and Global Development Finance [GDF]. http://databank.worldbank.org/ddp/html-jsp/QuickViewReport.jsp?RowAxis=WDI_Ctry~&ColAxis=WDI_Time~&PageAxis=WDI_Series~&PageAxisCaption=Series~&RowAxisCaption=Country~&ColAxisCaption=Time~&NEW_REPORT_SCALE=1&NEW_REPORT_PRECISION=0&newReport=yes&ROW_COUNT=1&COLUMN_COUNT=44&PAGE_COUNT=1&COMMA_SEP=true.)

consumption toward more efficient durable goods (e.g., appliances and electronics; ibid.).

To raise awareness, the government has instituted programs to spur the adoption of energy-efficient technologies. We detail here several, but not all, of these initiatives. First, the government has incentivized real-estate owners to boost energy efficiency by introducing an "energy certificate" program. This program, which has a legal basis in the Energy Saving Ordinance of 2007 (EnEV), stipulates that energy certificates be published for new construction projects and for existing buildings that are to be put on the market, detailing these buildings' energy usage rates (Bundesministerium für Wirtschaft und Technologie [BMWi], and BMU 2011, p. 22). Although the program does not set specific efficiency standards, it does allow potential buyers to easily compare different buildings' energy usage profiles. Since 2008, every building placed on the market has had such a certificate. Studies from the Association for Research of Renewable Energy have shown that buildings with better energy certificate ratings have recently increased in value relative to their peers (BMWi and Forschungs Verbund Erneuerbare Energien [FVEE] 2009). In this way, energy efficiency transparency has created a financial incentive for property owners and developers to increase their properties' energy efficiencies, helping the government to achieve the energy consumption reduction goals essential to the *Energiewende*.

To further encourage efficiency improvements, the German government has created the brand "Energy Efficiency: Made in Germany." This brand is intended to increase exports of efficient technologies by small and midsized businesses, helping them to lower costs through economies of scale (Kramer 2011, p. 5). The program provides businesses with free international advertising and access to energy efficiency experts, who help participants improve their products. The ultimate goal of the program is to further develop efficient technologies, making them cheaper and more attractive. The government hopes that, by opening new markets to German companies, the firms will achieve greater scale, making their products more affordable for both domestic and international customers. The initiative will help achieve the *Energiewende*'s goal of reducing new buildings' energy usage by 80% by 2050, whereas also increasing industrial energy efficiency.

Technological change outside the "Energy Efficiency: Made in Germany" program will also enable Germany to increase its energy-related efficiencies. In support of this notion, Dr. Stadermann notes that renewable energy production is not as efficient as ultimately achievable, meaning that, as technologies continue to improve, efficiency will increase (Stadermann 2011). Despite renewables' relative inefficiency when compared with fossil energy sources, Stadermann argues that the inefficiencies of wind or solar power production are offset by these power sources not emitting harmful pollutants. Hence, renewables should replace fossil fuels in the energy matrix (ibid.). Additionally, planned improvements to the German electrical grid will decrease renewables' transformation losses. These factors will allow the country to continue the trend of the past three decades, during which it surpassed many other countries in terms of energy efficiency. Figure 3.4 shows Germany's GDP per unit of energy consumed rising above the world average, as well as above the values of China and the United States (WBG 2011c).

We now offer a few additional examples of energy efficiency programs that the German government has been implementing. In 2010, the government published a new report, called the "Energy Concept," which laid down "measures for the development of renewable energy sources, power grids and energy efficiency" that will speed the *Energiewende* (BMWi and BMU 2011, p. i). For example, to foster the development of offshore wind turbines, the federal government promised to offer five billion Euros in loans to eligible programs (ibid., p. 8). The plan also detailed new measures

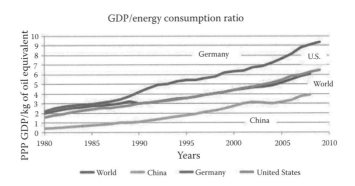

FIGURE 3.4

GDP per oil equivalent consumption in different countries. (Adapted from WBG. World Databank: World Development Indicators [WDI] and Global Development Finance [GDF]. http://databank.worldbank.org/ddp/html-jsp/QuickViewReport.jsp?RowAxis=WDI_Ctry~&ColAxis=WDI_Time ~&PageAxis=WDI_Series~&PageAxisCaption=Series~&RowAxisCaption=Country~&ColAxisCaption=Time~&NEW_REPORT_SCALE=1&NEW_REPORT_PRECISION=0&newReport=yes&ROW_COUNT=1&COLUMN_COUNT=44&PAGE_COUNT=1&COMMA_SEP=true.)

to improve energy efficiency. For example, the government promised to make energy efficiency "an important criterion for awarding public contracts" (ibid., p. 11). In addition, it will create an "Energy Efficiency Fund" that will support "the market introduction of highly efficient cross-application technologies (e.g., engines, pumps, refrigeration)," for the "optimisation of energy-intensive manufacturing processes" and for "especially innovative technologies to enhance energy efficiency" (ibid., p. 13). The last program to be addressed here is the energy efficiency–centric "Building Rehabilitation Programme," which aims to increase the rate of building renovation from "approximately 1% to 2% per annum" (ibid., p. 22). This program will support the aforementioned goal of reducing the primary energy demand of buildings by 80%, as three-fourths of all buildings in Germany were constructed before any substantial energy efficiency standards were established (ibid., p. 13). Combined, these efforts, along with others detailed in the "Energy Concept," are expected to double energy productivity in the country in the coming decades.

Moving away from energy efficiency in Germany, the rest of this section will focus on the current state of renewables. In 2010, renewables generated 10.3% of Germany's total energy supply, up from 9.3% in 2009. The effect of renewable energy was especially significant in gross electricity consumption, as renewables supplied 17% in 2010 (AE 2011). According to Eurostat (the statistics repository of the European Commission), in 2010, renewables also generated 9.5% of the thermal energy and 5.8% of the fuel energy consumed in Germany (AE 2011). Table 3.2 illustrates renewable energy's presence in each of the main sectors of the economy. The weighted sum of the percentages given in Table 3.2 yields the total renewable energy consumed in Germany, which amounted to nearly 10% of the total energy consumed in 2010 (AE 2011). Each of the different applications (electrical, thermal, and fuel) has different preferred primary energy sources. Thus, each application demands different renewable alternatives to replace its traditional fossil or nuclear energy sources. Table 3.3 illustrates this by using data from the BMU (2011a).

Something that cannot be seen in Table 3.3 is the fact that, until the 1990s, hydraulic power was the most prevalent renewable method of electricity generation. This changed toward the end of the century, however, with the installation of wind farms. Since then, the use of wind power has steadily increased, nearly doubling from 18,713 GWh in 2003 to 37,793 GWh in 2010 (BMU 2012a, p. 20). Solar energy has experienced a similar trend due to recent developments in photovoltaic (PV) technologies. The

TABLE 3.2

Consumption of End-Use Renewable Energy (by Sector and Form of Energy)

Usage	Industry (%)	Household (%)	Services (%)	Transport (%)
Electrical	6.0	4.3	6.8	0.3
Thermal	5.0	6.3	5.4	0.0
Fuel	0.7	0.5	0.2	5.7

Source: Adapted from AE. Auswertungstabellen 1990–2011. http://www.ag-energiebilanzen. de/viewpage.php?idpage=139.

TABLE 3.3

Shares of Renewable Energy Sources for the Three Main Energy Forms

Electrical (%)	Thermal (%)	Fuel (%)
Wind: 49	Biomass: 83	Biodiesel: 74
Photovoltaic: 31	Biogenic fuel: 9	Bioethanol: 24
Hydro: 9	Solar: 4	Vegetable oil: 2
Biomass: 9	Geothermal: 4	
Biogenic fuel: 3		

Source: Adapted from BMU. *Erneuerbare Energien in Zahlen* by F. Musiol, T. Nieder, T. Rüther, M. Walker, U. Zimmer, M. Memmler, S. Rother, S. Schneider, and K. Merkel. Paderborn, Germany: Bonifatius GmbH, 2011.

speed of development and introduction of this technology has been even more impressive than that of wind energy, as its capacity almost doubled from 9900 GWH in 2008 to 17,300 GWh in 2010. Figure 3.5a illustrates this and the development of the other renewable energy sources in electric energy production (BMU 2011a).

In Germany, thermal renewable energy applications are dominated by one source: biomass. Thermal energy is typically generated by small biogas plants, which, in 2011, had an average capacity of 390 kW each. These plants, which convert organic materials into methane through anaerobic digestion, are typically fueled either by energy crops (46% of their feedstock) or by manure (45% of their feedstock; Bundesministerium für Ernährung, Landwirtschaft und Verbraucherschutz [BMELV] and Fachagentur Nachwachsende Rohstoffe e.V. [FNR] 2012, p. 36). This methane gas is then combusted in combined heat and power units, which simultaneously produce electricity and heat very efficiently. Since 2001, biogas plants have demanded an ever-greater share of Germany's harvest, as the installed capacity of biogas plants soared from 111 MW in 2001

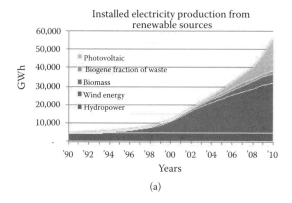

(a)

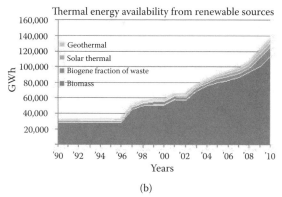

(b)

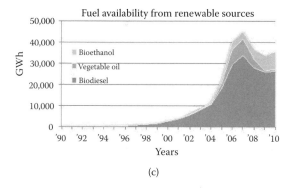

(c)

FIGURE 3.5
Production and availability of renewable energy for (a) electricity, (b) thermal, and (c) fuel production. (Adapted from BMU. *Erneuerbare Energien in Zahlen* by F. Musiol, T. Nieder, T. Rüther, M. Walker, U. Zimmer, M. Memmler, S. Rother, S. Schneider, and K. Merkel. Paderborn, Germany: Bonifatius GmbH, 2011.)

to 2728 MW in 2011. The most commonly used biomass crops include canola, sunflowers, corn, wheat, and potatoes (ibid., pp. 32, 34).

Biomass also dominates renewables' share of fuel production. Figure 3.5c illustrates the development of the biofuel market in Germany and shows that biodiesel is the dominant renewable fuel (accounting for 74% of all renewable fuels). It is refined mainly from a mixture of canola oil and methanol. Although Germany's biodiesel production expanded rapidly from 2000 until 2007, from 0.34 to 3.32 million tons, after 2007 it lost its tax-exempt status, leading to a decline in production. Beginning in 2008, the federal tax on biodiesel gradually increased, causing a major drop in demand due to higher prices. This caused production to decrease to 2.33 million tons as of 2011 (BMELV and FNR 2012, p. 22). At the same time, the production of bioethanol has increased steadily from 2009 to 2012, due in large part to incentives provided by the European Directive of 2009, which stipulated that its member countries should derive 10% of their transportation fuels from renewable sources, that is, biofuels (Deutsche Naturschutzring [DNR] 2012). However, as a result of 2012's poor harvests worldwide, on October 17, 2012, the European Union (EU) reversed course and proposed a new rule that would alter the original 10% target. Under the new rule, countries would be responsible for obtaining 5% of their transportation fuels from renewable sources. This measure is intended to slow the expansion of conventional biofuels, which are produced from crops that could otherwise be used for food production (e.g., corn, wheat, rapeseed, etc.). Under this new rule, the EU also hopes to encourage the production of second-generation biofuels, which are produced from plant materials that do not compete with food crops (such as the fast-growing, hardy jatropha bush or agricultural residues; ibid.). Given this new target and the intense debate over the use of food crops for biofuel production, the futures of bioethanol and biodiesel in Germany are uncertain. This could either hamper Germany's transition to renewables in the transportation sector or it could open the doors for new technologies such as electric vehicles. As of this writing, the future role of renewable energy sources in the transportation sector is still unclear.

In summary, the share of renewable sources on an application-by-application basis is shown in Figure 3.6. Biogenic fuels comprise approximately 70% of renewable energy production (biogenic fuels are renewable energy sources that stem from living organisms—e.g., methane produced in landfills or in cattle confines) (BMU 2011a). This is not surprising because

FIGURE 3.6

Structure of renewable energy supply in Germany. (Adapted from BMU. *Erneuerbare Energien in Zahlen* by F. Musiol, T. Nieder, T. Rüther, M. Walker, U. Zimmer, M. Memmler, S. Rother, S. Schneider, and K. Merkel. Paderborn, Germany: Bonifatius GmbH, 2011.)

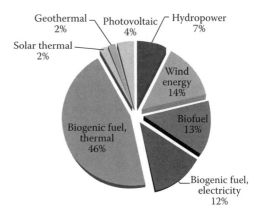

biomass is the most important source for heat production, and heat accounts for more than half of renewable energy usage (Figure 3.5b).

Despite the uncertain future of its biofuel industry, Germany stands at the forefront of Europe's renewable energy industry. According to a study by Eurostat, Germany was one of only two countries (the other was Hungary) in the EU that easily exceeded the 2010 objective (set in 2001) for energy-mix share attributable to renewable sources. Figure 3.7 displays the targeted and achieved shares of renewable energy for each EU country (European Environment Agency [EEA] 2008).

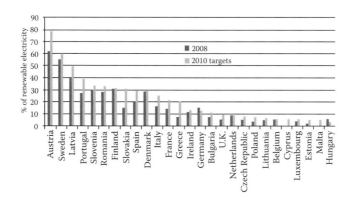

FIGURE 3.7

Share of renewable energy in electricity consumption. (Adapted from EEA. Renewable Electricity as a Percentage of Gross Electricity Consumption, 2008. http://www.eea.europa.eu/data-and-maps/figures/renewable-electricity-as-a-percentage-2.)

3.3 SUMMARY OF EVENTS CONTRIBUTING TO THE RISE OF RENEWABLES

A study by the Technische Universität Berlin synthesized the most important events leading to the current state of Germany's energy sector (BMU and AEE 2010). Table 3.4 is an extension of this study, as it also encompasses more recent events. These events are divided into three categories: (1) triggers for a new paradigm, (2) milestones for energy and climate politics, and (3) legislative actions.

TABLE 3.4

Chronology of Historical Facts Relevant to the *Energiewende*

	1972	Book published: The Limits to Growth
	1973–1974	First oil crisis
	1979–1980	Second oil crisis
	1986	Chernobyl nuclear accident
	1988–1990	Committee of inquiry: "Protecting the Earth's atmosphere"
	1991	Electric supply act (StrEG)
	1992	UN Framework Convention on Climate Change
	1992–1995	Committee of inquiry: "Protection of the atmosphere"
	1996	EU Directive on the internal electricity market
	1997	Kyoto Protocol
	2000	Renewable energy act (EEG)
	2001	EU Directive to promote electricity from renewable energy sources
	2002	Offshore wind strategy of the federal government
	2004	EEG revision 2004
	2005	National climate change strategy
	2007	Integrated energy and climate program of the federal government
	2008	Financial crisis
	2009	EEG revision 2009
	2009	EU directive to promote the use of energy from renewable sources
	2011	Fukushima nuclear accident
		Triggers for a new paradigm
		Milestones for energy and climate politics
		Legislative sanctions

Source: Adapted from BMU. Agentur für Erneuerbare Energien e.V., *20 Jahre Förderung von Strom aus Erneuerbaren Energien in Deutschland—eine Erfolgsgeschichte* by E. Bruns, D. Ohlhorst, and B. Wenzel. Berlin: Agentur für Erneuerbare Energien e.V., 2010.

Figure 3.8 displays the effect each event had on the production and consumption of energy in Germany. The chronological data presented in Table 3.4 were superimposed onto a timeline showing the evolution of Germany's energy mix (from 1978 to 2010) for clearer visualization. As can be seen in Figure 3.8, certain historical events caused different kinds of shifts in the energy paradigm. The oil crises of 1973 and 1979 and the financial crisis of 2008 are good examples of events that caused rapid changes, as described briefly in the introduction to this chapter. In most instances, however, effects manifested themselves more gradually. The sequence of events that took place throughout the 1990s and 2000s serves as an example of this slow change (AE 2011; BMU and AEE 2010). Most of these events were in the forms of laws and directives that took effect only gradually. Overall, it seems that unexpected events in Germany's history (such as crises and accidents) forced immediate reactions, whereas planned interventions led to creeping, long-term results.

Other factors (such as German culture) were not included in the Technische Universität's study because they have neither definitive starting dates nor exact ending dates. This chapter's analysis includes some of these factors and classifies all the included factors into the categories of

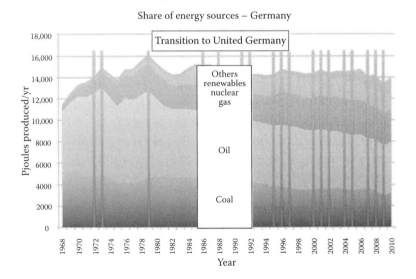

FIGURE 3.8
How the events (the bars on the chart) affected energy production and consumption in Germany (from 1970 until 2010). (Adapted from AE. Auswertungstabellen 1990–2011. http://www.ag-energiebilanzen.de/viewpage.php?idpage=139; and K. Borchardt. Zäsuren in der wirtschaftlichen Entwicklung. *Vierteljahrshefte für Zeitgeschichte* 61: 21–34, 1990.)

(1) historical events, crises, and endowments; (2) socioeconomic aspects; and (3) regulatory and political events. They range, respectively, from more exogenous factors (i.e., the German government had little control over them) to more endogenous factors (i.e., the state had a greater degree of control). The following sections will address the factors listed in Table 3.4 and provide an overview of some of the culturally significant historical developments that contributed to the realization of the *Energiewende*, as all these events are crucial to building a relatively complete understanding of the German willingness to undertake this project.

3.4 HISTORICAL CONTEXT AND ENDOWMENT FACTORS

The first factors relate to Germany's history and natural resource endowment, each of which has played a major role in determining the country's current energy policy. For centuries, German culture has revolved around the country's landscape, creating a strong bond between the people and the environment. This has led to a history rich in environmental activism, creating a base of support for the *Energiewende*. In addition, Germany's relatively small endowment of fossil fuels, with the exception of its relatively large coal reserves, has led the country to become dependent on oil and other energy imports, raising concerns about its energy security. This section will detail how these factors, combined with exogenous shocks, help explain the rise of the *Energiewende*.

3.4.1 Early History of the Environmental Movement in Germany, from the Roots of Sustainability to the National Socialists' "Blood and Earth"

Germany's historical identity has been associated with its environment—especially the German forest—for millennia. The German people embraced *Naturschutz* (the protection of nature) early in the modern period. This protectionism has been bolstered not only by environmental activists but also by entire cultural, artistic, and literary movements that, when taken as a whole, have created a rich historical narrative. This section will not attempt to discuss all of these different movements; rather, it will provide a brief summary of the cultural history surrounding German environmentalism by focusing on several focal points that helped define the country's relationship with the environment. We hope to convey a sense of the rich legacy on which the *Energiewende* is founded, but with the important caveat that this section does not provide a complete account of the factors that created the present-day concern for the natural world. Several moral principles have also played key roles during this history and, because of their influence on German thought today, will be discussed in this section and in Section 3.4.2.

The first well-known literary association of the German people with their environment stems from the writing of *Germania* by the Roman historian Tacitus. In this text, written around 100 AD, Tacitus described the German forest as a sinister, terrifying landscape (Hohls 2011, pp. 2–3). This established the foundation for the famed *Mythos Wald* (forest mythos) that conceptualized the forest as a mystical place filled with mysterious, even dangerous, creatures whose workings defy human comprehension. Tacitus focused on the negative aspects of this impenetrable forest, but this mythical incomprehensibility has also had positive connotations in the literary world. Nevertheless, for hundreds of years since *Germania*, the image of a foreboding forest would contribute to the German perception of the forest as a wild place that could either teach lessons or punish foolishness and wickedness. This is exemplified by many German myths and fables, such as those gathered and adapted by the Brothers Grimm (ibid.).

Out of this unique German relationship with the forest sprang the concept of *Nachhaltigkeit* (sustainability) in the eighteenth century. This concept strengthened the country's commitment to environmental preservation. Faced with a wood shortage in 1713, Hans Carl von Carlowitz enforced a policy of sustainable forest management in Freiburg, as the area's industries and mining activities had led to clear-cutting of much of the nearby wooded land (Schmidt 2012, p. 50). von Carlowitz recognized the importance of the resource wood and subsequently wrote *Sylvicultura Oeconomica* to outline his forest-management plans. The defining idea of this work comprised a system where only so much wood would be harvested as was expected to grow back in the same year. This was the first time the principle of sustainability had been applied consciously to any industry in Germany. von Carlowitz's idea quickly spread and would become a "cornerstone for the German forestry industry" (ibid., p. 51). This led, in turn, to the expansion of the movement throughout Germany, contributing greatly to the establishment of sustainable development.

Georg Ludwig Hartig, a forester famous during the eighteenth and nineteenth centuries, further helped develop the sustainability idea by providing it with a justification. He believed that sustainable forestry's purpose was to ensure that "posterity can extract at least exactly as much benefit (out of the forest), as the currently living generation dedicates to themselves" (Schmidt 2012, p. 50). This was the first time a German had expressed the idea of a *Generationenvertrag* (generational contract) that obligated living generations to preserve the natural environment so that future generations could benefit from it at least as much as current generations themselves had (ibid.). This contract is a subconcept of a broader theme in the German cultural consciousness and political sphere: the *Verursacherprinzip* (causative principle), which demands that whoever causes a problem also deal with its consequences. We will focus specifically on the generational contract concept in this chapter, but it should be noted that the broader *Verursacherprinzip* also plays a major role in the societal discourse about pollution, fossil fuels, etc., as evidenced by the German willingness to pay for the *Energiewende*. This demonstrates that the German people are largely willing to pay for the damage caused by fossil fuel usage now by creating a sustainable energy sector for the benefit of future generations. The costs of the *Energiewende*, therefore, stand as a massive indemnification paid to the unborn for damages caused by present consumption, honoring the *Generationenvertrag* and *Verursacherprinzip* simultaneously (ibid.). The two historical ideas of sustainability and the *Generationenvertrag* show how the environmental movement's early history contributed greatly to Germany's current social and political situations.

The next major epoch to be detailed in this section occurred during the nineteenth century, when the literary and artistic movement of the *Waldromantik* (forest romanticism) emerged. German romantics such as Ludwig Tieck and Joseph von Eichendorff embedded in the German cultural consciousness the notion that the forest and natural environment should be regarded as an "ideal world" (Hohls 2011, p. 4). Furthermore, they asserted that the environment could serve as an escape from the " 'hard cold' everyday life," which increasingly became a reality in an industrialized Germany (ibid.). Eichendorff went so far as to praise the forest, in his poem *Abschied*, as a "symbol for the beautiful and the good and eternal life" (ibid., p. 5). This school of thought would not inform widespread environmental movements until the turn of the twentieth century, but it did influence one of the first episodes of German *Naturschutz*.

In 1823, near the birth of the *Waldromantik*, it was decided that construction of the famous, albeit unfinished, Kölner Dom should be completed.

This required that stone from the nearby Drachenfels, which had served as the original quarry for the dome, be mined (Gudermann 2011). However, a grassroots movement arose to combat this further degradation of the Drachenfels, which had become an artistically treasured landmark. The protestors eventually triumphed and established the quarry as the first federally protected natural landmark in Germany. It stands to this day as a symbol for the genesis of German environmental protectionism and foreshadowed the romantic revolution that would sweep the country at the turn of the next century (Gudermann 2011).

The German people, beginning in the late 1800s, increasingly embraced a romanticized notion of the forest and of nature as a result of industrialization, even as Germany itself passed out of the romantic era. Beginning especially around 1900, the many people who had moved to cities looking for work began to travel in increasing numbers to heavily forested regions for vacation (Hohls 2011, pp. 6–7). They wanted to escape the rapid changes brought on by industrialization and to immerse themselves in a peaceful atmosphere they associated with their pasts. They sought this harmony, due in large part to the *Waldromantik* movement, in the forest. This resulted in an elevation in its importance and of the environment in general, a circumstance that persists to this day, as many families still regularly enjoy taking daytrips to or vacations in isolated wooded regions.

However, just as this love of the woods swept through the country, its very cause, that is, industrialization, endangered the existence of the traditional German forest. The enormous fuel needs posed by factories, shipbuilding, and household heating required tremendous amounts of wood. After an initial period of deforestation to meet this gluttonous demand, the German nobility interceded in many areas to establish a system of *Forstwirtschaft* (forest management), relying heavily on the sustainability principles developed during the eighteenth century (Hohls 2011, p. 7). This ensured the survival of the German forest. However, in the place of the traditional German linden and oak trees, new cultivated, man-made plantations of spruces and pines arose (ibid.). The German people were not satisfied with these synthetic forests and longed for the woods praised by the *Waldromantik*.

The longing to escape from the modern world into nature, and the resultant desire for environmental protection, raised an interesting paradox, which Professor Wolfgang Haber has identified. He argues that the spread of environmental protectionism had as its precondition the satisfaction of city dwellers' basic needs (Haber 2009, p. 23). These were the citizens who originally mobilized to defend the green, natural settings to which they could escape. However, to satisfy their needs, natural resources had to be extracted on a monumental scale, and this lust for resources was exactly what had endangered the environment in the first place. Thus, Haber concludes that the economic development that enabled citizens to take up the cause of environmental protection was itself the cause of the environment's distress (ibid.). This has resulted in a strong skepticism toward economic development in the minds of German environmentalists, as many associate any form of economic progress as necessitating harmful resource exploitation, whether in the form of extractive industries or in the form of pollution. The result has been a constant conflict between economic progress advocates and some environmental protectionists in Germany since before the turn of the twentieth century.

This mistrust of economic development and dissatisfaction with the new, forester-cultivated woods gave rise to the *Heimatschutzbewegung* (protection of home movement). This movement's goal was, and continues to be, the preservation of the authentic German *Heimat*. This word is defined by the Bund Heimat und Umwelt e.V. (BHU), the umbrella organization of the cultural–environmental movement, as "the region, the landscape, the city, the village, in a word the room (or place), in which we live, which we design (or shape) and to which we therefore have a special responsibility" (Gotzmann 2010b). The Bund Heimatschutz, the BHU's predecessor, was founded to protect the *Heimat* by music Professor Ernst Rudorff in 1904. It devoted itself to the care of historical landmarks and to the defense of the environment. This concern for the environment was driven by Rudorff's belief that an integral part of the German *Heimat* was the "poetic animal and lower (or more base or fundamental) plant world" (Gotzmann 2010a). He perceived this world as being endangered by industrialization's desire to put all land to rational use. Thus, he saw industrialization as a threat to both the environment and German cultural landmarks, leading him to mobilize support to halt the advance of development. His efforts, along with those of his followers, have expanded over the decades, allowing the BHU to claim a membership of 500,000, most of who operate in local organizations (ibid.). The longevity of this organization underscores how deeply rooted the German concern for the environment is in the nation's cultural consciousness. It also demonstrates that the desire to protect the *Heimat* drives many Germans to oppose intrusive development projects. One consequence of this is the strength of Germany's "not-in-my-backyard" (NIMBY) movement, which opposes the defacing of landscapes caused by (among many projects) the construction of power lines and wind turbines.

The people's connection to the traditional German environment did not always serve positive ends, however. When the National Socialists rose to power in 1933, they twisted the romantic perception of the natural environment as nearly holy and the protectionist *Heimatschutzbewegung* to their own purposes. They recognized the strong psychological influence the environment had on the German people and perverted it by spreading the notion that the German environment, like the German people, was the archetypical form of nature. This *Blut-und Bodenideologie* (blood and earth [or land, ground] ideology) envisioned the German people united with a perfect environment, which the Nazis undertook to construct. The Nazis discarded the longing for a natural forest, replacing it instead with their manufactured conception of an "ideal" environment (Hohls 2011, p. 8). They thus exploited Eichendorff's aforementioned notion that the forest was a symbol of eternal life, transferring the eternality of the German forest to their "thousand-year Reich." Hermann Göring, the Nazi forest and hunting minister, believed "eternal forest and eternal people, they belong together" (ibid., p. 10). The powerful influence the environment exerted on the minds of the German people was so pronounced that the National Socialists used every opportunity to exploit it.

3.4.2 Modern Developments in the History of Germany's Environmental Movement: A Turn to the Precautionary Principle

The institutionalized environmental protectionism established under the Nazi regime (the legal foundation of which will be discussed in Section 3.5.2) characterized German environmental policy until the emergence of the antinuclear movement in the 1970s. Despite the broad powers granted to the government under the Nazis' environmental laws, little protectionist action was taken because rebuilding the nation's economy took priority, as is evident from the story of atomic power in Germany. This section will provide an account of the early history of the German antinuclear movement while relating it to how the *Vorsorgeprinzip* (precautionary principle), in this case through environmental issues, extended its influence over German decision-making over the past four decades.

From 1955 (when West Germany regained its sovereignty from the Allies) to 1969, the peaceful use of atomic power enjoyed broad public support (Rieckmann 2000, p. 5). This was followed by a period of protest against the expansion of atomic power. The protests seem to have drawn strength primarily from opposition to atomic power in Switzerland and France, growing skepticism toward technological progress, and a lack of governmental response to local complaints about new nuclear construction projects (ibid., p. 10). The confluence of these factors led to what is considered by many to be the birth of the organized antinuclear movement in Germany: the occupation of the nuclear plant construction site at Whyl. This occupation peaked on February 23, 1975, when 28,000 protestors occupied the site in a mass act of civil disobedience. Their persistence paid off on March 14, 1977, when a court ruled that the Whyl plant could be completed only if an expensive fail-safe protection device was erected around it. The project then lost its financial viability, thus halting construction (Radkau 2011).

The Whyl event arose from "fears of ecological and atomic catastrophes... and state repression (of protests against atomic power)" and led to the first National Conference of the Anti-Nuclear Power Plant Movement in 1977, elevating the issue to the national level (Rieckmann 2000, p. 14). These drastic actions suggest that the *Vorsorgeprinzip* influenced thinking at the time. This principle is intuitive: take (precautionary) action now to minimize the chance of suffering (usually substantial) damages later. The success of the protests shows that this principle held its ground as a valid counterargument against a strictly myopic, profit-oriented view. Furthermore, the adoption of the *Energiewende* demonstrates that the *Vorsorgeprinzip* is still salient today. We believe that the antinuclear movement helped legitimize the *Vorsorgeprinzip* in Germany's political discourse where environmental issues are concerned, thereby contributing to the success of the *Energiewende*.

Another lesson of the antinuclear movement relates to the power of German grassroots movements. Those living in the immediate vicinity of power plant construction sites, especially local farmers and vintners, ignited the nationwide movement, eventually bringing enough attention to the issue to make a difference. Much like the Drachenfels protest detailed in Section 3.4.1, this demonstrates that local protests can transform into national debates in Germany. In fact, without local efforts, the drawbacks of atomic energy may never have entered the national political discourse.

Another tangible outcome of the Whyl event was that it contributed to Germany's Social Democratic Party (SPD) becoming the first major political party to officially oppose atomic power in 1977 (Rieckmann 2000, p. 6). The SPD was then joined by the German Green Party in 1983, when the Greens first achieved representation in the Bundestag. The Greens focused on environmental issues during their early years, broadening and enhancing the political discussion about atomic energy and environmental protectionism. Despite this strong political opposition aimed at atomic energy, the federal government continued to promote nuclear power and its expansion, as the pro–atomic energy Christian Democratic Party (CDU) and the Free Democrat Party (FDP) had come into power in 1982 (ibid., p. 7). This divided stance regarding nuclear power would persist until the Chernobyl accident in 1986, which, along with the contemporary antinuclear movement, will be addressed in Section 3.4.4.

Another issue shared the environmental spotlight with atomic power during the 1980s: the *Waldsterben* (tree extinction). This term refers to the predicted deaths of thousands to millions of acres of German forest due to acid rain, which was caused by sulfur oxide emissions from coal power plants. The national media pounced on the issue in the fall of 1981, even though acid rain's role in the observed damage had not yet been proven scientifically (Radkau 2011). Nevertheless, because of the aforementioned German affinity for the forest, this story precipitated a massive uproar. The government reacted quickly, forcing German coal plants to adopt new emissions standards, which were implemented by the late 1980s and reduced sulfur oxide emissions, thus supposedly saving the forests (as the alleged die-off tapered off during this time; Kandler 1993). This forceful response to an issue surrounded by uncertainty provides another example of the German willingness to support (expensive) precautionary measures, rather than face the potential consequences of inaction. As a result, we conclude that this voluntary adherence to the *Vorsorgeprinzip* can be considered a defining societal trait that contributes to the German people's readiness to undertake the *Energiewende*.

As detailed in this and in the previous sections, the environmental movement in Germany has a rich history. The German love for the landscape and the perceived obligation to future generations to preserve it have influenced private and public behavior for more than two centuries. The aforementioned actions on the part of policy makers—among many other actions—stand as testaments. Hence, we confidently conclude that the history of the German environmental movement has played a pivotal role in the *Energiewende*.

3.4.3 Club of Rome: *The Limits to Growth* and the Two Oil Crises

Moving to a more international historic perspective, in the 1960s, some of the world's major economic players began to look at the issue of sustainability with greater concern. In 1968, a group of scientists, industrialists, and public leaders conceptualized the Club of Rome. They established this organization as a platform to discuss the "future of humanity" (one of the current co-presidents of the Club of Rome, Dr. Eberhard von Koerber, is German; H+M Media AG 2012). In 1972, the NGO published the groundbreaking *The Limits to Growth*, which postulated that the planet's finite natural resources would eventually constrain economic growth (Meadows et al. 2008). The publication sold 30 million copies in more than 37 languages. Germany's baby-boomer generation, in particular, helped transform the book into a global bestseller. *The Limits to Growth* thus helped disseminate awareness of sustainability among the German population.

Another major international event that influenced Germany's energy policy was the first oil crisis in 1973, which resulted from a contraction of the oil supply by the OAPEC countries, the Arab members of OPEC (Williams 2011). This coordinated reduction in the oil supply was in retaliation against the American government's support of the Israeli military during the Yom Kippur War, when Arab countries attacked Israel on its holiest holiday. The drop in production caused oil prices to quadruple by 1974, increasing the price per barrel to more than US$5 for the first time, before prices ultimately topped out at more than US$12 per barrel (ibid.). This crisis interrupted economic growth in Germany, causing the nation's GDP growth rate to decrease from 11.4% in 1973 to 4.8% in 1975. It recovered to 8.4% in 1976, reflecting the economy's adjustment to higher oil prices (Destatis 2012, p. 14).

The oil embargo led to significant increases in unemployment and to a number of insolvencies nationwide. Restrictions on Sunday car traffic and reductions in the maximum highway speed limit followed; and it is speculated that the widespread adoption of daylight saving time may have been associated with this oil crisis (European Federation for Transport and Environment 2005). Although one might have expected that worldwide increases in oil prices would raise demand for coal (as an alternative to oil), this did not occur. Figure 3.8 shows that coal consumption actually declined during the crisis. Although decreases in coal consumption could be attributed to the crisis-induced drop in economic activity, demand for coal did not subsequently rebound after the recovery because, as the oil crisis persisted, Germany began to develop alternative energy sources. Hydraulic and nuclear power thus achieved considerable shares of Germany's energy matrix. In particular, the amount of nuclear energy tripled between 1973 and 1977 (Paeger 2012).

The phenomenal growth in the atomic energy industry in Germany continued as a result of the war between Iran and Iraq in 1979. Geopolitical uncertainty in the Middle East caused prices to soar to a record high of US$39 per barrel before they stabilized to less than US$30 in 1983 (Williams 2011). Although the price spike was temporary, its negative effects influenced Germany's energy mix substantially for at least 2 years after the onset of the crisis (Figure 3.8).

After the second oil crisis, the world experienced additional disruptions in oil supply and high prices during the Gulf War in 1990 and the Asian currency crisis of 1997 (Williams 2011). Germany, however, was not as vulnerable to these crises, due in part to the success of previous initiatives to reduce the country's dependence on imported fuels. Figure 3.8 shows how nuclear power and gas absorbed a fair amount of these oil supply shocks. In addition, Germany's GDP growth rates following these crises did not decline compared with the respective previous years, further supporting the assertion that the economy had become increasingly less dependent on oil (Destatis 2012, p. 14).

3.4.4 The Changing German Attitude toward Nuclear Energy

As mentioned previously, Germany began investing heavily in nuclear energy after the first oil crisis. The country became well-known for its nuclear technology, allowing it to export nuclear power station equipment and nuclear sector–specific knowledge. Until the late 1970s, all major German political parties supported the development of this technology. Nuclear energy seemed advantageous because of its ability to weaken Germany's oil addiction. However, the accident at the Soviet Chernobyl nuclear plant in 1986 provoked scrutiny of the previously ignored risks of nuclear power.

This accident sparked the societal discussion about the long-term elimination of nuclear energy. Like the *Waldsterben*, it raised fears about the potential consequences of nonrenewable energy sources. Moreover, the disaster justified and reinvigorated the antinuclear movement (Rieckmann 2000, p. 7). The catastrophe also made Germany's future use of atomic energy a central theme in the 1986 elections, with the SPD and Green Party calling for the abandonment of nuclear power. The ruling CDU and FDP, however, were still unconvinced that this step would serve Germany's best interests, as it could cause short-term energy shortages. Nevertheless, the last of Germany's nuclear power plants were constructed from 1986 to 1992, after which a "de facto expansion moratorium" reigned. Thus, the accident at Chernobyl has been labeled "a historic blow for the use of atomic power" (Rieckmann 2000, p. 7). This assertion is supported by Figure 3.8, which shows that nuclear power has consistently provided roughly a quarter of Germany's energy supply since the mid-1980s.

This trend now faces disruption because, after the March 2011 Fukushima accident, voter pressure on politicians to end the use of nuclear power has greatly increased. This accident, much like Chernobyl, has affected both the citizens and politicians in Germany. Fukushima, however, seems to be the catalyst that will complete the trend Chernobyl started. Before the Japanese incident, the German government had only been discussing the shutdown of 17 German nuclear plants. Fukushima forced the issue, accelerating the adoption of a final resolution. This can be seen from German Chancellor Angela Merkel's reversal of her stance regarding nuclear power after the accident. In particular, in 2010, she overturned a plan to shut down all German nuclear plants by 2022, but then revoked this decision after Fukushima (*The Economist* 2010). After the accident, she demanded the immediate shutdown of Germany's seven oldest nuclear plants and reinstituted the prior gradual shutdown schedule for those remaining, noting, "The nuclear accident in Fukushima is a turning point in the history of the industrial world" (*The Economist* 2011).

This willingness to back away from nuclear power because of its inherent (and, some would argue, minimal) risks represents another example of how deeply the *Vorsorgeprinzip* is embedded in the German psyche. That most other countries made minimal changes to their energy policies following the Fukushima disaster supports this.

Despite this distinctive adherence to the *Vorsorgeprinzip*, the transition away from nuclear power will not come easily. In the initial years, the country will face a gap in its energy supply. To fill it, German electrical utilities plan to construct additional coal and natural gas plants. This short-term increase in fossil fuel–caused CO_2 emissions will, however, be only temporary while renewables gain ground. Supporting this prediction, in addition to much legislation, is the decision by a Berlin complex to expand its research staff from 1000 to 3000 over the next 2 years. The complex anticipates the need for more staff to satisfy increased demand for new and improved renewable energy technologies (Stadermann 2011). In addition, government investment in renewable energy is expected to increase dramatically in the coming decade. The Fukushima accident has thus proven to be an important catalyst in the transition to renewable energy in Germany, as it precipitated the *Ausstieg aus der Atomkraft* (exit from nuclear power). This proves that the tragedy has had, and will continue to have for decades, an effect on the entire world.

The *"ungelöste Entsorgungsfrage"* (unresolved disposal problem) represents the last aspect of Germany's interaction with nuclear power to be addressed here (Rieckmann 2000, p. 7). This problem refers to what the federal government will do with the atomic waste generated over the decades by the country's nuclear power plants. In the interest of the *Generationenvertrag*, activists have called for the creation of a secure final resting place for this radioactive waste. However, the word "secure" here is highly significant, as this security must be ensured for at least one million years (Bojanowski 2011). The environmentalists' case for abandoning nuclear power has been strengthened because this waste will burden countless generations.

This section has provided an overview of Germany's contemporary attitude toward nuclear power. A number of important lessons should be emphasized. First, as shown by the case of Whyl, grassroots campaigns can achieve tangible results in Germany. Second, when faced with uncertain future outcomes, the German people are willing, under the right circumstances, to follow the *Vorsorgeprinzip*. Finally, the public's concerns for nuclear waste disposal and its consequences for future generations show the influence still exerted by the moral concept of the *Generationenvertrag*.

3.4.5 Fossil Fuel and Renewable Energy Endowments

Much like many historical occurrences, a country's government cannot control its resource endowments. Germany, for example, has almost no oil resources, forcing it to import nearly all its oil. All uranium for nuclear plants is also imported. In contrast, the country has significant reserves of coal. In 2005, brown coal and hard coal supplied, respectively, 42% and 20% of domestically produced primary energy (UBA, Bundesanstalt für Geowissenschaft und Rohstoffe [BGR], and Destatis 2007). For clarification, brown coal is a low-carbon coal containing 25% to 35% carbon, whereas hard coal contains 92% to 98% carbon. Coal was an important strategic asset in the 1950s during Germany's "economic miracle." During this period, coal was used to generate 89% of the country's energy (Cramer 2010, p. 2).

Germany's abundance of coal reserves could have been an obstacle to the development of renewable energy, but two factors hindered its over-exploitation, thereby encouraging green energy. First, even before the environmental movements began in the 1960s, Germans were concerned about depleting their finite coal resources (Haus der Essener Geschichte 2003). This was certainly influenced by the aforementioned established sustainability focus of other German economic sectors. Second, as environmental awareness increased, the country began to worry about the greenhouse gas effect of coal. This was especially pertinent to the German people, as the notion of the *Generationenvertrag*, defined in Section 3.4.1, is still very prominent in the contemporary consciousness. Because of the extensive environmental research of the past decades about the detrimental effects of global warming, the German people have identified the greenhouse effect as a danger to posterity, spurring them to reduce their country's coal consumption. Figure 3.9 compares hard and brown coal CO_2 emissions with other energy sources' emissions. Germans have come to regard coal as dangerous because it emits roughly twice as much CO_2 as natural gas and around 20 times more CO_2 than renewable energy sources (Wissenschaftliche Dienste des Deutschen Bundestages [WDDB] 2007).

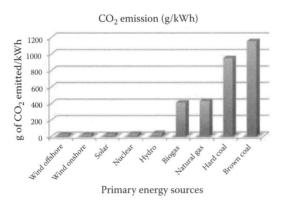

FIGURE 3.9
Comparison of different electricity sources' CO_2 emissions. (Adapted from Wissenschaftliche Dienste des Deutschen Bundestages, *Co2-Bilanzen verschiedener Energieträger im Vergleich* by D. Lübbert. Berlin: Wissenschaftliche Dienste des Deutschen Bundestages, 2007.)

In addition to its fossil fuel reserves, Germany possesses substantial endowments of some renewables such as wind (both onshore and offshore), but is lacking in other areas such as solar. According to estimates by the Risø National Laboratory of Denmark, most of the country experiences average wind speeds of 18 km/h (11 mph), reaching up to 31 km/h (19 mph) in the north (measurements taken at 50 m above ground; Troen and Petersen 1989). Among European countries, only the United Kingdom and the Scandinavian countries enjoy comparable wind speeds. Furthermore, Germany's access to the North and Baltic Seas is a strategic advantage, as these are high–wind speed regions (ibid.). Thus, the government is planning to construct large offshore wind farms (BMWi and BMU 2011, p. 5). However, some technical barriers must first be overcome, such as how to transport energy safely from offshore turbines to the continent. Moreover, a greater portion of Germany's economic activity occurs in the south, whereas the planned offshore wind farms will be in the north. Thus, additional power lines will need to be constructed to transfer the power not only to the continent but also from the north to the south.

Germany's development of solar energy faces even greater challenges. Its latitude and climate (characterized by cold and cloudy winters) create unfavorable conditions for solar energy production. For example, at the end of 2011, when the weather was particularly cloudy, more than 1.1 million households equipped with PV installations could not produce significant energy (Neubacher 2012). Moreover, the land available for PV energy production is limited, as solar panels are not permitted to compete with crops. Despite these disadvantages, Germany remains the country with the highest solar energy utilization, accounting (in 2010) for 43.5% of the world's installed solar capacity (Schünemann 2012).

3.5 REGULATORY AND POLITICAL CONTEXT

The political sector presents both general and idiosyncratic characteristics. In addition to the strong incentives created by the European directives and federal regulations, we observe through a more interpretative analysis that the German political mindset is another very important (and positive) differentiator. A number of guiding principles—such as the aforementioned *Vorsorgeprinzip* and the concept of the *Generationenvertrag*—steer politicians' decisions. Several other factors that comprise Germany's unique political context are discussed in the following sections.

3.5.1 The European Union

In 2009, the Energy Commission of the EU instituted a directive that affected the energy sectors of all the member countries. Paragraph 13 of Directive 2009/28/EC established "mandatory national targets consistent with a 20% share of energy from renewable sources and a 10% share of energy from renewable sources in transport in Community energy consumption by 2020" (European Parliament 2009). This directive is aimed at forcing compliance with the Kyoto Protocol, which was agreed upon in 1997 and took effect in 2005. In the EU context, a directive implies a legal obligation. However, unlike a regulation, it does not stipulate how the targets are to be achieved. This renewable energy directive has had a positive environmental effect in the European context. Since 1997, on average, the EU countries have nearly doubled the amount of renewable energy produced. Directive 2009/28/EC will ensure that this trend continues at a rapid pace.

Many individuals and organizations within Germany, however, have criticized the directive because of its lack of ambition. Although it has promoted the development of the renewable energy sector in the EU, many Germans feel that its goals are too lenient (Fell 2010). As illustrated in Figure 3.7, Germany and Denmark attained their goals far in advance of the 2020 deadline. According to German legislator Hans-Josef Fell, the emission-reduction goals are not substantial enough, and more aggressive goals should be implemented (ibid.). Similarly, Heiko Stubner, of the Bundesverband Erneuerbare Energie e.V. (National Association for Renewable Energy), believes that Germany could achieve more at the national and European levels in terms of emissions-reduction legislation. If the directives were stricter, he claims, Germany could profit more from its technological prowess by exporting technology, products, and services. The EU has acknowledged the current lack of ambition, and decision-makers are considering the possibility of instituting stricter directives. The German government is not alone in its push for more far-reaching goals. The Danish government has requested that the European goal for emissions reductions be intensified from 20% to 25% by 2020. The German government had not taken a position on this request as of this writing (Bauchmüller 2012).

3.5.2 German Governmental Structure

Moving away from the transnational, top-down influence of the EU, the German government has fostered the adoption of renewable energy. In part, as a response to the previously discussed environmental and antinuclear power movements, the structure of the German government and its constitutional mandates have facilitated the *Energiewende*. Article 20a of the German "Basic Law" (constitution) reads: "Mindful also of its responsibility toward future generations, the state shall protect the natural foundations of life and animals by legislation and, in accordance with law and justice, by executive and judicial action, all within the framework of the constitutional order" (Deutscher Bundestag 2010, p. 27). The German state is obligated by this provision to protect the natural environment due to this "responsibility toward future generations" (ibid.). The *Energiewende* definitely serves this purpose, as it will, with its goal of shifting Germany's energy mix to renewables, help combat global warming. The Intergovernmental Panel on Climate Change's (IPCC) 2007 *Fourth Assessment Report* found that the average of the estimates in the literature, as of 2005, was that each ton of CO_2 emissions averted will prevent US\$12 in damages to the environment that would come at the cost of future generations (IPCC 2007, p. 69). Consequently, this clause of the German constitution has played a major role in the conception and implementation of the *Energiewende*.

Although this obligation to protect the environment is anchored in the most recent German constitution, its origin lies in the constitution of the Weimar Republic, Article 150, which guaranteed the protection and care of the state to the landscape and monuments of nature. This original constitution was abolished by the National Socialists. The Nazis did, however, pass the first *Naturschutzgesetz* (nature protection law) in 1935, which required the government to protect "nature in all of its forms" (Haber 2009, p. 34). Despite the Nazis' atrocities, this law was maintained after the fall of the third Reich, as it gave the state effective tools for protecting the environment (Gudermann 2011). It was replaced in 1976 by the second *Naturschutzgesetz*, which expanded the previous law to protect the "diversity, unique character, and beauty as well as the value of nature and the landscape as a place for relaxation" (Haber 2009, p. 37). These laws and constitutional provisions provided the German government with the legal powers to set the *Energiewende* in motion and show by their very existence the importance the German people place on environmental protectionism.

Despite this progressive stance on environmental protection and energy-related development, the German government does not have an energy ministry. Rather, a group of ministries handle energy issues. Previously, all energy-related matters were dealt with by the Ministry of Economy and Technology (BMWi). In 2002, renewable energy issues were turned over to the BMU. As Andrea Meyer, an official at the BMU in Berlin, explained, the ministry had the sustainability knowledge required to tackle the *Energiewende* efficiently (Meyer 2011). This was one of the reasons for the transfer of authority. Another motivation was the government's goal of increasing public awareness about the *Energiewende*. The transfer of authority to the BMU has worked well, as the *Energiewende* has gained more system-wide attention. Under this new administrative structure, revisions to the *Erneuerbare-Energie-Gesetz* (EEG or Renewable Energy Act) of 2000 were approved, strengthening the law's incentive structure (ibid.).

The BMU has been growing significantly in recent years to deal with the challenges posed by the *Energiewende*. For example, its renewable energy department has expanded to collaborate with numerous energy agencies and associations to jointly develop several projects. One interesting aspect that helps to explain Germany's success is the seriousness and impartiality with which these relationships are built. Meyer stated that, "Although the associations have a vested interest in the development of renewable energy, we manage to maintain a very professional relationship with them and it is very clear to them what is considered an acceptable lobby and what is not" (Meyer 2011). What makes these relationships especially valuable and effective is that ministry officers have very close contact with legislative power. They provide suggestions to legislators and receive feedback as well. According to Andree Böhling—a Greenpeace environmental expert, former government officer, and contributor to the first draft of the EEG— the parliament participated significantly not only in the approval but also in the more technical aspects of the development of the EEG law. This process, which comprised numerous rounds of discussion, was influenced both by legislators and by the many ministries that would aid in its implementation, resulting in a program that was simultaneously representative of the people's interests and was practically enforceable (Böhling 2011). The next section will detail some of the results of this practical, collaborative law-making process, as a number of laws directly influence, and are indeed essential to, the functioning of the *Energiewende*.

3.5.3 *Stromeinspeisungsgesetz* and the *Erneuerbare-Energie-Gesetz*

Laws created to promote renewable energy are vital to the *Energiewende*. The *Stromeinspeisungsgesetz* (StrEG; Electricity Supply Act) was implemented in 1991 and mandated that utilities connect small, decentralized electricity producers to the grid so they could sell their excess energy on the open market. These small producers had previously been unsuccessful in the market because they could not afford to connect themselves to the grid. Thus, up until the early 1990s, hydraulic power was the only renewable energy source that had any chance of competing with fossil fuels. The StrEG changed this, as it defined a minimum price at which small producers' energy should be purchased. This measure provided an incentive for investment in decentralized electricity generation. A significant result of this new law was the boom in wind power, as other sources (e.g., solar energy, biomass) were not yet competitive with wind energy. The main limitation of the StrEG, therefore, was that it promoted primarily the most profitable renewable energy source—wind—because the incentives were designed to favor the most economical renewable energy source, rather than a diversified portfolio of sustainable energy sources. Wind's typical initial capital and production costs made it the most profitable renewable at the time and thus the best investment option. Despite its limitations, the StrEG laid the foundation for the EEG (Clearingstelle EEG and RELAW 2012).

The EEG, introduced in 2000, is arguably the most important piece of renewable energy legislation in German history, as it strengthened and expanded incentives for renewable energy enormously. Its main goal was to regulate the relationships among renewable electricity producers, power grid operators, and electricity consumers. At the time of this writing, it had been updated twice since its inception, in 2004 and 2009. It expanded the objectives of the StrEG along two dimensions. First, the EEG included provisions not only for electricity production but also for heat production. Second, and more importantly, it created incentives that promoted the production of renewable energy sources other than wind, such as biomass and solar. As its defining characteristic, the law set different minimum purchase prices for each type of renewable energy. As with the StrEG, it guaranteed fixed minimum prices (called feed-in tariffs) for every energy producer. However, it sought to make each source equally competitive from an investor's point of view, rather than promote only the most cost-effective source. The law accomplished this by mandating that utilities pay higher purchase prices for more-expensive-to-produce types of energy, such as solar, while paying lower prices per kilowatt-hour for energy generated from more cost-effective energy sources, such as wind. These differentiated feed-in tariffs thus made previously uncompetitive renewables economically feasible.

Table 3.5 provides a summary of the prices guaranteed to each type of energy producer. The complete EEG provides more details than displayed in the table, as it also considers a project's age, capacity (in kilowatts), purpose (private or commercial usage), and location (house roofs, buildings, etc.) in determining the specific feed-in tariff (BMU 2010a). It must be noted, therefore, that Table 3.5 provides only a simplified overview to give a glimpse into how the feed-in tariffs are structured.

TABLE 3.5

Prices (Feed-in Tariffs) Guaranteed to Producers of Various Types of Renewable Energy according to the EEG, as of 2009

Energy Source	Characteristics for which a Tariff is Guaranteed	Guaranteed Feed-in Tariff (€ cents/kWh)
Water	<500 kW	12.67
	<2 MW	8.65
	<5 MW	7.65
	<50 MW	4.34
	50 MW+	3.50
Biomass	<150 kW	11.67
	<500 kW	9.18
	<5 MW	8.25
	20 MW+	7.79
Geothermal	<10 MW	16.0
	10 MW+	10.5
Wind onshore	First 5 years	9.2
	Base price	5.02
Wind offshore	First 12 years	13.0
	Base price	3.5
Solar	On a construction site	33.0–43.01
	In a field	31.94
	For own usage	25.01

Source: Adapted from BMU. Vergütungssätze und Degressionsbeispiele nach dem neuen Erneuerbare-Energien-Gesetz (EEG) vom 31. Oktober 2008 mit Änderungen vom 11. August 2010. http://www.erneuerbare-energien.de/files/pdfs/allgemein/application/pdf/eeg_2009_verguetungsdegression_bf.pdf.

No discussion about the Pigouvian taxation elements inherent to free market–manipulating feed-in tariffs would be complete without a look at their corresponding pro- and counterarguments. After examining the EEG's feed-in tariffs, some might argue that conventional fuels have a significant cost advantage over renewables. Such libertarian-minded individuals contend that this cost advantage should determine which sources are actually used according to free market principles because cheaper energy sources will decrease the cost of a country's economic activity, making its industries more competitive. Cheaper energy will also allow the country to meet its citizens' requirements at a lower price. Opponents of feed-in tariffs could also argue that if the external costs of fossil fuel–based energy generation are actually relevant, that these external costs would already be included in the market prices of fossil fuels. This particular argument, however, assumes that the free market is perfect, which is clearly untrue, as can be seen, for example, from the overexploitation currently plaguing much of the world's fish stocks or from the hunting to near extinction of endangered species like rhinoceroses.

Thus, we feel that arguments against feed-in tariffs cannot stand up to pro-arguments, as conventional energy sources neither internalize their external costs to society nor to future inhabitants of the planet. Therefore, using the "cheapest" energy sources, as priced by the market, may not actually provide society with the cheapest goods and services, as there will be many indirect costs (e.g., environmental, health-related) that accompany the utilization of cheaper energy sources. Consequently, the prices of conventional energy sources cannot be compared properly with those of renewables, as the external costs of conventional energy sources are difficult to estimate (e.g., it is not easy to price the risk of a nuclear accident and its economic and social consequences into the cost of nuclear power). Delving into this issue further, nuclear energy prices were approximately €0.05/kWh in 2012. Nuclear was thus considerably cheaper than any renewable energy source. However, opponents of nuclear power generation argue that nuclear plants (1) have no insurance against potential nuclear meltdowns and hence their prices cannot be compared directly with those of renewable energy plants (which are usually insured) and (2) generate nuclear waste that must be dealt with for hundreds of thousands of years. Similarly, fossil energy sources do not internalize the cost of CO_2 emissions and are, thus, also not comparable.

Strengthening the case for renewables is the fact that renewable energy subsidy costs are, according to the text of the EEG, planned (rather than just expected) to decrease gradually, whereas the external costs of other energy sources will remain constant. For instance, the feed-in tariffs paid to solar energy projects installed on buildings have gradually decreased from €0.574/kWh, the price established in 2004. This tariff was expected to reach less than a third of its original value in 2013 (€0.1857/kWh; Handelsblatt 2012). In fact, in 2012, the government was accelerating the reduction in solar energy prices, given the technology's faster-than-expected progress (see Sections 3.6.2 and 3.6.3 for more details). Conversely, fossil energy prices are expected to increase in the long-term due to these sources' finite natures. Considering this, promoting renewable energy technologies now seems to be a way to ensure Germany's future energy security—not to mention a way to give the country a technological head start (and consequently, the potential for greater exports) when other countries' demand for renewable energy increases.

According to the BMU, more than 47 countries have used the EEG as a model to develop their own energy codes, as they have also recognized the benefits of green energy. Thus, it is considered the most successful law for renewable energy in the world. In Germany, at the very least, it has played an essential role in the development of wind, solar, biogas, wood heat, and vegetable oil heat energy generation, among other sources (Fell 2010). According to T. Wenzel, of the Deutsche Energie-Agentur (German Energy Agency), the law's greatest strength was the low level of regulation it imposed on the energy industry outside of enforcing the specified feed-in tariffs. It simply created market conditions that made the development of renewable energy sources economically feasible. These tariffs guarantee a fixed amount of revenue to investors for renewable energy generated, all within the framework of a private market. Investors range from families installing rooftop solar panels to large companies investing in offshore wind plants. Technology development has also been encouraged by this law, as demand and financial incentives have both increased constantly through 2011 (Wenzel 2011).

Overall, the EEG established the foundation for a successful renewable energy sector without the need for heavy government intervention, in stark contrast to the bidding or concession models used by many countries, which require much more government activity. These alternative models benefit primarily large companies in the energy sector because they can offer the lowest bid to the cost-conscious governmental agency offering the development rights to a project. Bidding models basically function as follows: the firm that offers to supply energy to consumers at the lowest price receives a private or government contract to develop an energy project. Typically, concession models are used to "delineate a monopoly franchise and auction it off to the bidder offering the best proposal" (Dosi and Moretto 2006). This may mean, again, that the firm promising to deliver the energy at the lowest price will win. It may also mean that the firm agreeing to pay the auctioneer the highest price for this "monopoly franchise" will win. The long-term monopolies created by these concessions tend to reduce incentives for innovation, unlike the EEG model, which encourages innovation to achieve cost reductions by gradually decreasing feed-in tariffs over time. Even in bid or concession model cases, in which there are incentives for competition, different renewables often compete against one another and, depending on the program's and country's characteristics, initially one renewable source will prove more profitable than (and therefore preferable to) the others (ibid.). As shown in Table 3.5, the EEG manages to avoid this cannibalization of renewable energy development by providing staggered feed-in tariffs to different energy sources based on their relative competitiveness. This ensures that a country will not become entirely dependent on a single energy source, which is important because some energy sources may start out being expensive but then prove cheaper in the long run as a result of technological breakthroughs. The EEG, therefore, avoids myopic decisions that may lead to expensive-path dependencies (BMU 2010a).

This avenue has had drawbacks, however, as energy has become more expensive. Average users, thus far, have ended up paying €2 more per month for energy than before the law. Legislator Fell argued that this higher price resulted in an additional three to four billion euros flowing into the energy sector, which has created jobs by stimulating investments in renewable energy (Fell 2009). Supporting this argument is the fact that, from 1998 to 2010, the number of jobs in the renewable energy industry jumped from 30,000 to 340,000 (ibid.).

Another advantage of the EEG lies in the fact that the costs of producing renewable energy technologies did not burden the state unduly. The higher cost of energy was transferred to grid controllers, who then transferred the costs to their customers. This had several advantages, including (1) consumers becoming more conscious of their energy consumption, which helped the government achieve its goal of encouraging energy conservation; (2) an easier passage of the law because the government budget was not affected by higher costs; and (3) producers expecting their renewable energy investments to yield positive net present values, meaning the private sector willingly invested in the industry. This last advantage began a virtuous cycle, as increased private investment led to technological breakthroughs and greater scale, both of which then reduced costs.

3.5.4 The German Politician's Mindset

These advantages, although important, did not guarantee the passage of the EEG. They contributed to the creation and success of the program, but the law's passage and the heretofore success of the *Energiewende* owe much more to Germany's aforementioned environmental movement and the pressure it applied to the country's political sphere. Before delving into this, however, we should describe the general political atmosphere in Germany. The German people (and politicians) have been celebrated for their austerity and focus. According to Transparency International's *Corruption Perception Index 2010*, Germany is ranked 15th in out of 178 countries terms of political transparency (Transparency International 2011). This indicates that Germany possesses a positive environment for change favoring the greater good. In addition, although administration changes from one party to another bear the risk that previously adopted programs and objectives will not be pursued, the German political structure is characterized by a results-oriented focus, step-by-step approaches, and long-term vision. These have been key elements in the success of renewable energy policies (Meyer 2011). Meyer adds that persistence, transparency, and communication were very important on a day-to-day basis. The government has followed a managerial approach, as the different administrations have constantly pressed officials for results. Government officials have had no fear of conflicts and have worked diligently toward concrete results, also taking the time to communicate adequately and to exchange ideas to build consensuses on important aspects of the program. Illustrating this point, Meyer states, in reference to government officials, that "they speak and they listen constantly" (ibid.). She also notes that the hierarchical structure within the BMU and the parliament are very conducive to transforming ideas into actions (ibid.).

The fact that serious discussions about environmental issues occurred relatively early on also helped the country along its path to the EEG. Mautz et al. (2008) recall that in 1982, the SPD administration increased the budget for renewable energy research from DM20 million to DM300 million. This event is seen as a milestone that positioned Germany to adopt renewable energy technologies on a national scale. Even as of this writing, in

2012, with a different party in power, the political and economic climates remain favorable for R&D activities.

History provided an additional incentive for the development of renewable energy because, during Gerhard Schröder's SPD administration, Green Party votes were necessary to obtain a legislative majority. In particular, when the reduction of taxes for efficient gas plants was under discussion, the SPD compromised with the Green Party. Under this compromise, the SPD agreed to pass the EEG, which was endorsed by the Green Party, whereas the Green Party agreed to support the law that would grant gas plants a favorable tax status (Umwelt- und Prognose-Institut e.V. [UPI] 1999). When a new coalition came into power in 2005 (comprising the German Free Democrat Party and the German Christian Democrat Party, the relatively libertarian and conservative parties of Germany, respectively), the law had already created a significant number of new jobs, and the industry had already reached such a size that dismantling it would have been very difficult politically. Some observers also consider important the fact that the current German chancellor, Angela Merkel, was a former Minister of the Environment from 1994 to 1998 (BMU 2012b, pp. 5, 30). Although there seems to be no indication of a disproportionate bias toward or against environmental issues, her experience, skill-set, and knowledge have presumably influenced her decision-making as chancellor (Deutsche Presse-Agentur [dpa] 2012).

3.5.5 Lobbying

Although lobbying is not a formal activity in Germany and is generally regarded as controversial, it has influenced the *Energiewende*. Because of industry deregulation, it is difficult to quantify its extent, but it is estimated that the country has more than 4500 lobbyists (the Ministry of Justice holds records of more than 2000 lobbying associations and agencies; Bundesministerium der Justiz [BMJ] 2012). The energy sector has one of the highest lobbying activity levels, with more than 100 associations. A sample of representatives from these institutions was interviewed for this chapter. When asked about their objectives, they invariably answered that it was to represent their members, to give politicians input, and to offer them suggestions regarding energy-related issues (Stadermann 2011; Stubner 2011; Timm 2011). Although some of the associations represented both renewable and fossil energy companies, there was a clear polarization of intentions. For example, although large, established energy companies such as RWE, EnBW, Vattenfall, and E.On allegedly have been supportive of renewable energy, historically, they have tried to extend the lives of their nuclear plants as much as possible because these plants represent steady profit streams. According to a 2010 documentary from the TV channel 3Sat, the aforementioned companies would realize a profit of more than €50 billion if they are allowed to operate their nuclear plants for an additional 8 years starting in 2010 (Skalski 2010). This indicates that the companies have a major economic incentive to engage in lobbying activities.

Citizens and activists complain that nuclear energy producers have an unfair lobbying advantage over renewable energy advocates. The sizes of these two groups are generally not comparable because the former are huge utility companies commanding massive resource pools and established ties to the government. Working to counteract the influence of the utilities is a significant number of so-called medium-sized companies that, combined, have some lobbying power. The Bundesverband Erneuerbare Energie (BEE, Federal Association of Renewable Energy), for example, is an umbrella organization encompassing 25 associations with more than 2000 member companies. As Heiko Stubner, spokesperson for the BEE, notes, although complete alignment of these many companies' interests was difficult, the association has managed to represent the interests of the renewable energy industry as a whole, thereby countering the fossil energy lobby (Stubner 2011). The BEE also received, and continues to receive, valuable feedback from the government, which helps the group to understand important trends and developments, thus aiding its lobbying efforts.

In conclusion, lobbying can work both in favor and against the renewable energy industry. From the research conducted for this chapter, it was unclear whether lobby intensification (perhaps with increased formalization akin to the U.S. Disclosure Act of 1995) would help or hinder the progress of renewable energy.

3.5.6 Citizen Participation

The German government has encouraged competition for influence against these lobbying efforts by fostering citizen participation in the *Energiewende*, which has helped bolster its popular support. Citizen participation is essential, according to economist and lawyer, Michael Zschiesche, as the nation cannot expect local citizens facing the construction of new projects in their areas to be consoled by talk of sacrificing for the greater good. Instead, Zschiesche states that locals are more concerned with issues such as decreases in "real estate prices, health risks, environmental damages, and dangers to the landscape image" (Netzwerk Bürgerbeteiligung 2012, pp. 5–6). This latter issue is especially important because the construction of wind turbines and power lines are essential to the *Energiewende*. These intrusive and highly visible infrastructure projects have sparked a massive NIMBY movement among the German people, who do not want new wind turbines or power lines to mar their local landscapes. They, therefore, vehemently defend their culturally important *Heimat* from a perceived *Verspargelung* (a uniquely German word referring to the "asparagus-izing" of the landscape, referring to how tall, skinny wind turbines rise prominently from the ground). To combat this NIMBY movement, the federal government has enacted legislation to ensure that the construction of major new projects can proceed only after citizen input has been considered (Germanwatch e.V., DAKT e.V., and Heinrich-Böll-Stiftung Thüringen 2012, pp. 11–12).

The first opportunity for citizens to provide feedback occurs during the so-called *Raumordnungsverfahren* (ROV, regional planning procedure) process, in which both the private and public sectors assess the suitability of a location for a new project. At that time, so-called "representatives of public concerns," such as citizen groups and environmentalist NGOs, take part in the creation of an *Umweltverträglichkeitsstudie* (environmental compatibility study; Germanwatch e.V., DAKT e.V., and Heinrich-Böll-Stiftung Thüringen 2012, pp. 11–12). This comprehensive study determines whether a project can be located in an area without causing untoward damage to the community or environment. The final report is then published so the community members can voice their objections. If significant enough, these objections can lead to a public hearing, which can influence the responsible regulatory agency's final decision on the matter (ibid., pp. 11–12).

This *Raumordnungsverfahren* is only the first step in the approval process for new renewable energy projects. The next, more important step is the *Planfeststellungsverfahren* (PFV, plan approval procedure), which, over a 4-week period during which the project is open to public scrutiny, seeks to consider an even broader range of interests before final approval is granted (ibid., p. 12). This affords every citizen the opportunity to file a complaint against the proposed project. In addition, the aforementioned "representatives of public concerns" must submit opinions about the published plans during this period. At the end of the 4 weeks, all objections and opinions must be addressed and answered by the organization proposing the project (typically a utility). Based on the community's response to these answers, the responsible regulatory agency may then schedule a hearing in which citizens can again voice their complaints. With this back-and-forth communication process, the project plan will, potentially, be changed until the community's objections have been addressed satisfactorily. Only then will the regulator grant permission for the project's construction (ibid., p. 12).

This basic outline of citizen participation has thus far touched only on regulatory procedures. The German government is also developing new measures to facilitate citizen investments in upgrading the German electrical grid. Private citizens can already invest in renewable energy installations (such as PV and wind energy projects). In fact, more than 50% of Germany's renewable energy capacity is owned by private individuals (BMU, Bundesministerium für Ernährung, Landwirtschaft und Verbraucherschutz [BMELV], and Agentur für Erneuerbare Energien e.V. [AEE] 2012, p. 15). New legislation would build on this success by empowering citizens to invest in billions of dollars of new electricity grid expansion projects. Minister of the Environment Peter Altmaier has suggested creating a system in which debt instruments of at least €500 yielding 5% interest would be offered to the public (Kröger and dapd Nachtrichtenagentur 2012). This program should encourage popular support for grid expansion by providing an attractive financial incentive to help overcome NIMBY objections.

Although these efforts do not provide a perfect means of alleviating all the inconveniences citizens face, they certainly help bring complaints to the attention of the German government. Without such opportunities for citizen participation, the *Energiewende* would distribute harm more unequally. Consequently, we surmise that these efforts have helped fuel the success of the *Energiewende*.

3.6 SOCIOECONOMIC CONTEXT

Despite the extensive citizen participation associated with the *Energiewende*, easy access to fossil energy represents a hurdle to the adoption of renewable energy. However, grassroots movements and a powerful environmental movement provide a robust counterbalance to the economic advantages of fossil fuels. The main socioeconomic factors characterizing these juxtaposed forces are detailed in this section.

3.6.1 The Entrenched Nonrenewable Energy Industry

One of the greatest obstacles to the *Energiewende*'s transition to renewable energy has been the existing nonrenewable energy–dependent utility companies. In terms of revenue, 15 of the 100 largest German companies are related to the energy sector. In 2010, the four largest energy producers—E.On, RWE, EnBW, and Vattenfall—were in positions 2, 5, 14, and 35, respectively (Forbes 2010). Table 3.6 displays their shares of each major energy source.

Three of the four companies listed in Table 3.6—Vattenfall, EnBW, and E.On—exceed or are close to the German average in renewable energy production as a percentage of their overall energy mixes (as stated previously, Germany produced approximately 10% of its energy from renewable sources in 2010). It is evident, however, that E.On, RWE, and EnBW are dependent on nuclear energy (E.On 2010; EnBW Energie Baden-Württemberg AG [EnBW] 2012; RWE AG 2010; Vattenfall 2009). Table 3.7 displays the costs of energy production for conventional sources in the United States to illustrate the cost-effectiveness of nuclear power.

Assuming German cost data are similar, it follows that nuclear energy production is much cheaper than other sources (Nuclear Energy Institute [NEI] 2012). Because consumers pay the same for all types of energy, nuclear energy producers would, in the absence of the EEG's feed-in tariffs, enjoy the highest margins. According to some accounts, when the legislation was being developed, established energy companies tried to fight it by claiming that renewable energy was undependable and too expensive. They also claimed that to ensure its affordability, the *Energiewende* would

TABLE 3.6

Share of Energy Sources for the Four Major Energy Producers in Germany

Energy Sources	E.On (%)	RWE (%)	EnBW (%)	Vattenfall (%)
Fossil	68	78	35	73
Nuclear	24	19	51	5
Renewable	8	3	11	22

Source: Adapted from E.On. Stromkennzeichnung gemäß §42 EnWG auf Basis der Daten von 2010. https://www.eon.de/de/westfalenweser/pk/services/Rechtliches_Veroeffentlichungspflichten/Energiemix/index.htm.; EnBW. Der Strommix der EnBW. http://www.enbw.com/content/de/der_konzern/enbw/neue_gesetzgebung/stromkennzeichnung/index.jsp; RWE AG. *2009 Jahresabschluss der RWE AG.* Bocholt, Germany: D+L Printpartner GmbH, 2010; and Vattenfall. Information zur Stromkennzeichnung. http://www.vattenfall.de/www/vf/vf_de/Gemeinsame_Inhalte/DOCUMENT/154192vatt/289100prod/289114priv/P02182175.pdf.

TABLE 3.7

Cost of Electricity Production from Different Fossil and
Nuclear Sources

Energy Sources	Production Cost of 1 kWh (U.S. 2009)
Nuclear	US$0.02
Coal	US$0.03
Gas	US$0.05
Oil	US$0.12

Source: Adapted from NEI. Monthly Fuel Cost to U.S. Electric Utilities (1995–2011). http://www.nei.org/resourcesandstats/
documentlibrary/reliableandaffordableenergy/graphics
andcharts/monthlyfuelcosttouselectricutilities/.

require cooperation across the entire EU and a massive grid reformula-
tion. This particular claim may prove true, demonstrating that utilities'
concerns should be taken into account, even if such a consideration is tempered by an acknowledgement of the companies' profit motives (Böhling
2011).

In recent years, these legacy companies have invested in renewable
energy and have learned to exploit its strategic benefits. For example, RWE
and E. On are supporting the DESERTEC initiative, which plans to install
wind and solar generators in desert areas around the globe (El-Sharif
2009). Furthermore, RWE has launched a new subsidiary, called Innogy,
which installs and operates wind farms in Germany and surrounding
countries (RWE Innogy 2012). Because of their strong balance sheets,
these large firms are able to invest in capital-intensive projects such as
offshore wind technologies. The presence of large institutions represents
an important complement to the development of onshore wind devices by
small investors and to the adoption of solar panels by households.

3.6.2 Technological Development

This section will address the position of German green technologies
in the world market. A study presented by the *Deutsches Institut für
Wirtschaftsforschung* (DIW, German Institute for Economic Research)
outlines Germany's importance as an actor in the worldwide markets
of wind, PV, and solar thermal technologies. The study points out that,
in 2011, Germany was the second largest exporter of wind technologies
after the United States and was the second largest exporter of PV devices

and components after China (Groba and Kemfert 2011). In 2010, German firms commanded 16.1% of the global wind turbine export market and 15.9% of the worldwide solar export market (ibid.). Since 2000, Germany increased its absolute volumes of exports in these two markets (wind and PV) by more than 900% and 700%, respectively. Undoubtedly, the world views Germany's products favorably in these markets, indicating that the country has made great strides in terms of research and development.

Germany is not the only country that increased its export volume in these product categories over the last 10 years. We will focus on the PV industry for the remainder of this section because it presents an interesting case of Germany losing its competitive advantage in a particular industry. J.H. Lange, of the HSH-Nordbank, recalled that, until 2009, Germany represented roughly 50% of the world's PV market (Lange 2011). At the same time, Janosch Ondraczek, of PwC Hamburg, pointed out that China "absorbed" the technology and improved production processes to achieve significant cost reductions (Ondraczek 2011). Chinese firms also benefitted from much lower labor costs and government subsidies that have allowed them to undercut Western competitors in the solar market. As a result, China became the main exporter of PV devices to Germany in 2011. Germany's enormous domestic demand, combined with the low prices offered by Chinese firms, actually caused the country to import more PV devices than it exported starting around 2005, indicating it had begun to lose its edge in this market (Groba and Kemfert 2011). This reversal in the trade balance of its PV industry shows how Germany's early efforts, which gave it a lead in the industry until the mid-2000s, were then tapped by foreign competitors, a trend that, as of this writing, had not yet spread to the wind energy industry, but could do so in the future.

Germany's early lead in technological development in the PV industry can be explained in part by a few factors that have also affected the development of other green technologies, such as wind power. First, in addition to beginning R&D in the renewable energy industries very early on, the country had a decentralized development process. Numerous small and medium-sized companies pursued independent R&D efforts, thereby ensuring a broad diversification of research efforts. This approach improved the chances of developing successful technologies simply because there were a significant number of projects. Second, Germany has a tradition of excellence in mechanical machinery technology (for decades, capital goods produced in Germany have been synonymous with high quality). This traditional leadership in high-quality manufacturing helped German companies to create superior green energy products. Third, specific pieces of legislation provided financial incentives to encourage investment in renewable energy. Finally, Germany has a fairly unique system in which the government provides financial support to competitors' joint efforts to develop new technologies, which are often supplemented by aid from German research institutes. According to the Bundesministerium für Bildung und Forschung (BMBF, Ministry of Education and Research), there were approximately 750 publicly funded research institutes, 506,000 employees in R&D, and 299,000 scientists and scholars working on research in 2008, illustrating the vast breadth of Germany's research network (BMBF and Deutscher Akademischer Austausch Dienst [DAAD] 2008). Furthermore, the country has a broad range of organizations in the field of applied research, providing a very strong research landscape geared toward industrial applications.

Despite the vast amount of research devoted to the PV industry, the recent plunge in the technology's costs, and the large global solar industry that has emerged, economists still express significant skepticism about solar energy. They argue that solar is too expensive and that the current subsidies are wasteful because other renewables are more economically competitive (Neubacher 2012). Despite these doubts, the German Ministries of the Economy and of the Environment agreed not to eliminate solar subsides and, as detailed in the last paragraph, the government has supported research in the field. Instead, they anticipate a reduction in solar energy subsidies of between 20% and 30%, depending on production conditions (Bauchmüller 2012). Although environmentalists see this as a step back (because subsidies will be reduced), the subsidy reduction was motivated by the recent high installation rate of new PV projects, which leads us to conclude that solar energy is gradually becoming competitive. Germany's development of this technology, therefore, seems to have contributed to the technology's global success, helping to justify subsidies for solar.

3.6.3 Private Sector Investment Incentives

Market demand for renewable energy has been driven, to a large extent, by grassroots investors. Small businesses, private citizens, and private capital in general have contributed much to the ongoing success of the *Energiewende*, complementing large companies and institutional investors. Mautz et al. (2008) supported this assertion, separating the development of renewable energy in Germany into two distinct phases. The first phase lasted until the 1980s and was predominantly a cultural movement. The expansion of nuclear plants increased public concern, which led citizens to react on different fronts. Many joined or supported NGOs, whereas others went further, building their own wind generators or biogas plants. These oppositionists did not wait for the government to alter its stance on nuclear and fossil energy. The independent researcher or "inventor," therefore, had an important role, as they helped give birth to what is today the state of the art in wind energy generators (ibid.). Many of the ventures centered on these independent researchers were once very small, independent private contractors, but some eventually evolved into mid-sized firms. Thus, we conclude that the first phase included a strong bottom-up component.

The second phase began with the introduction of the *Stromeinspeisungsgesetz* in 1991 (Mautz et al. 2008). This law created the means for institutions to invest in renewable energy because it guaranteed their projects would be connected to the grid, making them financially viable. At this point, the industry started to grow at a faster pace due to the construction of larger projects, which encouraged the evolution of more specialized green energy firms. Companies began to focus on different components of the value chain—including product development, manufacturing, construction, and investor relations—while consulting firms entered the process to support clients through each developmental phase. Ondraczek, who worked for PwC Hamburg supporting the renewable energy industry, stated that numerous projects were commissioned by both large and medium-sized companies during this period. In many instances, groups of investors seeking advice approached PwC because of the perceived profit opportunity (Ondraczek 2011). The projects often began with an assessment of location (domestic or international) and technology (onshore wind, offshore wind, solar, etc.). For onshore projects, investors would then lease the selected land (instead of purchasing it), allowing both farmers and investors to maximize expected returns. This was beneficial because (1) the landlords (farmers) could continue to exploit the land's agricultural potential and (2) the investors had much lower initial capital costs. Ondraczek also detailed current efforts in the offshore wind industry, stating that capital requirements are much greater, making bottom-up approaches impossible. Investor groups often work together with other capital providers, such as major banks, to overcome this hurdle (ibid.).

Despite the long set-up and operating periods of renewable energy projects, forecasting future revenues can be done quite precisely because the EEG clearly defines energy prices in most cases for at least a decade. This is highly consequential, as it usually takes 2 years for a project to reach the implementation phase. Some factors that contribute to the long lead time include identifying suitable land, negotiating with the landlord, settling technical details, creating financial projections, obtaining permits, and finalizing monetary agreements. The predictability provided by the EEG is therefore crucial to reduce the risks associated with these projects. For example, investing in onshore wind farms in Germany in 2011 provided an average rate of return of more than 6%, with very low risk, making such projects acceptable to pension funds.

At the same time, the volatility of oil and gas prices helps make renewable energy investments attractive. Germany constantly exports these projects but, because of less favorable and less precise regulations in other countries, the risks are usually higher, making international projects unfit for risk-averse investors. The EEG model in Germany, on the other hand, seems to appeal to all categories of investors. This is in sharp contrast to countries that follow the bid-and-concession model, as energy projects in those countries usually have large capital requirements, generally forcing small investors out of the market.

In addition to the two phases already referenced, we believe that a third wave began in the mid-2000s. As in the case of the first wave, this period has been defined by grassroots adoption of new technologies, especially PV cells, a technology that the government has prioritized by setting high feed-in tariffs for solar energy. Although PV cell technology has changed little in recent years, production processes and scale have improved significantly, especially since Chinese firms entered the market, allowing for significant cost reductions. Solar energy is attractive because installation can be done on almost any surface receiving sunlight, thus rendering PV cells very popular among households. Households' extra energy can be sold to the grid, requiring only that a current meter be installed. If the energy is consumed instead of sold, the household's energy bill is reduced.

This popularity has driven solar's growth. According to data from the BMU, solar energy generation has increased from 64 GWh in 2000 to 19,340 GWh in 2011 (BMU 2011a, p. 20). Furthermore, according to Lange (2011), nearly 80% of PV cells installed in Germany in 2011 were on the roofs of houses. The installed base could have been even greater if other sectors of the economy had invested extensively. In 2008, the government passed an amendment prohibiting the installation of solar devices in arable fields, which reduced opportunities for large investors but preserved Germany's food security.

3.6.4 German Culture

Germans, as exemplified by the *Heimatschutzbewegung* and the environmental clause of their constitution (among other aspects), have a special affection for nature. In addition, Germany's history has created a culture of wariness and skepticism, which has contributed to activism against nuclear power (Demandt 2008). This can be seen in German politicians frequently citing the *Vorsorgeprinzip* when debating issues. As discussed earlier, this principle demands that, when possible and reasonable, action should be taken before a problem arises in order to minimize the consequences. It plays a central role in the debate over nuclear power because the German people feel it is worth abolishing nuclear power to avoid nuclear accidents. Lessons from the world wars, in part, justify adherence to this principle. For example, when discussing the effects of the world wars on German society, Dr. Stadermann explains, "We became a pacifist people and now, for better or for worse, we don't want to do anything that might offer reason to condemn us later" (2011). He maintains that Germans still carry the burdens of the past and have made it their duty to foster peace, safety, and balance; hence, their reluctance to support anything (e.g., fossil fuels, nuclear power) that might endanger these values. He also states that the same concept is applicable to research, in which strict regulations are to be followed. According to data from the Bundesministerium für Naturschutz (BfN, Federal Office for Nature Protection), 89% of Germans consider protecting nature to be an important political task (BfN 2010). Furthermore, 80% of Germans support the administration's recent resolution to not extend the lives of the existing nuclear plants, and 70% are fearful that a nuclear accident such as Fukushima could happen in their country (Erdmann, Deutsche Press-Agentur [dpa], and dapd Nachtrichtenagentur 2011). Dr. Stadermann notes that Germans in general are not inconsequential and irresponsible but rather, are skeptical, apprehensive, and wary. This leads the German cultural mindset to be as focused on long-term survival as it is on short-term well-being (Stadermann 2011).

The environment-oriented mindset of Germany's population created a perfect scenario for the establishment of NGOs. Greenpeace was founded in 1971. Despite its Canadian origin, this NGO gained major importance and significance in Germany. More than 18% of the organization's current 2.8 million members worldwide are German (Greenpeace 2012), an impressive number considering that Germany's population represents only 7% of high-income countries' total population (WBG 2011b). In 2012, one of Greenpeace's largest offices, with 150 employees, was located in Hamburg. From research and interviews with Böhling, it became apparent that the organization's mission was not only to fight against nuclear power and create public support for renewable energy but also to participate broadly in related initiatives (2011). For example, the NGO provided a study, *The Plan* (most recently revised in 2011), to provide guidance for achieving a 100% renewable energy mix by 2050 (Greenpeace 2011). Greenpeace suggests replacing nuclear power temporarily with coal in the short-term, while high-efficiency gas plants and additional solar and wind installations are constructed to comprise the ultimate energy mix. In the long term, geothermal plants would also play a role in energy production (ibid.). The German Green Party referred to this plan while developing the EEG. Although Germany currently plans to generate 80% of its electricity from renewable energy sources by 2050, as opposed to the 100% suggested by Greenpeace, Böhling emphasizes that the existence of Germany's long-term goal is what is important. He is encouraged by the fact that this goal was set by a conservative government, indicating that it might be improved further by a more progressive administration (Böhling 2011).

3.7 DISCUSSION

3.7.1 Relationships among Factors

This chapter has outlined the main factors contributing to the success of renewable energy in Germany. Figure 3.10 illustrates some of the interactions among these factors, exemplifying the complexity of this sociocultural-political-economic *Energiewende*. We have identified each factor based on its appropriate category (social, economic, historical, political, or endowment). The arrows map the perceived relationships among these factors, along with the main directions in which influence flows between each linked pair. Figure 3.10 attempts to convey an understanding of the multiple underlying conditions and interrelationships leading to each listed factor, as well as an understanding of the outcomes of the

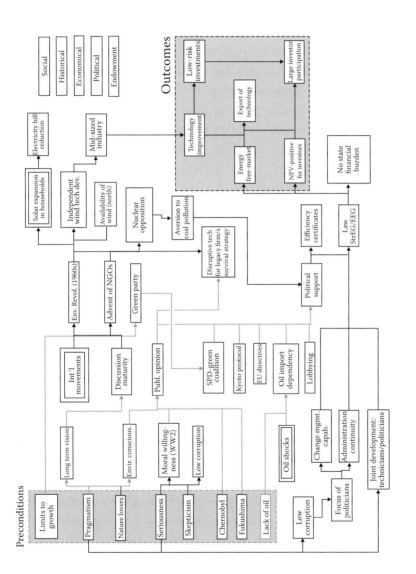

FIGURE 3.10

Flow chart of the factors leading to renewable energy industry development and their correlation.

entire network. Each of the two highlighted boxes contains a set of factors. The first set, called "preconditions," encompasses the primary causes or foundations that seem to have contributed the most to the successful development of renewable energy. The second set, labeled "outcomes," presents some of the results of the *Energiewende* that are currently affecting Germany. This diagram could be useful to other countries' policymakers as they attempt to transition to renewable energy.

Figure 3.10 shows that the main preconditions we identified here for the success of the *Energiewende* were key historical events (e.g., nuclear accidents, the international environmental movement) and some of the distinguishing characteristics of the German people (e.g., seriousness, skepticism, pragmatism). Most of the important outcomes, on the other hand, are related to economic advantages because, thanks to the unique structure of the *Energiewende*, the German renewable energy industry provides attractive financial returns to investors without compromising the government's budget.

Not all of the economic outcomes of the *Energiewende* are positive, however (at least when perceived from a purely short-to-medium-term economic perspective). This is because the *Energiewende* has increased energy prices in Germany since its inception. This outcome is related, through the EEG, to the expansion of renewable energy because higher feed-in tariffs must be paid for renewables (e.g., solar, wind, biomass) than for cheaper, conventional energy sources. As a direct result, as the share of renewables in the German energy matrix increases, replacing an ever-greater amount of cheaper fossil fuel capacity, energy prices will continue to increase, at least in the short-term to medium-term while renewable technologies become more cost-effective and the grid is upgraded. This overarching consequence of the *Energiewende* must be, according to the EEG, borne by consumers, as mentioned previously.

The significance of increasing energy prices, when compared with other, positive consequences of the *Energiewende,* such as lower carbon emissions and positive ROIs for investors, should not be understated. The other outcomes are dependent on consumers paying these higher prices through the renewable energy surcharge included in the price of each kilowatt-hour. This surcharge amounted to €0.036/kWh in 2012, and the main German utilities announced that this fee would increase to €0.0528/kWh in 2013 (Wiese 2012). These fees are used exclusively to finance the *Energiewende*'s expansion of the German electricity grid and to pay for the feed-in tariffs guaranteed to renewable energy sources. Considering that the average price paid by German households for electricity is €0.255/kWh, this fee is quite substantial, meaning that, although investors may enjoy profitable investments due to the *Energiewende*, households must foot the bill (ibid.).

These higher energy prices do not necessarily harm Germany's international competitiveness, however. Energy-intensive firms—such as cement producers, chemical companies, paper manufacturers, and mining companies—are largely exempted from taxes related to energy (Rosa Luxemburg Stiftung [RLS] and Arepo Consult 2012, p. 21). This tax relief is substantial, amounting to more than eight billion euros in 2011 and is estimated at more than nine billion euros in 2012 (ibid., p. 8). Thus, it is important to remember that, although the *Energiewende* has delivered great benefits, its costs have been borne by the people.

3.7.2 Ongoing Challenges

Nevertheless, the added fees paid by consumers are not going to waste, as the renewable energy industry in Germany is growing steadily. The proportion of renewables in the energy mix is expected to reach 35% by 2020 (Wiese 2012). Still, there are challenges ahead. Experts from various sectors (consulting, banking, government, nonprofit, etc.) agree that the primary challenges involve technological issues related to the reliability of renewable energy sources—in particular, how renewables will be able to meet peak power demand. Renewables, unlike fossil and nuclear energy sources, do not provide a constant supply of energy. Wind speed variability and the solar energy flux's seasonality represent unavoidable obstacles for a country that wants to source most of its energy from renewables, as these factors cause fluctuations in the energy supply. Fossil fuel plants have managed to absorb peak energy demand and to fill in when wind and solar electricity generation rates are low, but Germany will begin to face supply constraints as fossil fuels are phased out.

One theoretical solution for this supply issue lies in the development of large-scale energy storage devices that would store potential, electrical, or thermal energy when electricity is overproduced and then release it during periods of peak demand. Currently, however, creating additional storage capacity is prohibitively expensive. The most feasible method is to pump water to a higher elevation during periods of electricity oversupply. This elevated water can then be released through a dam to drive hydroelectric energy turbines during periods of demand. The problem with this solution is that such pumped energy reservoirs require specific geographic settings. Germany, unfortunately, has already exhausted its potential for developing such capacity. This solution, therefore, cannot be scaled up to satisfy the demands posed by the *Energiewende,* meaning a different energy storage option is needed.

Modernization of the grid presents another challenge. As previously mentioned, wind energy is presently produced primarily in the center and northern parts of the country, with offshore wind farms installed in the Baltic and North Seas. However, the greatest population density and most industrial activity are found further to the south. Distribution is thus an important precondition for the widespread use of renewable energy sources. Spurring investment in Germany's national grid will require strong incentives because transporting energy from offshore wind farms poses significant technical and environmental difficulties. For example, environmentalists argue that extending hundreds of kilometers of energy cables along the seabed is destructive because of both the pollution and dangers to sea life (Wind Energy Development Programmatic Environmental Impact Statement [WEDP EIS] 2009).

Wind energy faces further difficulties. In particular, many critics condemn this source because the generators are noisy and the windmill blades pose dangers to birds. Finding an optimal solution that is acceptable to society, economic actors, and environmentalists will involve conciliating tradeoffs. Fortunately, German society is already actively engaged in this process.

In addition to these technical and environmental obstacles, financing issues will also need to be tackled. Although financial players foresee the continuation of the EEG and its financial incentives, there is always the chance that the feed-in tariffs will be lowered in part because some see the subsidies created by the law as being deleterious to the market in the long term. These opponents contend that the law's subsidies rob producers of incentives to improve subsidized technologies and increase their competitiveness. However, the expansion of Germany's PV and wind exports, detailed in Section 3.6.2, seem to contradict this argument.

Thus, given the EEG's current success in expanding the role of renewables and German exports, an opportunity exists to address another challenge inherent in Germany's energy sector: the heat industry. A law similar to the EEG, but tailored to heating, could prove effective. Initially, large, established heat companies will likely object to such a measure, as they will seek to maintain their "cash cows," much like the established electrical utilities. Thus, an incentive structure similar to the EEG's feed-in tariffs will be needed to stimulate investment in green-heat sources, which could have a massive effect because, according to data from BMU, heating consumes almost 60% of the energy used in Germany (BMU 2011a). The German government has actually taken a first step with the 2009 *Erneuerbare-Energien-Wärmegesetz* (renewable energy heating law), which dictates that new buildings must derive set proportions of their heating requirements from renewable energy sources. However, this law does not lay down any requirements for existing buildings and is far more top-down than the EEG, leaving room for improvement (Stadt Mannheim 2013). A new heating law would perhaps face greater challenges than the EEG, considering that many older buildings in Germany house inefficient heating and insulation systems. Consequently, the new law would need to create financial incentives to spur improvement on two fronts: "greening" HVAC (heating, ventilation, and air conditioning) systems to reduce the environmental effect of heating/cooling and encouraging better insulation in new and existing structures.

3.7.3 Final Considerations

In 2011, Germany was the world's largest solar energy producer, third largest wind energy producer, and second largest exporter of both solar and wind equipment. The country is planning on generating 80% of its electricity from renewables whereas reducing overall energy consumption over the next four decades, goals put in place despite strong projected economic growth. In this chapter, we identified the main elements that helped Germany achieve such a remarkable position, ranging from exogenous factors (the historical context and endowments) to endogenous influences (cultural, political, economic, and social). Most notably, in terms of significant historical events, the oil shocks of 1973 and 1979 were important signals to Germany about the danger of external resource dependency. The nuclear accidents at Chernobyl and Fukushima then led the German population to believe that their nation had committed an error in embracing nuclear power. NGOs and the Green Party helped the environmental movement to capitalize on these horrific accidents, leading the country as a whole to regard the *Energiewende* favorably. This, along with the wary, pragmatic, and thoughtful mindset of German politicians, culminated in the passing of the EEG and the recent movement to abolish nuclear power.

The financial incentives for renewables supplied by this landmark piece of legislation have attracted the investments needed to create lasting change in Germany's energy mix.

Replicating this story in other countries has been fraught with difficulties. The cultural and historical factors that created such an expansive affinity for environmental protectionism among the German people are largely absent in other parts of the world. Other countries have not enjoyed inspired events such as the *Waldromantik* and *Heimatschutzbewegung.* Nor do other countries necessarily allow the *Vorsorgeprinzip* or the concept of the *Generationenvertrag* to factor into their politicians' decisions because of more short-term-oriented constituencies. Therefore, applying the successful model of the German renewable energy industry to other countries may not be straightforward, especially if political issues such as ideological misalignments among parties, frequent administration changes, and widespread corruption are present. A lack of popular support due to the financial burden posed by higher energy costs may also impede the adoption of an *Energiewende*-like program. Exogenous factors (such as economic shocks, nuclear disasters, or environmental catastrophes) could force change but are highly undesirable. Nevertheless, lacking such shocks, countries may be able to enact change in different ways. Ideally, they could learn from Germany's actions, adapting them to their own unique circumstances to transform their energy mixes.

We will now note some of the main recommendations that other countries can take away from the German experience and also provide some of our own personal suggestions. To begin with, an important component of the *Energiewende* was its compliance with free market principles after providing a framework favorable to renewables. The EEG's feed-in tariffs provided investors with reliable and positive economic returns without burdening the industry with other extensive regulations. We recommend that countries beginning their own programs consider this example, creating incentives characterized by two main criteria: (1) feed-in tariffs guaranteeing investors attractive annual returns over time spans of at least several years and (2) the ability to ensure a diversification of technologies.

The first criterion assures investors that their funds will not be lost due to fossil fuels undercutting renewables with respect to price. We would recommend, if at all possible, an approach similar to the German model, in which the ultimate cost is spread among final energy consumers, thereby embedding the tariffs into the structure of the free market. This is preferable to a program that is solely dependent on government financing, as politicians would be more likely to cut a government-dependent program in times of fiscal deficits. To minimize the cost burden to consumers, however, we believe that investments in renewables should be granted favorable tax treatment and guaranteed by the government. The former would boost investment returns through a mechanism other than higher energy prices, whereas the latter would reduce risk and, hence, the return demanded by investors. Both would reduce the need to increase retail energy prices. For example, a policy allowing individual investors to invest in renewable energy funds on a pretax basis, which would function like a 401(k) plan in the United States, would boost returns, as earnings would accumulate on a relatively larger, pre–tax principal. On another tax-related note, we recommend that renewable energy companies be allowed to organize themselves as flow-through entities similar to master limited partnerships or real estate investment trusts in the United States (a privilege already enjoyed by oil and gas companies). This would allow renewable energy companies to avoid the problem of double taxation that plagues C corporations in the United States, again boosting the net profit flow realized by investors. Such a shift in legal organization would also make investing in renewable energy projects much easier for the general public. Finally, in high corporate tax rate countries like the United States or Japan, we recommend either establishing a special, reduced corporate tax rate for renewable energy companies or applying a territorial international tax system to renewable energy companies (in contrast to the worldwide tax system in the United States, which taxes all repatriated earnings at the 35% corporate statutory rate, this territorial system would ensure that profits generated in foreign countries would only be taxed once at lower foreign rates). We feel that such preferential tax treatment for the industry is warranted because of the globally beneficial social good that renewable energy developers provide. Moreover, lower tax burdens would reduce the cost of doing business for these companies, making *Energiewendes* worldwide more affordable, ultimately benefitting taxpayers.

Moving to the topic of energy efficiency, consumers could be granted special rebates on their energy bills should they undertake efficiency-boosting improvements (e.g., furnaces, insulation, windows), thereby augmenting the returns from such improvements. We also recommend that countries supplement these measures with programs such as those detailed throughout this chapter (e.g., Germany's building certificate program and citizen financial participation in energy and grid projects).

The second criterion, diversification, will allow countries to exploit synergistic relationships among different renewables. For example, installing sizable biomass electricity-generation capacity will help smooth the aforementioned fluctuations associated with wind and solar sources, because biomass plants can accumulate large inventories of feedstock to increase electricity production when necessary. Through diversification, countries will also avoid path dependencies, allowing them to exploit future price declines in all renewable technologies. An enterprising country could even incorporate a public transportation clause into its renewable energy law, thereby reducing its dependence on oil without necessarily expanding renewables' capacity.

Germany's model also provides some guidance on how to mitigate possible social resistance to an *Energiewende*. We would suggest that politicians pushing their own versions of the EEG emphasize the law's benefits to future generations. This may need to be accompanied by a broad educational campaign and a push to encourage membership in environmental NGOs if a country's population is unaware of the consequences of climate change. Nevertheless, politicians should introduce the themes of the *Vorsorgeprinzip*, the *Generationenvertrag*, and energy security into their countries' societal discourses to help their constituents become familiar with these important topics. Whenever possible, countries should also incorporate citizen participation into their programs to mitigate the strength of NIMBY movements. In particular, participation needs to be encouraged in project site decisions and investment opportunities. We realize that encouraging broad civil support stands as a tremendous undertaking, especially considering the huge investor and commercial interests supporting the legacy fossil fuel industries. If this was not a problem, then the world would already be tackling climate change much more aggressively. Acknowledging this does not mean, however, that we must bow to it; it is part of our obligation to the future inhabitants of the planet to somehow counteract the damages wrought by fossil fuels, even if it means (and it almost certainly does) that the developed world must sacrifice economic benefits now.

We further recommend that countries exploit international expertise in renewable energy technologies and financing structures. Rather than reinventing the wheel, new players now have the ability to tap existing technologies and markets. They should capitalize on this. This could take the form of importing German wind turbines and Chinese PV panels. Countries could also partner with Germany's "Energy Efficiency: Made in

Germany" program to import efficient products. Ideally, these countries would, through their EEG-like programs, attract foreign direct investment by multinational utilities, renewable energy technology manufacturers, and energy efficiency–centric firms, allowing them to adopt the most cost-effective, state-of-the-art renewable technologies. Countries could even take steps to foster the adoption of the best technologies within each renewable type (e.g., solar, wind) by adjusting feed-in tariffs (not retrogressively) for specific energy sources based on annual evaluations of renewable technologies' costs. This would minimize the cost of an *Energiewende*.

In conclusion, an understanding of the German model could aid other countries in recognizing their weaknesses and strengths, thereby allowing them to structure programs tailored to their idiosyncratic characteristics and helping them to introduce effective renewable energy policies (laws, certificates, technology imports, etc.). It is our hope that the lessons emphasized here will prove valuable to countries seeking to combat climate change and fossil fuel addiction, and that the *Energiewende*'s success will spread across the globe, ensuring a safe and comfortable future fueled by green power.

REFERENCES

Arbeitsgemeinschaft Energiebilanzen e.V. (AE). Auswertungstabellen 1990–2011. Last modified July, 2011. Accessed May 12, 2012. http://www.ag-energiebilanzen.de/view page.php?idpage=139.

Bauchmüller, A. Solarförderung Sinkt Drastisch *Süddeutsche Zeitung*, February 23, 2012. http://www.sueddeutsche.de/wirtschaft/von-april-an-prozent-weniger-solarfoerder ung-sinkt-drastisch-1.1291104 (accessed December 27, 2012).

Böhling, A. Interview with A. Böhling, Greenpeace Energy Expert in Hamburg, Germany, by J.C. Thomaz. December, 2011.

Bojanowski, A. Streit über Endlager: Abschied vom Millionen-Jahre-Konzept. *Der Spiegel*, May 10, 2011. http://www.spiegel.de/wissenschaft/technik/streit-ueber-endlager-abschied-vom-millionen-jahre-konzept-a-760101.html (accessed March 10, 2013).

Borchardt, K. Zäsuren in der wirtschaftlichen Entwicklung. *Vierteljahrshefte für Zeitgeschichte* 61: 21–34, 1990.

Bundesamt für Naturschutz (BfN). *Pressehintergrund: Naturschutz/Biologische Vielfalt/ Umfrage/Naturbewusstsein* by Franz A. Emde. Bonn, Germany: Bundesamt für Naturschutz, 2010. http://www.bfn.de/fileadmin/MDB/documents/presse/BPK-Hintergrund-Naturbewusstsein_102010_Jessel_cs_final.pdf (accessed May 12, 2012).

Bundesministerium der Justiz (BMJ). *Bekanntmachung der öffentlichen Liste über die Registrierung von Verbänden und deren Vertretern.* Berlin: Bundesministerium der Justiz, 2012. http://www.bundestag.de/dokumente/lobbyliste/lobbylisteamtlich.pdf (accessed June 6, 2012).

Bundesministerium für Bildung und Forschung (BMBF), and Deutscher Akademischer Austausch Dienst (DAAD). *German Research Institutions at a Glance*. Köln, Germany: Moeker Merkur Druck GmbH, 2008. http://www.daad.org.uk/imperia/md/content/london/fundingdocs/german_research_institutions_at_a_glance.pdf (accessed May 12, 2012).

Bundesministerium für Ernährung, Landwirtschaft und Verbraucherschutz (BMELV), and Fachagentur Nachwachsende Rohstoffe e.V. (FNR). *Bioenergy in Germany: Facts and Figures*. Gützow, Germany: Fachagentur Nachwachsende Rohstoffe e.V., 2012. http://www.biodeutschland.org/tl_files/content/dokumente/biothek/Bioenergy_in-Germany_2012_fnr.pdf (accessed January 12, 2013).

Bundesministerium für Umwelt, Naturschutz und Reaktorsicherheit (BMU). Vergütungssätze und Degressionsbeispiele nach dem neuen Erneuerbare-Energien-Gesetz (eeg).vom 31. Oktober 2008 mit Änderungen vom 11. August 2010. Last modified November, 2010a. Accessed May 12, 2012. http://www.erneuerbare-energien.de/files/pdfs/allgemein/application/pdf/eeg_2009_verguetungsdegression_bf.pdf.

Bundesministerium für Umwelt, Naturschutz und Reaktorsicherheit (BMU). Struktur und Höhe des Primärenergieverbrauchs im Leitszenario 2010. Last modified 2010b. Accessed May 12, 2012. http://www.umweltbundesamt-daten-zur-umwelt.de/umweltdaten/public/document/downloadPrintImage.do?date=&ident=22569.

Bundesministerium für Umwelt, Naturschutz und Reaktorsicherheit (BMU). *Erneuerbare Energien in Zahlen* by F. Musiol, T. Nieder, T. Rüther, M. Walker, U. Zimmer, M. Memmler, S. Rother, S. Schneider, and K. Merkel. Paderborn, Germany: Bonifatius GmbH, 2011. http://www.erneuerbare-energien.de/files/pdfs/allgemein/application/pdf/broschuere_ee_zahlen_bf.pdf (accessed May 12, 2012).

Bundesministerium für Umwelt, Naturschutz und Reaktorsicherheit (BMU). *Erneuerbare Energien in Zahlen: Internet-Update ausgewählter Daten* by F. Musiol, P. Bickel, T. Nieder, T. Rüther, M. Walker, M. Memmler, S. Rother, S. Schneider, and K. Merkel. Paderborn, Germany: Bonifatius GmbH, 2012a. http://www.erneuerbare-energien.de/files/pdfs/allgemein/application/pdf/broschuere_ee_zahlen_bf.pdf (accessed May 30, 2013).

Bundesministerium für Umwelt, Naturschutz und Reaktorsicherheit (BMU). *Mit neuer Energie: 10-Punkte-Programm für eine Energie- und Umweltpolitik mit Ambition und Augenmaß von Bundesminister Peter Altmaier* by P. Altmaier. Frankfurt am Main, Germany: Zarbock GmbH & Co. KG, 2012b. https://secure.bmu.de/fileadmin/bmu-import/files/pdfs/allgemein/application/pdf/10_punkte_programm_bf.pdf (accessed March 10, 2013).

Bundesministerium für Umwelt, Naturschutz und Reaktorsicherheit (BMU), and Agentur für Erneuerbare Energien e.V. (AEE). *20 Jahre Förderung von Strom aus Erneuerbaren Energien in Deutschland—Eine Erfolgsgeschichte* by E. Bruns, D. Ohlhorst, and B. Wenzel. Berlin: Agentur für Erneuerbare Energien e.V., 2010. http://www.aee.ch/fileadmin/user_upload/Downloads/Downlaods/Wirtschaft/41_Renews_Spezial_20_Jahre_EE-Strom-Foerderung.pdf (accessed May 12, 2012).

Bundesministerium für Umwelt, Naturschutz und Reaktorsicherheit (BMU), Bundesministerium für Ernährung, Landwirtschaft und Verbraucherschutz (BMELV), and Agentur für Erneuerbare Energien e.V. (AEE). *Akzeptanz und Bürgerbeteiligung für Erneuerbare Energien: Erkenntnisse aus Akzeptanz und Partizipationsforschung* by C. Wunderlich. Berlin: Agentur für Erneuerbare Energien e.V., 2012. http://www.unendlich-viel-energie.de/uploads/media/60_Renews_Spezial_Akzeptanz_online_final.pdf (accessed December 25, 2012).

Bundesministerium für Wirtschaft und Technologie (BMWi). *Sichern Sie sich Ihren Vorsprung im Export!* by FLASKAMP AG. Berlin: Print Produktion Laube GmbH, 2009. http://www.efficiency-from-germany.info/EIE/Redaktion/PDF/infobroschuere-exportini-tiative-energieeffizienz,property=pdf,bereich=eie,sprache=de,rwb=true.pdf (accessed December 25, 2012).

Bundesministerium für Wirtschaft und Technologie (BMWi), and Bundesministerium für Umwelt, Naturschutz und Reaktorsicherheit (BMU). *The Federal Government's Energy Concept of 2010 and the Transformation of the Energy System of 2011.* Niestetal, Germany: Silber Druck oHG, 2011. http://www.bmu.de/files/english/pdf/application/pdf/energiekonzept_bundesregierung_en.pdf (accessed December 25, 2012).

Bundesministerium für Wirtschaft und Technologie (BMWi), and ForschungsVerbund Erneuerbare Energien (FVEE). *Themen 2008: Energieeffizientes und Solares Bauen.* Berlin: Hoch3 GmbH, 2009. http://www.fvee.de/fileadmin/publikationen/Themenhefte/th2008-1/th2008.pdf (accessed May 12, 2012).

Clearingstelle EEG, and RELAW—Gesellschaft für angewandtes Recht der Erneuerbaren Energien mbH. "StrEG." Last modified 2012. Accessed March 10, 2013. http://www.clearingstelle-eeg.de/streg.

Cramer, B. Bundesanstalt für Geowissenschaften und Rohstoffe (BGR) and Deutsche Rohstoffagentur (DeRA). Fossile Energieträger—Wie viel und wie lange noch? Last modified 2010. Accessed December 28, 2012. http://www.muenchner-wissenschaft stage.de/2010/upload/download/Cramer_Fossile_Energietraeger.pdf.

Demandt, A. *Über die Deutschen: Eine kleine Kulturgeschichte.* Bonn: Bundeszentrale für politische Bildung, 2008.

Deutsche Naturschutzring (DNR). Richtlinienvorschlag zu Biokraftstoffen. Last modified 2012. Accessed December 28, 2012. http://www.eu-koordination.de/umweltnews/news/klima-energie/1736-richtlinienvorschlag-zu-biokraftstoffen.

Deutsche Presse-Agentur (dpa). Klimaforscher: Merkels Klimapolitik doppelzüngig. *Focus,* July 17, 2012. http://www.focus.de/politik/deutschland/umwelt-klimaforscher-merkels-klimapolitik-doppelzuengig_aid_782941.html (accessed March 10, 2013).

Deutscher Bundestag. *Basic Law for the Federal Republic of Germany.* Berlin, Germany: Deutscher Bundestag, 2010. https://www.btg-bestellservice.de/pdf/80201000.pdf (accessed December 28, 2012).

Dosi, C., and M. Moretto. 2006. Concession bidding rules and investment time flexibility. Paper presented at the *10th Joint Conference on Food, Agriculture and the Environment, Duluth, MN, August 27–30.* St. Paul, MN: Center for International Food and Agricultural Policy, University of Minnesota. http://ageconsearch.umn.edu/bitstream/6630/2/cp06do01.pdf (accessed May 12, 2012).

The Economist. Germany's energy policy: Nuclear power? Um, maybe. *The Economist,* September 2, 2010. http://www.economist.com/node/16947258 (accessed December 27, 2011).

The Economist. Germany's energy policy: Nuclear power? No thanks (Again). *The Economist,* March 15, 2011. http://www.economist.com/blogs/newsbook/2011/03/german_energy_policy (accessed December 27, 2011).

El-Sharif, Y. Desertec-Projekt: Experten Zweifeln an Wüstenstrom-Wunder. *Der Spiegel,* July 13, 2009. http://www.spiegel.de/wirtschaft/desertec-projekt-experten-zweifeln-an-wuestenstrom-wunder-a-635811.html (accessed March 21, 2013).

EnBW Energie Baden-Württemberg AG (EnBW). Der Strommix der EnBW. Last modified 2012. Accessed May 12, 2012. http://www.enbw.com/content/de/der_konzern/enbw/neue_gesetzgebung/stromkennzeichnung/index.jsp.

E.On. Stromkennzeichnung gemäß §42 EnWG auf Basis der Daten von 2010. Accessed May 12, 2012. https://www.eon.de/de/westfalenweser/pk/services/Rechtliches_ Veroeffentlichungspflichten/Energiemix/index.htm.

Erdmann, L., Deutsche Press-Agentur (dpa), and dapd Natrichtenagentur. Deutsche wenden sich radikal von der Atomkraft ab. *Der Spiegel.* March 15, 2011. http://www. spiegel.de/panorama/umfragen-deutsche-wenden-sich-radikal-von-der-atomkraftab-a-750955.html (accessed December 18, 2011).

European Environment Agency (EEA). Renewable Electricity as a Percentage of Gross Electricity Consumption, 2008. Last modified August 9, 2011. Accessed May 12, 2012. http://www.eea.europa.eu/data-and-maps/figures/ renewable-electricity-as-a-percentage-2.

European Federation for Transport and Environment. Road Transport Speed and Climate Change. Last modified September 2005. Accessed March 10, 2013. http:// www20.gencat.cat/docs/dmah/Home/Ambits dactuacio/Atmosfera/Qualitatdelaire/ OficinaTecnicadePlansdeMillora/Enllacosdinteresiestudistecnicsrelacionats/ Enllacosidocs/2005-09_cars21_speed_co2.pdf.

European Parliament. Directive 2009/28/EC of the European Parliament and of the Council. *Official Journal of the European Union.* 140: 16–62, 2009. http://eur-lex.europa.eu/LexUriServ/ LexUriServ.do?uri=Oj:L:2009:140:0016:0062:en:PDF (accessed January 12, 2013).

Fell, H.J. Einspeisevergütung für erneuerbare Energien: Ein wirksames Konjunkturprogramm ohne staatliche Neuverschuldung. Last modified 2009. Accessed May 12, 2012. www. hans-josef-fell.de.

Fell, H.J. Finanzwirtschaft und klimaschutz: Eine gewinnbringende allianz. Last modified 2010. Accessed May 12, 2012. www.hans-josef-fell.de.

Forbes. The Global 2000: Germany. Last modified April 21, 2010. Accessed May 12, 2012. http://www.forbes.com/lists/2010/18/global-2000-10_The-Global-2000-Germany_ 10Rank_print.html.

Germanwatch e.V., DAKT e.V., and Heinrich-Böll-Stiftung Thüringen. *Energiewende und Bürgerbeteiligung: Öffentliche Akzeptanz von Infrastrukturprojekten am Beispiel der "Thüringer Strombrück,"* by K. Schnelle, and M. Voigt. Erfurt, Germany: Heinrich-Böll-Stiftung Thüringen, 2012. http://germanwatch.org/fr/download/4135.pdf (accessed December 25, 2012).

Gotzmann, I. Bund Heimat und Umwelt e.V. Geschichte. Last modified 2010a. Accessed December 28, 2012. http://www.bhu.de/bhu/content/de/ueberuns/geschichte/startseite. html?jid=1o2o6.

Gotzmann, I. Bund Heimat und Umwelt e.V. Standpunkte. Last modified 2010b. Accessed December 28, 2012. http://www.bhu.de/bhu/content/de/ueberuns/standpunkte/ startseite.html?jid=1o2o1.

Greenpeace. *Der Plan: Deutschland ist Erneuerbar!* Hamburg, Germany: EDP, 2011. http:// www.greenpeace.de/fileadmin/gpd/user_upload/themen/energie/DerPlan.pdf (accessed May 12, 2012).

Greenpeace. Greenpeace Start. Last modified 2012. Accessed May 12, 2012. http://www. greenpeace.de/.

Groba, F., and C. Kemfert. Erneuerbare Energien: Deutschland baut Technologie-Exporte aus. *DIW Wochenbericht.* 45: 23–29, 2011. http://www.diw.de/documents/publika tionen/73/diw_01.c.388573.de/11-45-4.pdf (accessed November 21, 2012).

Gudermann, R. NRW-Stiftung. Die Geschichte des Naturschutzes in Deutschland> So Kam der Naturschutz in Bewegung. Last modified 2011. Accessed December 25, 2012. http://www.nrw-stiftung.de/projekte/bericht.php?bid=3.

H+M Media AG. About the Club of Rome. Last modified 2012. Accessed March 9, 2013. http://www.clubofrome.org/?p=324.

Haber, W. Johannes Gutenberg-Universität Mainz. Naturschutz zwischen Wissenschaft und Praxis. Last modified 2009. Accessed December 25, 2012. http://www.studgen.uni-mainz.de/Dateien/Prof_Wolfgang_Haber_17-11-08.pdf.

Handelsblatt. Trotz Kürzungen bleiben Solaranlagen attraktiv. Last modified 2012. Accessed December 28, 2012. http://www.handelsblatt.com/politik/deutschland/solarfoerder ung-trotz-kuerzungen-bleiben-solaranlagen-attraktiv/6807250.html.

Haus der Essener Geschichte. Deckel auf den Pütt—Zechensterben begann vor 50 Jahren. Last modified 2003. Accessed March 10, 2013. http://www.essen.de/de/Rathaus/Aemter/Ordner_41/Stadtarchiv/geschichte/geschichte_einsichten_bergbaukrise.html.

Hohls, A. Der Deutsche Wald: Die Beziehung der Deutschen zum Wald im Verlauf der Zeitgeschichte. Last modified 2011. Accessed December 25, 2012. http://www.xn—waldpdagogik-kcb.de/pdf/produkte/themen/hausarbeit_wald.pdf.

Intergovernmental Panel on Climate Change (IPCC). *Climate Change 2007: Synthesis Report* by L. Bernstein, P. Bosch, O. Canziani, Z. Chen, R. Christ, O. Davidson, W. Hare, S. Huq, D. Karoly, V. Kattsov, Z. Kundzewicz, J. Liu, U. Lohmann, M. Manning, T. Matsuno, B. Menne, B. Metz, M. Mirza, N. Nicholls, L. Nurse, R. Pachauri, J. Palutikof, M. Parry, D. Qin, N. Ravindranath, A. Reisinger, J. Ren, K. Riahi, C. Rosenzweig, M. Rusticucci, S. Schneider, Y. Sokona, S. Solomon, P. Stott, R. Stouffer, T. Sugiyama, R. Swart, D. Tirpak, C. Vogel, G. Yohe, and T. Barker. Geneva, Switzerland: IPCC, 2007. http://www.ipcc.ch/pdf/assessment-report/ar4/syr/ar4_syr.pdf (accessed December 28, 2012).

Kandler, O. Food and Agricultural Organization (FAO). The Air Pollution/Forest Decline Connection: The "Waldsterben" Theory Refuted. Last modified 1993. Accessed December 25, 2012. http://www.fao.org/docrep/v0290e/v0290e07.htm.

Kramer, A. Bundesministerium für Wirtschaft und Technologie. The Export Energy Efficiency Initiative in Germany. Last modified 2011. Accessed December 25, 2012. http://www.gaccny.com/fileadmin/ahk_gaccny/Consulting/Green_Corner/GACC_Presentation_Kramer.pdf.

Kröger, M., and dapd Natrichtenagentur. Altmaier will Bürgerbeteiligung schnell umsetzen. *Der Spiegel*, November 11, 2012. Accessed December 25, 2012. http://www.spiegel.de/politik/deutschland/altmaier-verspricht-gesetz-zur-buergeranleihe-fuer-starkstromleitungen-a-866530.html.

Lange, J.H. Interview with J.H. Lange, Director of the Energy Department at HSH-Nordbank, by J.C. Thomaz. December 22, 2011.

Mautz, R., A. Byzio, and W. Rosenbaum. *Principles of the Innovation. auf dem Weg zur Energiewende: Die Entwicklung der Stromproduktion aus Erneuerbaren Energien in Deutschland*. Göttingen: Universitätsverlag Göttingen, 2008.

Meadows, D.H., D.L. Meadows, J. Randers, and W. Behrens. *The Limits to Growth*. Universe Books, New York, 2008.

Meyer, A. Interview with A. Meyer, Officer at the BMU—Bundesministerium für Umwelt, Naturschutz und Reaktorsicherheit, by J.C. Thomaz. December 13, 2011.

Netzwerk Bürgerbeteiligung. *Energiewende, Netzausbau und Öffentlichkeitsbeteiligung—Wie geht das zusammen?* by M. Zschiesche. Bonn, Germany: Netzwerk Bürger beteiligung, 2012. http://www.netzwerk-buergerbeteiligung.de/fileadmin/Inhalte/PDF-Dokumente/newsletter_beitraege/beitrag_zschiesche_121005.pdf (accessed December 25, 2012).

Neubacher, A. Energie: Verblendet. *Der Spiegel*, January 16, 2012. http://www.spiegel.de/spiegel/print/d-83588339.html (accessed May 26, 2012).

Nuclear Energy Institute (NEI). Monthly Fuel Cost to U.S. Electric Utilities (1995–2011). Last modified 2012. Accessed May 12, 2012. http://www.nei.org/resources andstats/documentlibrary/reliableandaffordableenergy/graphicsandcharts/monthlyfuelcosttouselectricutilities/.

Ondraczek, J. Interview with J. Ondraczek, Consultant for Energy Projects at PwC—Pricewaterhouse and Coopers in Hamburg, Germany, by J.C. Thomaz. December 21, 2011.

Overgaard, S. Statistics Norway. Issue Paper: Definition of Primary and Secondary Energy. Last modified 2008. Accessed December 25, 2012. http://unstats.un.org/unsd/envac counting/londongroup/meeting13/LG13_12a.pdf.

Paeger, J. Eine kleine Geschichte der Atomkraft. Last modified 2012. Accessed December 25, 2012. http://www.oekosystem-erde.de/html/atomkraft.html.

Radkau, J. Bundeszentrale für politische Bildung. Eine kurze Geschichte der deutschen Antiatomkraftbewegung. Last modified 2011. Accessed December 25, 2012. http://www.bpb.de/apuz/59680/eine-kurze-geschichte-der-deutschen-antiatomkraftbewegung?p=all.

Rieckmann, M. Die Anfänge der Anti-Atom-Bewegung in der Bundesrepublik Deutschland. Last modified 2000. Accessed December 25, 2012. http://basisgruen.gruene-linke.de/fachbereiche/energie/atom/00-09—rieckmann—antiakw-bewegung.pdf.

Rosa Luxemburg Stiftung (RLS), and Arepo Consult. *Befreiung der Energieintensiven Industrie in Deutschland von Energieabgaben* by S. Rieseberg, and C. Wörlen. Berlin, Germany: Arepo Consult, 2012. http://www.rosalux.de/fileadmin/rls_uploads/pdfs/Themen/Nachhaltigkeit/RLS-Studie_Energieintensive_Industrie.pdf (accessed February 28, 2013).

RWE AG. *2009 Jahresabschluss der RWE AG.* Bocholt, Germany: D+L Printpartner GmbH, 2010. http://www.rwe.com/web/cms/mediablob/de/389550/data/239998/3/rwe/investor-relations/berichte/2009/Jahresabschluss-RWE-AG-2009.pdf (accessed May 12, 2012).

RWE Innogy. About RWE Innogy. Last modified 2012. Accessed March 10, 2013. http://www.rwe.com/web/cms/en/87202/rwe-innogy/about-rwe-innogy/.

Schmidt, O. Bayerische Landesanstalt für Wald und Forstwirtschaft. Von den Wurzeln der Nachhaltigkeit: Vor drei Jahrhunderten erfand die Forstwirtschaft das Prinzip der Nachhaltigkeit. Last modified 2012. Accessed December 25, 2012. http://www.lwf.bayern.de/waldoekologie/standort-bodenschutz/aktuell/2012/43671/linkurl_1.pdf.

Schünemann, C. Regenerative-Zukunft.de. Photovoltaik. Last modified 2012. Accessed December 28, 2012. http://www.regenerative-zukunft.de/erneuerbare-energien-menu/photovoltaik.

Skalski, T. Umstrittene Laufzeiten von Kernkraftwerken: Es tobt eine heftiger Kampf um Interessen. Last modified August, 2010. Accessed May 12, 2012. http://www.3sat.de/page/?source =/scobel/147197/index.html.

Stadermann, G. Interview with G. Stadermann, Director at the FVEE—Forschungsverbund Erneuerbare Energien, by J.C. Thomaz. December 19–20, 2011.

Stadt Mannheim. EEWärmeGesetz. Last modified 2013. Accessed January 11, 2013. http://www.mannheim.de/buerger-sein/eewaermegesetz.

Statistisches Bundesamt (Destatis). *Volkswirtschaftliche Gesamtrechnungen Inlandsprodukts-berechnung Lange Reihen ab 1970.* Wiesbaden, Germany: Statistisches Bundesamt, 2012. https://www.destatis.de/DE/Publikationen/Thematisch/Volkswirtschaftliche Gesamtrechnungen/Inlandsprodukt/InlandsproduktsberechnungLangeReihen PDF_ 2180150.pdf?__blob=publicationFile (accessed December 25, 2012).

Stubner, H. Interview with H. Stubner, Leader for Political Issues at BVEE—Bunderverband Erneuerbare Energien, by J.C. Thomaz. December 19, 2011.

Timm, M. Interview with M. Timm, Representative for Energy Affairs at BDEW— Bundesverband der Energie-und Wasserwirtschaft, by J.C. Thomaz. December 20, 2011.

Transparency International. Corruption Perceptions Index 2010. Last modified 2011. Accessed May 12, 2012. http://www.transparency.org/cpi2010/results.

Troen, I., and E.L. Petersen. *European Wind Atlas*. Roskilde: Risø National Laboratory, 1989.

Umwelt- und Prognose-Institut e.V. (UPI). Wie Milliarden-Subventionen die Köpfe Vernebeln. Last modified November, 1999. Accessed March 10, 2013. http://www. upi-institut.de/oeskohlegas.htm.

Umweltbundesamt (UBA). Primärenergieverbrauch. Last modified 2011. Accessed May 12, 2012. http://www.umweltbundesamt-daten-zur-umwelt.de/umweltdaten/public/ theme.do?nodeIdent=2326.

Umweltbundesamt (UBA), Bundesanstalt für Geowissenschaft und Rohstoffe (BGR), and Statistiches Bundesamt (Destatis). *Umweltdaten Deutschland: Nachhaltig Wirtschaften—Natürliche Ressourcen und Umwelt Schonen*. Berlin: KOMAG, 2007. http://www.umweltdaten.de/publikationen/fpdf-l/3244.pdf (accessed May 12, 2012).

U.S. Energy Information Administration (U.S. EIA). Voluntary Reporting of Greenhouse Gases Program Fuel Emission Coefficients. Last modified 2011. Accessed May 12, 2012. http://www.eia.gov/oiaf/1605/coefficients.html.

Vattenfall. Information zur Stromkennzeichnung. Last modified 2009. Accessed May 12, 2012. http://www.vattenfall.de/www/vf/vf_de/Gemeinsame_Inhalte/DOCUMENT/ 154192vatt/289100prod/289114priv/P02182175.pdf.

Wenzel, T. Interview with T. Wenzel, Energy Expert, by J.C. Thomaz. December 19, 2011.

Wiese, T. Tagesschau.de. Fragen und Antworten: Die EEG-Umlage. Last modified 2012. Accessed December 28, 2012. http://www.tagesschau.de/wirtschaft/faq-oekostro mumlage100.html.

Williams, J. WTRG Economics. Oil Price History and Analysis. Last modified 2011. Accessed December 28, 2012. http://www.wtrg.com/prices.htm.

Wind Energy Development Programmatic Environmental Impact Statement (WEDP EIS). U.S. Department of the Interior, Bureau of Land Management. Wind Energy Development Environmental Concerns. Last modified 2009. Accessed March 10, 2013. http://windeis.anl.gov/guide/concern/index.cfm.

Wissenschaftliche Dienste des Deutschen Bundestages (WDDB). *Co2-Bilanzen verschiedener Energieträger im Vergleich* by Daniel Lübbert. Berlin: Wissenschaftliche Dienste des Deutschen Bundestages, 2007. http://www.bundestag.de/dokumente/analy sen/2007/CO2-Bilanzen_verschiedener_Energietraeger_im_Vergleich.pdf (accessed May 12, 2012).

The World Bank Group (WBG). OECD Members. Last modified 2011. Accessed January 11, 2013. http://data.worldbank.org/country/OED.

The World Bank Group (WBG). World Databank: World Development Indicators (WDI) and Global Development Finance (GDF). Last modified 2011. Accessed May 12, 2012. http://databank.worldbank.org/ddp/html-jsp/QuickViewReport. jsp?RowAxis=WDI_Ctry~&ColAxis=WDI_Time ~&PageAxis=WDI_Series~&PageA xisCaption=Series~&RowAxisCaption=Country~&ColAxisCaption=Time~&NEW_ REPORT_SCALE=1&NEW_REPORT_PRECISION=0&newReport=yes&ROW_ COUNT=1&COLUMN_COUNT=44&PAGE_COUNT=1&COMMA_SEP=true.

4

China's Energy Profile and the Importance of Coal

Julia Zheng and Xiaoting Zheng

CONTENTS

4.1 INTRODUCTION

China's energy production and consumption have increased continuously over the past four decades, especially during the first decade of the twenty-first century. This dramatic growth has been driven by the country's swift economic development, which increased its gross domestic product (GDP) to the second largest in the world (World Bank Group [WBG] 2011). A notable feature of China's energy sector is the extraordinarily large proportion of coal in the energy mix. Coal is the country's most abundant energy resource and has played a major role in China's industrialization and urbanization, mainly through its use in power generation and industrial plants. Presently, coal alone satisfies more than two-thirds of the country's total energy demand, a figure that is not expected to change much in the foreseeable future (U.S. Energy Information Administration [U.S. EIA] 2011).

The extensive use of coal, one of the "dirtiest" energy sources, has led to major environmental consequences. Carbon emissions have expanded drastically, and China is now the world's largest emitter of greenhouse gases. One by-product of burning coal, SO_2, also causes acid rain and poses a serious threat to local ecosystems. Other external environmental and social problems include coal gangue, coal dust, water pollution, air pollution, land subsidence, and mining accidents (Greenpeace 2008).

China has recognized the drawbacks of coal and has taken an active role in developing technologies to overcome some of the resource's inherent flaws. The country is leading efforts in researching and testing advanced power generation methods that use this resource more efficiently. It is also investigating the possibility of carbon capture and sequestration (CCS), so that it can safely store the carbon produced from coal combustion. In addition to seeking ways to make coal usage cleaner over time, China is developing alternative energy sources, which will be discussed in Chapter 5.

4.2 ENERGY CONSUMPTION OVERVIEW AND FUTURE FORECASTS

China's energy production and consumption have increased dramatically over the last four decades. As shown in Figure 4.1, total energy production increased from 627.7 million tons of coal equivalent (Mtce) in 1978 to 2600 Mtce in 2008. The unit tce represents the energy obtained from the combustion of one metric ton of coal, equivalent to 29.39 GJ (gigajoules), 27.78 Btu, and 8.14 MWh. Total energy consumption grew from 571.44 Mtce in 1978 to 2850 Mtce in 2008, with an annual compounded growth rate of 5.6%. China's energy consumption accounted for 20% of the world's total energy consumption in 2010. Moreover, nearly 40% of

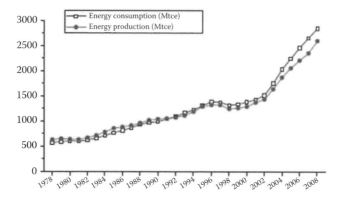

FIGURE 4.1

Energy production and consumption, 1978–2008. (Adapted from Zhang N. et al., *Energy* 36:3639–3649, 2011.)

the total increase in energy consumption since 1990 can be attributed to China (U.S. EIA 2011). The country's energy consumption from 2003 to 2006 exceeded the sum of energy consumed in the previous 25 years (Fang 2009). Consumption first exceeded production in 1992, and this gap has been widening steadily ever since.

Energy demand exceeded supply in China in 2003 and 2004, causing power interruptions during periods of peak usage. These shortages drove the increase in renewable energy–generating capacity as a proportion of the total energy mix beginning around 2004 (IBISWorld 2013). Also, as a result of these shortages, the Chinese government set ambitious energy-saving targets in the 11th Five-Year Plan, aiming for change during the 2006 to 2011 timeframe. Although the government had intended for non–energy-intensive sectors to make up a greater percentage of the economy, through 2012 the share of these sectors had actually declined. The service sector made up only 40% of GDP in 2005, compared with 41.5% in 2002 (Wang et al. 2011).

China's energy consumption is dominated by coal, which is the country's most ample energy resource and has played a major role in its recent rapid development. Currently, coal provides 70% of the total energy used in the country, followed by oil liquids, hydroelectric energy, and other sources (ibid.). China's energy consumption is projected to increase in the foreseeable future without significant changes to its energy matrix, as shown in Figure 4.2.

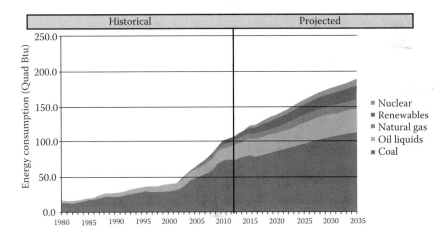

FIGURE 4.2
China's historical and future forecasted energy consumption. (Adapted from U.S. EIA. *International Energy Outlook 2011*. Washington, DC: U.S. Department of Energy and U.S. Energy Information Administration, 2011.)

4.2.1 Economic Development

China's energy consumption growth has been driven largely by the economic boom that followed the opening of the economy in 1978. The 13th Central Committee of the Communist Party of China Conference, held in October 1987, proposed a three-step economic development strategy. The first step, achieved by the end of the 1980s, involved doubling the 1980 gross national product (GNP) and ensuring that the population's basic needs were served (i.e., that "people had enough food and clothing"; Feng 2008). The second step was to quadruple the 1980 GNP by the end of the twentieth century. This goal was attained in 1995. The third step is to bring per capita GNP up to the level of medium-developed countries by 2050.

In 2010, China surpassed Japan to become the nation with the second-largest GDP in the world. From 1980 to 2010, its GDP expanded by an average annual rate of 10%, resulting in a current figure that is nearly 30 times greater than the 1980 level. As a percentage of the world's total, China's GDP grew from 2% in 1980 to 9% in 2010 (Figure 4.3) and is expected to continue to increase in the foreseeable future, although perhaps at a slightly slower rate. It is projected to reach US$21.4 trillion (in 2010 US$) in 2035, 13% of the world's total, compared with US$5.7 trillion in 2010 (U.S. EIA 2011).

An increasing GDP does not necessitate greater energy consumption. Exceptions to the general rule of a positive correlation between energy consumption and GDP have occurred in specific situations, such as in Denmark.

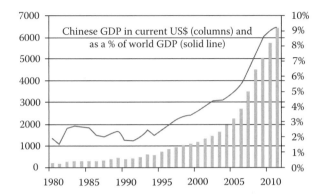

FIGURE 4.3
China's historical GDP and percentage of world GDP. (Adapted from IMF. World Economic Outlook Database. http://www.imf.org/external/pubs/ft/weo/2010/02/weodata/download.aspx [accessed November 19, 2012].)

In China, however, climbing energy consumption levels have accompanied GDP growth. In fact, the country's energy consumption per unit of GDP (energy intensity) is more than twice that of both the United States and the world average (Figure 4.4), although it has been decreasing continuously over the past three decades. This high energy intensity reflects a low degree of energy efficiency, which results from the large percentage of coal in China's energy supply. However, the country's energy consumption per capita is still low compared with those of developed countries. This indicates that its aggregate energy consumption is likely to increase even more in the future with continued modernization and higher living standards (Zhang 2011).

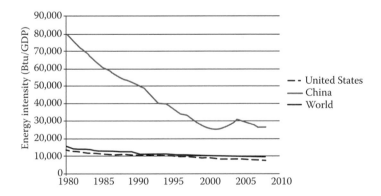

FIGURE 4.4
Energy intensities of the U.S., China, and the world. (Adapted from U.S. EIA. International Energy Statistics. http://www.eia.gov/cfapps/ipdbproject/IEDIndex3.cfm [accessed November 19, 2011].)

4.2.2 Carbon Emissions

The unprecedented expansion of China's industrial sector has required a large amount of energy and natural resources, in part because two-thirds of the primary energy consumed in the country is derived from coal, the least efficient of traditionally used energy sources. Its growing, coal-dominated energy matrix has exacerbated the magnitude of its carbon emissions. Carbon dioxide (CO_2) is a major greenhouse gas and a critical factor in global climate change. Carbon emissions have negative environmental consequences on the entire planet, affecting global temperature, water cycles, ocean circulation, and the biosphere (Intergovernmental Panel on Climate Change [IPCC] 2007).

China's annual carbon emissions increased from 1.45 billion tons in 1980 to 7.71 billion tons in 2009, accounting for 25.42% of the world's total that year (U.S. EIA 2012b). Of its total carbon emissions, 84% come from coal. The average annual growth rate over the past decade has been a dramatic 11.7%, as shown in Figure 4.5.

China's CO_2 emissions per capita began in 1980 below the non-OECD average, surpassed the non-OECD average in the mid-1990s, and exceeded the world average by around 2005 (Figure 4.6). Its per capita emissions were 5.82 tons in 2009, above the world average of 4.47 tons, but still well below the OECD's average of 10.36 tons (U.S. EIA 2012a).

Based on assumptions of continued economic growth, further population expansion, and increased energy consumption, China's CO_2 emissions are projected to increase significantly in the next few decades, with a total increase of more than six billion tons by 2035 (Figure 4.7; U.S. EIA 2011).

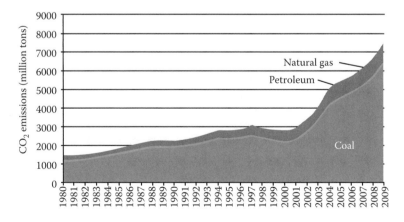

FIGURE 4.5

China's CO_2 emissions by source over time. (Adapted from U.S. EIA. International Energy Statistics. http://www.eia.gov/cfapps/ipdbproject/IEDIndex3.cfm [accessed November 19, 2011].)

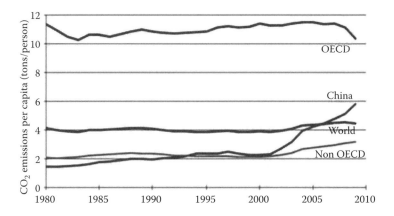

FIGURE 4.6
CO_2 emissions per capita of China, OECD, non-OECD, and world over time. (Adapted from U.S. EIA. International Energy Statistics. http://www.eia.gov/cfapps/ipdbproject/ IEDIndex3.cfm [accessed November 19, 2011].)

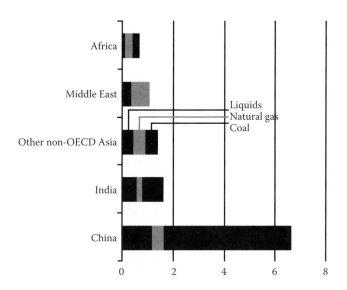

FIGURE 4.7
Projected increases in annual carbon emissions by fuel type for regions with the highest forecasted economic growth rates, 2008–2035 (GtC). (Adapted from U.S. EIA. *International Energy Outlook 2011.* Washington, DC: U.S. Department of Energy & U.S. Energy Information Administration, 2011.)

As a result of structural, industrial, and technological improvements and the increased adoption of low-carbon and noncarbon resources, China's carbon intensity (carbon emissions per unit of GDP) decreased dramatically from 6.7 tons per thousand US$ in 1980 to 2.01 tons per thousand US$ in 2000, with an impressive annual rate of decline of 5.8% (Figure 4.8; U.S. EIA 2012a). During this period, the faster growth rate of GDP relative to carbon emissions accounted for the decrease in carbon intensity. Energy consumption grew by a factor of 2.5 from 1978 to 2001, although GDP increased by a factor of 8.2 (National Bureau of Statistics of China 2012). This unusually large decline in carbon intensity can be attributed to improvements in the country's formerly exceedingly poor energy efficiency, as well as to efforts to continually decrease the share of coal in total energy production and consumption. It can also be attributed to policy measures enacted throughout this period, especially between 1994 and 2000, to commercialize the economy and improve energy efficiency (Transatlantic Academy 2012). Nevertheless, despite the dramatic decrease, China's carbon intensity is still more than six times the OECD average and more than three times the world's average (Zhang et al. 2011).

From 2001 to 2004, however, China's carbon intensity increased at an annual rate of 6.7%. Fixed asset investments as a share of total investments were 36% in 2002 and 47% in 2005, reflecting an expansion of the

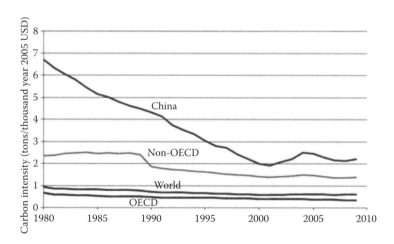

FIGURE 4.8
Carbon intensities over time for China, OECD, non-OECD, and the world using market exchange rates. (Adapted from U.S. EIA. International Energy Statistics. http://www.eia.gov/cfapps/ipdbproject/IEDIndex3.cfm [accessed November 19, 2011].)

industrial sector, particularly of energy-intensive industries such as steel, cement, chemicals, and heavy manufacturing. China became the world's largest producer of cement, steel, and flat glass; and its exports of energy-intensive products also increased dramatically. Most of this periods' GDP growth can be traced to the expansion of these energy-intensive industries. Consequently, during this time, carbon emissions grew more rapidly than GDP (Zhang et al. 2011). The annual growth rate of carbon emissions was 14.4%, whereas GDP grew at an annual rate of 10% (U.S. EIA 2012b).

Carbon-intensive products such as cement, power, steel, nonferrous metals, glass, paper, and chemicals are produced disproportionately by small-scale, energy-inefficient plants. For example, more than 5000 small plants were producing cement in 2005, whereas the top 10 producing enterprises accounted for just 13% of the national cement production. Energy-efficient rotary-type kilns accounted for only 40% of the production (Zhang et al. 2011). Carbon intensity dropped in 2006 after the adoption of the National Energy-Saving Policy (ibid.). However, it recently climbed again by 4% in 2009 over the level in 2008.

4.3 ENERGY SUPPLY

4.3.1 Energy Resources

At the end of 2007, China had recoverable reserves of 176.8 billion tons of coal, 21.2 billion tons of crude oil, 14.3 billion tons of nonconventional oil, 22.03 trillion cubic meters of natural gas, and 400 GW of hydropower (Wang et al. 2011). It has the third-largest supply of coal reserves in the world and the largest hydropower reserves (Zhang et al. 2011). Therefore, it has large energy reserves in absolute terms (Table 4.1). When considered on a per capita basis, however, the country's energy resources per capita are very low compared with world averages (Wang et al. 2011). For example, its per capita energy resources of coal and hydropower are only 50% of the world average level, and the per capita oil and gas resources are only approximately 7% of the world average (Zhang et al. 2011). Furthermore, many of China's renewable and nonrenewable energy resources are located far from major demand centers. Innovations in transmission and transportation, therefore, are crucial to its ability to deploy energy resources optimally (Wang et al. 2011).

TABLE 4.1

China's Energy Resources

Primary Source	Raw Coal (billion tons)	Crude Oil (billion tons)	Nonconventional Oil (billion tons)	Natural Gas (trillion cubic meters)	Hydraulic (GW)
Remaining recoverable reserves	176.8	21.2	14.3	22.03	400
Primary source	Onshore wind (GW)	Offshore wind (GW)	Geothermal ($\geq150°$) (GW)	Wave power generation (GW)	Biomass (Mtce/year)
Technically feasible potential	380	700	6	28	350

Source: Adapted from Wang, Y. et al., *Energy Policy* 39:6745–6759, 2011.

4.3.2 Energy Production

In recent years, primary energy production has had two defining characteristics: a major change in output and only slight changes in composition. In 2007, China produced 2.23 billion tce (equal to 1.56 billion tons of oil equivalent, or toe) of primary energy, 6.5% more than in the previous year. When analyzing the evolution of energy production in China (Figure 4.9),

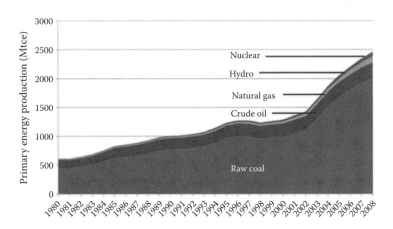

FIGURE 4.9

China's energy production by type over time. (Adapted from National Bureau of Statistics of China. *China Energy Statistical Yearbook 2011*. Beijing: China Statistics Press, 2012.)

it is important to recall that the government closed many inefficient and out-of-date enterprises in the northeast region during the reform of the Chinese economic system around 2000. These closures, coupled with a lack of accurate output data for small coal mines, resulted in the recorded primary energy production being lower than in 1997.

4.3.3 Power Distribution

The power grid configuration is very important when considering the national energy supply. In 2010, China's national power grid was connected to all the provinces except Hainan Island, Tibet, and Xinjiang. The Southern Power Co., Ltd. and the State Power Co., Ltd. are the two power grid operators. Southern Power covers Guangdong, Guangxi, Yunnan, Guizhou, and Hainan, and State Power serves the rest. In 2009, Hainan was connected to the main grid by a 500 kV transmission line across the strait of Qiongzhou. Tibet had four independent local grids, and Xinjiang had an independent power grid. At the end of 2007, China's transmission lines spanned 1,106,345 km, with 100,000 km of 500 and 750 kV lines and 95,855 sets of transformers with a total capacity of 2424 million kVA (Wang et al. 2011).

4.3.3.1 The Geographic Imbalance between Demand and Supply and the Smart Grid

China's power production and consumption centers exhibit a significant geographic discrepancy (Figure 4.10). In the eastern coastal regions, electricity demand exceeds electricity supply; in the northern regions, the opposite is true.

In March 2010, Premier Wen Jiabao outlined a plan to build a unified smart grid system by 2020. Through this, he hoped to integrate advanced information technology into the power system. With this technological aid, smart grids allow grid operators to monitor generation, transmission, and storage to better meet fluctuations in demand. Smart grids aid in load-shifting and peak-shaving and also employ smart meters to allow energy consumers to respond to fluctuating energy prices. The construction of this system is included in China's 12th Five-Year Plan. The Southern Grid Company (SGC) has its own three-phase plan for the construction of a national smart grid system based on a core of an ultra-high voltage (UHV)

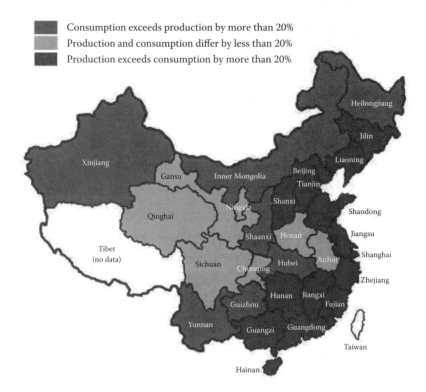

FIGURE 4.10

The geographic imbalance of power production and consumption in China. (Adapted from ChinaMaps.org. China blank map. http://www.chinamaps.org/china/china-blank-map.html.; and China Wind Power Center. Wind Energy Resource Characteristics and Development Potential. http://www.cwpc.cn/cwpc/en/node/6295 [accessed November 19, 2012].)

network. This project will require investments totaling US$40 billion by 2020 (International Energy Agency [IEA] 2011).

The first step of this plan is to accelerate the construction of UHV transmission networks throughout the country. A 1100-kV UHV line has higher electricity transmission capability than a 500-kV line due to lower levels of power loss (ibid.). In 2009, a state investment of US$112 billion (50.9% of the total investment in the power sector) in the transmission system exceeded the investments in new plant construction for the first time, illustrating that the government has recognized that its infrastructure is lagging its energy production capacities. Investment in UHV transmission power lines accounted for 2.5% of the total investment in the power transmission system (ibid.).

SGC and the China State Grid Corporation (CSG) have begun working on this first step. In January 2009, a 1000 kV UHV AC power line (Shanxi–Hubei) was put into commercial operation. In June 2009, an 8000 kV DC line (Yunan–Guangdong) was added. Additionally, SGC has announced plans to construct a "three horizontal, three vertical, and one circle" UHV AC transmission line system, which it plans to complete by 2015. The "three horizontal" component refers to the following UHV lines: western Inner Mongolia–Weifang, Shandong; Jinzhong, Shanxi–Xuzhou, Jiangsu; and Ya'an, Sichuan–Wannan, Anhui. The "three vertical" consists of Xilingol, Inner Mongolia–Nanjing, Jiangsu; Zhangbei, Hebei–Nanchang, Jiangxi; and Shanbei, Shaanxi–Changsha, Hunan (Xinhuanet.com 2011).

4.3.3.2 *Weak Interregional Connections*

Currently, China's grid system can be broken down into six different regions, operating separately under the control of three different transmission and distribution companies (Figures 4.11 and 4.12). The CSG manages

FIGURE 4.11

Regional grid clusters in China. (OECD/IEA 2011; Reprinted with permission from International Energy Agency. *Integration of Renewables: Status and Challenges in China,* by Kat Cheung. Paris: OECD/IEA, 2011. http://www.iea.org/publications/freepublications/publication/Integration_of_Renewables.pdf.)

FIGURE 4.12
Regional power grid clusters in China. (OECD/IEA 2011; Reprinted with permission from International Energy Agency. *Integration of Renewables: Status and Challenges in China*, by Kat Cheung. Paris: OECD/IEA, 2011. http://www.iea.org/publications/freepublications/publication/Integration_of_Renewables.pdf.)

the northeast, northwest, east, and central grids, as well as the part of the north region to the right of the black line in Figure 4.11; the SGC manages the south grid, and the Western Inner Mongolia Grid Corporation, an independent company, manages the part of the north grid to the west of the black line. Each company is responsible for its own profits and losses, so there is little incentive for intercompany cooperation (ibid.).

Interregional grid connections are currently very weak. In 2009, the cross-regional trade of electricity amounted to 158 TWh, representing only 4% of the total electricity flow. This amount of trade, although small, was up 11.4% from the previous year (ibid.).

Strengthening the interregional grid connections will be vital once China begins to incorporate more variable renewable electricity sources, such as solar and wind, into its energy mix. Because electricity cannot be stored in large quantities in an economically feasible way, the grid system and its ability to move electricity around the country will prove to be an important tool for balancing power supply and demand. Variable renewable sources' power output cannot be easily predicted or scheduled. As a result, incorporating a greater amount of renewable energy into the grid leads to less predictable load variation.

Currently, regional grid companies try to balance supply and demand from variable energy sources, such as wind, in three ways. First, they try to accommodate the supply. If that fails, they try to curtail the surplus. Their third option, transmitting the surplus to neighboring power systems, is limited due to the weakness of the interregional grid connections. Therefore, strengthening these connections represents a way to broaden available alternatives for reducing load volatility. SGC is planning to strengthen the connections among the north, central, and east grids (IEA 2011).

4.3.3.3 Demand-Side Management

Demand-side management (DSM) refers to the management of utility activities to encourage energy consumers to modify their patterns of electricity usage behavior. Currently, DSM is employed sporadically and only at the provincial level. Jiangsu Province provides an example in which the local government launched a system of differentiated peak and valley electricity prices, thereby encouraging greater consumption of energy during off-peak hours. However, this method does not allow for real-time adjustments to demand–supply variations (ibid.).

4.3.3.4 Testing the Preferential-Dispatch Method

In 2007, the National Development and Reform Commission (NDRC) began testing its Regulation on Energy Conservation Power Generation Dispatching in five Chinese provinces, ranking deployment of power generation technologies in the following order:

1. Wind, solar, and ocean- and run-river hydro plants
2. Adjustable hydro, biomass, geothermal, and solid waste plants
3. Nuclear plants
4. Coal-fired CHP plants
5. Natural gas– and coal gasification–based combined cycle plants
6. Other coal-fired plants, including cogeneration plants without head load, and
7. Oil-based power plants (ibid.)

The program attempts to prioritize dispatching to variable renewable sources, ensuring that, as long as the reliability and the stability of the supply are secure, renewables will run at full capacity. System operators adjust production of the rest of the generation infrastructure according to this ranking.

Although Chinese officials have not yet concluded whether this mandated prioritization is successful, this set of rules could conceivably lead coal-fired plants to run at less than full loads. This, in turn, would cause them to operate for insufficient periods of time, thus negatively affecting their useful lifetimes. This legislation affects the revenues of electric utilities, which are linked directly to the number of hours their plants run per year. Consequently, unofficial reports indicate that pockets of resistance to the dispatching program exist (ibid.).

4.4 ENERGY DEMAND

4.4.1 Energy Consumption by Type and Sector

Figure 4.13 shows the historical and projected energy consumption by sector in China. The industrial sector was responsible for 74% of energy consumption in 2010, followed by the transportation sector, which accounted for 13%; the residential sector, which used 9%; and the commercial sector,

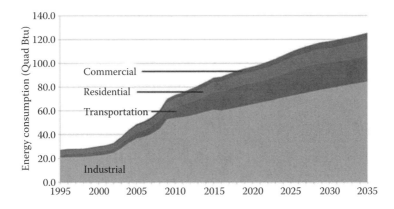

FIGURE 4.13
China energy consumption by sector, 1995–2035. (Adapted from U.S. EIA. *International Energy Outlook 2011*. Washington, DC: U.S. Department of Energy and U.S. Energy Information Administration, 2011.)

which consumed the remaining 4%. By 2035, the industrial sector's share is expected to decrease slightly to 67%, with the transportation sector increasing to 18%, the residential sector rising to 11%, and the commercial sector remaining roughly the same (U.S. EIA 2011).

Table 4.2 provides a detailed breakdown of final energy consumption in 2007 compared with 2000. The industrial sector comprises mining and quarrying, manufacturing, and utilities. The utilities subsector encompasses the production and distribution of electric power and heat, gas, and water.

In 2007, manufacturing accounted for 88% of the total industrial final energy consumption (117 Mtce). Smelting and pressing of ferrous metals, manufacturing of raw chemical materials and chemical products, manufacturing of nonmetallic mineral products, and processing of petroleum, coking, and nuclear fuels comprised the top four energy-consuming sectors. Their total energy consumption made up 76% of manufacturing energy consumption and 67% of total industrial energy consumption. Each of these sectors had annual consumptions of more than 100 Mtce that year. Thus, there was tremendous growth in the heavy industrial sectors, especially smelting and pressing of ferrous metals, resulting in substantial growth in China's output of cement, steel, aluminum, and other energy-intensive products from 2000 to 2007 (Wang et al. 2011).

TABLE 4.2

Final Energy Consumption by Sector: A Comparison of Year 2007 and Year 2000 Statistics

	Final Consumption (10 ktce)			Share of Total (%)	
	2007	**2000**	**Growth (2007/2000)**	**2007**	**2000**
Total	183,546	92,518	1.98	100.0	100.0
Agriculture	6084	4138	1.47	3.3	4.5
Industry	127,090	61,701	2.06	69.2	66.7
Non-energy use	9745	7280	1.34	5.3	7.9
Construction	3344	1053	3.17	1.8	1.1
Transport	19,054	8885	2.14	10.4	9.6
Service	3878	1923	2.02	2.1	2.1
Residential	18,188	10,686	1.70	9.9	11.6
Urban	11,458	6501	1.76	6.2	7.0
Rural	6730	4185	1.61	3.7	4.5
Other	5909	4132	1.43	3.2	4.5

Source: Adapted from Wang, Y. et al., *Energy Policy* 39:6745–6759, 2011.

Energy consumption in the buildings sector (broadly defined as including wholesale and retail trade, hotels and restaurants, other services, the government, and residential sectors) accounted for 15% of total consumption in 2007. This sector consumes a large percentage of many types of energy, such as coal briquettes (83%), LPG (76%), natural gas (29%), gasoline (31%), electricity (20%), and heat (26%). It also consumes a large proportion of China's renewable energy. Much of the renewable energy use is not formally documented, but unofficial data indicate that roughly 335 Mtce of renewable energy, including biomass burned directly (agricultural waste, 340 million tons; and firewood, 182.2 million tons) was used by this sector in 2007. If these unofficial data are accurate, approximately 55% of the energy consumed by this sector comes from renewables (mostly biomass). As incomes rise and urbanization increases, demand for commercial fuels such as oil, gas, coal, and electricity will outstrip and replace the demand for "noncommercial" fuels such as biomass (Wang et al. 2011).

Figure 4.14 displays each sector's energy consumption by type of energy using 2010 data. For the industrial sector, 64% of total consumption is supplied directly by coal, followed by 19% by electricity and 13% by liquids. Transportation is supplied almost entirely by oil liquids. The residential

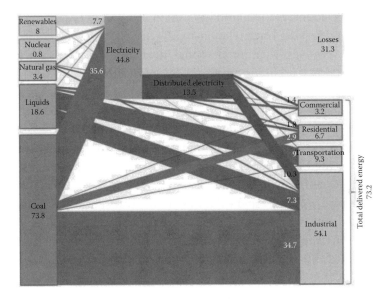

FIGURE 4.14
China energy consumption by type and sector, 2010 (unit: quadrillion Btu). (Adapted from U.S. EIA. *International Energy Outlook 2011.* Washington, DC: U.S. Department of Energy and U.S. Energy Information Administration, 2011.)

sector also consumes a fair amount of coal and electricity, and the commercial sector is supplied mostly by electricity and oil liquids. Electricity as an intermediate energy source is fueled primarily by coal, which accounts for approximately 80% of total generation (U.S. EIA 2011).

4.4.2 Energy Imports and Exports

Throughout the 1980s and the first half of the 1990s, China was entirely self-sufficient in terms of its primary energy supply (Zhang et al. 2011). In fact, energy products—especially oil, coal, and coke—once accounted for the majority of the country's export revenues (Wang et al. 2011). This has changed dramatically: although the oil industry's growth rate has continued at 1% to 3% per year since the mid-1980s, oil consumption has increased at annual rates of 5% to 8% (Zhang et al. 2011). Additionally, as the manufacturing industry developed, domestic demand for energy skyrocketed (Wang et al. 2011). China became a net oil product importer in 1993, a net crude oil importer in 1996, and a net primary energy importer in 1997. Net imported primary energy increased continually from 23 Mtce in 1997 to 201 Mtce in 2006, corresponding to an increase in energy import reliance from 1.7% in 1997 to 8.2% in 2006 (Zhang et al. 2011). However, the country still maintains a small volume of energy exports for the sake of preserving trade relationships with neighboring countries and to generate revenues when international buyers are willing to pay higher energy prices than domestic buyers (Wang et al. 2011).

For strategic purposes, the government tries to keep the energy supply reliant on mostly domestic sources. Currently, more than 90% of energy is supplied domestically (Zhang et al. 2011). Net imports of petroleum, however, have swollen from 12.2 million tons in 1995 to nearly 200 million tons in 2008 (ibid.). Imported oil, as a result, accounted for more than 50% of China's oil supply in 2007 and an estimated 52% in 2008 (Wang et al. 2011). Coupled with the rapid expansion of the transportation sector, this growing dependence on foreign oil is viewed as a threat to national energy security. Thus, the government has taken measures to raise fuel consumption standards for vehicles, to introduce alternative fuels, and to develop new engine technologies (Zhang et al. 2011). The strategic implications of increasing energy dependence on foreign oil are a key driver of government support for alternative energy sources. Although coal-based fuel stands as a convenient substitute for oil on a large scale in the near term, the negative environmental effect and low energy conversion efficiencies associated with coal-to-liquid fuel production have led the government to consider other alternatives, including renewable energy.

4.5 POLICY

4.5.1 Supply-Side Policy

Given the rapid growth of energy demand, pollution from energy consumption, global warming concerns, and volatile oil prices, China's supply-side energy policies focus on securing supplies of key fuels and supplying cleaner and lower-carbon energy (Wang et al. 2011).

4.5.1.1 Renewables

Increasing carbon emissions and higher levels of energy consumption, along with economic growth, have led the Chinese government to subsidize the development of renewable energy. Since 2004, there has been a substantial increase in energy consumed from renewable sources as a proportion of total energy consumption. The alternative power generation sector, however, currently faces many challenges. High initial investments, low reliability, and more costly energy sources due to incipient technologies and smaller scale result in renewable power generation methods that are less competitive than conventional sources, sometimes even when government subsidies are accounted for (ibid.).

The *Medium- and Long-Term Development Plan* of 2007 included an objective of reaching a total investment in renewable energy development of two trillion yuan (US$0.26 trillion) by 2020. The plan set goals of having 10% of the energy demand satisfied by renewable sources by the end of 2010 and 15% by 2020. Despite all the international attention this plan received, it was relatively conservative because the target included large hydropower, which accounted for 65% of the total investment China had earmarked for renewables (ibid.).

Table 4.3 displays specific targets for different types of renewable energy sources under the 2007 plan. Apart from large hydropower, mini-hydropower developments dominated the goals. The installed capacity for grid-connected wind power reached 6 GW in 2007, thus exceeding the target for 2010. An updated plan changed the windpower target for 2020 to 150 GW and the photovoltaic capacity target to 20 GW (ibid.).

TABLE 4.3

Targets for Renewable Energy Sources

	2010	2020
Mini-hydro	50 GW	75 GW
Biomass for Power Generation	5.5 GW	30 GW
Biomass Pellets	1 million tons	50 million tons
Biogas	19 billion m^3	44 billion m^3
Ethanol	2 million tons	10 million tons
Biodiesel	0.2 million tons	2 million tons
Wind Power	5 GW	150 GW
PV and Solar Thermal Power	0.3 GW	20 GW
Geothermal	4 Mtce	12 Mtce
Solar Water Heater	150 million m^2	300 million m^2

Source: Adapted from Wang, Y. et al., *Energy Policy* 39:6745–6759, 2011.

The 2005 Renewable Energy Law contained provisions for renewable portfolio standards, feed-in tariffs for biomass, government-guided prices for wind power, the obligation for utilities to purchase all renewable power generated, new financing mechanisms and guarantees, and other provisions (Li and Martinot 2010). Implementation regulations included laws guiding the use of renewable energy for power generation, the *Medium- and Long-Term Renewable Energy Development Plan*, the 11th Five-Year Plan for Renewable Energy Development, and regulations setting out the responsibilities for power utilities to purchase electricity from renewable resources. These regulations provided guiding principles to be considered while developing renewable energy, development targets, and codes and norms intended to improve the technical performance of these projects along with regulations geared toward making renewable energy projects more competitive (Wang et al. 2011). The regulatory environment in China has proven to be very dynamic. Since the enactment of the Renewable Energy Law in 2005, the government has issued more than 20 regulations, policies, and norms to guide its implementation (ibid.).

4.5.2 Demand-Side Policy

The energy conservation law took effect in China on April 1, 2008. Policies on the demand side focus on energy efficiency improvements in various sectors. Efforts have been concentrated on the industrial sector, which consumes 70% of the energy. Attention has also been given to the buildings and transportation sectors (ibid.).

4.5.2.1 Industry

The government regulates energy efficiency in the industrial sector by monitoring large energy consumers. For the "Top 1000 Enterprises Program," 997 large, mostly industrial enterprises were selected to report their energy consumptions and energy efficiencies annually. On average, in 2009, each firm's individual annual energy consumption was more than 180,000 tce. After examining these data, the government set norms for energy consumption per unit produced for steel, cement, and 18 other types of products. For aluminum production, for example, these norms specify a maximum electricity usage rate for existing plants in China of 14,400 kWh per ton smelted, whereas the maximum consumption rate for new facilities is set at 13,800 kWh per ton, which is nearly equal to the international best practice standard of 13,500 kWh per ton. Construction of a new plant is allowed only if it is designed to meet appropriate energy intensity requirements through an independent evaluation. Furthermore, since April 2009, all organizations that consume more than 10,000 tce of energy annually must report their previous year's energy consumption and energy efficiency by April 2011 (ibid.).

4.5.2.2 Buildings

Energy-efficiency regulations for the buildings sector focus on controlling space-heating efficiencies and electricity consumption by appliances, lights, and office equipment. Based on the 1980 design code for civilian buildings, China adopted a 30% savings code in 1986 and a 50% savings code in 1996. The country is currently developing a code for 65% savings. Regions have different codes based on weather conditions. For example, Beijing's average space-heating load, according to the 1980 code, was 31.7 W/m^2. The number decreased to 25.3 W/m^2 in 1986 and to 20.6 W/m^2 in 1996, and is projected to soon decrease to 11.1 W/m^2. Unfortunately, the code only controls designed energy use in newly constructed buildings, and there is a shortage of supervision to enforce the code. Therefore, the efficiency of space heating in China remains low despite attempts at regulation (Wang et al. 2011).

Regulations passed in October 2008 provide guidance to builders, operators, and owners of buildings for meeting efficiency requirements.

For example, the Chinese government encourages the use of efficient technologies and materials in new construction projects. It also requires operators to manage buildings efficiently to ensure that the buildings do not exceed the aforementioned maximum energy consumption norms. Finally, the government encourages owners to retrofit their properties to meet the building code (ibid.).

The Chinese government has adopted two measures to help owners and builders reduce their structures' electricity consumption. First, it requires and encourages the production of efficient appliances through the implementation of efficiency standards and labels, creating a supply of such products. Second, it subsidizes customers' purchases of these appliances (ibid.).

4.5.2.3 Transportation

China is making great strides in creating an efficient national transportation system. For example, the country constructed the most efficient railway system in the world in terms of energy consumption per ton-kilometer and per person-kilometer. Further efficiency improvement efforts are focused on automotive fuel economy and mass transport systems in cities (ibid.). In 2007, the Chinese government signed an agreement with the United States to develop energy efficient cars, with special emphases on electricity-powered, hybrid, fuel cell, and alternative fuel vehicles. The government is also considering adopting efficiency labels for vehicles. As of 2010, cars with engines smaller than or equal to 1.6 L are subject to a 5% consumption tax, whereas cars with larger engines are subject to a 10% tax (ibid.). Local governments have also begun to subsidize public transport facilities significantly to encourage the use of mass transit. As a result, many large cities in China are building their first new subway lines in years (ibid.).

4.5.3 Climate Change Policy

China's Policies and Actions for Addressing Climate Change was published in October 2008, and *China's National Climate Change Program* was published in June 2007 (Information Office of the State Council of the People's Republic of China [IOSCPRC] 2007, 2008). They stand as the two most important documents that describe China's current position on climate change and also summarize the main mitigating measures and actions that can be found in related official documents. In addition, they detail the country's plans in the areas of energy efficiency and renewable energy development. China is also trying to control and mitigate greenhouse gas emissions, but has not adopted a greenhouse gas emissions cap. Policy makers thus far have shied away from a fixed target for emissions (Wang et al. 2011).

4.6 COAL SUPPLY

4.6.1 Coal Reserves

China has extensive coal resources, whereas its petroleum and natural gas reserves are relatively low. As of January 2011, the country had 20.35 billion barrels of proven oil reserves, making its reserves the 14th largest in the world. These reserves accounted for only 1.38% of the world's total (Central Intelligence Agency [CIA] 2011b). China's natural gas reserves were 3.03 trillion cubic meters, ranked 13th, accounting for 1.58% of the world total (CIA 2011a). As shown in Table 4.4, the world's total recoverable coal reserves are estimated at 948 billion tons, and China possesses 13% of this amount, following the United States and Russia (U.S. EIA 2011). As a result, coal dominates the country's energy infrastructure and plays a key role in its economic development.

4.6.1.1 Geographic Problems

Even though China has enormous coal reserves, these resources are not distributed equally. Most are located in the northern part of the country, especially in Shanxi Province, whereas coal from the southern mines tends to have higher sulfur and ash contents and is not suitable for many applications. The detailed map in Figure 4.15 shows the distribution of coal reserves across the country.

This geographic imbalance poses a considerable logistical problem for supplying energy to the more heavily populated areas. Transporting coal from the largest coal-producing region, Inner Mongolia, to seaports along China's coast, for instance, has overloaded highways such as China National Highway 110, resulting in chronic traffic jams and delays. In many cases, the shipping companies that handle coal shipments by rail or sea garner higher profits than the producers do (Greenpeace 2008). For example, the Daqin Railway Co., Ltd. in Shanxi Province (a leading coal-producing region) runs the Daqin Railway, which connects Datong

TABLE 4.4

World Recoverable Coal Reserves as of January 2010

| Region/Country | Recoverable Reserves by Type of Coal | | | | 2008 Production | Reserves-to-Production Ratio (years) |
	Bituminous and Anthracite	Sub-Bituminous	Lignite	Total		
World total	445.7	287.0	215.3	948.0	7.5	126.3
United States	119.2	108.2	33.2	260.6	1.2	222.3
Russia	54.1	107.4	11.5	173.1	0.3	514.9
China	68.6	37.1	20.5	126.2	3.1	40.9
Other non-OECD Europe and Eurasia	42.2	19.1	40.1	101.4	0.3	291.9
Australia and New Zealand	40.9	2.5	41.4	84.8	0.4	191.1
India	61.8	0.0	5.0	66.8	0.6	117.5
OECD Europe	6.2	0.8	54.3	61.3	0.7	94.2
Africa	34.7	0.2	0.0	34.9	0.3	123.3
Other non-OECD Asia	3.9	3.9	6.8	14.7	0.4	34.4
Other Central and South America	7.6	1.0	0.0	8.6	0.1	95.8
Canada	3.8	1.0	2.5	7.3	0.1	97.2
Brazil	0.0	5.0	0.0	5.0	0.0	689.5
Other	2.6	0.6	0.1	3.4	0.0	184.5

Source: Adapted from U.S. EIA. *International Energy Outlook 2011.* Washington, DC: U.S. Department of Energy and U.S. Energy Information Administration 80, 2011.

FIGURE 4.15
Map of coal reserves and industry distribution in China. (Reprinted with permission from Li, Z. et al., *Energy Policy* 48:93–102, 2012.)

(a coal-mining city in Shanxi) to Qinhuangdao in Hebei Province. With a total length of 653 km, it is *the* major coal transportation railway in China and the busiest railway in the world (Daqin Railway 2006). Its annual coal shipment volume is about one-third of China's total.

4.6.2 Coal Production

China is the largest coal producer in the world. The country's coal extraction reached 3.47 billion tons in 2011 and is expected to continue to increase, as shown in Figure 4.16 (World Coal Association 2012). This huge extraction volume has generated profits for China's largest coal enterprises by more than ¥67 billion (US$8.75 billion). The growth rate in coal production really exploded after 1980, before which it followed a trend marked by more modest yearly growth. This eruption in mining, therefore, seems to have started at about the same time as the opening of

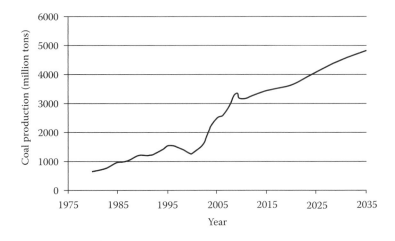

FIGURE 4.16

China's coal production history and projections by year. (Adapted from U.S. EIA. *International Energy Outlook 2011*. Washington, DC: U.S. Department of Energy and U.S. Energy Information Administration, 2011.)

TABLE 4.5

Coal Production by Region in 2009

Region	Coal Production (million tons)	Percentage (%)
Inner Mongolia	600.6	20.4
Shanxi	593.5	20.1
Shaanxi	296.1	10.0
Henan	230.2	7.8
Shandong	143.8	4.9
Other	1086.3	36.8

Source: Adapted from National Bureau of Statistics of China. *China Energy Statistical Yearbook 2011.* Beijing: China Statistics Press, 2012.

China's economy (Tu 2007). Table 4.5 illustrates China's coal production by region.

4.6.2.1 Shanxi Province

Shanxi Province, in northern China, covers 157,000 km². More than 35% of this area contains coal reserves, with deposits totaling 265.5 billion tons, accounting for 26% of the country's total reserves (Pu 2007; SXGOV.cn 2009). Its annual extraction comprises more than 20% of China's total. Of this production, more than half is transported to 26 provinces and cities in China, accounting for 80% of China's total transprovincial coal trade (SXGOV.cn 2009). Thus, Shanxi's economy is focused on heavy industries such as coal and chemical production, power generation, and metal refining.

4.6.2.2 Inner Mongolia

The Inner Mongolia Autonomous Region has been expanding its coal production rapidly and recently moved ahead of Shanxi Province as the largest coal-producing region in China. In 2010, extraction in the region reached 782 million tons, accounting for 24% of China's total production, followed closely by Shanxi's 742 million tons (Xinhua 2011). In 2011, production expanded to 979 million tons. The largest open-pit coal mine in China, the Haerwusu mine, is located in the middle of the Zhungeer Coalfield in Inner Mongolia. This mine covers an area of 67 km², with total reserves estimated at 1.73 billion tons, consisting of mostly low-sulfur steam coal. Construction of the mine began in May 2006, and the total investment cost topped seven billion yuan (US$0.91 billion; China.org.cn 2008). The Shenhua Group began mining in October 2008. The mine has an annual extraction capacity of 20 million tons, with an effective life estimated at 79 years (ibid.).

4.7 COAL CONSUMPTION

4.7.1 Coal Consumption Overview

China is not only the largest coal producer in the world, but also its largest consumer (Figure 4.17). The country consumed 3.2 billion tons in 2010, more than double the rate of a decade earlier (Figure 4.18). In the first three quarters of 2011, it consumed 2.28 billion tons, representing an increase of 10.3% year-over-year (Zhang 2011). Coal supplies 70% of the energy consumed in the country. This proportion is drastically greater than the world average of 40% (Figure 4.19). China's annual coal production target for 2015 is 3.6 to 3.8 billion tons (Zhang 2010).

The U.S. EIA, in its *International Energy Outlook 2011* (IEO 2011), estimated that coal consumption in China will continue to increase. In fact, the IEO 2011 projects that the country's coal consumption growth will

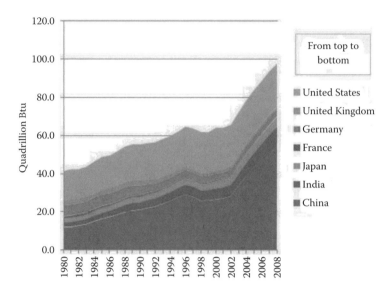

FIGURE 4.17
Consumption of seven of the largest coal-consuming countries, 1980–2008. (Adapted from Brown, L.R. Earth Policy Institute. Plan B 4.0: Mobilizing to Save Civilization: Supporting Data. http://www.earth-policy.org/books/pb4/pb4_data [accessed November 20, 2012].)

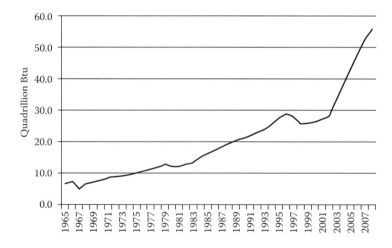

FIGURE 4.18
Coal consumption in China. (Adapted from Lester R. Brown. Earth Policy Institute. Plan B 4.0: Mobilizing to Save Civilization: Supporting Data. http://www.earth-policy.org/books/pb4/pb4_data [accessed November 20, 2012].)

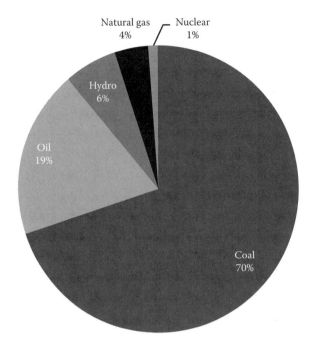

FIGURE 4.19
Total energy consumption in China by type in 2009. (Adapted from U.S. EIA. International Energy Statistics. http://www.eia.gov/cfapps/ipdbproject/IEDIndex3.cfm [accessed November 19, 2011].)

account for 76% of the world's net growth in coal usage through 2035. This forecast is based on a strong projected annual economic growth rate of 5.7%.

Table 4.6 shows that the regions in China that consume the greatest quantities of coal are the four provinces in northeastern China, south of Beijing: Shandong, Shanxi, Hebei, and Henan. These regions are heavily populated: Henan and Shandong are the most and third-most populous provinces, respectively. They are also highly industrialized, with each having a heavy focus on manufacturing. This industrialization largely drives their massive coal consumption.

The National Coal Trade Fair handles 48% of China's entire coal trade volume, whereas 44% is processed by regional trading centers. Coal is also traded through individual transactions among producers, consumers, and traders (Greenpeace 2008).

In 2009, for the first time in history, China became a net importer of coal. Imports totaled 125.83 million tons, more than triple the amount imported in the previous year (Zhang 2010). Imports continued to grow

TABLE 4.6

Coal Consumption by Region in 2009

Region	Coal Consumption (million tons)	Percentage of Total Consumption (%)
Shandong	348	9.9
Shanxi	278	7.9
Hebei	265	7.6
Henan	245	7.0
Inner Mongolia	241	6.8
Jiangsu	210	6.0
Liaoning	160	4.6
Guangdong	137	3.9
Zhejiang	133	3.8
Anhui	127	3.6
Other	1370	39

Source: Adapted from National Bureau of Statistics of China. *China Energy Statistical Yearbook 2011.* Beijing: China Statistics Press, 2012.

significantly in 2010. This recent surge in imports has been attributed to higher domestic coal prices, increasing domestic demand, and supply shortages (which may have spurred some coal users to stockpile the resource). In the first three quarters of 2011, China imported 123 million tons of coal, an increase of 1.9% year-over-year, and exported 12.12 million tons (Zhang 2011). Its coal imports and exports during the first half of 2011 are shown in Table 4.7. According to the National Development and Reform Commission, in 2010, Indonesia was the largest coal exporter to China, followed by Australia, Vietnam, Mongolia, and Russia. These five countries supplied 84% of the country's coal imports (Zhao 2011).

TABLE 4.7

China's Import and Export of Coal in the First Half of 2011

	Amount of Coal (million tons)	Change year-over-year (%)	Total Value (US$ billion)	Change year-over-year (%)	Average Price (US$ per ton)	Change year-over-year (%)
Export	8.752	−13.7	1.59	51.7	182	75.8
Import	70.62	−10	7.73	0.4	109.5	11.5

Source: Adapted from E-to-China.com. China's Coal Imports and Exports Down in Quantity in First Half of 2011. http://www.e-to-china.com/2011/0913/97058.html (accessed November 20, 2012).

According to the IEO 2011, China will remain a net coal importer through 2035 and is even expected to surpass Japan in 2015 to become the world's largest importer of coal. Its coal imports are estimated to reach approximately 300 million tons in 2035 (U.S. EIA 2011).

4.7.2 Coal Consumption by Sector

Nearly half of China's coal consumption is used to generate electricity. The second largest coal-consuming sector is manufacturing, within which the leading uses are oil refining, production of ferrous metals, manufacturing of nonmetallic products, and chemical creation (National Bureau of Statistics of China 2012). Table 4.8 provides a detailed breakdown of coal consumption by sector from 2005 to 2009.

4.7.2.1 Electricity Generation

China is the second largest electricity producer in the world, generating 3.7 trillion kWh in 2009; coal supplied approximately 80% of this power. The country's total electricity generation has more than tripled since 1995 (Figure 4.20; U.S. EIA 2012a).

Coal's share of electricity generation is projected to decline to 66% by 2035 as a result of the expanding use of other energy sources. Despite this proportionate decline, coal use in China's electricity sector is projected to increase in absolute terms at an average annual rate of 3.0% over the next 20 years. In comparison, coal use in the United States is projected to grow at a rate of only 0.2% per year (U.S. EIA 2011). Driving this absolute growth, in part, is the fact that electricity's share of total industrial energy consumption is expected to increase from 18% in 2008 to 26% in 2035. China's installed electricity generation capacity exceeds 900 million GW, of which coal-based capacity accounts for 496 million GW. This capacity is predicted to expand to 1233 GW by 2035 (Xinhua 2010). Figure 4.21 compares this growth to that in other regions of the world.

Thermal power generation efficiency in China has improved significantly over the last two decades (Figure 4.22). The average efficiency was 32% in 2009, 8% lower than the OECD average (ABB 2011).

Government efforts to develop larger and more advanced thermal power plants while closing small stations with capacities less than 100 MW are the main drivers of this improvement in efficiency. New plants include supercritical (SC) and ultra-supercritical (USC) units with capacities of

TABLE 4.8

Coal Consumption (Million Tons) by Sector

	2005			2006			2007			2008			2009		
	Amount	Growth (%)	Share (%)	Amount	Growth (%)	Share (%)	Amount	Growth (%)	Share (%)	Amount	Growth (%)	Share (%)	Amount	Growth (%)	Share (%)
Total consumption	2318.51	—		2550.65	10.0		2727.46	6.9		2810.96	3.1		2958.33	5.2	
Power production and supply	1062.9		45.8	1210.6		47.5	1336.5		49.0	1367.3		48.6	1448.8		49.0
Manufacturing	920.2		39.7	990.8		38.8	1020.0		37.4	1081.8		38.5	1120.1		37.9
Mining and quarrying	158.3		6.8	168.8		6.6	189.7		7.0	195.0		6.9	217.3		7.3
Residential	100.4		4.3	100.4		3.9	97.6		3.6	91.5		3.3	91.2		3.1
Others	76.7		3.3	80.0		3.1	83.6		3.1	75.5		2.7	81.0		2.7

Source: Adapted from National Bureau of Statistics of China. *China Energy Statistical Yearbook 2011.* Beijing: China Statistics Press, 2012.

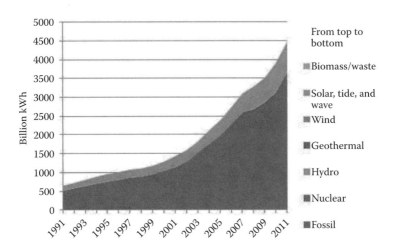

FIGURE 4.20
China's electricity generation by type, 1991–2011. (Adapted from U.S. EIA. International Energy Statistics: Electricity Net Generation by Type [billion kilowatthours]. http://www.eia.gov/cfapps/ipdbproject/iedindex3.cfm?tid=2&pid=alltypes&aid=12&cid=CH,&syid=2007&eyid=1991&unit=BKWH [accessed November 20, 2012].)

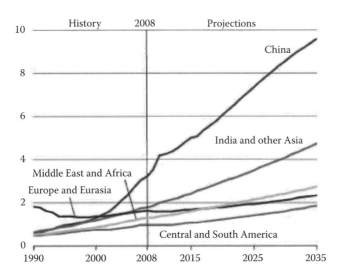

FIGURE 4.21
Non-OECD net electricity generation by region, 1990–2035 (trillion kWh). (Reprinted under the U.S. EIA's media resources usage policy: U.S. EIA. *International Energy Outlook 2011.* Washington, DC: U.S. Department of Energy and U.S. Energy Information Administration, 2011.)

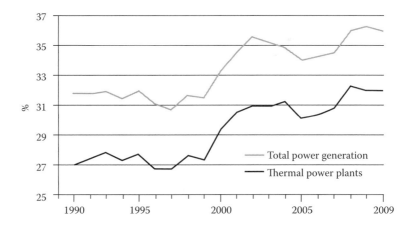

FIGURE 4.22
Efficiency of power generation and thermal power plants. (Reprinted with permission from ABB. *Trends in Global Energy Efficiency 2011: China: Energy Efficiency Report.* Zurich: ABB, 2011. http://www05.abb.com/global/scot/scot316.nsf/veritydisplay/63246e 62080610aec12578640050f217/$file/china.pdf.)

more than 600 MW and large combined cycle units (Chen and Xu 2010). The percentage of thermal power generation plants of more than 100 MW increased from 62.5% in 1995 to 72.4% in 2005 (ibid.). China's first domestically developed SC unit began operation in 2004 (Sun 2010). In 2006, the country added an additional 18 GW of SC plants, expanding capacity to 30 GW. In 2011, contracts were in place to install approximately 100 GW of capacity, which implies that the percentage of SC plants in new installations would increase significantly, improving overall efficiency. As a result, the IEA projects that thermal power generation efficiency in China will increase to 39% by 2030 (ibid.). The country's first USC unit, the Yuhuan Power Plant of the Huaneng Group, went online in 2006. A summary of China's thermal generation plants is provided in Table 4.9.

In 2006, virtually all thermal power generation units in China installed dust-cleaning facilities. Most of these plants were also discharging waste water that met national pollution standards at the time (IOSCPRC 2007). As of 2010, China had also achieved its 5-year goals of reducing pollutant emissions by 10% and reducing energy consumption per unit of GDP by approximately 20% (Lu 2011). As a part of the most recent 12th Five-Year Plan, by 2015, China aims to reduce energy intensity by 16% and to reduce carbon emissions by 17%. This will put the country on track to achieve its 40% to 45% reduction targets by 2020 (Ohshita and Price 2011).

TABLE 4.9

China's Thermal Power Generation Units

	2002		2005	
	Number of Units	Capacity (GW)	Number of Units	Capacity (GW)
600+ MW	26	15.6	84	55.5
300–600 MW	278	86.0	450	144.1
200–300 MW	248	51.3	255	53.0
100–200 MW	456	54.4	573	70.9
6–100 MW	5026	87.5	5619	100.6
Total	6034	294.8	6981	424.1

Source: Adapted from Sun, G. *Coal in China: Resources, Uses, and Advanced Coal Technologies.* Arlington, VA: Pew Center on Global Climate Change, 2010. http://www.pewclimate. org/docUploads/coal-in-china-resources-uses-technologies.pdf.

In 2012, the price of coal in China shot up 40% to 50%. However, the government did not increase electricity prices due to fears of inflation. Electricity producers, therefore, faced both supply shortages and price ceilings. This is a chronic issue plaguing the industry, as demonstrated by a similar situation occurring in 2011. After coal prices increased 15% during the winter of 2011, electricity companies experienced difficulties in meeting demand because they could not pass coal price increases along to consumers. Consequently, power shortages occurred. Such a situation in 2010 also caused China's top five thermal power companies to post a combined loss of ¥13.7 billion (US$2.1 billion) in their thermal generation businesses, again showing the longevity of the problem. These companies also posted a loss of ¥35 billion (US$5.5 billion) in 2011 (Wei and Juan 2011).

Furthermore, China's power transmission system is still underdeveloped, exacerbating power supply problems. As mentioned above, there are several regional grids rather than a national grid. The lack of long-distance power transmission capacity means that power cannot be shared easily among regions with power surpluses and those with power shortages, compounding the severity of regional power shortage problems. This is especially acute for northern areas that experience shortages in the winter due to increased heating demands, and for eastern and southern areas that experience shortages in late spring and early summer due to higher air conditioning demands.

4.7.2.2 Steel Industry

China's steel industry has expanded steadily since the economic reform in 1978, becoming the world's largest in 1996. In 2008, the country produced more than 500 million tons of steel, accounting for 37.6% of world production (Reserve Bank of Australia 2010). Figure 4.23 depicts the growth since 1995.

China has been a net steel exporter since 2006. Its largest steel trading partner is South Korea, which receives 20% of China's exports, followed by India and Thailand. Figures 4.24 and 4.25 illustrate these relationships.

China's steel industry is highly decentralized. The leading steelmakers are large groups owned by the local, provincial, and central governments. The rest are small- to midsized firms, a significant proportion of which are privately held. In 2008, there were more than 660 steel producers, with the top 10 companies accounting for less than half the total output, and with the next 75 accounting for only an additional 30% (ibid.). Both Beijing and Hebei produce more than 100 million tons of steel annually. They are followed by Liaoning, Shandong, and Jiangsu, each producing between 30 and 100 million tons. These coastal provinces account for approximately

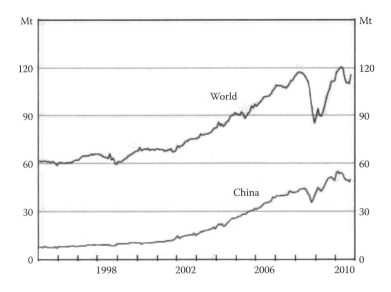

FIGURE 4.23

China's crude steel production over time. (Reprinted with permission from Reserve Bank of Australia. *China's Steel Industry*, by Holloway, J., I. Roberts, and A. Rush. Sydney, Australia: Reserve Bank of Australia, 2010. http://www.rba.gov.au/publications/bulletin/2010/dec/pdf/bu-1210-3.pdf.)

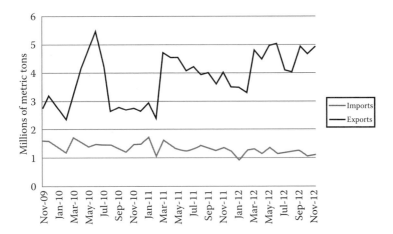

FIGURE 4.24
China's imports/exports of steel mill products. (Reprinted with permission from U.S. Department of Commerce. International Trade Administration. *Steel Industry Executive Summary*: January 2013. Washington, DC: Import Administration, 2013. http://hq-web03.ita.doc.gov/License/Surge.nsf/webfiles/SteelMillDevelopments/$file/exec%20summ.pdf?openelement.)

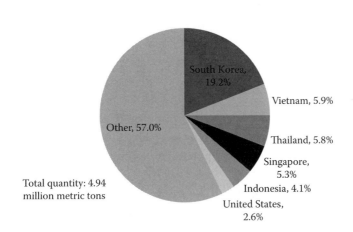

FIGURE 4.25
Chinese exports of steel mill products by trading partner. (Adapted from U.S. Department of Commerce. International Trade Administration. *Steel Industry Executive Summary*: November 2012. Washington, DC: Import Administration, 2012. http://hq-web03.ita.doc.gov/License/Surge.nsf/webfiles/SteelMillDevelopments/$file/exec%20summ.pdf?openelement.)

65% of total crude steel production, with the remainder spread across the country (ibid.).

The industry's lack of consolidation negatively affects its economies of scale. Furthermore, overcapacity problems have been known to occur as resources are not always used most efficiently (Hong and Mu 2010). More than 90% of the crude steel in China is produced with traditional blast furnaces, utilizing the basic oxygen converter method, which requires more iron and coal and is much less efficient than more advanced methods (Reserve Bank of Australia 2010).

The Chinese government has created plans to spur consolidation in the steel industry, both through mergers and acquisitions and through closures of small- to midsized producers. In June 2010, the State Council unveiled its goal to increase the production share of the top 10 steelmakers from 44% in 2009 to 60% by 2015, with the ultimate aim of cultivating three to five large, globally competitive iron and steel conglomerates (ibid.).

4.7.2.3 Residential Sector

Residential coal use makes up only a small percentage of the total coal consumption in China, but its negative effects on health and the environment are disproportionately high. Burning coal is banned in cities (the practice, however, still exists but is on the decline), but is nearly ubiquitous in rural areas. Many Chinese households rely on solid fuels such as coal and biomass for energy. Honeycomb coal briquettes are the most common form of coal used in the residential sector and are typically burned in coal stoves. These stoves are a major source of pollution due to their large number, their suboptimal combustion conditions, and their lack of emission filters (Bi et al. 2008). Thus, they represent a significant source of greenhouse gas emissions and environmental degradation.

Furthermore, the stoves fill rooms with high levels of toxic materials, leading to poor indoor air quality. Consequently, the stoves are responsible for more than 380,000 premature deaths annually in China (Impact Carbon 2013). Food cooked over coal can also absorb some of the toxic substances emitted, including arsenic, fluorine, polycyclic aromatic hydrocarbons, and mercury. This can lead to major health complications, such as endemic fluorosis, arsenic poisoning, and lung cancer (ibid.).

4.7.3 External Costs

4.7.3.1 Greenhouse Gas Emissions

China's growing, coal-dominated energy mix is causing ever-higher carbon emissions, resulting in consequences at a global level. Coal is a highly inefficient and dirty energy source and accounts for 84% of China's total CO_2 emissions. The country's use of coal has contributed to increased greenhouse gas levels in the atmosphere. These gases have been linked to various climate change concerns that can affect humans significantly (IPCC 2007).

Thus, as the largest carbon emitter in the world, China faces immense international pressure to reduce its emissions in absolute terms. The country does not dispute its status, instead choosing to stress that its carbon emissions per capita are much lower than those of the United States or Europe. Nevertheless, the country is working to make its economy less carbon-intensive by reducing its emissions per unit of GDP. Its progress toward this goal, however, owes more to technological advancement than to policy measures. To continue this decarbonization, therefore, China needs to pursue efforts to achieve a low-carbon economy on all fronts— technological and political (Friedman and ClimateWire 2012).

Coalbed gas represents another dangerous greenhouse gas released as a result of China's coal dependency. This gas is primarily methane, which has a greenhouse effect more than 20 times that of CO_2 over a 100-year period (U.S. Environmental Protection Agency [U.S. EPA] 2012). The Chinese government, recognizing this threat, has made significant efforts to reduce these emissions. The total volume of coal mine methane released reached 7.35 billion cubic meters in 2010. Of this, 2.5 billion cubic meters were utilized for energy production (China Coal Information Institute [CCII], and China Coalbed Methane Clearinghouse [CCMC] 2011).

4.7.3.2 Acid Deposition

Acid deposition, caused largely by burning coal, emerged as a serious environmental problem in the late 1970s (Larssen et al. 2006). As shown in Figure 4.26, China's SO_2 emissions, the leading cause of this hazard, have increased dramatically over the past five decades. The average sulfur content in coal in China is 1.1%, but it can be as high as 4% in some heavily industrialized areas in the southwest (ibid.). The country's annual SO_2 emissions peaked at 34 million tons in 2006 and then decreased to

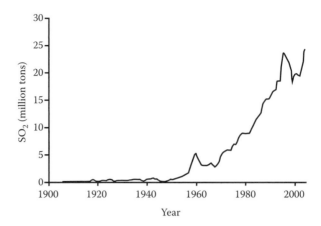

FIGURE 4.26
China's SO_2 emissions by year. (Reprinted with permission from Larssen, T. et al., *Environmental Science & Technology*, 418–425, 2006. http://pubs.acs.org/doi/pdfplus/10.1021/es0626133.)

30.8 million tons in 2010 due to the wide application of flue gas desulfurization (FGD) equipment in power plants (Lu and Streets 2011).

Approximately 30% of China is afflicted by coal-related acid deposition (Greenpeace 2008). Although the atmospheric SO_2 concentration is high in most areas of the country, the alkaline dust occurring naturally in the north's deserts largely neutralizes these acids. Thus, acid deposition is a problem mainly in the southern and the southwestern regions, as shown by the pH levels provided in Figure 4.27 (Larssen et al. 2006).

4.7.3.3 Coal Gangue

Coal gangue, the solid waste that results from coal washing, is one of the major sources of industrial waste in China. Statistics show that the amount of coal gangue discharged in the country annually is at least 200 million tons, sometimes reaching more than 300 million tons (Zhao 2010). More than five billion tons of the substance is currently stored throughout the country. Coal gangue represents a substantial problem because, during spontaneous combustion, it emits harmful gases such as SO_2, CO_2, and CO, which lead to serious air pollution problems, destroy surrounding ecosystems, and harm the health of local residents. In response to these problems, China plans to create infrastructure to utilize its coal gangue

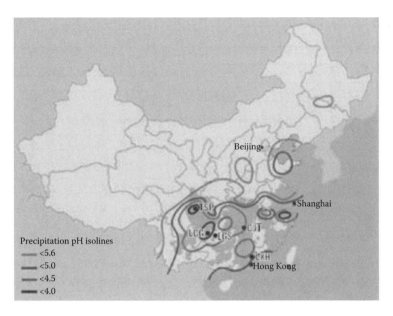

FIGURE 4.27
Precipitation pH distribution in China. (Reprinted with permission from Larssen, T. et al., *Environmental Science & Technology*, 418–425, 2006. http://pubs.acs.org/doi/pdfplus/10.1021/es0626133.)

more safely. The country plans to recover more than 0.39 billion tons, of which 0.2 billion tons will be burned in low-calorific-value-fuel power plants, 0.09 billion tons will be converted into coal gangue bricks, and 0.1 billion tons will be used in reclamation (e.g., to create new land), road building, and backfilling (which entails putting the gangue back into the ground, sometimes by filling in former coal mines; UNDP/Spanish MDG Achievements Funds 2009).

4.7.3.4 Coal Dust

China has approximately 6000 coal storage facilities that lack dust protection, which results in the release of 10 million tons of coal dust each year (Radical Geography 2007). This represents not only a significant loss of coal resources, but also a serious source of pollution. In particular, this dust causes coal worker's pneumoconiosis (CWP). The most severe health issue these workers face, this disease causes lacerations in the lung that result from dark coal deposits; hence, it is also known as "black lung disease" (Finkelman et al. 2002). At later stages, patients suffer from airway

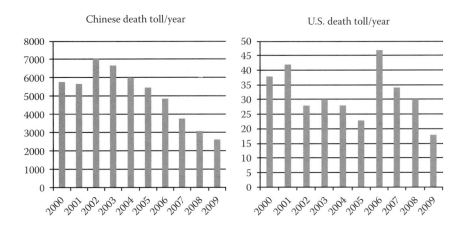

FIGURE 4.28
Coal mining–related deaths, a comparison of the United States and China, 2000–2009. (Adapted from Wei, G., *International Journal of Business Administration* 2:82–86, 2011.)

obstruction, reduced lung capacity, and diminished absorption of oxygen into their bloodstreams. CWP, which is accompanied by high morbidity and mortality rates, accounts for 39% of pneumoconiosis cases in China, making it the most serious occupational lung disease. More than 600,000 cases of pneumoconiosis had been recorded by 2000, with 10,000 to 15,000 new cases being reported every year since (Christiani and Wang 2003).

In addition to causing this severe health problem, coal dust has also been responsible for many deadly coal mine explosions, resulting in thousands of deaths over the last half century. It is estimated that 87.32% of major coal mines in China are at risk for these explosions (Feng et al. 2009). Figure 4.28 illustrates the consequences, comparing the total number of coal-mining deaths in the United States and China from 2000 to 2009.

4.7.3.5 Coal Transportation

As mentioned previously, because of its unequal geographic distribution, China's coal must be transported from the northern and western regions to the southern and eastern provinces. Rail is the dominant means of transportation, carrying more than 60% of coal freight. Coal, in turn, accounts for 40% of China's railway freight. In 2007, the total mass of coal transported amounted to 1.54 billion tons. It is much more expensive to transport coal via trucks, but is often necessary to move the coal short

TABLE 4.10

External Costs of Coal Transportation by Different Means in China, 2002

	Railway	Road	Waterway
Freight (RMB/millions)	32,248	3300	7009
External costs (RMB/ton)	2.08	2.1	4.5

Source: Adapted from Greenpeace. *The True Cost of Coal* by Mao Yushi, Sheng Hong, and Yang Fuqiang. Hong Kong: Greenpeace, 2008. http://act. greenpeace.org.cn/coal/report/TCOC-Final-EN.pdf.

distances from railway terminals to end users. Coal is also transported via water to coastal cities (Tian 2008).

This intensive coal transportation also contributes to the country's serious coal dust problems. Every year, railway and road transportation generate more than 11 million tons of the dust. Trucks have proven especially harmful, as they spread coal dust to residents' homes while traveling through densely populated cities. The dust is also released at coastal ports during loading and unloading. In addition, because the equipment used in this process becomes coated with the dust, cleaning it can cause water pollution (China Coal Information Institute [CCII] 2003). Table 4.10 provides estimates of the external costs of coal transportation.

4.7.3.6 Water Resources

The coal extraction industry consumes a large amount of water. Coal mining, processing, combustion, and coal-to-chemical industries are responsible for 22% of China's total water consumption, second only to agriculture (Cho 2011). Washing one ton of coal requires 4 to 5 m³ of water, and approximately 40 million cubic meters of water are used to wash coal annually. Wastewater from the coal industry accounts for 25% of China's total wastewater output (Greenpeace 2008). This wastewater contains large amounts of coal sediment and heavy metals. Of the 30 million cubic meters of wastewater discharged into river systems from coal mines each year, only 15% of this volume is treated, meaning that the 85% flows freely into rivers, carrying with it hundreds of thousands of tons of coal residues. The wastewater also has higher levels of salt and sulfate than allowed by national irrigation standards, causing damage to local agriculture (Wei 2003).

The extraction of coal also alters water tables, which sometimes causes groundwater sources to dry up. Thus, coal mining can create water shortage problems. Of the 96 major, state-owned coal mines, 71% face water shortages, with 40% facing severe shortages. The damage to water resources is particularly acute in the major coal production areas of Shanxi, Shaanxi, and the western part of Inner Mongolia (Greenpeace 2008). This damage, between 2004 and 2009, resulted in Inner Mongolia and Xinjiang losing 46.8 million and 95.5 million cubic meters of fresh water, respectively (Cho 2011).

To meet water and energy demands in the northern and western regions, China is constructing the US$62 billion South-to-North Water Diversion Project, the largest such project ever attempted. When completed in 2050, it will link the Yangtze, Yellow, Huaihe, and Haihe Rivers and divert 44.8 billion cubic meters of water every year from the south to the north (ibid.). As part of the 12th Five-Year Plan, China also plans to combat water shortages by reducing water usage per unit of industrial output by 30% and reducing water pollution 8% by 2015.

4.7.3.7 Land Subsidence

When coal is extracted, the weight of the land on top of the mine can cause the overlying ground's surface to collapse, causing mining subsidence (Pennsylvania Department of Environmental Protection [PDEP] 2012). It is estimated that, for every ton of coal extracted, 0.2 ha of land are subjected to subsidence (State Administration of Coal Mine Safety [SACMS] 2006). The total area affected by subsidence exceeds 400,000 ha and is increasing at more than 20,000 ha annually (Hu et al. 1997). As of December 2006, the damages were estimated at ¥50 billion (US$6.53 billion), caused by subsidence's effects on roads, railways, bridges, and electric power transmission lines (Greenpeace 2008). In addition, large amounts of coal reserves are located under existing buildings, railways, and water bodies, leading, in some cases, to expensive displacements. For example, in mining areas with dense populations, on average, approximately 2000 people have to be relocated for every 10 million tons of coal extracted (Guo et al. 2009).

4.7.3.8 Additional Air Pollution Considerations

Heavy industrialization has left China with severe air pollution problems. The country monitors air quality by computing the air pollution index (API), which takes into account different pollutants, including CO_2, NO_x, and particulate matter. Lower API values correspond to higher air quality levels. Based on the API, the air quality in different regions of China is rated on a scale ranging from standard I (the best) to standard V (the worst). During the first half of 2012, among the 120 major cities monitored in China, fewer than half reached standard II more than 95% of the time (Xinhuanet.com 2012). Only one city, Haikou, reached standard I (Fu 2012). This lack of high-quality air detrimentally affects the more than 650 million people living in urban areas. In January 2011, the Institute of Public and Environmental Affairs (IPE) and the Renmin University School of Law published the Air Quality Transparency Index (AQTI), which rated and compared the availability of public information on air quality in 20 Chinese cities and 10 international cities. The results, summarized in Figure 4.29, show that the monitoring system in China is still underdeveloped and more transparency is needed (IPE and Renmin University School of Law 2011).

Coal is key in the debate over air pollution, standing as the largest source of air pollution in China. It is responsible for 75% of SO_2 emissions, 85% of NO_2 emissions, 60% of CO emissions, and 70% of total suspended particulate emissions (Greenpeace 2008). The absolute levels of these pollutants are also on the rise. For example, China's total SO_2

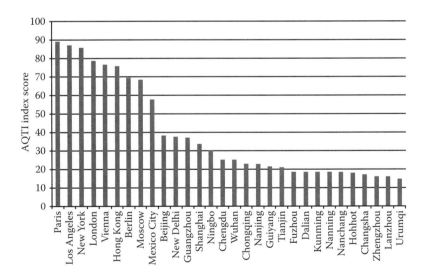

FIGURE 4.29
Comparison of AQTI scores between Chinese and international cities. (Adapted from IPE and Renmin University School of Law. Air Quality Transparency Index. http://www.ipe.org.cn//Upload/Report-AQTI-CH.pdf [accessed January 18, 2011].)

emissions increased from 14 million tons in 1981 to 33.2 million tons in 2006 (Lu et al. 2010). Geographically, emissions in northern China increased by 85% from 2000 to 2006, whereas emissions in the southern part of the country grew by only 28% (ibid.). Limited improvements are occurring, however, as SO_2 emissions per kilowatt-hour are 40% lower than 1990 levels due to more common installations of FGD systems at power plants. However, by the end of 2005, the total power generation capacity equipped with desulfurization technologies reached only 53 GW, or less than 15% of China's total power generation capacity (Chen and Xu 2010; Lu and Zhang 2010).

One promising trend is that China's soot (black carbon) emissions decreased from 14.5 million tons in 1981 to 11.82 million tons in 2005 (Chen and Xu 2010). These emissions are second only to carbon dioxide in accelerating global warming. They absorb incoming solar energy while also producing a greenhouse effect worse than that of CO_2. On the other hand, soot has a much shorter life and stays in the atmosphere for only a week to 10 days (Skirble 2011). Nevertheless, it is very dangerous because small carbon particles can penetrate human lungs and cause serious health problems. Soot emissions in power generation plants have been largely controlled in China due to the wide use of electric dust removal systems with efficiencies as high as 99.6% (Chen and Xu 2010).

A serious environmental event that also contributes to pollution, particularly in northern China, is the occurrence of sandstorms, which have become increasingly severe for several reasons, including deforestation and drought. Winds from the Gobi Desert pick up sand, dust, and toxic particles from coal-burning factories and plants and carry them southeast to populated centers such as Beijing. The toxic pollutants transported include mercury, lead, cadmium, fluoride, sulfur, selenium, and antimony, all of which can cause serious health problems (Wang 2011b).

Among these toxins, mercury is especially hazardous. On average, Chinese coal has a high mercury content of 0.145 to 0.280 mg/kg, depending on the type (ibid.). Nationwide mercury emissions in 2008 were estimated at approximately 74 tons, with the highest emissions in Jiangsu, Henan, and Shandong. Annual emissions have declined in recent years due to numerous installations of FGD systems and to the closures of smaller power plants. Mercury is toxic and causes soil pollution problems, food poisoning, and damage to the central and peripheral nervous systems, with harmful effects on the digestive and immune systems, lungs, and kidneys (World Health Organization [WHO] 2012).

TABLE 4.11

Death Rate for Each Million Tons of Coal Produced in China

Year	2000	2001	2002	2003	2004	2005	2006
Death rate	4.95	4.38	5.27	4.28	3.08	2.69	1.91

Source: Adapted from Homer, A.W., *Journal of Contemporary Asia* 39:424–439, 2009.

4.7.3.9 Mining Accidents and Deaths

China's coal-mining industry is both the largest and the deadliest in the world. In 2010, 2433 Chinese workers died in coal mines (Wang 2011a), compared with only 48 in the United States (Huber 2010). Although China's coal-mining mortalities have decreased significantly over the years, as exhibited in Table 4.11, when compared with the United States's death rate of 0.05 deaths per million tons of coal, China's mines are still considerably more dangerous. In fact, during 2007, China produced only one-third of the world's coal, but had a staggering four-fifths of coal fatalities.

A key reason for this high death rate is that 95% of the coal mines in China are permeated by high levels of hazardous gases such as poisonous hydrogen sulfide and explosive coalbed methane (Greenpeace 2008). Leaks of these gases can have serious consequences, especially when emergency exits are not well-maintained. Coalbed methane is highly flammable and, thus, poses a major threat to miners. In 2007, 1084 of 3770 coal mining fatalities in China were caused by gas explosions (China Daily 2008). Most of the accidents occurred in poorly constructed and inadequately maintained small mines, which account for 90% of all the mines in the country. The State Administration of Work Safety announced that China will ban the new construction of such coal mines with annual production capacities of less than 300,000 tons and close more than 4000 small mines in an effort to improve coal production safety (ibid.). Coal plants create numerous other health problems in China. For example, in addition to the CWP detailed earlier, emissions cause chronic obstructive pulmonary disease. In 2010, more than 38 million people in China suffered from this disease (Shan 2010).

4.8 CLEAN COAL TECHNOLOGIES

Faced with a consistently high demand for coal and the environmental effect that its use entails, China has placed a strong emphasis on developing clean coal technologies. In the 11th Five-Year Plan, the country focused on coal processing, high-efficiency combustion, coal conversion, pollution control, and waste management technologies to make its coal industry cleaner. Under the 12th Five-Year Plan, it continues to research and develop clean coal technologies by incorporating the industry's latest international advancements (DoNews 2011).

4.8.1 Advanced Power Generation Technologies

4.8.1.1 Supercritical and Ultrasupercritical Systems

The thermodynamic efficiency of a Rankine cycle, which is used to generate electricity from coal, increases with higher steam temperatures and pressures. SC describes the thermodynamic state of a substance where there is no clear distinction between its liquid and gaseous phases. SC and USC plants utilize higher steam temperatures and pressures so the steam reaches a SC state. As a result, conversion efficiencies are greatly improved (Beér 2009). Figure 4.30 compares the efficiencies of multiple conversion technologies.

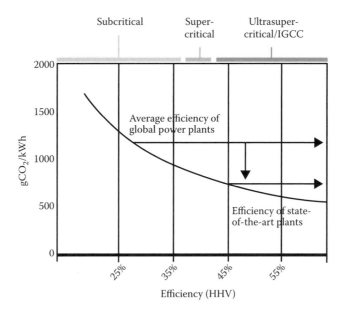

FIGURE 4.30
Energy and environmental efficiencies of SC and USC systems. Note: 1% increase in efficiency = 2%–3% decrease in emissions. (Reprinted with permission from IEA. *Focus on Clean Coal*, by Brian Ricketts. Paris: OECD/IEA, 2006. http://www.iea.org/publications/freepublications/publication/focus_on_coal.pdf.)

Power plants can use energy more efficiently by recycling used turbine steam in their boilers. Once back in the boiler, the plants reheat it and then reintroduce it into the turbine. The performance of SC and USC plants depends on steam conditions that are usually defined by three parameters: pressure, main steam temperature, and recycled steam temperature, which could be denoted as "30 MPa/600°C/600°C," for example. If steam is reused twice, the second reheat or recycled steam temperature will be listed after the first recycled steam temperature (ibid.).

SC power plants operate at steam conditions of 30 MPa/600°C/600°C, allowing them to reach a heat efficiency of 44% or higher, whereas subcritical plants, operating at only 455°C, have an average efficiency of 28%. A USC system further increases both temperatures to as much as 760°C while raising pressure to as much as 37 MPa. USC plant efficiencies exceed those of SC systems by about one percentage point for every 20°C increase in temperature (ibid.).

The key to achieving high steam pressure and temperature conditions, and consequently high efficiency, lies in the materials used to construct the plants. SC steam conditions can be achieved using steel with 12% chromium content. Austenite, also known as gamma phase iron, is an allotrope of iron generated when the iron is heated to a very high temperature (912°C–1394°C). At this temperature, the iron undergoes a phase transition, allowing it to withstand steam conditions of up to 31.5 MPa/620°C/620°C. Nickel-based alloys permit conditions of 35 MPa/700°C/720°C, yielding efficiencies of up to 48%. The successful research and development of new alloys may have a significant effect on the futures of SC and USC plants (Power4Georgians, LLC 2012).

By increasing energy efficiency, SC and USC plants reduce fuel consumption and CO_2, NO_x, SO_x, and particulate emissions. The extra investment, compared with subcritical plants, can be offset by savings in fuel and operating costs. Such upgrades cost less than many other measures to reduce coal's environmental effect and can be fully integrated with CO_2 sequestration systems. Consequently, SC power plants stand as a leading clean coal technology with widespread applications globally (National Energy Technology Laboratory [NETL] 2003).

Since the 1990s, China has introduced SC units with capacities of 350, 600, and 900 MW and has developed the technologies rapidly (Zhu and Zhao 2008). In 2000, the construction of a 600-MW SC thermal power generation project was included as a top priority in the country's development plan. This project was developed at the Huaneng Qinbei Power

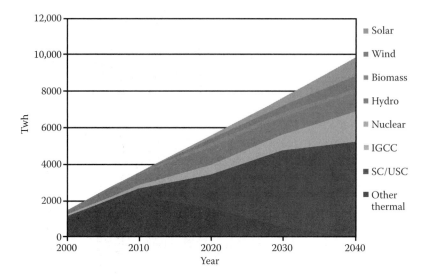

FIGURE 4.31

Power generation projection of China in low carbon scenario. (Reprinted with permission from Li, Y., *Energy Policy*, 51:138–142, 2012. doi:10.1016/j.enpol.2011.06.012.)

Plant with parameters of 24.2 MPa/566°C/566°C, and the plant was put into operation in 2004 (Li 2012). In 2006, two 1000-MW USC plants at the Huaneng Yuhuan Power Plant were put into operation in October and December, respectively (ibid.). As a result of this success, China required that all future power plants with capacities of 600 MW or greater must utilize SC technology.

Thus, China is now a world leader in SC and USC clean coal power plants. By the end of June 2007, the country had more than 180 SC units with capacities of 600 MW or greater, 30 1000-MW USC units, and 34 USC units with capacities of 600 or 660 MW in operation or under construction (ibid.). It is expected that SC units will account for 30% of the total power capacity in China by 2020 (Chen and Xu 2010). As shown in Figure 4.31, in a low carbon scenario, SC and USC power plants are projected to be the leading thermal power generation technologies, replacing all subcritical plants by 2040 (ibid.).

4.8.1.2 Circulating Fluidized Beds

A fluidized bed combustion (FBC) system refers to a technology in which the bed materials, including the fuel and sorbent, are suspended by primary combustion air drafts that come from below the combustor floor.

The particles move freely in the air, causing them to be fluidized. A pressurized fluidized bed combustion (PFBC) system is an FBC system that operates at an elevated pressure. There are two fluidized bed designs: bubbling fluidized and circulating fluidized (CFB). A major advantage of the CFB is its ability to utilize a wide range of fuels, including any type of coal, petroleum coke, and biomass (U.S. EPA 2010).

Under the CFB process, high-pressure air comes in through air nozzles located at the bottom of the bed and keeps the bed materials and crushed coal in suspension. The exposed surface area of the coal is thereby maximized, allowing it to undergo rapid and efficient combustion as the turbulent mixing of fuel and air results in more effective chemical reactions and heat transfer. As a result of the greater efficiency, the furnace is able to operate at a lower temperature, which limits the formation of NO_x. In addition, by mixing coal with sulfur-absorbing material such as limestone, more than 95% of the sulfur content can be captured inside the boiler, significantly reducing air pollutants. Fine particles of coal and ash enter the cyclone, which separates the particles from the gases. The solids then fall into the hopper and are recycled into the furnace. This process recirculates almost all of the bed materials. The remainder of the hot gas exits the cyclone, passes through a heat exchanger (which enhances the process's efficiency), and is then emitted as flue gas (ibid.). Figure 4.32 provides a diagram of the process.

This process can be scaled up easily and is characterized by low emissions. These favorable traits, combined with the ability to utilize low-quality coal, make CFB plants an attractive option in China for power plants with capacities between 300 and 600 MW (Chen and Xu 2010). The Sichuan Baima Power Plant, constructed by the French power company Alstom, is the first CFB demonstration unit in China and has a capacity of 300 MW. The plant was commissioned in 2005 (Alstom Power 2012). Since then, several more 300-MW units have gone into operation, including the Huaneng Kaiyuan Power Plant, the Datang Honghe Power Plant, and the Huadian Xunjiansi Power Plant. Many more 300 MW CFB projects are planned (ibid.).

Nevertheless, to enter the electricity market at capacities higher than 600 MW and to meet efficiency and capacity requirements, the steam parameters of CFB plants need to be increased to SC conditions. The SC CFB plant is the next generation of CFB technology and can have a power generation efficiency that is 42% higher than previous generations (Wuhu Huicheng Boiler Chemical Equipment Manufacturing Co., Ltd. 2009).

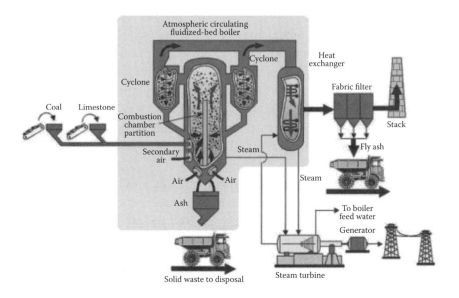

FIGURE 4.32

Simplified diagram of circulating fluidized beds power plant. (Reprinted under the U.S. Department of Energy's media resources usage policy from NETL. CCPI/Clean Coal Demonstrations: Nucla CFB Demonstration Project. http://www.netl.doe.gov/technologies/coalpower/cctc/cctdp/project_briefs/nucla/nuclademo.html [accessed November 20, 2012].)

China's National Development and Reform Commission approved a 600-MW SC CFB demonstration project and chose the Sichuan Dongfang Boiler Group as the provider. In 2012, the project proceeded to the construction phase. If successful, this will be the largest CFB plant in the world (Chen and Xu 2010).

4.8.1.3 Coal Gasification and the Integrated Gasification Combined Cycle

Coal gasification is a process that converts coal into a synthetic gas consisting of carbon monoxide and hydrogen through reactions of fuels at high temperatures with oxygen and steam. The vessel where the reaction occurs, called a gasifier, carefully controls the amount of oxygen in the reaction chamber so the majority of the feedstock undergoes partial oxidation, which causes it to be optimally chemically dissociated. The resulting synthetic gas (syngas or coal gas) is combustible and can be used as an energy source. This gasification process is a very efficient method for extracting energy from coal and facilitates the control of common pollutants from coal-burning plants. For example, during gasification, the sulfur content in coal is converted to hydrogen sulfide (H_2S) and carbonyl sulfide (OCS), both of which can be extracted easily. The lack of oxygen in the gasifier prevents the formation of NO_x. Instead, ammonia (NH_3) is generated and can be removed easily (U.S. Department of Energy [U.S. DOE] 2011). This ability to achieve extremely low sulfur, nitrogen, and particle emissions provides significant environmental benefits and makes this process very attractive.

Gasifiers have three major designs according to their flow geometry: entrained flow, fluidized bed, and moving bed gasifiers (U.S. EPA 2010). The details of each design are listed in Table 4.12 and explained in the following paragraphs.

The most common design is the entrained flow gasifier, in which pulverized coal particles and gases flow concurrently at high speed. As a result, coal reacts with steam and air while suspended. Coal enters the process dry in a transport gas (typically nitrogen) or wet in a water slurry. This process operates at high temperatures of 1200°C to 1600°C and pressures of 2 to 8 MPa. The advantage of this gasifier is its ability to accept both solid and liquid fuels. Its disadvantage is that its high temperature requires expensive construction materials and more maintenance because the high-temperature slag, which contains melted ash, can damage the gasifier's components (Minchener 2005).

TABLE 4.12

Comparison of Gasifier Designs

	Temperature	Pressure (MPa)	Feedstock Type	Main Advantage(s)	Main Disadvantage(s)
Entrained flow	1200°C–1600°C	2–8	Dry or wet	Accepts solid and liquid fuels	High temperature causes high costs
Fluidized bed	900°C–1050°C	3	Dry	Operates at variable loads. Avoids ash melting	Low temperature causes incomplete conversion
Moving bed	1300°C–1800°C	3	Dry	Eliminates the need for syngas cooler	Requires good bed permeability

Source: Adapted from Minchener, A.J., *Fuel* 84:2222–2235, 2005.

In a fluidized bed gasifier, an upward flow of gas transports coal and fluidizes the bed, allowing the coal to react and combust efficiently. It operates at lower temperatures of 900°C to 1050°C, which are not hot enough to cause the ash to melt. This process's main advantage is its ability to operate at variable loads, as operators can quickly increase or decrease the fluidized bed gasifiers' electricity outputs. However, the process's low temperature can result in incomplete carbon conversion, leading to lower efficiency levels. It is also suitable only for solid fuels (ibid.).

The moving bed gasifier operates only with solid fuels and can process coal with relatively high ash content. In this system, gases flow upward and coal is fed from the top via a lockhopper system. There are two types of moving bed gasifiers: dry ash and slagging. The former uses a much higher ratio of steam to oxygen and, consequently, has a much lower reaction temperature (1300°C compared with 1500°C–1800°C for the latter; ibid.). Because the two streams of materials flow against each other, the produced syngas is cooled upon its exit from the process, eliminating the need for expensive syngas coolers.

Both entrained flow and fluidized bed gasifiers are being developed in China. Tests have also been conducted on coal water slurry entrained flow gasification. In 2005, a multinozzle striking flow coal slurry-feed gasifier was developed successfully and put into operation (Chen and Xu 2010). A test facility with a two-stage, dry-feed pressurized gasification plant was also completed.

An integrated gasification combined cycle (IGCC) system uses syngas as fuel to generate power. The sulfur, nitrogen, and particle contents in the syngas are removed before the combustion process. The syngas is burned in a gas turbine, akin to a natural gas turbine, to generate electricity. The exhaust heat of the turbine and some heat from the gasification process are then used to generate steam, which drives a steam turbine, creating more electricity. Due to this dual source of electricity, the technology is called "combined cycle" (U.S. EPA 2010). This system can reach efficiency above 40% Higher Heating Value (the amount of heat released during combustion of the substance, including the latent heat of water vapor), much higher than conventional thermal plants. IGCC technologies have various advantages in addition to high efficiency, as they can also utilize a wide range of low-quality and high-quality coals. The technology also effectively controls sulfur, nitrogen, and particulate emissions because these pollutants are removed during the gasification process. In addition, it generates less solid waste and uses less water than SC plants (ibid.). Finally, when oxygen is used in the gasifier, the CO_2 produced is in a concentrated gas steam, which facilitates CCS. Some disadvantages, compared with SC technology include additional plant complexity, higher construction costs, and poorer performance at high-altitude locations (ibid.). Figure 4.33 provides a simplified diagram of an IGCC plant.

IGCC was selected as one of the key technologies for clean coal power generation in China's 15-year National Program for Medium-to-Long-term Scientific and Technological Development (2005–2020), along with the SC and USC CFB technologies (The Central People's Government of the People's Republic of China 2006). China's first IGCC project is the US$420 million Tianjin project. Its goal is to construct and operate a 250-MW IGCC power plant with an annual electricity output of 1470 GWh (Asian Development Bank [ADB] 2010). It was to be completed in 2012 and is the first of the three phases of the GreenGen program that aims to develop a near-zero emission coal power plant by 2015 (ibid.).

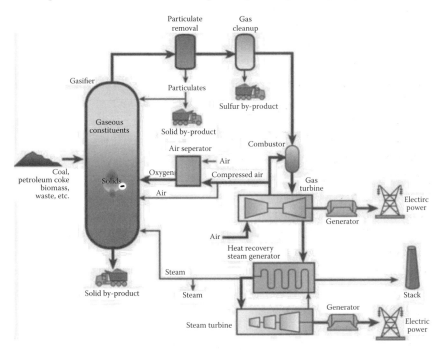

FIGURE 4.33

Simplified diagram of IGCC plant. (Diagram courtesy of the National Energy Technology Laboratory; reprinted under the U.S. Department of Energy's media resources usage policy from NETL. U.S. DOE. *Gasification—An Overview*, by Jenny B. Tennant. Pittsburgh, PA: NETL/U.S. DOE, 2011. http://www.netl.doe.gov/technologies/coalpower/gasification/pubs/pdf/110401%20GasificationTrainingToledo%20FINAL.pdf.)

TABLE 4.13

Cost Comparison of Clean Coal Power Generation Technologies

	SC	USC	CFB	IGCC
Capacity (MW)	300	600	300	250
Unit cost (¥/kW)	3919	3924	4799	7751
Reference electricity cost (¥/MWh)	321.0	310.7	354.8	504

Source: Adapted from Chen, W., and R. Xu., *Energy Policy* 38:2123–2130, 2010.

Table 4.13 provides a cost comparison among the clean coal power generation technologies mentioned thus far. IGCC still has the highest cost among the options. However, if other benefits are taken into account, this option effectively becomes cost-competitive.

4.8.2 Carbon Capture and Sequestration

CCS is a process that seizes CO_2 from large emission sources, such as power plants, and stores it in isolation from the atmosphere. This technology is being developed to minimize atmospheric carbon emissions.

Carbon capture technologies are divided into postcombustion and precombustion. Postcombustion carbon capture means removing CO_2 from the flue gas emitted from power plants. Due to the low CO_2 intensity in the flue gas, this process faces significant thermodynamic challenges. However, it has the highest short-term potential because it can be installed at existing power plants (Figueroa et al. 2008). The postcombustion capture technologies are listed in Table 4.14, along with their respective advantages and disadvantages. Precombustion carbon capture removes CO_2 before the fuel is burned. Technologies under development are presented in Table 4.15.

The last technological option that has been proposed for CCS is oxy-combustion, in which nearly pure oxygen (mixed with recycled flue gas) is injected into the combustion process in place of regular air. This produces flue gas that consists of water (which can be separated out easily) and very high concentrations of CO_2.

In all these methods, captured CO_2 is transported through pipelines to deep underground formations. The depth must be at least 1 km for pressure to be high enough to liquefy the gas. This liquefaction is vital because this makes it difficult for the gas to escape. In a process referred to as enhanced oil recovery, the captured CO_2 displaces oil and gas from mostly exhausted reservoirs and is then stored there permanently. Highly pressurized heated CO_2 can dissolve in oil; injecting it into the reservoirs can

TABLE 4.14

Postcombustion Carbon Capture Technologies

Technology	Material	Advantage	Disadvantage
Amine-based	Amines	Removes CO_2 effectively	Equilibrium-limited
Carbonate-based	Soluble carbonates	Little energy required for regeneration	Expensive
Ammonia-based	Ammonia	Low heat of reaction Potentially high CO_2 capacity Lower degradation Tolerance for oxygen Low cost	Flue gas must be cooled to react Ammonia can be lost during regeneration
Membranes	Membranes with amine, or inorganic silica membranes	Reduced loss of amine Higher loading differential	Still small-scale and underdeveloped
CO_2 capture sorbents	Solids that react with CO_2	Able to capture and regenerate easily Forms a stable solid after reaction	Solids are more difficult to implement than liquids
Metal organic frameworks (MOFs)	Hybrid material built from metal ions	High storage capacity Low heat recovery	Need to determine stability
Enzyme-based	Enzymes that mimic respiratory process	Low heat of absorption Fast reaction rate	Still small-scale and underdeveloped
Ionic liquids	Salts with organic cation and inorganic anion	High temperature stability Low heat for regeneration	Still small-scale and underdeveloped

Source: Adapted from Figueroa, J.T. et al., *International Journal of Greenhouse Gas Control* 2:9–20, 2008.

thus help to extract more oil from the rocks. This method has also dominated existing CCS projects because the injection pipelines are already in place at the reservoirs (Blunt 2010).

China has been active in developing CCS technologies. CCS is highlighted in the National Program for Medium-to-Long-term Scientific and Technological Development (2005–2020) as a frontier technology in reducing carbon emissions. The National High-tech R&D Program (863 Program) focused on three research areas: carbon capture absorption, carbon capture adsorption, and carbon storage technologies (Chen and Xu 2010). Various projects are being conducted to develop technologies

TABLE 4.15

Precombustion Carbon Capture Technologies

Technology	Method	Advantage	Disadvantage
IGCC	CO_2 is separated in the shift converter	High CO_2 concentration of flue gas and high pressure	Technical and economic uncertainty
Physical solvent	Solvent selectively absorbs CO_2	No chemical reaction	Energy intensive heat transfer
Membranes	Polymer-based membranes	Less energy-intensive No phase change Low maintenance	Still small-scale and underdeveloped
Sorbents	Lithium silicate-based sorbent	High CO_2 removal rate High regenerability	Still small-scale and underdeveloped
Chemical looping combustion and gasification	Oxygen is supplied by a solid oxygen carrier	High CO_2 concentration of flue gas	Still in its early stages
Improved auxiliary process	Improved auxiliary process in IGCC	Improves IGCC availability and economics	Underdeveloped

Source: Adapted from Figueroa, J.T. et al., *International Journal of Greenhouse Gas Control* 2:9–20, 2008.

that use CO_2 for oil recovery and for coal bed methane exploitation. China has also participated in international projects, such as the China–U.K. Near-Zero Emissions Coal initiative and the China–Australia Geological Storage project (NZEC 2009).

Petro China started China's first CO_2 usage and storage project at the Jilin Oil Field in 2006. In July 2008, the Huaneng Group opened a postcombustion carbon capture facility that can recover more than 85% of CO_2 from the power plant with a purity of 99.99%, capturing 3000 tons of CO_2 annually (Chen and Xu 2010). The previously mentioned GreenGen project also plans to install CCS technology at its IGCC plant during the project's second stage.

4.9 CONCLUSION

The Chinese energy mix is currently characterized by a large proportion of coal as well as by a widening gap between consumption and supply. The rapid growth of China's energy production and consumption over the past 30 years has positioned China as a major player and driver of the energy industry. The widening gap between the nation's consumption and production is regarded as a serious problem by Chinese policymakers. One of the primary motivations driving the Chinese government is its desire to preserve its political power. There are two types of threats: external and internal.

The widening gap between the nation's consumption and production has led to a rapidly increasing share of energy imports. Policymakers have especially focused on the increasing oil imports as a potential external threat. Shortly after China became a net oil importer in 1993, the first article with "energy security" as a keyword emerged from the "China economic news library" of China INFOBANK, a database of Chinese periodicals. When China's oil imports doubled in 2000, the first article with a title containing "energy security" appeared (Leung 2011). Thus, it seems that the country's shift to a net oil importer has increased oil-related concerns. Historically, after the Western powers subjected China to an oil embargo during the Korean War, the country turned to the Soviet Union and Eastern Europe for oil products. But after the Sino–Soviet split in the 1960s, the Soviet Union curtailed this trade relationship, and China's transport sector and military were left with dangerous fuel shortages (ibid.). Apart from coal, fossil fuels make up the largest component of the energy mix. As a result, oil security is a key motivation behind the government's desire to develop alternative energy sources.

To defend itself from internal threats (i.e., domestic unrest), the Chinese government must promote economic development. Some argue that the government and the country's populace have an implicit contract: the people grudgingly accept the party's rule in exchange for growing prosperity and personal freedom (Pei 2004). In the 1980s and 1990s, the government adopted a "first development, then environment, first pollute, then cleanup" approach to economic development, reflecting its prioritization of economic growth over the environment. Recently, however, policy and academic documents have reflected a greater concern about the long-term stability of its political power. The 12th Five-Year Plan is often cited as the definitive signal of the growing effort to transition from China's old growth-at-all costs model to a new growth model subject to economic and social constraints (Boyd 2012).

China's old development model created unsustainable patterns of energy use, many of which have come under fire from academics and central government documents (Boyd 2012). Economic growth has been fueled by energy consumption growth, but the current and historical domination of coal, one of the dirtiest energy resources, in the energy mix presents a threat to the sustainability of economic development. This is shown by the numerous problems outlined in this chapter, such as dramatic increases in carbon emissions. In fact, China is now the world's largest producer of greenhouse gases, acid deposition, coal gangue, coal dust, water pollution, other forms of air pollution, land subsidence, and mining accidents.

In particular, the *National Assessment Report on Climate Change* and China's *Climate Change White Paper* highlight the following potential threats to China from climate change: (1) the damage that extreme and unpredictable weather will pose to already declining crop yields and live-stock; (2) large-scale biodiversity loss; (3) melting permafrost and glaciers, resulting in an overall decline in ice cover in the Tibetan Plateau, with major effects on the water resources in nearly all of China's major river systems; (4) greater incidence and severity of both floods and droughts, but with an overall decline in already stressed water resources; (5) rising sea levels threatening China's populous and rich coastal areas; and (6) increasing social and economic costs because of the widespread poverty and underdevelopment that will exacerbate the difficulties of adapting to climate change. As such, the government is encouraging the development of technologies to make the country's energy mix cleaner (Boyd 2012).

Furthermore, because coal makes up such a large portion of China's energy mix, the government views maintaining a reliable supply as a "foundation of China's national wealth and energy security" (ibid.). As we have seen in the chapter, however, China has already become a net importer of coal due to the geographical disparity between the supply of coal and the demand for energy and a strained infrastructure. The growing gap between energy supply and demand is an especially troubling issue for the government, and has also led to electricity shortages that have increased the importance of transitioning to a more sustainable growth model.

China's energy situation, however, presents more than threats to the country. The search for alternative energy sources currently pursued by the country will serve both as a way to transition to a more sustainable model of economic growth and, according to some academics and poli-cymakers, as a historic opportunity for China to position itself as an economic and technological leader. For example, Hu Angang, of Tsinghua University, has argued that, historically, countries that are able to exploit key technologies before others have succeeded in changing the balance of power in the international system. He argues that green energy technology will be linked to the next wave of modernization and that China should position itself at the forefront of this technological revolution. Others have argued that China is well-situated to position itself in this way because its economy is already in a state of transition from an energy-intensive export industry to a less energy-intensive economy based on increasing productivity and higher-quality goods and services. The State Council has echoed these sentiments, stating that China can gain a "first mover advantage" and "seize a historic opportunity" by using the development of low-carbon technologies to forge a new comparative advantage in international trade (ibid.).

As an industry that requires high capital investment and that is not yet economically competitive, the alternative energy industry relies heavily on government support. China's rapid GDP growth can be attributed in part to its uninhibited pursuit of increasing economic gains. This growth has been accompanied, however, by increasing energy consumption and a widening gap between the production and consumption of energy. The heavy utilization of cheap and dirty coal, geographic disparities between supply and demand centers, and the lack of adequate infrastructure in China have led to visible environmental and social damage, energy supply shortages, and threats to the country's energy security. All these problems threaten the sustainability of its economic growth model and have posed enough of a danger to China's economic growth that the government has taken notice. These considerations have fueled attempts to transition from a growth-at-all-costs model to a model of economic growth subject to environmental and social constraints.

The following chapter will demonstrate that the abundance of such a cheap energy resource, coal, presents another barrier to the development of cleaner energies: the attractiveness of an investment is partially determined by the attractiveness of alternative investments. The search for alternative energy sources and the current status of the alternative energy industry is discussed in greater detail in the following chapter.

REFERENCES

ABB. *Trends in Global Energy Efficiency 2011: China: Energy Efficiency Report*. Zurich: ABB, 2011. http://www05.abb.com/global/scot/scot316.nsf/veritydisplay/63246e62080610 aec12578640050f217/$file/china.pdf (accessed December 13, 2012).

Alstom Power. *Circulating Fluidised Bed Technology: Clean, Efficient, & Fuel-Flexible*. Baden, Switzerland: Alstom Power, 2012. http://www.alstom.com/Global/Power/Resources/

Documents/Brochures/circulating-fluidised-bed-boiler-technology-coal-oil-power. pdf (accessed January 10, 2013).

Asian Development Bank (ADB). *People's Republic of China: Tianjin Integrated Gasification Combined Cycle Power Plant Project*. Manila, Philippines: ADB, 2010. http://www. adb.org/Documents/PAMs/PRC/42117-01-prc-pam.pdf (accessed January 10, 2013).

Beér, J.M. *Higher Efficiency Power Generation Reduces Emissions: National Coal Council Issue Paper*. Cambridge, MA: MIT Energy Initiative, 2009. http://mitei.mit.edu/ system/files/beer-emissions.pdf (accessed January 10, 2013).

Bi, X., B.R.T. Simoneit, G. Sheng, and J. Fu. Characterization of molecular markers in smoke from residential coal combustion in China. *Fuel* 87:112–119, 2008.

Blunt, M. *Grantham Institute for Climate Change Briefing Paper No 4: Carbon Dioxide Storage*. London: Imperial College London, 2010. http://www.co2storage.org.uk/ Blunt_2010_CC_Storage.pdf (accessed January 12, 2013).

Boyd, O.T. *China's Energy Reform and Climate Policy: The Motivating Change*. Action, Australia: Crawford School of Public Policy, 2012. http://ideas.repec.org/p/een/ ccepwp/1205.html (accessed April 15, 2013).

Brown, L.R. Earth Policy Institute. Plan B 4.0: Mobilizing to save civilization: Supporting data. Last modified 2009 (accessed November 20, 2012). http://www.earth-policy. org/books/pb4/pb4_data.

Central Intelligence Agency (CIA). The World Factbook, Country Comparison: Natural Gas—Proved Reserves. Last modified January 1, 2011a (accessed November 19, 2012). https://www.cia.gov/library/publications/the-world-factbook/rankorder/2179rank. html.

Central Intelligence Agency (CIA). The World Factbook, Country Comparison: Oil— Proved Reserves. Last modified January 1, 2011b (accessed November 19, 2012). https://www.cia.gov/library/publications/the-world-factbook/rankorder/2178rank. html.

The Central People's Government of the People's Republic of China. China National Program for Medium-to-Long-term Scientific and Technological Development (2005–2020). Last modified February 9, 2006 (accessed January 10, 2013). http:// www.gov.cn/jrzg/2006-02/09/content_183787_3.htm.

Chen, W., and R. Xu. Clean coal technology development in China. *Energy Policy* 38:2123– 2130, 2010.

China Coal Information Institute (CCII). *The Environmental Effects of Coal Production and Utilization in China*. Beijing: CCII, 2003. http://www.wwfchina.org/wwfpress/ publication/climate/mtyxbg.pdf (accessed January 11, 2013).

China Coal Information Institute (CCII), and China Coalbed Methane Clearinghouse (CCMC). *Methane to Markets Partnership in China Summary Report* by S. Huang, W. Liu, G. Zhao, F. Sang, X. Liu, and L. Huang. Beijing: CCII and CCMC, 2011. http:// www.epa.gov/outreach/cmop/docs/MethaneToMarketsChina_summaryreport.pdf (accessed December 13, 2012).

China Daily. China to ban small coal mines for improving pit safety record. *China Daily*, August 15, 2008. http://www.chinadaily.com.cn/bizchina/2008-08/15/content_ 6940070.htm (accessed January 10, 2013).

ChinaMaps.org. China blank map. Accessed May 12, 2013. http://www.chinamaps.org/ china/china-blank-map.html.

China.org.cn. China's largest open-cast coal mine starts production. *China.org.cn*, December 18, 2008. http://www.china.org.cn/china/photos/2008-12/18/content_16972032_2. htm (accessed November 20, 2012).

China.org.cn. Lang subsidence forms giant hole in mining area. *China.org.cn*, February 15, 2011. http://www.china.org.cn/china/2011-02/15/content_21926898.htm (accessed December 15, 2012).

China Wind Power Center. Wind Energy Resource Characteristics and Development Potential. Last modified December 28, 2009 (accessed November 19, 2012). http://www.cwpc.cn/cwpc/en/node/6295.

Cho, R. The Earth Institute at Columbia University. How China is dealing with its water crisis. Last modified May 5, 2011 (accessed December 15, 2012). http://blogs.ei.columbia.edu/2011/05/05/how-china-is-dealing-with-its-water-crisis/.

Christiani, D.C., and X.R. Wang. Occupational lung disease in China. *International Journal of Occupational and Environmental Health* 4:320–325, 2003.

Daqin Railway Co., Ltd. Company Introduction. Last modified June 30, 2006 (accessed November 20, 2012). http://www.daqintielu.com/2006-06-30/00002176.shtml.

DoNews. Closer look at the 12th five-year plan: The changed and the unchanged in the new energy industry. *DoNews*, March 21, 2011. http://www.donews.com/it/201103/398632.shtm (accessed January 9, 2013).

E-to-China.com. China's Coal Imports and Exports Down in Quantity in First Half of 2011. Last modified September 13, 2011 (accessed November 20, 2012). http://www.e-to-china.com/2011/0913/97058.html.

Fang, R. China Association of Sciences and Technology (CAST). Kuangdi Xu calls for low-carbon economy in response to climate change. Last modified September 9, 2009 (accessed November 19, 2012). http://www.cast.org.cn/n35081/n35473/n35518/11485489.html.

Feng, X. China Economy Net. 1987: The Great Implications of the "Three-Step" Economic Development Strategy. Last modified December 29, 2008 (accessed April 18, 2013). http://views.ce.cn/fun/corpus/ce/fx/200812/29/t20081229_17823822.shtml.

Feng, C.G., P. Huan, G.X. Jing, X.J. Li, Z.Y. Liu, X.M. Qian, and Y.P. Zheng. A statistical analysis of coal mine accidents caused by coal dust explosions in China. *Journal of Loss Prevention in the Process Industries.* 22:528–532, 2009.

Figueroa, J., T. Fout, S. Plasynski, H. McIlvried, and R.D. Srivastava. Advances in CO_2 capture technology—The U.S. Department of Energy's carbon sequestration program. *International Journal of Greenhouse Gas Control* 2:9–20, 2008.

Finkelman, R.B., W. Orem, V. Castranova, C.A. Tatu, H.E. Belkin, B. Zheng, H.E. Lerch, S.V. Maharaj, and A. Bates. Health impacts of coal and coal use: Possible solutions. *International Journal of Coal Geology* 50:425–443, 2002.

Friedman, L., and ClimateWire. China greenhouse gas emissions set to rise well past U.S. *Scientific American*, February 3, 2012. http://www.scientificamerican.com/article.cfm?id=china-greenhouse-gas-emissions-rise-past-us (accessed January 11, 2013).

Fu, X. 33 Cities Have Air Quality Lower than Minimum Standard, Beijing PM10 Level Ranked Third to Last. Last modified 2012 (accessed December 15, 2012). http://city.china.com.cn/index.php?m=content&c=index&a=show&catid=89&id=25906138.

Greenpeace. *The True Cost of Coal* by Mao Yushi, Sheng Hong, and Yang Fuqiang. Hong Kong: Greenpeace, 2008. http://act.greenpeace.org.cn/coal/report/TCOC-Final-EN.pdf (accessed December 21, 2011).

Guo, W., Y. Zou, and Y. Liu. Current status and future prospects of mining subsidence and ground control technology in China. In *Coal: Coal Operators' Conference: University of Wollongong & the Australasian Institute of Mining and Metallurgy*, 2009, 140–145.

Homer, A.W. Coal mine safety regulation in China and the USA. *Journal of Contemporary Asia* 39:424–439, 2009.

Hong, Y., and Y. Mu. *China's Steel Industry: An Update, EAI Background Brief No. 501.* Singapore: National University of Singapore, 2010. http://www.eai.nus.edu.sg/BB501. pdf (accessed December 13, 2012).

Hu, Z., F. Hu, J. Li, and H. Li. Impacts of coal mining subsidence on farmland in Eastern China. *International Journal of Mining, Reclamation and Environment* 11:91–94, 1997.

Huber, T. U.S. coal mine deaths: 2010 deadliest year since 1992. *The Huffington Post,* December 30, 2010. http://www.huffingtonpost.com/2010/12/30/us-coal-mine-deaths-in-20_n_802790.html (accessed December 16, 2012).

IBISWorld. China Industry Reports: Alternative Energy—Industry at a Glance: Executive Summary. Last modified 2013 (accessed April 24, 2013). http://clients1.ibisworld. com/reports/cn/industry/ataglance.aspx?entid=719.

Impact Carbon. Household Health and Energy in China. Last modified 2013 (accessed April 24, 2013). http://impactcarbon.org/our-projects/stoves-in-china/.

Information Office of the State Council of the People's Republic of China (IOSCPRC). *China's Energy Conditions and Policies.* Beijing: IOSCPRC, 2007. http://en.ndrc.gov. cn/policyrelease/P020071227502260511798.pdf (accessed December 13, 2012).

Information Office of the State Council of the People's Republic of China (IOSCPRC). *China's Policies and Actions for Addressing Climate Change.* Beijing: IOSCPRC, 2008. http://www.ccchina.gov.cn/WebSite/CCChina/UpFile/File419.pdf (accessed December 13, 2012).

The Institute of Public and Environmental Affairs (IPE) and Renmin University School of Law. Air Quality Transparency Index. Last modified January 18, 2011 (accessed January 18, 2011). http://www.ipe.org.cn//Upload/Report-AQTI-CH.pdf.

Intergovernmental Panel on Climate Change (IPCC). *Climate Change 2007: Synthesis Report* by Lenny Bernstein, Peter Bosch, Osvaldo Canziani, Zhenlin Chen, Renate Christ, Ogunlade Davidson, William Hare, Saleemul Huq, David Karoly, Vladimir Kattsov, Zbigniew Kundzewicz, Jian Liu, Ulrike Lohmann, Martin Manning, Taroh Matsuno, Bettina Menne, Bert Metz, Monirul Mirza, Neville Nicholls, Leonard Nurse, Rajendra Pachauri, Jean Palutikof, Martin Parry, Dahe Qin, Nijavalli Ravindranath, Andy Reisinger, Jiawen Ren, Keywan Riahi, Cynthia Rosenzweig, Matilde Rusticucci, Stephen Schneider, Youba Sokona, Susan Solomon, Peter Stott, Ronald Stouffer, Taishi Sugiyama, Rob Swart, Dennis Tirpak, Coleen Vogel, Gary Yohe, and Terry Barker. Geneva, Switzerland: IPCC, 2007. http://www.ipcc.ch/pdf/assessment-report/ar4/ syr/ar4_syr.pdf (accessed December 28, 2012).

International Energy Agency (IEA). *Focus on Clean Coal* by Brian Ricketts. Paris: OECD/ IEA, 2006. http://www.iea.org/publications/freepublications/publication/focus_on_ coal.pdf (accessed April 23, 2013).

International Energy Agency (IEA). *Integration of Renewables: Status and Challenges in China* by Kat Cheung. Paris: OECD/IEA, 2011. http://www.iea.org/publications/ freepublications/publication/Integration_of_Renewables.pdf (accessed November 19, 2012).

International Monetary Fund (IMF). World Economic Outlook Database. Last modified 2010 (accessed November 19, 2012). http://www.imf.org/external/pubs/ft/ weo/2010/02/weodata/download.aspx.

Larssen, T., E. Lydersen, D. Tang, Y. He, J. Gao, H. Liu, L. Duan, H.M. Seip, R.D. Vogt, J. Mulder, M. Shao, Y. Wang, H. Shang, X. Zhang, S. Solberg, W. Aas, T. Økland, O. Eilertsen, V. Angell, Q. Liu, D. Zhao, R. Xiang, J. Xiao, and J. Luo. Acid rain in China. *Environmental Science & Technology,* 40(2):418–425, 2006. http://pubs.acs.org/doi/ pdfplus/10.1021/es0626133 (accessed December 13, 2012).

Leung, G. China's energy security: Perception and reality. *Energy Policy* 39:1330–1337, 2011.

Li, J. and E. Martinot. Renewable Energy World. Renewable Energy Policy Update for China. Last modified July, 2010 (accessed November 19, 2012). http://www.renewableenergyworld.com/rea/news/article/2010/07/renewable-energy-policy-update-for-china.

Li, Y. Dynamics of clean coal-fired power generation development in China. *Energy Policy.* 51:138–142, 2012. doi:10.1016/j.enpol.2011.06.012.

Li, Z., P. Liu, L. Ma, and L. Pan. A supply chain based assessment of water issues in the coal industry in China. *Energy Policy* 48:93–102, 2012.

Lu, H. China meets 5-year energy-saving goal: NDRC. *Xinhua News.* Last modified January 6, 2011 (accessed December 13, 2012). http://news.xinhuanet.com/english2010/business/2011-01/06/c_13679329.htm.

Lu, Z., and D.G. Streets. Sulfur dioxide and primary carbonaceous aerosol emissions in China and India, 1996–2010. *Atmospheric Chemistry and Physics* 11:20267–20330, 2011.

Lu, Z., D.G. Streets, Q. Zhang, S. Wang, G.R. Carmichael, Y.F. Cheng, C. Wei, C. Chin, T. Diehl, and Q. Tan. Sulfur dioxide emissions in China and sulfur trends in East Asia since 2000. *Atmospheric Chemistry and Physics* 10:6311–6331, 2010.

Lu, Y. and S. Zhang. China's coal-fired power plant desulfurization technology problems and countermeasures. Last modified September 27, 2010 (accessed December 15, 2012). http://eng.hi138.com/?i257821_Chinas_coal-fired_power_plant_desulfurization_technology_problems_and_countermeasures.

Minchener, A.J. Coal gasification for advanced power generation. *Fuel* 84:2222–2235, 2005.

National Bureau of Statistics of China. *China Energy Statistical Yearbook 2011.* Beijing: China Statistics Press, 2012.

National Energy Technology Laboratory (NETL). CCPI/Clean Coal Demonstrations: Nucla CFB Demonstration Project. Last modified 2003 (accessed November 20, 2012). http://www.netl.doe.gov/technologies/coalpower/cctc/cctdp/project_briefs/nucla/nuclademo.html.

National Energy Technology Laboratory (NETL). U.S. Department of Energy. *Gasification— An Overview* by Jenny B. Tennant. Pittsburgh, PA: NETL/U.S. DOE, 2011. http://www.netl.doe.gov/technologies/coalpower/gasification/pubs/pdf/110401%20GasificationTrainingToledo%20FINAL.pdf (accessed April 23, 2013).

NZEC Carbon Capture & Storage. *CCS Activities in China.* Work Package 5, 2009. http://www.nzec.info/en/assets/Reports/CCS-Activities-in-China.pdf (accessed January 10, 2013).

Ohshita, S., and L. Price. The Network for Climate and Energy Information. Targets for the Provinces: Energy Intensity in the 12th Five-Year Plan. Last modified April 18, 2011 (accessed December 13, 2012). http://www.chinafaqs.org/blog-posts/targets-provinces-energy-intensity-12th-five-year-plan.

Pei, M. Beijing's social contract is starting to fray. *The Financial Times*, June 3, 2004. http://yaleglobal.yale.edu/content/beijings-social-contract-starting-fray (accessed April 15, 2013).

Pennsylvania Department of Environmental Protection (PDEP). What is mine subsidence? Last modified 2012 (accessed April 22, 2013). http://www.dep.state.pa.us/msi/whatisms.html.

Power4Georgians, LLC. Supercritical power plants. Last modified 2012 (accessed January 10, 2013). http://power4georgians.com/supercritical.aspx.

Pu, Y. Shanxi: Total coal reserves account for 26% of national reserves, annual production 866 million tons. *Xinhua News.* Last modified October 31, 2007 (accessed November 20, 2012). http://news.xinhuanet.com/local/2007-10/31/content_6976322.htm.

Radical Geography. Coal, China—A case study of growing resource exploitation on the people and the environment of an LEDC country. Last modified 2007 (accessed December 15, 2012). www.radicalgeography.co.uk/chinathoughts.doc.

Reserve Bank of Australia. *China's Steel Industry* by J. Holloway, I. Roberts, and A. Rush. Sydney, Australia: Reserve Bank of Australia, 2010. http://www.rba.gov.au/publications/bulletin/2010/dec/pdf/bu-1210-3.pdf (accessed December 13, 2012).

Shan, J. 38 million people suffering COPD in China. *China Daily.* Last modified 2010 (accessed January 9, 2013). http://www.chinadaily.com.cn/china/2010-11/15/content_11616546.htm.

Skirble, R. Study: Reducing soot is fastest way to slow climate change. Last modified September 1, 2011 (accessed December 15, 2012). http://www.voanews.com/content/study-reducing-soot-is-fastest-way-to-slow-climate-change-129070528/169690.html.

State Administration of Coal Mine Safety (SACMS). *China Coal Industry Year Book.* Beijing: China Coal Industry Publishing House, 2006.

Sun, G. *Coal in China: Resources, Uses, and Advanced Coal Technologies.* Arlington, VA: Pew Center on Global Climate Change, 2010. http://www.pewclimate.org/doc-Uploads/coal-in-china-resources-uses-technologies.pdf (accessed December 13, 2012).

SXGOV.cn. Introduction to Coal Resources in Shanxi. Last modified March 8, 2009 (accessed November 20, 2012). http://www.sxgov.cn/cultrue/cultrue_content/2009-03/08/content_44122.htm.

Tian, S. Overview of China's Coal Transportation. Last modified 2008 (accessed January 11, 2013). http://blog.sina.com.cn/s/blog_51bfd7ca0100aq4s.html.

Transatlantic Academy. *China's Long Road to a Low-Carbon Economy: An Institutional Analysis* by Philip Andrews-Speed. Washington, DC: Transatlantic Academy, 2012. http://www.transatlanticacademy.org/sites/default/files/publications/AndrewsSpeed_China%27sLongRoad_May12_web.pdf (accessed January 27, 2013).

Tu, J. Coal mining safety: China's Achilles' heel. *China Security* 3:36–53, 2007.

UNDP/Spanish MDG Achievements Funds. *China Climate Change Partnership Framework—Sector Study and Project Pilot Identification on Promoting the Adoption of Heat Recovery Power Generation in Coal Gangue Brick Making Sector. Final Report* by Energy and Environmental Development Consulting Limited. Beijing: Energy and Environmental Development Consulting Limtied, 2009. http://www.unido.org/fileadmin/media/documents/pdf/Procurement/Notices/1008/16002201/Final%20report_EN-090108.PDF (accessed January 10, 2013).

U.S. Department of Commerce. International Trade Administration. *Steel Industry Executive Summary: November 2012.* Washington, DC: Import Administration, 2012. http://hq-web03.ita.doc.gov/License/Surge.nsf/webfiles/SteelMillDevelopments/$file/exec%20summ.pdf?openelement (accessed December 13, 2012).

U.S. Department of Energy (U.S. DOE). How coal gasification power plants work. Last modified January 6, 2011 (accessed January 10, 2013). http://fossil.energy.gov/programs/powersystems/gasification/howgasificationworks.html.

U.S. Energy Information Administration (U.S. EIA). *International Energy Outlook 2011.* Washington, DC: U.S. Department of Energy and U.S. Energy Information Administration, 2011.

U.S. Energy Information Administration (U.S. EIA). International Energy Statistics. Last modified 2012a (accessed November 19, 2011). http://www.eia.gov/cfapps/ipdbproject/IEDIndex3.cfm.

U.S. Energy Information Administration (U.S. EIA). International Energy Statistics: Total Carbon Dioxide Emissions from the Consumption of Energy (Million Metric Tons). Last updated 2012b (accessed November 19, 2012). http://www.eia.gov/cfapps/ipdbproject/iedindex3.cfm?tid=90&pid=44&aid=8&cid=CH,&syid=1980&eyid=2009&unit=MMTCD.

U.S. Energy Information Administration (U.S. EIA). International Energy Statistics: Electricity Net Generation by Type (billion kilowatthours). Accessed November 20, 2012. http://www.eia.gov/cfapps/ipdbproject/iedindex3.cfm?tid=2&pid=alltypes&aid=12&cid=CH,&syid=2007&eyid=1991&unit=BKWH.

U.S. Environmental Protection Agency (U.S. EPA). *Available and Emerging Technologies for Reducing Greenhouse Gas Emissions from Coal-Fired Electric Generating Units.* Research Triangle Park, NC: U.S. EPA, 2010. http://www.epa.gov/nsr/ghgdocs/electricgeneration.pdf (accessed November 19, 2012).

U.S. Environmental Protection Agency (U.S. EPA). Greenhouse Gas Emissions. Last modified June 14, 2012 (accessed December 13, 2012). http://epa.gov/climatechange/ghgemissions/gases/ch4.html.

Wang, H. China Coal Mine Deaths Fall 'but Still Remain High.' Last modified February 26, 2011a (accessed December 16, 2012). http://www.asiaone.com/News/Latest+News/Asia/Story/A1Story20110226-265428.html.

Wang, S. United Nations Environmental Program. Mercury emissions from coal-fired power plants in China. Last modified January 23, 2011b (accessed December 15, 2012). http://www.unep.org/hazardoussubstances/Portals/9/Mercury/Documents/INC2/technical%20briefing%20presentations/updated/2_China_Coal_Presentation.pdf.

Wang, Y., A. Gu, and A. Zhang. Recent development of energy supply and demand in China, and energy sector prospects through 2030. *Energy Policy* 39:6745–6759, 2011.

Wei, W. Current issues of China's coal industry: The case of Shanxi. In *Proceedings of the 15th Annual Conference of the Association for Chinese Economics Studies Australia (ACESA), Melbourne.* Melbourne, Australia: RMIT University, 2003. http://mams.rmit.edu.au/9tqqzgfj9oks1.pdf (accessed December 15, 2012).

Wei, G. Statistical analysis of Sino-U.S. coal mining industry accidents. *International Journal of Business Administration* 2:82–86, 2011.

Wei, H., and D. Juan. Power shortages on the way. *China Daily*, November 12, 2011. http://www.chinadaily.com.cn/bizchina/2011-11/12/content_14082975.htm.

World Bank Group (WBG). GDP (current US$). Last modified 2011 (accessed April 17, 2013). http://data.worldbank.org/indicator/NY.GDP.MKTP.CD?order=wbapi_data_value_2011+wbapi_data_value+wbapi_data_value-last&sort=desc.

World Coal Association. Coal statistics. Last modified 2012 (accessed November 20, 2012). http://www.worldcoal.org/resources/coal-statistics/.

World Health Organization (WHO). Mercury and Health. Last modified April 2012 (accessed April 22, 2013). http://www.who.int/mediacentre/factsheets/fs361/en/.

Wuhu Huicheng Boiler Chemical Equipment Manufacturing Co., Ltd. Circulating fluidized bed technology in China the status of clean coal technology. Last modified 2009 (accessed January 10, 2013). http://www.hcboiler.com/en/display.asp?id=90.

Xinhua. China's power generation capacity leaps above 900 mln kilowatts. *People's Daily Online.* Last modified September 20, 2010 (accessed November 20, 2012). http://english.peopledaily.com.cn/90001/90778/90862/7146272.html.

Xinhua. Inner Mongolia's coal output hits 782 mln tonnes in 2010. *People's Daily Online.* Last modified January 17, 2011 (accessed November 20, 2012). http://english.people.com.cn/90001/90778/7263140.html.

Xinhuanet.com. China's 120 Major Cities Air Quality Ranking in the First Half of 2012. Accessed December 15, 2012. http://forum.home.news.cn/detail/100874279/1.html.

Xinhuanet.com. China power corp announced UHV transmission line construction plan, complete three-horizontal three-vertical one-circle grid by 2015. Last modified January 5, 2011 (accessed May 22, 2013). http://news.xinhuanet.com/fortune/2011-01/05/c_12948645.htm.

Zhang, A. China top energy official: Not favor expanding coal output to 4 bln T by 2015. *Xinhua News Agency.* Last modified October 28, 2010 (accessed November 20, 2012). http://www.istockanalyst.com/article/viewiStockNews/articleid/4618521.

Zhang, N., N. Lior, and H. Jin. The energy situation and its sustainable development strategy in China. *Energy* 36:3639–3649, 2011.

Zhang, X. China's coal consumption rises 10.3 Pct in first three quarters. *Xinhua News.* Last modified October 23, 2011 (accessed November 20, 2012). http://news.xinhuanet.com/english2010/china/2011-10/23/c_131207356.htm.

Zhao, L. China continues with coal gangue projects. *The Electricity Forum.* Last modified September 10, 2010 (accessed December 13, 2012). http://www.electricityforum.com/news/sep10/Chinacontinueswithcoalgangueprojects.html.

Zhao, T. China's coal imports up 31% in 2010. *China Daily.* Last modified January 27, 2011 (accessed November 20, 2012). http://www.chinadaily.com.cn/business/2011-01/27/content_11926703.htm.

Zhu, B. and Y. Zhao. Development of ultra-supercritical power generation technology in China. *Huadian Technology.* Last modified February 1, 2008 (accessed January 10, 2013). http://www.hdpower.net/qikan/manage/wenzhang/20080201.pdf.

5

China's Search for Cleaner Electricity Generation Alternatives

Julia Zheng and Xiaoting Zheng

CONTENTS

5.1 INTRODUCTION

The two largest components of China's energy mix are coal (70%) and oil (18%). Their negative implications (e.g., environmental damage, health consequences, supply shocks due to a lack of diversification, and energy insecurity) have propelled government support for other energy sources. The long-term negative environmental effect and low energy conversion efficiency of coal described in Chapter 4, in particular, have led the government to explore alternative energy sources, including renewables.

The Chinese government's support for alternative energy sources has already yielded noticeable results, as demonstrated by the rapid growth of the alternative energy industry in China over the past few years. The country has seen significant progress in the use of natural gas, which is more energy-efficient and environmentally friendly than coal and oil. This strong demand has recently made China a net importer of natural gas (U.S. Energy Information Administration [U.S. EIA] 2013). In addition, China is now the world leader in installed wind capacity and installed hydropower capacity (Chinese Renewable Energy Industries Association [CREIA], Global Wind Energy Council [GWEC] and Greenpeace 2010; Engelsiepen 2013). It is also the world's leading manufacturer of solar photovoltaic (PV) panels, as a result of aggressive production policies, and has recently begun to take an interest in developing domestic installation of these panels as well. Although China's wind, solar, and water resources are not exceptionally large on a per capita basis, they are very abundant on an absolute basis. Resources for further development are even available in mature industries such as hydropower. The country is also investigating advanced technologies in nuclear power generation, even though large-scale deployment has been limited by low domestic uranium reserve levels.

Despite government support for developing the clean energy industry, many challenges still prevent the sources from being adopted on a large scale. Chief among these difficulties are a geographical mismatch between resource supply centers and energy demand centers, a lack of adequate infrastructure to transport the generated power, unintended effects of policy, misaligned incentives, and large costs despite government support. Furthermore, coal generation facilities are characterized by long ramp-up times and therefore provide little flexibility to grid operators in balancing the variability of renewable sources. This will be an especially large problem for China in its future integration of renewables. Increasing the proportion of natural gas and hydropower, two sources with short start-up times, can help balance out variability as more renewables are added to the mix.

Careful policy design and implementation are also important. In any industry that is not yet cost-competitive, government support helps nascent efforts and shapes outcomes. Many policies currently create unintended effects, as exemplified by the current incentive misalignment between power generators and grid operators and between provincial and regional governments. Even assuming that policies have their intended effects when implemented, enforcement is also crucial to the continued development of alternative energy in China.

This chapter serves as a continuation of Chapter 4 and will provide an overview of the cleaner energy alternatives that China currently supports. Each section will detail a specific energy source and will outline dominant technologies, the role that the source is expected to play in the national energy mix, idiosyncratic advantages and disadvantages, and how the government is currently fostering or planning to promote the source. The chapter concludes with a discussion of the current status of the alternative energy industry in general and its prospects.

5.2 NATURAL GAS

Natural gas is much more energy-efficient and causes much less environmental problems than other fossil fuels, such as coal and oil. The capital costs to build natural gas power plants are also much lower than those of coal, and these plants can be started up and shut down faster than their coal counterparts. Due to these reasons, China views natural gas as a significant player in improving its energy mix and has spent extensive effort encouraging expanded utilization across the country, especially in residential use in urban areas. These efforts have proven very effective in increasing the adoption of natural gas, and it is expected that this resource will capture a growing share of China's energy matrix going forward.

5.2.1 Natural Gas Supply

China's known natural gas reserves are estimated at 56 trillion cubic meters (Wang et al. 2011). However, different entities report varying figures based on whether the reserve is proven, economically recoverable, or theoretically present (Brennand 2001). According to the EIA, China's proven reserves in 2011 were 107 trillion cubic feet, ranking 13th in the world (U.S. EIA 2013). The historical trend of estimated reserves is provided in Figure 5.1.

The reserves are located mostly in offshore basins—such as Bohai Bay, the Pearl River Delta, the East China Sea, and the Gulf of Tonkin—and in inland basins in Xinjiang, Inner Mongolia, Qinghai, and Sichuan. The 13 largest basins contain more than 80% of China's total natural gas reserves (China Baike 2011).

Commercial natural gas is a mixture of hydrocarbon and nonhydrocarbon gases, in which the combustible share of the gas exceeds 85%. It usually occurs as an associated gas together with oil reservoirs or as a nonassociated gas in large deposits on its own (Neiva and Gama 2010).

In response to the increasing demand for natural gas as a cleaner energy alternative, the production of natural gas in China has increased rapidly, from 11 million cubic meters in 1949 to 85 billion cubic meters in 2009 (Dai et al. 2008; National Bureau of Statistics of China [NBSC] 2010). As noted

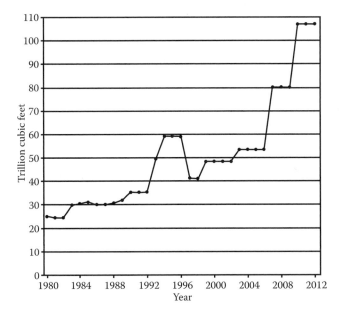

FIGURE 5.1
The development of China's natural gas reserves over time. (Reprinted with permission from U.S. EIA. China Country Analysis Brief, 2013. http://www.eia.gov/countries/country-data.cfm?fips=CH#ng.)

in Figure 5.2, most of this increase has occurred in recent years. There are about 60 natural gas companies in China, but the market is dominated largely by the "big three"—China National Petroleum Corporation Group (CNPC), China Petroleum and Chemical Corporation Group (Sinopec), and China National Offshore Oil Corporation Group (CNOOC).

Although known natural gas resources are abundant, the fact that many of these reserves are difficult to exploit represents one of the major challenges for China. Another barrier is the geographical disparity between where the gas is located and where the energy is demanded, as well as the lack of transportation infrastructure. The vast majority of China's natural gas reserves (84%) are located in its central (30%) and western (28%) regions, far from major markets in the more developed and populated southern and eastern areas of the country (China Center for Energy and Development [CCED] 2012). In addition, natural gas has a low energy density compared with coal and oil, which makes transportation difficult and requires that supply and demand sites be connected directly by pipelines.

Some pipelines are already in place or under construction to connect major cities to supply centers, as shown in Figure 5.3. The West–East Gas Pipeline, from the Tarim Basin in Xinjiang to Shanghai, which passes through 10 major provinces, is a multibillion dollar 4200-km project that transports 20 billion cubic meters of gas annually to support the Yangtze River Delta area in eastern China (Brennand 2001). However, in many small-sized to medium-sized cities, the demand for natural gas is

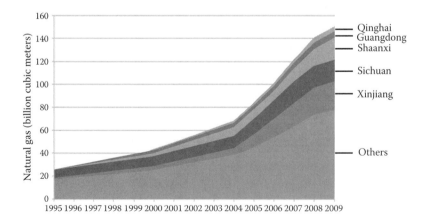

FIGURE 5.2
Chinese natural gas production by region. (Adapted from NBSC. *China Energy Statistical Yearbook 2010.* Beijing: China Statistics Press, 2011.)

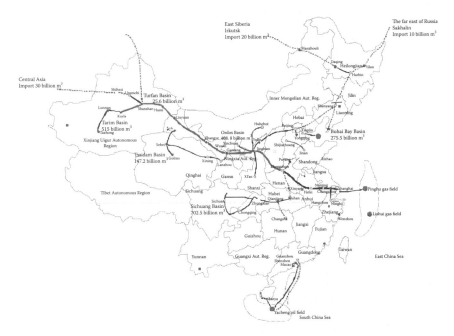

FIGURE 5.3
Existing and planned natural gas pipelines in China. (Reprinted with permission from Institute of Energy Economics Japan (IEEJ). *Natural Gas in China*, by Kaoru Yamaguchi and Keii Cho. Tokyo: IEEJ, 2003. http://eneken.ieej.or.jp/en/data/pdf/221.pdf.)

not great enough to make pipeline construction economically feasible; consequently, the infrastructure remains inadequate there.

5.2.2 Natural Gas Demand

Natural gas is considered a cleaner, more efficient energy resource than other fossil fuels because it generates less greenhouse gas and causes less environmental pollution. Methane's carbon–hydrogen ratio—which measures the ratio of the mass of carbon to the mass of hydrogen in a molecule—is 3.0, whereas it is 95 for coke, 31 for anthracite, 8.3 for wood, and 6.5 for diesel. Methane's calorific value—the amount of energy released when a defined volume of gas is burned under specified conditions—is double that of coal gas and eight times that of producer gas (Brennand 2001). Natural gas emits 58% less CO_2, 79% less NO_x, and much less SO_2 than coal. In addition, constructing natural gas power plants costs only two-thirds as much as equal-sized coal plants. Thus, natural gas is a promising alternative energy source (Zhou 2010).

These advantages related to improved health and environmental effects have led the Chinese government to promote the use of natural gas, especially as the gas of choice for urban settings. The government has encouraged city residents to use natural gas in daily energy-consuming activities such as cooking. In addition, the *Policies on Natural Gas Use*, issued in August 2007 by the National Development and Reform Commission, identified urban natural gas use as its most favored expansion area (National Development and Reform Commission [NDRC] 2007). As a result, many municipalities have begun to replace traditional residential energy sources, such as coke, with natural gas. As illustrated in Figure 5.4, residential gas consumption has increased dramatically, from only 1.9 billion cubic meters in 1995 to 17.8 billion cubic meters in 2009. Use of natural gas in the industrial sector has also grown—although not as drastically in percentage terms—from 15.4 billion cubic meters in 1995 to 57.8 billion cubic meters in 2009. In the industrial sector, natural gas is used primarily as a raw material for fertilizer production. This particular use accounts for more than half of the natural gas consumption in the industrial sector, which is followed by the electricity generation sector, which accounts for 23%. China is the fourth-largest natural gas consumer in the world after the United States, Russia, and Iran (NBSC 2011).

Due in large part to this governmental support, the proportion of natural gas in China's energy mix has grown from 2.0% in 1990 to 4.3% in 2008. This share is expected to reach 10% by 2020, representing a total annual consumption of 200 billion cubic meters (Shi et al. 2010).

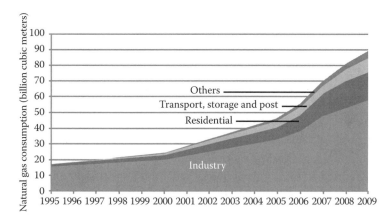

FIGURE 5.4

Natural gas consumption by sector over time. (Adapted from NBSC. *China Energy Statistical Yearbook 2010*. Beijing: China Statistics Press, 2011.)

The adoption of natural gas is also widespread because most regions have experienced a dramatic increase in natural gas consumption over the last two decades. Beijing, for example, began to switch from coal to natural gas in the 1990s, and total consumption grew from 0.08 billion cubic meters in 1990 to 6.94 billion cubic meters in 2009—an 80-fold increase. Guangdong, Shanghai, and many other regions that did not consume natural gas at all in 1990 are now the top-consuming areas in China. This growth can be attributed to the discovery of new natural gas reserves and the construction of pipeline infrastructure, as well as the government's encouragement of greater usage of natural gas (Shi et al. 2010). It is expected that natural gas consumption in China will continue to play an increasingly significant role. The largest natural gas consumption regions are shown in Table 5.1.

To date, the Chinese government has been controlling the price of natural gas. Thus, an open market has not yet developed. To encourage the adoption of natural gas, the government has been setting its domestic price at a level lower than the international market price. To facilitate a more mature and developed natural gas industry, China will gradually transition to an open market, allowing prices to be determined by supply and demand (Ministry of Commerce of the People's Republic of China: Department of Foreign Trade [MOFCOM: DOFT] 2013).

Until recently, China's natural gas consumption was fueled by domestic producers such as Sichuan Province. However, with the aforementioned growth in demand, the country became a net importer of natural gas in 2007, with net imports growing exponentially from 1.42 billion cubic meters in 2007 to 16.61 billion cubic meters in 2010 (MOFCOM: DOFT

TABLE 5.1

Regional Breakdown of China's Natural Gas Consumption (2010)

Region	Natural Gas Consumption (billion cubic meters)	Total (%)
Sichuan	12.70	9
Guangdong	11.29	8
Beijing	6.94	5
Xinjiang	6.79	5
Jiangsu	6.34	5
Others	92.44	68
Total	136.50	

Source: Adapted from National Bureau of Statistics of China. *China Energy Statistical Yearbook 2010.* Beijing: China Statistics Press, 2011.

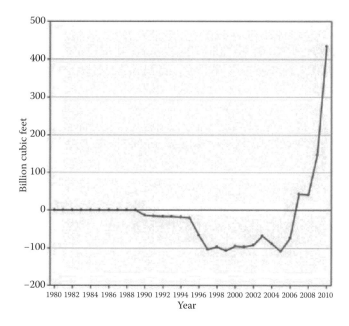

FIGURE 5.5

China's net natural gas imports over time. (Reprinted with permission from U.S. EIA. China Country Analysis Brief, 2013. http://www.eia.gov/countries/country-data.cfm? fips=CH#ng.)

2011; Figure 5.5). In 2012, China imported 40.8 billion cubic meters, an increase of 29.9% over the previous year (MOFCOM: DOFT 2013). Turkmenistan and Kazakhstan provide the bulk of China's natural gas imports through the Central Asia Gas Pipeline (CAGP, Turkmenistan-Uzbekistan-Kazakhstan-China; Cutler 2011).

5.2.3 Liquefied Natural Gas

Liquefied natural gas (LNG) is the liquid form of natural gas, which is produced by lowering the temperature of the gas to −162°C while subjecting it to intense pressure. LNG's volume is only 1/600 that of normal natural gas, presenting an effective solution to natural gas' low energy density problem. This allows for the storage and transportation of natural gas. LNG is also an attractive option for importing natural gas because it bypasses the restriction of cross-border pipelines, allowing buyers to choose from a wider range of suppliers (Figure 5.6; Shi et al. 2010).

In light of the growing demand for natural gas, China has expanded its LNG industry significantly, despite the product's relatively high cost. The

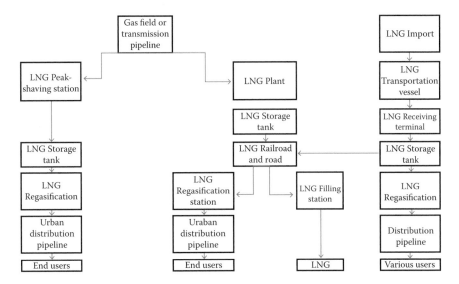

FIGURE 5.6
China's LNG value/transportation chain. (Adapted from Shi, G.H. et al., *Energy Policy* 38: 7457–7465, 2010.)

ability to transport LNG using a number of different modes (e.g., ships, trains, and trucks) frees this energy source from expensive pipelines, representing a major advantage over conventional natural gas. LNG can be transported via railway tankers or road tankers and then gasified and distributed to individual customers through local pipelines (Shi et al. 2010). As a result of this advantage of flexibility, it is now of increasing importance to a range of Chinese cities and plays a central role in improving their energy mixes. Existing LNG terminals are all located in coastal areas (as illustrated in Figure 5.7) and are responsible for 90% of China's total LNG imports (CNPC 2010).

The Dapeng LNG receiving terminal in Shenzhen, Guangdong formed a 25-year contract with Australia in June 2006 through the North West Shelf Venture, under which the terminal imports 3.7 million tons of LNG annually (North West Shelf Venture 2010). After the LNG is received and gasified, 65% is used for power generation and the rest is consumed as urban gas (Shi et al. 2010). CNOOC owns a 3% stake in the venture and is the largest shareholder. CNOOC is also operating the receiving terminal in Putian, Fujian, where it has been importing 2.6 million tons of LNG annually from Indonesia since 2008 for both power generation and residential purposes. Shanghai's terminal at the Yangshan deepwater port imports 1.1 million tons per year of LNG from Malaysia, supplying natural gas to the city's gas

LNG Terminals
In operation
1. Dapeng terminal
2. Putian terminal
3. Shanghai terminal
Building
4. Dalian terminal
5. Jiangsu terminal
Proposed
6. Tangshan terminal
7. Shandong terminal
8. Zhejiang terminal
9. Zhuhai terminal
10. Hainam terminal
LNG Plants
In operation
11. Shanghai plant
12. Zhongyuan plant
13. Guanghui plant
14. Fushan plant
15. Jianwei plant
16. Tianli plant
17. Weizhou plant
18. CNOOC plant
19. Taianplant
20. Xining plant
21. Longquan plant
22. Erdos plant
23. Shuntai plant
24. Shitai plant
Building
25. Anyang plant
26. Hefei plant
27. Jincheng plant
28. Yinchuan plant
29. Shuntianda plant
30. SK plant
31. Kelin plant
32. Ansai plant
33. Lanzhou plant
34. Hanas plant

FIGURE 5.7
Sites of LNG terminals and plants in mainland China. (Reprinted with permission from Shi, G.H. et al., *Energy Policy* 38: 7457–7465, 2010.)

grid. Additional projects are under development to import more LNG from Australia and Qatar. Sinopec, for example, is participating in the Australia Pacific LNG project and plans to import 4.3 million tons of LNG per year beginning in 2015 (Reuters 2011). It is estimated that China will consume 31 million tons of LNG per year by 2015 (Bloomberg News 2011).

5.2.4 Unconventional Gas Resources

5.2.4.1 Coalbed Methane

China is also looking into tapping its reserves of unconventional gas resources. The country hopes that these resources will improve the environmental friendliness of its energy mix while simultaneously contributing to greater energy security. The recent rapid escalation in the demand for natural gas, in particular, has provided the Chinese government with an incentive to add unconventional reserves to its existing exploitable resources.

The unconventional gas source to be discussed in this section, coalbed methane (CBM), is natural gas trapped in China's coal mines. China has 36.8 trillion cubic meters of CBM at depths of less than 2 km, of which 10.9 trillion cubic meters are readily recoverable at depths of less than 1.5 km (Standard Chartered Bank 2010). This level of reserves is the third-largest in the world. The resource is located mostly in northern and northwestern China, with a substantial portion in the Ordos Basin (China United Coalbed Methane Corporation, Ltd [CUCBM] 2009). CBM may not only provide a clean and viable energy source but may also reduce coal mine accidents caused by gas explosions, as well as greenhouse effects due to less CBM escaping into the atmosphere. China's CBM production was 11.5 billion cubic meters in 2011 and was expected to increase to 15.5 billion cubic meters in 2012 (China Daily 2012), with a planned increase to 21 billion cubic meters by 2015 (Zhu 2011).

5.2.5 Interim Conclusion

Most of the expansion of China's natural gas production occurred in the last decade. Before the construction of the west–east gas pipeline in 2004, the capital costs of adopting natural gas as an energy source were prohibitive. In addition, due to natural gas' much smaller domestic supply, it has historically been more expensive than coal. Recently, however, the government's encouragement of the industry and the construction of natural gas pipelines have led to its adoption on a greater scale.

In fact, the annual growth rate of gas consumption in China has exceeded GDP growth, demonstrating the growing importance of natural gas in the energy mix. The residential sector has been the main contributor to this growth in consumption, and increasing penetration in this sector is currently a government priority. Thus, this sector will likely continue to drive growth. By 2020, increasing demand is expected to result in a supply–demand gap of 40 to 80 billion cubic meters (International Energy Agency [IEA] 2009).

Price reform will be crucial to ensure that this demand can be met in a non–market-distorting manner. Domestic gas prices should align with international prices—a change that will promote energy efficiency and conservation in China. It will also ensure that the country's energy sector can import gas where economical and cover its cost of doing so, rather than having to purchase expensive foreign gas and selling it at artificially depressed domestic prices. An alignment of prices will also accelerate the development of domestic production and make the natural gas market more attractive for exporters. Currently, a falling natural gas price in the international market has provided the perfect opportunity for price reform (IEA 2009).

On a separate note, China seems to be accelerating its investments in and acquisitions of energy resources abroad, taking advantage of its large reserves of foreign currency and the depressed prices of assets resulting from the global economic slowdown. The country's investments in oil and gas projects in geopolitically sensitive areas could contribute to an increase in international tensions (IEA 2009). Thus, the vast reserves of CBM also offer a promising domestic opportunity for China to reduce carbon emissions and increase energy security going forward.

The low cost of coal, the fluctuating price of oil, and the lack of adequate infrastructure continue to present barriers to development. However, the attractive characteristics of natural gas make it a viable alternative energy resource. Therefore, expanding natural gas as a proportion of the energy mix can lower overall emissions and facilitate the integration of various renewables into the energy grid. It is expected that natural gas will play a significant role in China's future development.

5.3 HYDRO

Hydropower is a mature, commercially proven technology. In 2010, hydropower already made up the largest proportion (17%) of China's energy mix after fossil fuels (Deutsche Bank Group 2011). Unfortunately, the hydropower industry is still beleaguered by a number of environmental, economic, technological, diplomatic, and social concerns. Hydropower can function both as base-load power and as storage. Thus, it can be very useful for the expansion of the clean energy sector in China by aiding in the integration of variable renewables into the energy grid.

5.3.1 Characteristics of Hydropower

Hydropower is an exceedingly reliable and flexible power source. Unlike variable wind and solar power, it can provide base-load power (Deutsche Bank Group 2011). It can also adjust its power output level to meet load fluctuations minute-by-minute, and hydropower dams with large storage reservoirs can be used to store energy to meet peak demand—over days, weeks, months, seasons, or even years, depending on the size of the reservoir (International Renewable Energy Agency [IRENA] 2012a). These capabilities make it particularly useful for allowing the large-scale penetration of variable power sources, such as wind and solar. When the source of variable power is abundant, reservoir levels build up and allow energy to be stored for future use, such as when the wind stops blowing. In addition, when large changes in supply are needed due to fluctuations in variable power generation, hydroelectric generating units can start up very quickly and operate efficiently almost instantly. Hydropower offers the only large-scale and cost-efficient storage technology available today. Thermal plants take several hours to start up, and their response time is not nearly as fast. Because of these useful properties, systems with significant shares of large-scale hydropower and reservoir storage potential will be able to integrate higher levels of variable renewables at a lower cost than other systems (ibid.).

In China, hydropower technology provides flood control and irrigation services in addition to power generation. Although hydropower is generally CO_2-free, there are greenhouse gas (GHG) emissions associated with dam construction, silting in the reservoirs, and the decomposition of organic material in tropical regions caused by the creation of new reservoirs. In addition, hydropower plants can often affect river flow, water quality, and biodiversity, while also displacing local populations and affecting fish migration. Therefore, ideally, social and environmental impact assessments must be conducted and negative effects on local populations, ecosystems, and biodiversity should be mitigated in the project plan (IRENA 2012a).

5.3.2 Hydropower Technology

Hydropower is based on the simple concept of using the energy of flowing water. Figure 5.8 shows the structure of a typical low-head hydropower plant with storage ("head" refers to the vertical distance the water falls; Brown et al. 2011). The main components of a conventional hydropower plant are the dam, intake, surge chamber, turbine, generators, transformer, transmission lines, and outflow. The dam creates a water reservoir that can be used as storage by holding back the water. Some dams also incorporate a desilter to cope with sediment buildup behind the dam. The dam's gates

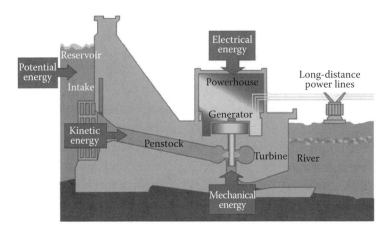

FIGURE 5.8

Typical "low head" hydropower plant with storage. (Reprinted with permission from IRENA. *Renewable Energy Technologies: Cost Analysis Series: Hydropower.* United Arab Emirates: IRENA, 2012. http://www.irena.org/DocumentDownloads/Publications/RE_Technologies_Cost_Analysis-HYDROPOWER.pdf.)

open and gravity conducts water through the penstock to the turbine. Sometimes there is a headrace before the penstock and a surge chamber or tank is used to reduce surges in water pressure that could potentially damage or lead to increased stresses on the turbine. The turbine is attached to a generator by a shaft, and converts energy from falling water to rotating shaft power (IRENA 2012a).

As the turbine blades turn, the rotors inside the generator also turn, and electric current is produced as magnets rotate inside the fixed-coil generator to produce alternating current (AC). The transformer in the powerhouse then converts the AC voltage into a higher-voltage current for more efficient long-distance transport, resulting in lower losses. The transmission lines send the electricity to a grid connection point or directly to a large industrial consumer, where the electricity is converted back to a lower-voltage current to be used by consumers.

The used water itself is carried out through pipelines, called tailraces, and re-enters the river downstream. The outflow system may also include "spillways" that allow the water to bypass the generation system and be "spilled" at times of good or very high inflows and reservoir levels (IRENA 2012a). Even though the water is returned to the river, this process may still have a significant effect on the river system in the area (ibid.).

The three main types of hydropower schemes are run-of-river, storage, and pumped storage. Run-of-river schemes use the natural flow of a river, with electricity generation depending on the timing and size of river flows. Storage schemes use dams to decouple generation from hydro inflows (IRENA 2012a). Pumped storage schemes involve two reservoirs. At times of low demand and low electricity prices, electricity (produced by outside sources) is used to pump water from the lower basin to an upper basin. The water is released to create power when demand and prices are high, improving storage capacity and providing grid flexibility (Brown et al. 2011).

5.3.3 Construction and Maintenance

Hydropower plants typically last between 30 and 80 years. The most time-consuming and effort-intensive components of a hydropower project to build are the dam, water intake, headrace, surge chamber, penstock, tailrace, and powerhouse (IRENA 2012a). There are two types of turbines: "impulse" and "reactionary." The former extracts energy from the momentum of flowing water, whereas the latter extracts energy from the pressure of the water head (ibid.). The most widely used reactionary turbine is the Francis turbine, as it is suitable for a variety of head and flow rates. The Kaplan turbine is another reactionary type that is used for lower heads (between 10 and 70 m). The most commonly used type for higher heads is the Pelton turbine, an impulse turbine. Another impulse turbine that is less efficient, but also less dependent on discharge, is the cross-flow turbine, also referred to as the Banki–Mitchell or Ossberger turbine (ibid.).

5.3.4 Water Supply

China has the largest supply of hydropower resources in the world, with estimates for total possible capacity of approximately 400 GW and an annual generating capacity ranging from one to two million terawatt hours (TWh). To put these numbers into context, China's hydropower resources are estimated to account for one-sixth of the world's total (Mirae Asset Financial Group 2012). Figure 5.9 illustrates the global hydropower technical resource potential, which is calculated as the proportion of the theoretical potential that could be converted to hydropower with today's technologies (IRENA 2012a).

These resources are concentrated primarily in southwest China, specifically in Sichuan, Yunnan, Tibet, and Guizhou. There are also ample resources in South Central China and in North China (Huang and Zheng 2009). Figure 5.10 displays the distribution of these resources, as well as

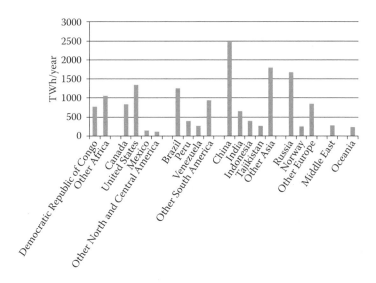

FIGURE 5.9

The world technically feasible hydropower resource potential. (Adapted from IRENA. *Renewable Energy Technologies: Cost Analysis Series: Hydropower.* United Arab Emirates: IRENA, 2012. http://www.irena.org/DocumentDownloads/Publications/RE_Technologies_Cost_Analysis-HYDROPOWER.pdf.)

the provinces with highest energy demand, and highlights the geographic mismatch between supply and demand centers.

As of 2008, only 21.5% of China's total hydropower resources were being exploited, much lower than the worldwide average level of development. Therefore, although hydropower already made up 20.36% of the total installed electricity generation capacity in the country in 2007, there is still enormous room for this proportion to expand (Huang and Zheng 2009).

5.3.5 Hydropower Adoption and Growth

The first hydropower plant was built in China as early as 1912, but hydropower resources were not utilized consistently until after the People's Republic of China was established in 1949 (Huang and Zheng 2009). The pace of development picked up even more after the liberalization of the economy around 1970 (Figure 5.11). Table 5.2 shows a tremendous annual average growth rate of 12.4% in installed capacity from 0.163 GW in 1949 to 145.26 GW in 2007, and an annual average energy generation growth rate of 11.9% from 0.71 TWh in 1949 to 486.7 TWh in 2007. The share of hydropower in total installed capacity increased from 8.8% to 20.36% between 1949 and 2007, but decreased from 16.5% to 14.95% in total energy

The People's Republic of China

FIGURE 5.10
The distribution of economically exploitable hydro resources in China. (Reprinted with permission from Deutsche Bank Group: DB Climate Change Advisors. *Hydropower in China: Opportunities and Risks*, by M. Carboy, C. Sharples, L. Cotter, and J. Cao. Frankfurt am Main, Germany: Deutsche Bank AG, 2011. http://www.banking-on-green.com/flash/Hydropower_in_China-Opportunities_and-Risks.pdf.)

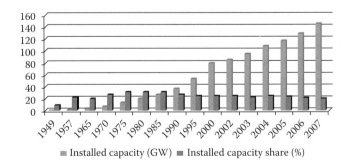

FIGURE 5.11
Growth of hydropower installed capacity. (Adapted from Huang, H. and Y. Zheng, *Renewable and Sustainable Energy Reviews* 13: 1652–1656, 2009.)

TABLE 5.2

Growth of Hydropower in China during the Past 58 Years

Year	Installed Capacity (GW)	Installed Capacity Share (%)	Energy Generation (TWh)	Energy Generation Share (%)
1949	0.163	8.8	0.71	16.5
1957	1.019	22	4.82	24.9
1965	3.02	20	10.41	15.4
1970	6.235	26.2	20.46	17.7
1975	13.428	30.9	47.63	24.3
1980	20.318	30.8	58.21	19.4
1985	26.42	30.4	92.4	29
1990	36.05	26.1	126.7	20.4
1995	52.18	24	186.8	18.6
2000	79.35	24.9	243.1	17.8
2002	84.56	24	271	16.5
2003	94.9	21.25	281.3	14.76
2004	108.26	24.57	328	15
2005	116.52	22.9	395.2	16
2006	128.57	20.67	416.7	14.7
2007	145.26	20.36	486.7	14.95

Source: Adapted from Huang, H. and Y. Zheng, *Renewable and Sustainable Energy Reviews* 13:1652–1656, 2009.

generation. By 2005, the country boasted the world's largest installed capacity and energy generation of hydropower (Huang and Zheng 2009).

5.3.6 Large Hydropower Plants

Large hydropower plants offer both flexibility to and storage capabilities for the energy grid. Typically, these plants are classified by generating capacity and their type of generation scheme (run-of-river, storage, or pumped storage). Classifying hydropower by head is useful because head determines water pressure on turbines, which, together with discharge, is the most important parameter for determining what type of hydraulic turbine to use. There are no standard definitions, but a general guiding definition for large hydropower is a hydropower plant with 100 MW or more of capacity feeding into a large electricity grid (IRENA 2012a).

Large hydropower plants with storage can decouple the timing of hydropower generation from variable river flows. Combined with pumped storage stations, such plants represent the only low-cost, large-scale energy storage mechanism. To operate as a base-load plant, the plant needs to be designed with a very large reservoir relative to the size of the plant. If hydropower capacity exceeds the amount of reservoir storage, the plant becomes a peaking plant, and can serve the function of generating large quantities of electricity to meet peak electricity demand. Hence, large hydropower plants can be very flexible tools to meet electricity demand quickly and to offset variable electricity generation. The economic lifetime of a large hydropower plant is generally between 40 and 80 years (IRENA 2012a).

Prior to 2005, China had 21 hydropower plants in operation with capacity exceeding 1000 MW, for a gross installed capacity of 39.73 GW, representing approximately 34.2% of China's cumulative hydropower capacity. Currently, 1182 large-scale and middle-scale hydropower plants are under construction with a gross installed capacity of 92.5 GW. China has planned 13 hydropower bases with a cumulative capacity of 275.77 GW. Based on factors such as hydropower resources, submergence, and construction costs, China gives priority to the development of the projects on the Yellow River Upper Reaches, the Hongshuihe River, the Yangtze River Upper Reaches, and the Wujiang River (Huang and Zheng 2009).

During the 11th Five-Year Plan period, China planned to focus on hydropower bases on the Yangtze River, the Yellow River, the Dadu River, and the Mekong River. By the end of 2010, these four bases were 78.7%, 34.4%, 25.6%, and 23.3% completed, respectively, with the total development level across all hydropower bases at approximately 25.3%, as shown in Table 5.3. During the 12th Five-Year Plan period, China plans to focus on hydropower development in southwestern areas with abundant resources, including the Jinshajiang River, the Yalongjiang River, the Dadu River, and the Mekong River. These bases are projected to reach development levels of 29.0%, 57.3%, 68.7%, and 57.4%, respectively, bringing the total development level of hydropower bases up to 49.3%. During the 13th Five-Year Plan period, China plans to concentrate on developing bases on the Jinshajiang River, the Mekong River, and the Jujiang River, with the development level of these bases increasing to 54.8%, 76.6%, and 33.6%, respectively, and further increasing total exploitation to 70.3%. During the 14th and 15th Five-Year Plan periods, hydropower construction is expected to be concentrated in Tibet and Xinjiang (Ming et al. 2013).

The flagship of the Chinese hydropower projects is the Three Gorges Project, the largest hydropower project in the world. China began construction in 1994 after about 70 years of debate and deliberation (Huang

TABLE 5.3

Development Plans of 13 Hydropower Bases in China

Hydropower Base Name	Hydroelectric Potential	2010		2015		2020	
		Exploited	Development Levels (%)	Exploited	Development Levels (%)	Exploited	Development Levels (%)
Jinshajiang River	5858	180	3.1	1700	29	3210	54.8
Yalongjiang River	2531	340	13.4	1450	57.3	1850	73.1
Daduhe River	2460	630	25.6	1690	68.7	2140	87
Wujiang River	1079	358	33.2	850	78.8	1010	93.6
Yangtze River upper reaches	3320	2612	78.7	2750	82.8	2830	85.2
Nanpanjiang River and Hongshuihe River	1431	498	34.8	918	64.2	1192	83.3
Lancangjiang River	2560	597	23.3	1470	57.4	1960	76.6
Yellow River, upper reaches	2003	690	34.4	1250	62.4	1400	69.9
Yellow River, main	641	163	25.4	343	53.5	579	90.3
West Hunan	590	176	29.8	310	52.5	519	88
Fujian and Zhejiang and Jiangxi	1092	330	30.2	567	51.9	845	77.4
The Northeast	1869	373	20	802	42.9	1131	60.5
Nujiang River	2142	18	0.8	300	14	720	33.6
Total	27,576	6965	25.3	14,400	49.3	19,386	70.3

Source: Adapted from Ming, Z. et al., *Renewable and Sustainable Energy Reviews* 20: 169–185, 2013.

and Zheng 2009), and completed the project in 2009. The dam was a success, as annual generation reached the projected designed capacity of 84.7 billion kilowatt-hours (kWh; IBISWorld 2011). The site is in Sandouping, Yichang, in Hubei Province. The massive project comprised three parts: the construction component, the reservoir component, and the transmission component. The construction component included the dam, the powerhouses of the hydropower plant, the ship lock, and the ship lift. The normal pool level for the reservoir is 175 m, with the dam crest level at 185 m. Its total storage capacity is 39.3 billion cubic meters, of which 22.15 billion cubic meters is the flood control capacity. The hydropower plant has 26 hydroturbine generating units with 700 MW of nominal capacity each, totaling to an installed capacity of 18.2 GW and an annual average energy generation of 84.68 TWh. The total investment in the project was roughly ¥180 billion (US$29 billion) (Huang and Zheng 2009).

Large hydropower plants use advanced technologies and equipment. Leading international suppliers provide much of the technology, equipment, computer software, and construction contracts. Although foreign suppliers are still more competitive in technology and equipment manufacturing, recently, domestically manufactured hydropower generators of large capacity (>500 MW) have been purchased. Domestic suppliers that have started to become competitive with foreign ones include Harbin Electrical Machinery Works and Dongfang Electrical Machinery Works (IBISWorld 2011).

5.3.7 Small Hydropower Stations

The Chinese government has identified small hydropower as not only a way to reduce GHG emissions, but also as a means of alleviating poverty through installation in rural areas (Wang and Tseng 2012). Approximately one-third of China's counties rely on small-scale hydropower as their primary source of electricity (IBISWorld 2011). Definitions of what constitutes a small hydropower plant differ by country, as shown in Table 5.4 for countries with the greatest hydropower generation.

In China, any plant with less than 50 MW of generating capacity is considered small. Small hydropower is very cost-effective when located close to energy consumers or existing transmission lines. Although the construction of small hydropower proceeds more quickly than that of large-scale hydropower, the planning and approval processes are often

TABLE 5.4

Definition of Small Hydropower by Country (in Megawatts)

	Small Hydropower Definition (MW)
Brazil	≤30
Canada	<50
China	≤50
European Union	≤20
India	≤25
Norway	≤10
Sweden	≤1.5
United States	5–100

Source: Adapted from IRENA. *Renewable Energy Technologies: Cost Analysis Series: Hydropower.* United Arab Emirates: IRENA, 2012. http://www.irena.org/DocumentDownloads/ Publications/RE_Technologies_Cost_Analysis-HYDROPOWER.pdf.

very similar. Despite long lead times, the costs of very small hydropower projects seem to be lower than those of PV systems, making them a viable option for off-grid electrification. Two types of generators can be used in small hydropower plants: asynchronous (induction) and synchronous. The former are generally used for microhydroprojects, whereas the latter are used for larger small hydropower projects. Microhydropower projects typically provide power for small communities or rural industries in remote areas. Larger projects (1–20 MW) usually feed into a grid (IRENA 2012a). The typical economic lifetime for such projects is 40 years, but is sometimes less (ibid.).

The technically exploitable capacity of small hydropower in China is estimated at 128 GW, representing the ability to generate, on average, 450 TWh/ year. In the small hydropower station segment, western China accounts for 67.7% of total capacity, central China accounts for 16.8%, and east China accounts for 15.6%. These stations are spread out in more than 1600 mountainous counties across the country. Development of small hydropower stations in these rural areas has flourished as the result of rural economic growth and governmental support, such as tax incentives (Huang and Zheng 2009). In 2000, China initiated the Western Development Program, an important goal of which is to develop hydropower resources in remote areas such as those along the Daduhe River, the Lancangjiang River, and the Yalongjiang River (ibid.). Since 1994, small hydropower plants have enjoyed a favorable value-added tax (VAT) rate of 6% (incurred by the end user), compared with large-sized and medium-sized hydropower plants that face a VAT of 17%. In some regions, no tax on small hydropower plants is required during the first 2 years of operation and a limited amount of tax can be deducted during the next 3 years. In other areas, the tax is levied first at a normal rate but is then either partially or fully rebated later for further investment (IBISWorld 2011). By the end of 2005, small hydropower projects had an installed capacity of 38 GW and generated 130 TWh, which was approximately 32.9% of the total hydropower generation that year. At that time, more than 40,000 small hydropower plants had been built and 653 rural counties had achieved preliminary electrification (Huang and Zheng 2009).

Despite the large number of small hydropower plants, however, smaller plants use poorer technology and have lower efficiency levels when compared with larger projects. These plants use inefficient operation models, face problems in high-water and low-water seasons, provide only intermittent electricity rather than constant base-load power, are accident-prone, and frequently endure suspensions of operations during maintenance. Relatively poor technology contributes to the high electricity-generating costs of these plants, leading to lower profitability levels. This has resulted in a large number of small plants suffering financial losses, despite their favorable tax statuses (IBISWorld 2011).

This profitability problem is also driven by small hydropower's higher average investment cost per kilowatt when compared with large plants, particularly for plants with capacities of less than 1 MW in which the specific electromechanical costs per kilowatt can dominate total installed costs. The investment costs per kilowatt of small hydropower plant projects will be lower if the plants have higher head and capacity potentials (IRENA 2012a).

Another disadvantage of small hydropower stems from its environmental effect. Competition among firms to demarcate rivers for building small stations has caused serious environmental damage (Wang and Tseng 2012) because these small dams can block species' migration routes, disrupt the natural flow of water and local ecosystems, and lead to harmful run-off during construction.

Some attribute the disadvantages of small hydropower to the reform of the Chinese electricity industry. This industry has evolved since the early 1990s into a "dual system." The system consists of China's centralized national power at the core and a decentralized group of government organizations and private enterprises owning the generation system at the periphery (Cherni and Kentish 2007). The crux of the argument states that the reform of the electricity industry bound the interests of state-owned enterprises (SOEs) with those of the provincial governments while binding the interests of lower levels of local governments with those of privately owned small power companies.

The SOEs and provincial governments built large dams along major rivers, whereas the local governments and privately owned small power companies built smaller dams along branches of major rivers. Without solid centralized planning, the rivers were cut into pieces to serve the divergent interests of individual parties, who did not necessarily seek the most effective and least environmentally harmful use of the rivers. Furthermore, the small hydropower plants in operation needed to be hooked up to national or regional grids controlled by SOEs. Unfortunately, the SOEs generally had little interest in buying unstable electricity generated by small hydropower stations and, therefore, did not buy it at all or bought it at a price point lower than cost. This last development has driven many companies out of the market in recent years (Wang and Tseng 2012).

5.3.8 Pumped Storage Stations

At the opposite end of the hydropower spectrum, pumped storage stations are large-scale energy storage devices that facilitate grid stability and the integration of variable renewables. Along with large-scale hydropower plants with reservoirs, they represent the most viable large-scale electricity storage options (IRENA 2012a). Nevertheless, these plants are more expensive to build than large hydropower plants with reservoirs, and it is also more difficult to find suitable sites for them.

Pumped storage plants generally operate by using off-peak electricity to pump water from a river or lower reservoir up to a higher reservoir for storage until its release during peak demand time. They can also use electricity at other times when additional generation helps reduce grid costs or improve system security. For example, in some cases, pumped hydropower demand can enable thermal power plants to generate in a more optimal load range, reducing the costs of providing spinning reserve (IRENA 2012a). By 2005, there were 11 pumped storage power stations in 10 provinces in China. Their gross installed capacity reached 6.4 GW, amounting to approximately 1.3% of the total installed capacity for electric power. As of 2009, 10 additional pumped storage power stations were under construction (Huang and Zheng 2009).

5.3.9 Hydropower Cost

Building hydropower requires substantial investments of capital and time due to the immense amount of effort needed for planning, design, and civil engineering works. The two major cost components for hydropower projects are the civil works and the electromechanical equipment. Civil works costs include infrastructure development outlays to access the site and project development costs. Infrastructure development expenses can contribute significantly to total costs if the site is far from the existing transmission lines or roads. Project development encompasses planning and feasibility assessments, environmental impact analyses, licensing, fish and wildlife conservation measures, development of recreational amenities, historical and archaeological mitigation, and water quality monitoring and mitigation. Civil works also include the expenditures associated with dam and reservoir construction; tunneling and canal construction; powerhouse construction; site access infrastructure; grid connection; engineering, procurement, and construction; and developer costs (IRENA 2012a).

Electromechanical equipment costs include the costs of turbines, generators, transformers, cabling, and control systems. As equipment costs do not depend on site characteristics, there is very little variation. There is, however, great variability in total investment costs due to the site-specific nature of infrastructure and development costs and the cost structure of the local economy. The lowest investment costs are associated with adding capacity to existing hydropower schemes or capturing energy from existing dams. The development of an entirely new scheme is typically more expensive (IRENA 2012a). Globally, large-scale hydropower project costs range from a low of US\$1000/kW to approximately US\$3500/kW. It is not unusual, however, for project costs to fall outside this range, so these numbers should be viewed as only a rough estimate (ibid.).

As the equipment and civil works become eroded by water flow or constant use, it becomes economical to refurbish plants to reduce increasing operation and maintenance (O&M) costs and to restore or boost generation capacity to or above its original level. Refurbishing costs fall into two categories. Under the first category, life extension, equipment is replaced on a "like-for-like" basis. These refurbishments will generally yield a 2.5% gain in capacity due to the worn-out state of the equipment prior to replacement. Under the second category, upgrades, increased capacity and efficiencies are incorporated into the refurbishment. These might lead to capacity gains of 10% to 30%, depending on the extent of the upgrades (IRENA 2012a).

5.3.10 Factors that Restrict Hydropower Development

Hydropower's expansion in China faces many challenges. First, because of extremely uneven temporal and spatial distributions of precipitation, river flows vary dramatically within individual years from the wet to dry seasons (and, by extension, between continuously or abnormally dry and wet years). As a result, the amount of reservoir storage capacity required to regulate river flows is large, causing significant investment and resettlement problems. Furthermore, uneven precipitation leads to the concentration of hydropower energy far from major urban centers, which entails a need for long-distance and extra–high-voltage transmission lines to move electricity from the west to China's energy-hungry east (Huang and Zheng 2009).

Also, the exploitation of hydro resources is often criticized for the effects of submergence of lands and the resettlement necessary to create reservoirs. Criticism has also been aimed at hydropower's negative effects on ecosystems (Huang and Zheng 2009). Sedimentation is a problem common to hydropower schemes with reservoirs. The sediments that build up behind dams can, over time, take up a significant amount of the original storage capacity, and soils cannot continue to refresh the river system downstream of the dam, which often has adverse effects on sustainable riparian vegetation and the continued use of lands for agriculture. Periodic flushing or dredging from reservoirs has been implemented successfully in China as a solution to this problem (IRENA 2012a).

Another environmental consequence of installing hydropower capacity is that it changes water quality. Dam installation can alter the amount of dissolved oxygen in the water, increase total dissolved gases, modify nutrient levels, change temperatures, and affect heavy metal levels both upstream and downstream. These problems are rarely serious enough to pose significant barriers to viability, and numerous mitigation measures can be utilized. For example, multilevel draw off works can ensure higher-quality water near the surface for usage, and autoventing turbines can oxygenate the water (International Hydropower Association [IHA] 2000).

Hydropower projects built on rivers that flow through multiple countries can also cause international disputes. For example, the Chinese dam on the Mekong (Lancang) River has had adverse effects on the river further downstream, generating discontent from Laos, Thailand, Cambodia, and Vietnam. More than 80 million people depend on this river for drinking water, fishing, transportation, and irrigation. The Chinese developed hydropower plants on the Mekong unilaterally and have declined to become members of regional institutions such as the Mekong River Basin Commission. The countries that the river flows through downstream of the Mekong have serious concerns about the dam but are restricted from serious counteraction because China is the uppermost riparian state and is the most politically powerful country in the basin. Detrimental environmental effects include the increased frequency and magnitude of landslides and earthquakes caused by construction, a reduction in nutrient and sediment deposits, sedimentation, and harm to aquatic life (Institute of Defence and Strategic Studies Singapore [IDSS] 2004).

Hydropower projects have a significant effect on fish and fishermen as well. Installation of hydropower capacity creates alterations in habitat quality and availability, changes in flow regime, and hindrances to fish passage. In particular, hydropower plants often store water during periods of high flow to use for energy generation later in the year. Unfortunately, this alteration of the natural river cycle can affect habitat stability during periods of spawning and incubation. Furthermore, fish will inevitably enter the water that passes through the turbines, particularly during migration season. These problems can be mitigated with behavioral systems that include methods and equipment to direct fish through bypass channels. These methods can keep more than 90% of the fish away from the dangerous hydropower turbines for some species. Nevertheless, native fish cannot always tolerate the changes in water quality and can experience increased mortality rates as a result (IHA et al. 2000).

5.3.11 Future Prospects

The National Development and Reform Commission of China *issued the Medium- and Long-term Renewable Energy Development Plan* in September 2007. The plan's goal is to reach 300 GW of gross installed capacity of hydropower by 2020. Of that, 225 GW will come from large-scale and middle-scale hydropower, whereas 75 GW will come from small hydropower (Huang and Zheng 2009). The plan actually sets forth a goal of adding 120 GW of installed capacity by 2020, implying a total installed capacity of 335 to 340 GW (Deutsche Bank Group 2011).

The Chinese government has also included specific targets for investment in its plans. To achieve its 2020 target of an additional 120 GW of installed capacity, the Ministry of Water Resources has decided to double the funds invested in the sector. No detailed background is included in the plans, but from 2011 to 2015, China will invest ¥1.8 trillion (US$281 billion), and by 2020 it will have invested a total of ¥4 trillion (US$625 billion). Expected annual expenditures will double from an average ¥200 billion (US$31 billion) to ¥400 billion (US$62.5 billion). The investments will focus on establishing early warning flood systems in 1100 counties, reinforcing 5400 to 6595 reservoirs, and harnessing 5000 or more rivers by 2021. The government plans to fund these projects by funneling 10% of the land sale taxes collected in the eastern provinces to the western provinces, which have smaller tax bases, even though some have commented that this funding mechanism is nothing but a first step because land availability is limited (Deutsche Bank Group 2011).

5.3.12 Interim Conclusion

China intends to exploit its untapped water resources further in the coming years. Beneath the successes heretofore, however, lie many challenges for the hydropower industry. As noted previously, dam construction can have serious environmental and social consequences. Nevertheless, hydropower's abilities to provide both base-load power and storage capacity can help integrate variable renewables into the energy grid, thereby presenting a strong argument for continued development.

Going forward, small hydropower capacity is expected to increase whereas investment in large-scale projects moderate (IBISWorld 2011). Currently, small hydropower suffers from high capital investment and poor technology. Furthermore, the lack of centralized analysis on how large and small projects taken together affect surroundings leads to dam construction that neither maximizes power generation and nor minimizes negative social and environmental impact. These mistakes will prove difficult to remedy. Furthermore, the attractiveness of small hydropower projects diminishes as wind and solar projects require less initial investment and can enable similar benefits to rural communities located far from the electricity grid. Future development relies on increased government support and more favorable market conditions (IBISWorld 2011).

Hydropower is a cumbersome industry, with high capital investment and many serious problems that are difficult to remedy. The government will continue to support the development of the industry going forward, with special encouragement for small hydropower projects. Nonetheless, hydro's ability to provide base power, mitigate the negative effects of other renewables' intermittency, and negligible carbon emissions pose benefits that are too great to ignore. Hydropower is currently the success story of clean energy in China, and can serve as a great base as China continues to move toward environmentally friendly and variable energy sources such as wind and solar.

5.4 WIND

Along with water, wind has been used since antiquity to generate power. It was an essential source until the invention of the steam engine in the nineteenth century. Unlike large hydropower but similar to other renewable forms of energy generation, wind power is highly variable, making integration into the grid challenging. Also, like other types of renewable energy, wind power generation requires substantial capital investment but entails little-to-no fuel cost. Wind is a very attractive power source from an environmental perspective but its generation intermittency places an implicit extra cost on every kilowatt-hour. In China, its adoption faces the hurdles of an inadequate power grid, little domestic innovation, and unintended policy consequences.

5.4.1 Overview of Wind Energy Technology

Modern wind power technology converts the kinetic energy of wind to mechanical power. The wind turns the wind turbine, which is connected to the drive shaft that provides mechanical energy to power the generator. A typical utility-scale wind turbine has three blades, a diameter of 80 to 100 m, and a capacity of 0.5 to 3 MW, and is part of a wind farm of 15 to 150 turbines, all connected to the grid. Wind farms are collections of wind turbines, along with roads for site access, buildings, and the grid connection point (IRENA 2012b).

Different types of wind power technologies can be classified by axis orientation (horizontal or vertical) and whether they are located onshore or offshore. The total generation is determined by wind speed, turbine height, and rotor diameter, and is constrained by the capacity of the turbine. Most modern large-scale wind turbines are horizontal axis wind turbines (HAWT) with three blades. HAWT are more aerodynamic than vertical axis wind turbines (VAWT) and consequently are used more frequently (IRENA 2012b).

The most common HAWT design concept is a three-bladed, stall- or pitch-regulated, horizontal axis machine operating at a near-fixed rotational speed. Wind turbines will typically begin generating electricity at a wind speed of 3 to 5 m/s, reach maximum power at 15 m/s, and generally cut out at a wind speed of approximately 25 m/s (IRENA 2012b).

There are two main methods for controlling the power output from the rotor blades: pitch control and stall control. The former involves adjusting the angle of the turbine's blades with a control system that has built-in braking. The latter utilizes the inherent aerodynamic properties of the blade to determine power output. Under this method, the twist and thickness of the rotor blades are designed so that turbulence occurs behind the blade whenever the wind speed becomes too high. The downside of this design is that the blade also minimizes the power output at higher wind speeds. Stall control machines also have brakes at the blade base (IRENA 2012b).

Available energy increases as a cubic function of wind speed, giving developers an incentive to site wind farms in areas with high average wind speeds. In addition, a fivefold increase in the height of a wind turbine can result in twice as much power because wind speeds are typically faster higher above the ground. "Smoother" air is considered more desirable, as turbulent air reduces output and can increase stress on structure and equipment, increasing structural fatigue and thus O&M costs. The maximum energy that can be harnessed by a wind turbine is proportional to the swept area of the rotors. So, by doubling rotor diameter, power output can increase by a factor of four. The average wind turbine's size has continued to grow for onshore turbines because larger turbines provide greater efficiency and economies of scale. They are, however, more complex to build, transport, and deploy (IRENA 2012b).

Typically, blades are manufactured from fiberglass-reinforced polyester or epoxy resin. Recently, carbon fiber has been incorporated into the design to reach the strength-to-weight ratio needed for the larger wind turbine blades now being developed. The turbine rotor and hub assembly spin at a rate of about 10 revolutions per minute (rpm), depending on turbine size and design. The hub is generally attached to a low-speed shaft connected to the turbine gearbox. The gearbox and all the main turbine components are housed in the nacelle, which is the main structure of the turbine and is made out of fiberglass. The gearbox converts the low-speed, high-torque rotation of the rotor to high-speed rotation (~1500 rpm) with low torque for input to the generator. The generator then converts the mechanical energy from the rotor into electrical energy. The controller monitors and controls the turbine and collects extensive operation data. A yaw mechanism ensures the turbine constantly faces the wind. Control systems are becoming increasingly advanced, and new developments have the potential to improve energy output substantially. The main turbine components are elevated by the tower, which is generally made of steel. The transformer steps up the voltage output from the generator to between 1 and 35 kV, depending on the requirements of the local grid. The transformer is often housed inside the tower (IRENA 2012b).

5.4.2 Policy

Because wind power is not cost-competitive with conventional power generation, the modern wind energy industry would not exist without government policies. Therefore, policy has shaped and continues to heavily influence this industry in China. The major elements of wind energy policy in China stem from several key pieces of legislation.

5.4.2.1 Medium- and Long-Term Development Plan

The *Medium- and Long-term Renewable Energy Development Plan*, passed in 2007, outlined goals for wind capacity installation. It called for installed grid-connected wind capacity of 5 GW by 2010 and 30 GW by 2020. By the end of 2010, China had already exceeded its 2020 goal. The Chinese government report, *Development Planning of New Energy Industry*, estimated that China's wind installed capacity would reach 200 GW by 2020 (GWEC 2010). To achieve this goal, wind energy installation will need to grow at a rate roughly equivalent to 16% per year. Very few countries are pursuing wind development on this scale (Wang et al. 2012).

5.4.2.2 The Renewable Energy Law

The Renewable Energy Law was enacted in February 2005 and took effect on January 2006. Its primary function was to establish an overarching legal framework for promoting renewable energy in China and a basis for setting national renewable energy targets, as informed by provincial energy plans (Wang et al. 2012). According to the law, the NDRC determines energy pricing and planning, the Standardization Administration of China administers technical standards and codes, and the Ministry of Finance regulates economic incentive mechanisms, such as tax breaks, subsidized loans, and research and development encouragement (China Wind Power Center [CWPC] 2013). Much of the law's language is deliberately vague, providing only guidelines and general principles to be applied in the more specific "implementation regulations" to be released later. This can be seen in Figure 5.12, which outlines the broad goals and framework of the law. Reportedly, the law was designed this way to allow for greater flexibility. Nevertheless, a few sections have been directly implementable. Power grid companies are required to connect all licensed renewable energy projects to the grid and to buy their generated energy. The additional costs associated with this requirement are covered in part by a renewable energy surcharge. The burden of the surcharge is distributed among consumers, excluding agricultural and other low-income communities (IEA 2011).

This connection and purchasing requirement is not strictly enforced. Currently, when local grids are saturated and cannot easily transmit the electricity to neighboring grids, companies will curtail electricity generated by wind. They choose to do so because on-grid prices for coal-fired plants are lower and because they do not wish to pay more to integrate wind power while increasing power system variability (IEA 2011). The Renewable Energy Law also authorized the establishment of feed-in tariffs (FiT) for renewable electricity, set the groundwork for cost-sharing mechanisms through the establishment of a special fund for renewable energy development, and called for much-needed national surveys of available renewable energy resources (Wang et al. 2012).

Renewable energy targets			
10% Renewable energy by 2010 15% Renewable energy by 2020		10 GW Wind by 2010 30 GW Wind by 2020	

Grid companies	Mandatory grid connection	Mandatory purchase of electricity	RE Surcharge levied on consumers

Project developers	Mandatory market share requirements	Tender-based pricing	Reduced VAT on power generation; Reduced income tax

Wind turbine manufacturers	R&D support	Reduced tax rates	Tax rebate on import of components

FIGURE 5.12

The present policy framework governing the wind power sector in China. (Adapted from CWPC. Renewable Energy Law of the People's Republic of China, 2012. http://www.cwpc.cn/cwpc/en/node/6548.)

The law was amended in 2009 to address three major problems that arose from the rapid development of the wind energy market. These challenges include (1) the lack of coordination among national and provincial level governments and among grid enterprises and power generation enterprises, (2) the inefficiencies resulting from renewable energy targets being formulated in terms of installed power generation capacity rather than electricity actually generated, and (3) the failure of grid enterprise to provide timely grid connections to renewable energy power generation facilities due to a lack of financial incentives and poor enforcement of existing mandatory grid connection regulations.

The amendment sets out more specific plans and requirements for the development and utilization of renewables. The amendment further mandates that the State Council energy authorities and national power supervisory institutions have the plans on record. These measures aim to improve coordination nationwide.

Furthermore, to address the second problem of inefficiencies resulting from misaligned goals and incentives, the 2009 amendment added a new provision requiring authorities to set a goal for the quantity of renewable electricity generated as a percentage of total generation. This important revision shifts the focus of renewable energy targets from energy capacity to energy generated (CWPC 2013).

Finally, to address the third problem of failure to establish grid connection, renewable energy development and utilization plans are now required to include a plan for the construction of supplementary electricity grid transmission capacity so that less energy is wasted. The amendment also mandates that renewable energy projects' grid connections conform to special technical standards and that power enterprises coordinate with grid enterprises to ensure the stability of the power grid. The hope is that these provisions will improve the utilization rate of renewable energy while supporting the grid companies so they will eventually come to promote renewables (CWPC 2013). Although the language in the amendment is strongly worded, the actual measures for enforcement are not well detailed (ibid.).

5.4.2.3 Support for Technological Advancements

Various tax benefits have been bestowed on wind technology equipment manufacturers in China over the past decade. In addition, the Ministry of Science and Technology (MOST) has subsidized wind energy research and development expenditures at various levels over time. A prototype machine developed through this research was approved at the national level. The MOST is now fostering the development of megawatt-size wind turbines, which entails support for variable-pitch rotor technologies and for variable-speed generators as part of the 863 National High Tech R&D Program. *The 11th Five-Year Development Plan of Science and Technology (2006–2010)* included support for the commercialization of 2- to 3-MW–sized wind turbines. Furthermore, the Ministry of Finance's Interim Measures on Management of Special Project Funds for the Industrialization for Wind Power Generation Equipment also provided funding support for the commercialization of wind power generation equipment (Wang et al. 2012).

5.4.2.4 Pricing Policy

The wind power pricing system in China changed dramatically with the NDRC's "Notice on Policy to Improve Grid-Connected Power Pricing for Wind Power Generation" issued in July 2009. Previously, wind power prices had been determined in a variety of ways, including a bidding process for government-selected concession projects and tariffs determined on a project-by-project basis for other large-scale endeavors. Each wind concession project received the tariff price bid for the first 30,000 load hours of the project (and the tariff received by desulfurized coal plants for any subsequent load hours). Five rounds of wind concessions produced 18 wind projects ranging from 100 MW to 300 MW in size, totaling 3350 MW of new wind installations. Tariffs under the wind concession program between 2003 and 2007 ranged from ¥0.42/kWh (US$0.066) to ¥0.551/kWh (US$0.086; Wang et al. 2012).

Although imperfect and plagued by reports of gaming in the bidding system, the wind concession projects were successful in helping the government determine the current price for wind power in China, and set the groundwork for the establishment of FiTs. The notice established a unified nationwide pricing standard and a standard return on investment by categorizing the country into four resource areas based on the quality of wind resources, with Category I having the highest quality resources and Category IV having the lowest quality resources. Each category was assigned a FiT (Figure 5.13; Wang et al. 2012).

The "Outline Measures for the Administration of Offshore Wind Power Development" was introduced in early 2011, and is the first policy

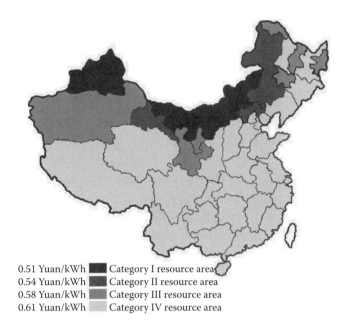

0.51 Yuan/kWh ▇ Category I resource area
0.54 Yuan/kWh ▇ Category II resource area
0.58 Yuan/kWh ▇ Category III resource area
0.61 Yuan/kWh ▢ Category IV resource area

FIGURE 5.13

Regional divisions for feed-in-tariffs in China. (Adapted from ChinaMaps.org. Blank China Map, 2012. http://www.chinamaps.org/china/china-blank-map.html; and CREIA, GWEC, and Greenpeace. *2010 China Wind Power Outlook*, edited by Li Junfeng, Shi Pengfei, and Gao Hu. Beijing: Greenpeace, 2010. http://www.greenpeace.org/eastasia/Global/eastasia/publications/reports/climate-energy/2010/2010-china-wind-power-outlook.pdf.)

governing offshore wind development in China. In May 2010, a concession program led to the development of four offshore wind projects in Jiangsu Province—located in Binhai (300 MW), Sheyang (300 MW), Dafeng (200 MW), and Dongtai (200 MW) (Wang et al. 2012).

5.4.2.5 Clean Development Mechanism

The Clean Development Mechanism (CDM) is a component of the Kyoto Protocol that facilitates the financing of clean energy projects in relatively poorer developing countries using support from richer nations. China has taken advantage of this. The United Nations has approved 869 Chinese projects under the CDM, representing 38.7% of all registered projects. There is uncertainty about the magnitude of approvals in the future. The uncertainty stems from challenges to the way Chinese projects have interpreted the rule that any CDM project must be "additional" to those that would be developed whether the CDM existed or not. In addition, it is not known if the CDM will continue unchanged after the expiration of the Kyoto Protocol emissions reduction period in 2012 (CREIA, GWEC, and Greenpeace 2010).

5.4.3 Offshore Wind Power

Offshore wind power is a more recent technological development in the wind power industry that offers many advantages, including higher average wind speeds and being able to utilize larger turbines. These benefits allow for greater capacity factors but also result in higher capital costs.

Higher capital costs result primarily from the complex foundations required for offshore wind turbines. These foundations must be designed to survive harsh marine environments and the impact of large waves. The challenges of higher installation costs and higher grid connection costs mean that they are significantly more expensive than land-based systems. There are three types of foundations: single-pile structures, gravity structures, and multi-pile structures. The choice depends on seabed conditions, water depth, and estimated costs. Currently, most offshore wind turbines use single-pile structures and are sited in shallow water not exceeding 30 m. The average capacity of offshore wind turbines was 3.4 MW in 2011, up from 2.9 MW in 2010. Most recently, wind farms have utilized 3.6 MW turbines, but 5.0 MW or larger-capacity turbines are available (IRENA 2012b).

China has begun to build offshore wind farms only recently. Shanghai's Donghai Bridge Wind Farm, with a total project investment of ¥2.365 billion (US$0.3695 billion) and total installed capacity of 102 MW, is located near the Donghai Bridge in waters that have an average depth of 10 m with annual average wind speeds of 8.4 m/s at the turbines' height of 90 m. This wind farm comprises 34 wind turbines, each with a capacity of 3 MW. It was designed to yield an annual electricity output of 267 GWh, which will meet the annual demand of 200,000 families and lead to 246,058 tons of CO_2 emission reductions per year (Chen 2011).

Four additional offshore wind power bases will be constructed in wind-rich Jiangsu Province. Located along the eastern coast, the average wind speed at 50 m can be more than 6.4 m/s, making the effective wind energy there more than 150 W/m², whereas the average annual hours of effective wind energy amounts to 5400. The 11th Five-Year Plan and "Outlook 2020 of Jiangsu's Wind Power" have set a goal to reach 10,000 MW of installed wind capacity by 2020, comprising 3000 MW of offshore wind power and 7000 MW of onshore wind power. In the long term, the wind power installed capacity is planned to reach 21,000 MW, 3000 MW of which will be onshore wind power and 18,000 MW of which will be inshore wind power (Chen 2011).

The four bases will be in Rudong, Dongtai, Dafeng, and Qidong. The "Master Plan of Wind Farms in Inshore and Tidal Zones in Rudong County" provides a scheme for the development of wind farms in offshore and tidal zones. Rudong's 1000-MW scale wind farms will be the first pilot demonstration base in Jiangsu Province. Upon completion, Rudong will have a total installed capacity of 4220 MW, which will include 1720 MW of onshore and tidal wind farms and 2500 MW of offshore wind farms. Currently, 790 MW of capacity at a cost of 10 billion yuan (US$1.56 billion) have already been installed, generating 1.2 billion kWh (Chen 2011). Dongtai is also currently under construction. The first project phase, comprising 200 MW, is already in operation; and it is estimated that total capacity will reach 1000 MW by 2016, with an annual electricity generation of more than two billion kilowatt-hours (ibid.).

5.4.4 Small Wind Power

Small wind turbines can be defined roughly as turbines with capacities of 100 kW or less, although there is no official cutoff. Because small wind turbines generally have higher capital costs per unit capacity and achieve lower capacity factors, their most important function is to satisfy unmet electricity demands. These key uses include off-grid electrification, rural electrification, and satisfying the electricity needs of individual homes, farms, and small businesses. According to a conservative estimate, there are approximately 200,000 small wind turbines currently in operation in China. If the average power of each is about 500 W, then installed capacity is approximately 100 MW, accounting for 0.01% of wind power capacity. This low number makes small wind power generally too insignificant to mention in reports on green energy in China (Zhang and Qi 2011). Worldwide, small wind turbines accounted for only 0.14% of capacity in 2010 (IRENA 2012b).

Nearly all small wind turbines use permanent magnet generators, direct drive passive yaw control, and two to three blades. Some turbines use four to five blades to reduce the rotational speed and increase torque. Siting is a critical issue because collecting accurate wind measurements is not economically feasible due to the cost and time required relative to the small size of the total investment. Siting must be based on experience and expert judgment. As a result, many systems perform poorly and can even suffer accelerated wear and tear from bad siting (IRENA 2012b).

Another issue with small wind turbines is low towers, which have low capacity factors and often expose the turbines to excessive turbulence. Unfortunately, increasing tower height is extremely costly. As a result, there has been increasing interest recently in vertical-axis technologies because they are less affected by turbulent air than horizontal-axis turbines, have lower installation costs for the same height as horizontal-axis, require lower wind speeds (which increases their ability to serve areas with lower-than-average wind speeds), and rotate at one-third to one-fourth the speed of horizontal-axis turbines (IRENA 2012b).

5.4.5 Growth of Wind

Wind turbine construction began in China as early as the 1980s but did not begin to develop rapidly until after the Renewable Energy Law took effect in 2006. Figure 5.14 illustrates that total capacity there doubled every year between 2006 and 2009, and expanded more than 100-fold from the end of 2001 to the end of 2010. By the end of 2010, China had the world's largest cumulative installed capacity. Its 44.7 GW accounted for 22.7% of the global total, as shown in Table 5.5 (CREIA, GWEC, and Greenpeace 2010). In addition, China installed the largest amount of new capacity in 2010, accounting for 49.5% of global growth, as shown in Table 5.6.

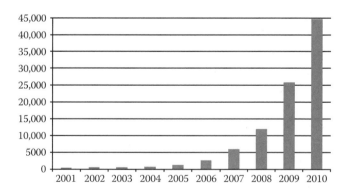

FIGURE 5.14
Cumulative installed wind capacity. (Adapted from CREIA, GWEC, and Greenpeace. *2010 China Wind Power Outlook*, edited by Li Junfeng, Shi Pengfei, and Gao Hu. Beijing: Greenpeace, 2010. http://www.greenpeace.org/eastasia/Global/eastasia/publications/reports/climate-energy/2010/2010-china-wind-power-outlook.pdf.)

TABLE 5.5

Cumulative Installed Wind Power Capacity as of 2010

Country	MW	Total (%)
China	44,733	22.7
U.S.	40,180	20.4
Germany	27,214	13.8
Spain	20,676	10.5
India	13,065	6.6
Italy	5797	2.9
France	5660	2.9
U.K.	5204	2.6
Canada	4009	2
Denmark	3752	1.9
Rest of the world	26,749	13.6
Total top 10	170,290	86.4
World total	197,039	100

Source: Adapted from CREIA, GWEC, and Greenpeace. *2010 China Wind Power Outlook*, edited by Li Junfeng, Shi Pengfei, and Gao Hu. Beijing: Greenpeace, 2010. http://www.greenpeace.org/eastasia/Global/eastasia/publications/reports/climate-energy/2010/2010-china-wind-power-outlook.pdf.

TABLE 5.6

New Wind Capacity Installed during 2010 by Country

Country	MW	Total (%)
China	18,928	49.5
U.S.	5115	13.4
India	2139	5.6
Spain	1516	4
Germany	1493	3.9
France	1086	2.8
U.K.	962	2.5
Italy	948	2.5
Canada	690	1.8
Sweden	604	1.6
Rest of the world	4785	12.5
Total top 10		87.5
World total		100

Source: Adapted from CREIA, GWEC, and Greenpeace. *2010 China Wind Power Outlook*, edited by Li Junfeng, Shi Pengfei, and Gao Hu. Beijing: Greenpeace, 2010. http://www.greenpeace.org/eastasia/Global/eastasia/publications/reports/climate-energy/2010/2010-china-wind-power-outlook.pdf.

China's 18.9 GW of newly installed wind energy accounted for nearly half of the global wind power installed in 2010. With total demand for installed electricity capacity in China at 1500 to 1600 GW (CREIA, GWEC, and Greenpeace 2010), wind energy capacity covered approximately 3% of the total energy demand at the end of 2010.

5.4.6 Wind Supply

Wind resources in China are most plentiful during the autumn, winter, and spring. Because the rainy season generally runs from March or April to June or July, the seasonal distribution of wind energy generally complements that of hydropower. The China Meteorological Administration, the United Nations Environmental Programme, and the National Climate Center have conducted surveys on total wind resources in China. Their conclusions vary based on differences among exploitable capacity estimates at different heights and between theoretically and technically exploitable capacity definitions. Exploitable capacity numbers are higher for studies assuming greater heights. Theoretically exploitable capacity describes the total power that could be generated if all the available wind technologies and resources were used. Technically exploitable capacity refers to the power that could be generated consistently with readily available technology (UNESCO 2012).

At a height of 10 m above ground level, the total technically exploitable capacity is approximately 700 to 1200 GW, including both land-based and offshore wind energy (for more details about how these numbers were measured and calculated, see the Third National Wind Energy Resources Census or the SWERA web site; CREIA, GWEC, and Greenpeace 2010). Figure 5.15 illustrates the annual average wind energy density at a height of 70 m above ground level. The map shows that wind resources are most abundant in the north and west, far from the demand centers on the eastern coast. Because the present electricity grid network is weak, it will need to be strengthened to facilitate further development of wind power (ibid.).

5.4.7 Wind Power Bases

The regions with the highest annual average wind density are located in northwestern, northern, and northeastern China (Figure 5.15). In 2008, the National Energy Board designated the construction of seven wind

(WPD ≤300 W/m², 70 m height)

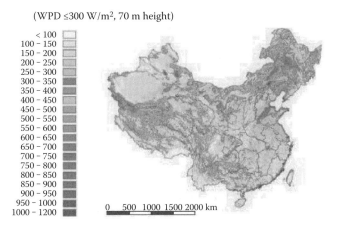

FIGURE 5.15

Distribution map of China's average wind energy density at 70 m above ground level. (Reprinted with permission from IEA and ERI. *Technology Roadmap: China Wind Energy Development Roadmap 2050*, by Zhongying et al. Paris: OECD/IEA, 2011. http://www.iea. org/publications/freepublications/publication/china_wind.pdf.)

power bases, each with at least 10 GW of wind power capacity—in Gansu, Xinjiang, Hebei, the eastern and western parts of Inner Mongolia, Jilin, and the coastal area of Jiangsu. The installed and planned capacities for each of these regions are detailed in Table 5.7 (CREIA, GWEC, and Greenpeace 2010).

At the end of 2010, there was a total of 22.7 GW of wind energy capacity installed in these bases, accounting for a little more than half of the total wind energy capacity installed in China. By 2020, there should be 138 GW of total capacity installed in these regions, accounting for approximately 69% of the total wind energy capacity planned for 2020 (CREIA, GWEC, and Greenpeace 2010). Figure 5.16 depicts the location of the wind power bases and shows the schematic of electricity delivery from the main wind power bases. The power generated by the northeastern wind power base is absorbed by the northeastern grid, whereas most of the power from the onshore and offshore wind power bases in the coastal Jiangsu region is absorbed locally. The Hebei base and the eastern Inner Mongolian base will soon be connected to the northeastern grid and the north China grid, allowing them to divert some of their electricity to China's east. The power bases in Jiuquan and Kumul will also soon be connected to the northwestern grid and the central China grid (ibid.).

TABLE 5.7

Wind Power Bases (in Megawatts)

Wind Power Base	2010 (Installed)	2015 (Planned)	2020 (Planned)
Hebei	4160	8980	14,130
Eastern Inner Mongolia	4211	13,211	30,811
Western Inner Mongolia	3460	17,970	38,320
Jilin	3915	10,115	21,315
Jiangsu	1800	5800	10,000
Jiuquan	5160	8000	12,710
Kumul	0	5000	10,800
Total	22,706	69,076	138,086

Source: Adapted from GWEC. *Global Wind Report: Annual Market Update 2010*. Brussels: GWEC, 2010. http://gwec.net/wp-content/uploads/2012/06/GWEC_annual_market_update_2010_-_2nd_edition_April_2011.pdf.

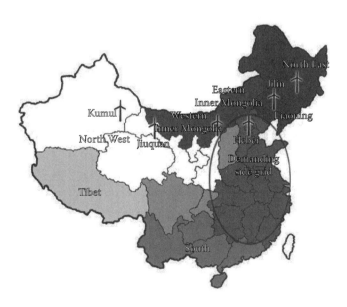

FIGURE 5.16

Wind power installations in China. (Adapted from ChinaMaps.org. Blank China Map, 2012. http://www.chinamaps.org/china/china-blank-map.html; and CREIA, GWEC, and Greenpeace. *2010 China Wind Power Outlook*, edited by Li Junfeng, Shi Pengfei, and Gao Hu. Beijing: Greenpeace, 2010. http://www.greenpeace.org/eastasia/Global/eastasia/publications/reports/climate-energy/2010/2010-china-wind-power-outlook.pdf.)

5.4.8 Cost of Wind Power

Installing wind power capacity requires large initial capital investments, but no fuel costs, a characteristic common to many renewable energy sources. The installed cost of a wind power project is dominated by the cost of turbines, as shown in Figure 5.17, which illustrates that the three largest cost components of a turbine—the rotor blades, the tower, and the gearbox—account for nearly half its cost. The generator, transformer, and power converter make up another 13%. Overall, capital investment accounts for 64% to 84% of project costs, with the grid connection, civil works, and other costs comprising the rest. Globally, increasing commodity prices led to the increase in wind project prices up to high points in 2007 and 2008. The subsequent price declines can also be explained by lower costs for steel, copper, and cement (IRENA 2012b).

Figure 5.18 shows that the cost of wind turbines is lowest in China, and that costs declined noticeably in every country except Japan from 2008 to 2010. To get a better sense of the structure and costs of a turbine, Figure 5.19 illustrates the cost breakdown of a 5 MW offshore wind turbine.

The tower and the blades account for 26% and 22%, respectively, and represent nearly half of the total cost. The gearbox is the third-largest component, accounting for 13%. O&M costs for gearboxes are actually

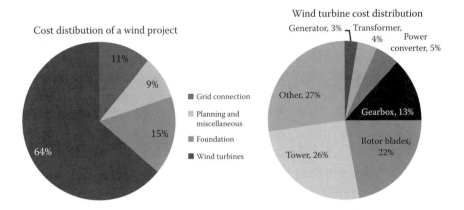

FIGURE 5.17

Breakdown of capital costs of a wind power project. (Adapted from IRENA. *Renewable Energy Technologies: Cost Analysis Series: Wind Power.* United Arab Emirates: IRENA, 2012. http://www.irena.org/DocumentDownloads/Publi cations/RE_Technologies_Cost_Analysis-WIND_POWER.pdf.)

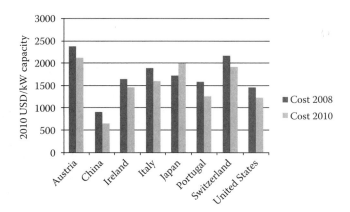

FIGURE 5.18

Wind turbine cost in selected countries, 2008 and 2010. (Adapted from IRENA. *Renewable Energy Technologies: Cost Analysis Series: Wind Power*. United Arab Emirates: IRENA, 2012. http://www.irena.org/DocumentDownloads/Publications/RE_Technologies_Cost_Analysis-WIND_POWER.pdf.)

significant, as they can require extensive maintenance. Figure 5.19 displays the cost structure of an offshore wind turbine; outlays for the tower and blades will occupy a smaller proportion of the total cost for onshore wind turbines (IRENA 2012b).

O&M costs typically make up 20% to 25% of the total levelized cost of electricity, which is the price at which electricity must be sold for the project to prove profitable over its entire lifetime for current wind power systems. They increase over time because of a growing probability of component failure as turbines age and of failure being outside the manufacturers' warranty period. Fixed O&M costs include insurance, administration, fixed grid access fees, and service contracts for scheduled maintenance (IRENA 2012b).

Variable O&M costs include scheduled and unscheduled maintenance not covered by fixed contracts, replacement parts and materials, and other labor costs. The costs for offshore wind farms involve costlier maintenance on turbines, cabling, and towers that are needed because of the harsher marine environment and the higher expected failure rates of some components (IRENA 2012b).

Grid connection also contributes to the overall cost, making up 1% to 14% of the initial capital cost of onshore wind farms and 15% to 30% of offshore wind farms. Wind farms can be connected to electricity grids via

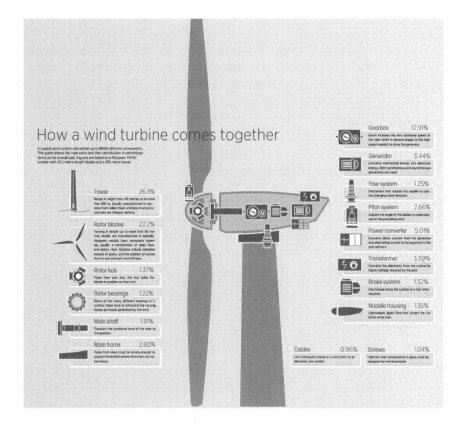

FIGURE 5.19
How a wind turbine comes together. (Reprinted with permission from IRENA. *Renewable Energy Technologies: Cost Analysis Series: Wind Power.* United Arab Emirates: IRENA, 2012. http://www.irena.org/DocumentDownloads/Publications/RE_Technologies_Cost_Analysis-WIND_POWER.pdf.)

the transmission network or the distribution network. When connected to the former, transformers are required to bring up to higher voltages the power provided by turbines. This increases costs. In addition, planners must choose between using a high-voltage alternating current (HVAC) and a high-voltage direct current (HVDC). This decision depends on the distance from the wind farm to the grid connection point. HVDC reduces losses during transmission but generates losses in converting to direct current and back again to alternating current. Thus, HVDC is most suitable for longer distances, whereas HVAC is better for shorter distances. Grid connection costs for investors are also affected by how costs are distributed (IRENA 2012b).

Transportation and installation costs are also substantial, but have been trending downward as a share of total cost with the increasing size of turbines. As wind turbine size increases, the absolute cost per turbine increases as well, whereas transport and installation costs have remain relatively fixed. Offshore costs are much higher than onshore costs, however (IRENA 2012b).

5.4.9 Technological Advancements

China's wind turbine component supply chain is already sophisticated, and key components can now be produced domestically at scale. As a result, many firms have been able achieve price reductions (Wang et al. 2012). However, even though China is now producing advanced megawatt-class wind turbines, firms still need to catch up in terms of the quality of the technology they are manufacturing. Prior to 2004, only one domestically owned Chinese firm was offering to install its own commercial wind turbines on the Chinese market. Most competitors were foreign-owned, and very few had the capability to produce megawatt-class wind turbines. Now, there are more than 30 manufactures that have deployed their wind technology in China, the majority of which are Chinese-owned (Figure 5.20; IRENA 2012b).

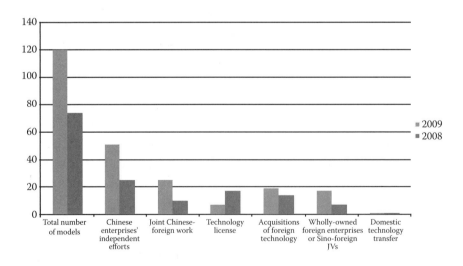

FIGURE 5.20

Sources of wind turbine technology in China by turbine models available in 2008 and 2009. (Adapted from Wang, Z. et al., *Energy Policy* 51: 80–88, 2012.)

Technological progress is difficult to measure, but others have used turbine size, origin of technological innovation, and amount of intellectual property to determine countries' technological levels. The average wind turbine size being installed annually can be used as a proxy, as this has increased consistently and globally over time. In addition, because the majority of China's wind power installed in recent years has come from Chinese manufactures (>80% in 2009), looking at the average turbine size of new capacity over the last few years can give us an idea of where Chinese firms stand internationally in terms of technical prowess. Overall, China is currently installing smaller wind turbines than the world average, suggesting that Chinese firms may still have some catching up to do (IRENA 2012b).

The average wind turbine installed in China in 2009 had a capacity of 1362.7 kW; 86.3% of these were megawatt-class turbines (1 MW or greater in size), and 0.56% were multiple megawatt-class turbines (2 MW or greater in size). Currently, 10 Chinese firms are capable of producing 2.5 and 3 MW wind turbine models, primarily for offshore wind farms. Fifteen Chinese enterprises, led by Sinovel and Xiangdian, are trying to develop 5 and 6 MW offshore turbines (IRENA 2012b).

Measuring technical progress by examining the origins of the technological innovations and intellectual property utilized by Chinese firms corroborates our analysis of turbine size. Prior to 2008, most Chinese firms interested in developing wind power technology acquired wind turbine designs through licensing arrangements with foreign firms. Since then, the amount of independent innovation and R&D, both in developing new designs and in assimilating and building on licensed foreign designs, has increased. Among the 120 types of wind turbines available in the Chinese market in 2009, 51 were developed independently by domestic Chinese enterprises, up from 25 in 2008. Another 25 models were developed jointly by Chinese and foreign firms, up from 10 in 2008 (IRENA 2012b). Figure 5.20 depicts the sources of wind turbine models available in China for 2008 to 2009.

Because of China's specific resource conditions, wind turbines developed for other markets need to be modified. Both Envision and Goldwind have introduced 1.5 MW turbine designs with 93-m rotor diameters, and Sinovel has introduced a 1.5-MW turbine with an 89-m rotor diameter, all of which are optimized for China's Category IV regions, which have low wind speeds. Other designs have been developed for wind farms located on plateaus (Wang, Jin, and Lewis 2012). Although the number of Chinese firms capable of conducting pioneering independent R&D is increasing, there are still only a few firms in the country capable of world-class innovation, and quality control remains a problem (ibid.).

5.4.10 Risks and Opportunities for Future Development

The biggest obstacles to and opportunities for improvement in the Chinese wind power industry are related to transmission and integration challenges, unintended policy consequences, and technical flaws. Because of the geographic discrepancy between the wind resource centers and electricity demand centers, transmission is crucial to the wind power industry. For example, two of the wind power bases located in Taonan and Qian'an, in the Tongyu area of the Jilin Province, are 150 km away from the nearest 220 kV line and 300 km away from the nearest city. Wind power's variability means that the integration of this renewable resource creates another enormous challenge. This variability is especially problematic for China due to its reliance on coal generation facilities. As mentioned, these facilities are characterized by long start-up times and thus provide little flexibility to grid operators dealing with the intermittency of wind power (Wang, Jin, and Lewis 2012).

The importance of policy to the renewable energy industry also makes the effects of unintended consequences of legislation on the structure of the wind industry particularly strong. Despite the fact that many policies related to this industry have been in place for only a short time, their consequences have still emerged. For example, the VAT was changed on January 1, 2009. China introduced a "Consumption Type VAT" regime on this date, which meant that input VAT would be included in the purchase price of fixed assets and could be credited against output VAT when calculating VAT payable. This reform thus eliminated the problem of "double taxation" on fixed assets, reducing the overall taxes on fixed assets. It also reduced the tax liability faced by wind power developers, thus boosting investment in wind power. At the same time, however, it reduced the direct benefit that local governments gain from hosting wind projects. As a result, many local governments now find that encouraging local manufacturing and using other means of reaping direct benefits from wind farm development in their regions can be a more attractive option. At worst, local manufacturing can be used primarily as a bargaining tool that local officials use in approving wind projects rather than as a strategic choice that maximizes benefits to the local region. As a result, excess manufacturing facilities have been reported across the country, even when they make little economic sense (Wang, Jin, and Lewis 2012). In addition, in 2011, approximately 16% of the electricity generated by wind farms in northern China went unused. This amounted to 12.3 billion kWh, and caused a loss of ¥6.59 billion (US$1.03 billion), resulting primarily from incentives for local governments to increase the installed capacity of wind turbines without needing to consider how much electricity they ultimately generate (Yang 2012).

There have been no systematic wind equipment failures in China, but such failures present a risk to the industry. Within a very short time, the country has installed an extraordinary amount of wind power. Most of the turbines were not designed specifically for local conditions, as many manufacturers relied on foreign designs that assumed other wind and environmental conditions (Yang 2012).

5.4.11 Interim Conclusion

China's more than 100-fold increase in wind power capacity over the past decade is an impressive achievement that demonstrates the country's commitment to renewable energy. This vast capacity expansion, however, masks underlying problems that must be addressed going forward to continue to increase the proportion of wind energy in the energy mix.

The growth in wind power capacity resulted from strong government incentives that had a number of unintended consequences. Because of its focus on the installation of wind power capacity, Chinese policy neglected to incentivize the creation of a supporting infrastructure that would facilitate the integration of all the new capacity into the grid. China also failed to first promote the design of wind turbines specially suited to China's wind environment, leading to a greater threat of technological failure.

The variable nature of wind power also hampers the adoption of wind energy as a larger share of the energy. This problem is compounded by the large proportion of coal in the energy mix; the long start-up times for coal generation facilities allow for little flexibility in smoothing out the variability of wind power. The intermittency of wind power will ultimately limit its importance in the energy mix, although there is still the potential for future development.

To increase wind energy as a proportion of its energy mix, China has taken some steps to revise policies to focus on incentivizing actual electricity generation rather than just capacity installation. China may also consider encouraging R&D for wind turbines designed specifically for the country. Although innovations have been increasing, the majority of wind turbines available in the Chinese market are still foreign-developed. However, as noted in the previous chapter, China is already investing substantially in the electricity grid infrastructure, which will help it utilize its wind capacity more fully.

5.5 SOLAR POWER

Solar power is clean, variable, and renewable. Solar output cycles with energy demand, naturally peaking during the day. Due to its variable nature, becoming a dominant proportion in the energy mix is highly unlikely, but solar power holds potential for development nonetheless, and can serve both as a great supplement to grid-integrated power and help in rural electrification. China's resources are abundant, and it is currently the largest producer of solar PV panels in the world. However, domestic installations make up a tiny proportion of total PV production. China has recently implemented policy measures to boost domestic utilization of solar energy. The impetus for this change comes from the desire to attain renewable energy goals in the wake of Japan's 2011 nuclear crisis as well as dwindling demand from its major customer, Europe. Given its importance in the solar industry, increased government support signals an increase in domestic installations.

5.5.1 Policy

Solar power industries in every country have relied on extensive government support. Without beneficial policies, the high costs of substrate materials and low efficiencies of solar cells make development very unattractive. According to one comparison, solar power is approximately 520% more expensive than coal (Haley and Schuler 2011).

There are two broad categories of government support policies: consumption and production. Consumption policies assume that installation demand will drive mass production of increasingly more cost-effective solar equipment, whereas production policies assume that economies of scale will drive down the price of solar power to grid parity. Consumption policies include consumption subsidies, FiTs, renewable portfolio standards, tax credits, and concessionary financing. Production policies include production subsidies, technology transfer, publicly funded R&D, research deployment, and industry research consortia. China's policies have focused largely on production assistance (Haley and Schuler 2011).

5.5.2 Solar Resources

China's solar resources are concentrated in the western, northern, and southwestern regions of the country (Figure 5.21). The regions have been classified into four zones, based on solar resource abundance. Table 5.8 breaks down the regions into zones for greater detail and shows the annual solar radiation for each zone. Note that zones I, II, and III account for more than 96% of the total.

More than two-thirds of the country receives more than 2000 annual sunlight hours, and the theoretical total solar energy per year reaches

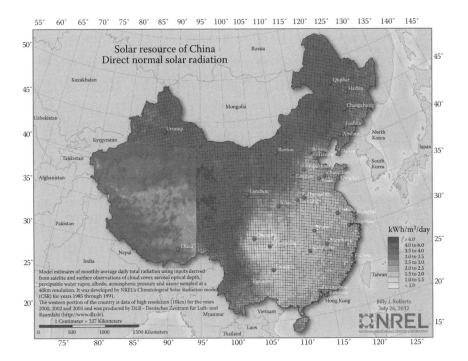

FIGURE 5.21

Distribution of solar resources in China. (Reprinted under the U.S. DOE's media resources usage policy: Roberts, B.J. Solar Resource of China: Direct Normal Solar Radiation, 2012. http://www.nrel.gov/analysis/staff/billy_roberts.html.)

TABLE 5.8

Solar Resource Distribution in China

Classification	Zone	Annual Solar Radiation (kWh/m²)	Share of the National Total (%)	Area
Most abundant	I	≥1750	17.4	Tibet, South Xinjiang, Qinghai, Gansu, and West Inner Mongolia
Very abundant	II	1400–1750	42.7	North Xinjiang, Northeast China, East Inner Mongolia, Huabei, North Jiangsu, Huangtu Plateau, East Qinghai and Gansu, West Sichuan, Hengduan Mountain, Fujian, South Guangdong, and Hainan
Abundant	III	1050–1400	36.3	Hill areas in the southeast, Hanshui River Basin, Guangxi
Normal	IV	<1050	3.6	Sichuan and Guizhou

Source: Adapted from Greenpeace, EPIA, and WWF. *2007 China Solar PV Report*, edited by J. Li, S. Wang, M. Zhang, and M. Lingjuan. Beijing: China Environmental Science Press, 2007. http://www.wwfchina.org/english/downloads/ClimateChange/china-pv-report-en.pdf.

49 million picojoules (Zhang et al. 2012). China's solar resources are comparable in size to those of the United States and greater than those of Japan and Europe (Greenpeace, European Photovoltaic Industry Association [EPIA], and World Wildlife Fund [WWF] 2007). Only a very small percentage of China's solar resources are currently being exploited, meaning that enormous potential exists for increasing installed capacity in China.

5.5.3 History of the Chinese Solar Industry

The Chinese solar industry can be divided into two broad sectors. The first is the solar–thermal industry, which refers mainly to solar water-heating systems in China. The industry was valued at ¥13 billion (US$2.03 billion) in 2005, and Chinese production and use of solar water heaters accounted for more than half the world's total. These heaters are used primarily to meet domestic water-heating needs in many small- and medium-sized towns and cities. This industry is mature, with installations in more than 30 million Chinese households (Ng 2011).

The second sector is the PV industry. PV refers to the particular hardware that converts solar energy directly into electricity (Ng 2011). Terrestrial application of solar cells began in China as early as 1974 (Marigo 2007). In the late 1970s, three semiconductor plants were transformed into PV monocrystalline silicon cell manufacturers, establishing the first manufacturing base. Three more manufacturers were established in the 1980s (ibid.). Industry growth prior to 1981 was inhibited, however, by a lack of government support (Ng 2011). Beginning in the period of the 6th (1981–1985) and 7th (1986–1990) Five-Year Plans, the government began to promote PV development in specialized industries and rural areas (Greenpeace, EPIA, and WWF 2007).

In 2002, the National Development and Planning Commission of the Township Electrification Program introduced the use of PV electricity in more than 700 townships in seven western provinces (Tibet, Xinjiang, Qinghai, Gansu, Inner Mongolia, Shanxi, and Sichuan). PV systems installed in China, primarily in distant areas and for special business purposes, exceeded 70,000 kW by 2005, and the production capacity of solar cells reached 300 MW/year. By 2008, this capacity reached 1.78 GW, accounting for 26% of the global production capacity (Ng 2011).

Global solar energy demand, driven largely by Germany and Japan, has been growing at a rate of approximately 25% per year over the past 15 years (Marigo 2007). Both of these countries focus heavily on consumption policies for the solar industry. Chinese companies have ramped up production capacity in response (ibid.). China currently boasts the world's largest solar PV production capacity as well as the largest actual production numbers (Ng 2011). In 2010, the country was responsible for more than 50% of the global production of solar cells and modules (Marigo 2007). The government has set targets to continue this trend (Ng 2011). Figure 5.22 shows the tremendous growth in PV installed capacity in China, and Table 5.9 provides a breakdown of cumulative installed PV power for the top 10 countries in 2010.

5.5.4 Production and Installation Gap

Despite the rapid increase in the production capacity of solar PV modules in China and the country's abundant solar resources, the installation of solar power has lagged behind solar cell production (Zhang et al. 2012). The industry's structure is a direct result of China's focus on production policies. In 2008, only about 1% of annual domestic PV production was

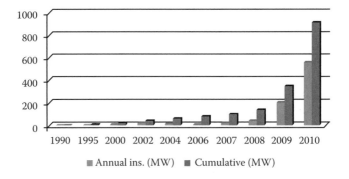

FIGURE 5.22
PV installed capacity in China. (Adapted from Zhang, D. et al., *Energy* 40: 370–375, 2012.)

TABLE 5.9

Cumulative Solar Installed Capacity Breakdown for the Top 10 Countries in 2010

Country	MW	Total (%)
Germany	17,320	43.5
Spain	3892	9.8
Japan	3617	9.1
Italy	3502	8.8
US	2520	6.3
Czech Republic	1953	4.9
France	1025	2.6
China	893	2.2
Belgium	803	2.0
South Korea	573	1.4
Rest of the world	3680	9.3

Source: Adapted from BP p.l.c. *BP Statistical Review of World Energy: June 2011.* London: BP p.l.c., 2011. http://www.bp.com/sectionbodycopy.do?categoryId=7500&contentId=7068481.

installed in the domestic market (Zhang et al. 2012). Prior to 2009, domestic PV installation occurred almost exclusively through government-financed, off-grid rural electrification projects (Marigo 2007). Moreover, polysilicon (a major input for solar cells) demand has greatly outstripped supply. As a result, imports of polysilicon totaled 7991 tons in February 2013, whereas exports amounted to only 376 tons, despite the decline in production volume that currently plagues the industry (Loo 2013).

In 2007, more than 99% of China's manufactured solar modules were exported, whereas only 100 MW of solar power were utilized domestically, compared with approximately 513,000 MW of coal (Ng 2011). From 1980 to 2010, the rate of PV installations grew by approximately 17%, much lower than the 30% to 40% global rate. The largest application of generated PV power was in rural electrification, which consumed 41.3% of domestic solar electricity in 2006 (ibid.). From an environmental standpoint, the gap between production and installation has resulted in high energy consumption and pollution in China and energy savings and emissions reductions abroad (Zhang et al. 2012). This gap represents one of the strongest criticisms of China's PV industry.

As a result of the PV industry's export orientation and the country's production policies, up to 95% of China's output in recent years has been exported. The majority of the PV panels are exported to Europe (~60%) and North America. In the United States, for example, solar panels from China account for about half of the market, whereas domestic panels make up less than a third. China is home to about 700 solar panel manufacturers, with a combined annual production capacity of 40 GW of electricity. Chinese companies have been able to grow and cut costs quickly because of large loans from government-owned banks (Bradsher and Wald 2012).

Tensions between China and Europe and the United States escalated in early 2012 after Chinese firms dropped their prices by 30%. European Union countries accused China of dumping artificially low-cost panels onto the European market because of the Chinese government's aforementioned production policies. As a result, a group of 25 European companies, led by Germany's Solar World, filed a complaint with the European Commission in September 2012, claiming Chinese rivals were benefitting unfairly from illegal subsidies (Bradsher and Wald 2012).

The geographical distribution of PV-producing firms has also been criticized. Polysilicon manufacturing facilities should be established in areas of China with abundant energy and cheap electricity, but new projects are being built in eastern and central China instead. These areas have high electricity prices and few energy resources, suboptimal conditions for the energy-intensive production of polysilicon (Ng 2011). Figure 5.23 shows the distribution of PV industry clusters in China.

5.5.5 Cost Structure

Solar cells comprise two layers of semiconductor material, usually silicon. One layer is positively charged and the other is negatively charged. When sunlight passes through the material, it creates a flow of electrons between

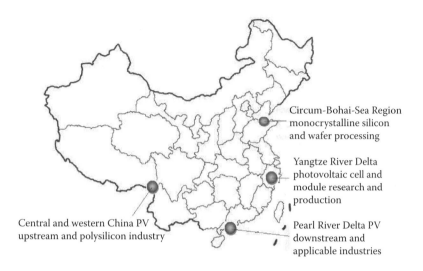

FIGURE 5.23

Photovoltaic industry clusters in China. (Adapted from ChinaMaps.org. Blank China Map. 2012. http://www.chinamaps.org/china/china-blank-map.html; and Ng, M. Economic impact of the photovoltaic industry in China after the financial crisis of 2009. *The Chinese Economy* 44(3): 22–44, 2011.)

the two layers, generating a DC current. PV systems are made up of PV cells, modules, and inverters. The cells gather sunlight and are composed primarily of silicon wafers. The inverter then converts the energy generated into a form conducive for day-to-day use (Greenpeace, EPIA, and WWF 2007).

In 2007, the vast majority of solar cells were made from monocrystalline and polycrystalline silicon (Ng 2011). Monocrystalline ingots require more sophisticated manufacturing processes, higher conductivity, and command higher prices (Haley and Schuler 2011). The cost of silicon accounted for the majority of the cost of the entire production process (56%). Figure 5.24 displays the cost breakdown of a typical PV panel. A shortage of polycrystalline silicon in 2006 created a bottleneck in the industry. With the growth of global PV demand, stemming largely from favorable consumption policies in Germany and Spain, the price of the feedstock increased from US$25/kg in 2003 to more than US$200/kg in 2006 on the black market. Black market prices approximate the value of the silicon if it were to be traded in an unregulated market. Because of small production scale, Chinese PV firms could not achieve adequate cost reductions to offset the increase in material prices. There is a strong correlation between production scale and manufacturing costs (Ng 2011).

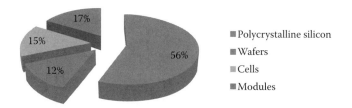

FIGURE 5.24
Percentage cost of each stage in the manufacturing process. (Adapted from Greenpeace, EPIA, and WWF. *2007 China Solar PV Report*, by Li, J., S. Wang, M. Zhang, and M. Lingjuan. Beijing: China Environmental Science Press, 2007. http://www.wwfchina.org/english/downloads/ClimateChange/china-pv-report-en.pdf.)

5.5.6 Supply Chain

Solar PV supply chains comprise three parts: polysilicon (upstream); wafer, cell, and module (midstream); and installation (downstream). Table 5.10 presents detailed descriptions of each of these stages. Currently, Chinese firms dominate polysilicon, cell, wafer, and module production, but account for a relatively small proportion of installation (Haley and Schuler 2011).

5.5.7 Effects of the Financial Crisis

In 2008, China's PV export value had reached ¥150 billion (US\$23.4 billion), making the country the world's largest exporter of PV cells. After the financial crisis, overseas demand for polysilicon plummeted by 40% in the first quarter of 2009 alone. This substantial decline in demand, paired with the rapid capacity expansion and a flurry of IPOs in the industry between 2006 and 2008, spawned overcapacity problems in the Chinese PV industry. Polysilicon output capacity was approximately 20,000 tons in 2008, whereas the actual output was only 4000 tons, reflecting the direness of the situation. Furthermore, production capacity under construction was more than 80,000 tons at the time. In response, in September 2009, the government proposed strict measures to handle overcapacity in the industry (Ng 2011).

The financial crisis, however, put downward pressure on the international price of polysilicon feedstock, which declined from US\$400/kg in mid-2008 to about 10 dollars/kg by the end of 2008. The price of this product—the major cost component in the manufacturing process of PV cells—thus became very favorable for firms operating in the downstream capacity (Ng 2011).

TABLE 5.10

Solar PV Supply Chain

Product	Process	Industry Characteristics	Technology	Generic Strategies
Polysilicon	Quartz silica changed into silicon ingots	• Oligopolistic • 5–10 companies • High entry barriers • Ample supply of inputs	• Vapor to liquid deposition	• Set price ceilings
Wafer	Silicon ingots cut into wafers	• About 50 companies • Medium entry barriers due to high investment • High dependence on polysilicon suppliers	• Monocrystalline • Multicrystalline • String ribbon	• Access low-cost financing • Develop proprietary technology • Integrate midstream operations
Cell	Circuitry put on wafer	• Highly competitive • About 100 companies • Low entry barriers • Essential component of silicon-based power • Boom–bust exposure	• Crystalline • Thin film (CIGS, CdTe, a-Si)	• Establish proprietary technology • Integrate midstream operations
Module	Cells placed on glass and made into panels	• Highly competitive • More than 400 companies • Low entry barriers due to low investment hurdle • Boom–bust exposure	• Low technology	• Differentiation
Installation	Solar panels installed	• Fragmented • More than 5000 companies • Requires financing and connections	• Low technology	• Price • Nonmarket strategies

Source: Adapted from Haley, U.C.V. and D.A. Schuler. Government policy and firm strategy in the solar photovoltaic industry. *California Management Review.* HBS Case Study, 2011.

5.5.8 Future Prospects

Like many other industries that are not economically competitive, the solar PV industry in China is buttressed by government policy. As noted previously, the Chinese government subsidizes manufacturers through preferential loans, tax incentives, R&D support, central government planning, and local and province policies. It also spurs demand through central government planning and goals, direct subsidies, a FiT, and local and province policies (The George Washington University Trachtenberg School of Public Policy and Public Administration [The Trachtenberg School] 2011).

Preferential loans are provided through lower real interest rates and the greater availability of credit. In particular, the availability of credit allowed firms to access funds during economic uncertainty. Under the Renewable Energy Law, eligible PV companies are exempt from VATs and customs duties, and business income taxes may be reduced to 15%. The Chinese MOST also provides monetary assistance to companies and public research institutions for R&D. China provided US$18.5 million for PV R&D in the 11th Five-Year Plan. Through the selective distribution of these programs' benefits, central government planning plays a role in picking winners in an effort to foster consolidation (and thereby economies of scale) in the industry (The Trachtenberg School 2011).

China has also created numerous consumer-centric incentives. In terms of direct subsidies, the government has rolled out a number of programs to stimulate growth. Two major programs are the building-integrated PV program (BIPV) and the Golden Sun Program. Table 5.11 details the major components of these two initiatives (The Trachtenberg School 2011). In addition, after years of taking advantage of overseas demand, China recently implemented its own FiT to develop its domestic solar market (Liu 2011). A FiT typically guarantees a fixed price for all renewable energy connected to the grid. The FiT for China is approximately US$0.18/kWh for projects completed by December 31, 2011, and US$0.15/kWh for projects approved by July 2011 but not completed before the end of that year (The Trachtenberg School 2011). The FiT guarantees payback in 7 years and attractive cash yields for nearly another two decades. The effects of this policy have already manifested themselves in Qinghai, where the local labor pool was exhausted by solar project developers shortly after the FiT was issued (Liu 2011).

The motivation behind this new development stems from the need to compensate for nuclear development setbacks while fulfilling alternative

TABLE 5.11

Major Terms of the BIPV and Golden Sun Programs

	BIPV Program	Golden Sun Program
Applications	Grid-connected rooftop and BIPV systems	Grid-connected rooftop, BIPV, and ground-mounted systems. Off-grid systems in rural areas
System size	≥50 kW	≥300 kW
Subsidy	¥15/W for rooftop systems ¥20/W for BIPV systems	50% of total cost for on-grid systems, 70% total cost for off-grid systems
Other terms	Conversion efficiency requirement: 16% for monocrystalline, 14% for multicrystalline, 6% for thin film	For grid-connected systems, on-site consumption is encouraged. Excess electricity will be sold to utility. Buyback rate is based on local benchmark coal-fired grid price

Source: Adapted from Ng, M., *The Chinese Economy* 44(3): 22–44, 2011.

energy development goals as well as for dwindling overseas demand. After Japan's nuclear disaster in 2011, China halted approval for new projects due to safety concerns. The slowdown in nuclear power development, however, had to be compensated by other clean energy sources to stay on track to obtaining 15% of the country's power supply from non–fossil fuels by 2020. Furthermore, Chinese manufacturers face dwindling demand as Italy, Spain, Germany, and other European countries, which make up the majority of the market for Chinese solar panels, slashed subsidies for solar energy (Liu 2011).

Nevertheless, the government is exhibiting some caution in developing the solar energy market. Wu Dacheng, the vice chairman of the Photovoltaic Committee of the Chinese Renewable Energy Society, an influential organization in Beijing, claimed that China is likely to set a ceiling for solar energy growth due to cost concerns (Chapter 6 provides a more detailed overview of these cost concerns, as it addresses Spain's experience with its "tariff deficit"). He also attributed this to the government's desire to avoid the mistakes made in the wind power sector. As noted in the section on wind power, China issued a similar FiT scheme for wind projects in 2009, which contributed to the rapid development of the industry in China. Now, however, due to inadequate grid infrastructure, wind-generated electricity is being wasted; in the first half of 2010, enough energy to supply 10 million people for an entire year never made it to consumers. Solar power will face similar grid infrastructure problems as solar resources are concentrated in areas far from energy demand centers (Liu 2011).

5.5.9 Interim Conclusion

Like wind power, solar power is variable, and thus the inflexibility of the primary component of the energy mix, coal, presents a serious problem to grid integration. The rapid growth of the manufacturing component of this industry has propelled China to become the largest manufacturer of solar PV cells; however, the legality of its subsidies is being questioned, and the quality of its solar panels relies on foreign, as opposed to domestic, innovation. Furthermore, the majority of these solar panels are exported, leading to low domestic installations.

Recently, in response to nuclear setbacks as well as dwindling demand from its major customer, Europe, China implemented a FiT similar to the one used to promote the rapid development of the wind power industry. Analysts believe this FiT will incentivize similarly strong growth in the solar power industry. It seems, however, that the government plans to exercise more caution in its support of the domestic solar power industry to prevent the problems of overinstallation and waste found in the wind power industry from occurring in the solar energy market.

Despite this caution, most analysts agree that domestic installations of solar panels will proceed rapidly as long as the government maintains incentives. Solar power is clean and renewable, can serve as a great supplement to grid-integrated power, and can help in rural electrification. In particular, solar power will complement China's energy-intensive heavy industry, which historically has demanded more energy during the day, when solar output naturally peaks. This will help China cope with peak electricity demand periods. Ultimately, although the growth of the proportion of solar power in the energy mix will be limited due to the inherent variability of this source, China's vast resources still hold huge potential for development.

5.6 NUCLEAR POWER

China's electricity demand has been growing at an annual rate of 10% and is expected to continue to increase over the next few decades. However, faced with the significant environmental and health effects caused by extensive coal use, the country is taking action to diversify its energy supply to meet its drastically increasing demand. Oil and natural gas are not abundant in China, and these resources' international prices have consistently trended upward. Many clean renewable energy sources, such as wind and solar, are variable. Nuclear power is clean, efficient, and able to overcome many of the barriers that other resources face. Thus, it is considered a crucial part of China's future energy solution (Zhou 2010).

5.6.1 Uranium Resources

Recoverable reserves are defined as reserves with extraction costs under US\$130/kg of uranium. The total recoverable uranium reserves in the world amounted to 5.4 million tons as of 2009 (Figure 5.25). This is lower than previous estimates mainly due to increases in extraction costs. Australia holds a substantial amount of these total reserves (31%), followed by Kazakhstan (12%), Canada (9%), and Russia (9%). In addition, the majority of the uranium reserves in Australia are very economically exploitable, with production costs lower than US\$80/kg of uranium. China has only 171,000 tons of uranium reserves, 3% of the world's supply (World Nuclear Association [WNA] 2012).

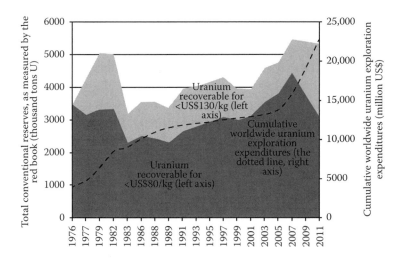

FIGURE 5.25

Known uranium resources and exploration expenditures. (Adapted from NEA and IAEA. *Forty Years of Uranium Resources, Production and Demand in Perspective: The Red Book Retrospective.* Paris: OECD Publishing, 2006; NEA and IAEA. *Uranium 2005: Resources, Production and Demand.* Paris: OECD Publishing, 2006; NEA and IAEA. *Uranium 2007: Resources, Production and Demand.* Paris: OECD Publishing, 2008; NEA and IAEA. *Uranium 2009: Resources, Production and Demand.* Paris: OECD Publishing, 2010; NEA and IAEA. *Uranium 2011: Resources, Production and Demand.* Paris: OECD Publishing, 2012.)

There are 15 types of uranium deposits, classified by geological conditions and origins of mineralization. Sandstone deposits, for example, are contained in sandstones in marine sedimentary environments. In China, there are approximately 200 uranium deposits, mostly small- to medium-sized and of low grade because they contain only 0.005% to 0.3% U (50–3000 ppm; Yang and Huang 2010) with very few of 0.5% grade (5000 ppm; Zhou and Zhang 2010). The major types of deposits present in China are granite, volcanic, and sandstone (OECD Nuclear Energy Agency [NEA] and International Atomic Energy Agency [IAEA] 2006), as shown in Table 5.12. Both granite and volcanic deposits are in hard-rock areas where exploration can be difficult and costly, posing yet another challenge for the country's uranium production.

5.6.2 Uranium Production

China's uranium production has remained stable over the last decade, amounting to 827 tons in 2010 (Figure 5.26). Currently, all production is used to satisfy domestic nuclear power plant needs. Nuclear power is a minor part of China's energy supply, and plants are relatively small.

The uranium extraction technologies in China include the following four types: heap leaching, *in situ* leaching, stope or block leaching, and conventional leaching. Heap-leaching technology involves mining

TABLE 5.12

Uranium Deposit Types in China

Uranium Deposit Types	Total Uranium Reserves in China	ppm (U)	Distribution Region
Granite	38	60–300	Guangdong Guidong zone, Zhuguang Mount Northern China, Tao Mount Jiangxi, Jilin NW China
Volcanic	22	400–40,000	Xiangshan Jiangxi, Xianqiuyuan Zhejiang, Baiyanghe Xinjiang
Sandstone	19.5	400–4000	Yili Basin, Xunwu Basin, Henyang Jianchang Basin, west of Yunnan
Carbonaceous-siliceous pelitic rock	16	<100	Huangcai, Laowolong zone, Chanziping zone, Ruoergai zone

Source: Adapted from Yang, G. and W. Huang, *Energy Policy* 38: 966–975, 2010.

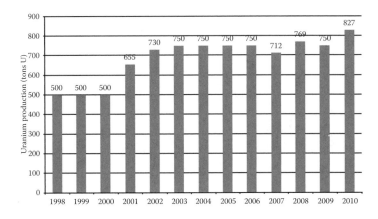

FIGURE 5.26

China's uranium production. (Adapted from WNA. Supply of Uranium. 2012. http://www.world-nuclear.org/info/inf75.html.)

uranium ore from the deposit, piling it on the surface, and using sulfuric acid or other chemicals to extract the uranium. This technique is most suitable for oxide ore deposits, that is, those that are close to the surface and have reacted with oxygen through natural processes. Hard-rock deposits in southeastern China are generally exploited using this method (NEA and IAEA 2006).

In situ leaching is very similar to heap leaching. The difference is that the uranium remains underground until the minerals are dissolved in a solution. The solution is then pumped to the surface for recovery. This process involves little alteration to the surface or to the geological formation, but is viable only where the deposit is permeable to the extracting solution. This method is also limited to situations where there is no risk of contaminating local water bodies. China has used this technology mainly to recover sandstone deposits in the northern part of the country (NEA and IAEA 2006).

The growth of China's nuclear power industry is constrained significantly by its uranium resources and production. The expansion and implementation of large-scale nuclear power plants would have to be fueled by uranium imports.

5.6.3 Nuclear Power Demand and Production

Nuclear power is much more efficient than coal in terms of energy output per unit of input mass. A 1000 MW coal power plant consumes approximately two million tons of coal equivalent every year, whereas a similar capacity nuclear power plant consumes only 190 tons of uranium, over 10,000 times more energy-efficient than coal by mass of feedstock (WNA 2013). China has expanded its nuclear industry over the last decade, but it is still significantly behind the United States, Europe, and Japan in terms of installed capacity. In 2011, China had 14 nuclear power units running with an aggregate installed capacity of 11.7 GW, over 25 units under construction, and more than 100 planned and proposed (WNA 2013; Zhou and Zhang 2010). There were 442 nuclear power plants in the world with total installed capacity of 385 GW, which, when combined, consumed approximately 67,000 tons of uranium per year (Yang and Huang 2010).

China's nuclear power output has quintupled since 1995 (Figure 5.27). However, it still provides only about 2% of the country's electricity output, much lower than the world average of 16%, and less than 1% of total energy consumption (NBSC 2011). According to China's *Medium- and Long-term Nuclear Power Development Plan (2005–2020)*, China plans to increase its nuclear power capacity to 40 GW by 2020, accounting for 4% of total electricity generation (Zhou 2010). To meet this target, a significant number of nuclear plants would have to be brought online in the next few years.

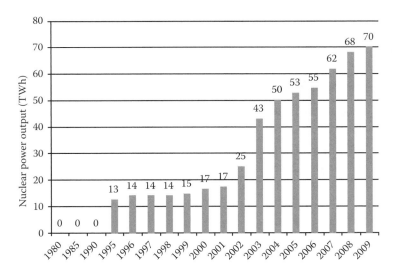

FIGURE 5.27
China's nuclear power output. (Adapted from NBSC. *China Energy Statistical Yearbook 2009*. Beijing: China Statistics Press, 2010.)

Nuclear power plants involve higher initial investment than most other power plants. Thus, this development plan relies on the continuous growth of the Chinese economy to provide the necessary capital to fund the construction of nuclear plants, as well as to finance the R&D for new technologies. In the long run, however, these plants can prove cost-competitive, mostly because they do not require as much input material as coal plants, which reduces transportation costs. Studies have also shown that nuclear power is cost-competitive with hydropower in the country's eastern coastal region (Wu and Siddiqi 1995).

China's current uranium production supports about half of its nuclear industry's needs, with the remaining portion imported from Australia, Kazakhstan, Russia, and Namibia. China signed the "Sino-Australian Cooperation in the Peaceful Use of Nuclear Energy" agreement with Australia in 2006 to import uranium. In 2010 and 2011, the country imported 17,135 and 16,126 tons of uranium, respectively (Ding and Liu 2012). In the future, as it expands its nuclear power capacity, it will need a significant amount of uranium resources. It is estimated that to support the 40-GW nuclear power target in 2020, China will require approximately 10,000 additional tons of uranium each year, much more than what domestic reserves can supply (Zhou 2010). Inevitably, China will have to become even more reliant on uranium imports.

Currently, spent nuclear fuel management is not a huge problem, as China has only a small amount of nuclear waste. However, as uranium consumption grows, waste disposal will pose a significant challenge.

5.6.4 Nuclear Power Plants and Technologies

China's nuclear plants utilize mostly pressurized water reactor (PWR), generation II technology. The power plants are clustered in three bases: the Guangdong Daya Bay Nuclear Power Plant Base, the Zhejiang Qinshan Nuclear Power Base, and the Jiangsu Tianwan Nuclear Power Plant Base. All are located in coastal areas with high economic activity and energy demand, but few natural resources. A map of the plants is shown in Figure 5.28. A detailed listing of nuclear power plants in China is provided in Table 5.13.

5.6.4.1 Daya Bay Reactors

The Daya Bay reactors in Guangdong Province are two identical French PWR units, constructed by Électricité de France, with a combined

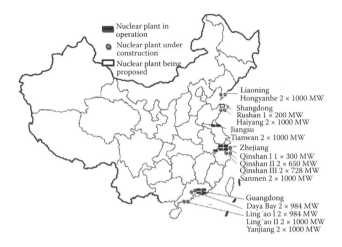

FIGURE 5.28

Nuclear power in operation and under construction in China. (Adapted from ChinaMaps. org. Blank China Map. 2012. http://www.chinamaps.org/china/china-blank-map.html; Yang, G. and W. Huang, *Energy Policy* 38, 966–975, 2010.)

TABLE 5.13

Summary of Nuclear Power Plants in China

Name of Reactor	Net Generating Capacity (MW)	Gross Generating Capacity (MW)	Type	Operation Began
Qinshan-I	278	300	PWR	1991
Daya Bay-1	944	984	PWR	1993
Daya Bay-2	944	984	PWR	1994
Qinshan-II-1	610	650	PWR	2002
Qinshan-II-2	610	650	PWR	2004
Qinshan-II-3	610	650	PWR	2010
Qinshan-III-1	665	728	PHWR	2002
Qinshan-III-2	665	728	PHWR	2003
Ling'ao-I-1	935	985	PWR	2002
Ling'ao-I-2	935	985	PWR	2003
Ling'ao-II-1	935	1000	PWR (CPR-1000)	2010
Ling'ao-II-2	935	1000	PWR (CPR-1000)	2011
Tianwan-1	1000	1060	PWR	2006
Tianwan-2	1000	1060	PWR	2007
Total: 14	11,066	11,764		

Source: Adapted from Yang, G. and W. Huang, *Energy Policy* 38, 966–975, 2010; WNA. Nuclear Power in China. 2013. http://www.world-nuclear.org/info/inf63.html.

capacity of 1.968 GW. The station is based 45 km from the center of Shenzhen and 50 km from Hong Kong. Commercial operation began in 1994. The plant supplies more than 14 billion kWh of electricity annually, of which 70% is transmitted to Hong Kong, with the remainder transported to Guangdong Province (Hong Kong Nuclear Investment Company Limited [HKNIC] 2010).

5.6.4.2 Ling'ao Reactors

The Ling'ao reactors are located on the Dapeng Peninsula, which is also in Guangdong Province. Ling'ao-I includes two 985-MW PWR units based on the French 900-MW three–cooling loop design. The reactors began operation in 2002 and 2003. Ling'ao-II includes two 1000-MW CPR-1000 units. CPR-1000 is a Chinese improved version of the PWR unit, a generation II+ technology designed and operated by the China Guangdong Nuclear Power Company (CGNPC). Operation began in 2010 and 2011 (China National Nuclear Corporation [CNNC] 2011).

5.6.4.3 Qinshan Reactors

The Qinshan reactors are located in Qinshan Town in Zhejiang Province. The Qinshan-I 300 MW reactor was the first unit designed and constructed by China. It is a PWR unit that began operation in 1991. Qinshan-II, comprising four 650-MW PWR units with a combined capacity of 2600 MW, was also designed, constructed, and managed by China itself. Currently, three of the four units have begun operation, with the first brought online in 2002. The last unit was scheduled to be completed by 2012. Qinshan-III includes two units utilizing CANDU-6 technology, each with a capacity of 728 MW. It was designed by Atomic Energy of Canada Ltd., and the construction was completed in 2003, 112 days ahead of schedule. The plant generates more than 15 billion kWh annually (CNNC 2011).

5.6.4.4 Tianwan Reactors

The Tianwan plant is located in Lianyungang City in Jiangsu Province. It comprises two model AES-91 1000-MW PWR units constructed by Russia Atomstroyexport. Operation began in 2007. The site has capacity for a total of eight 1000-MW units (CNNC 2011).

5.6.4.5 Plants Under Construction

The Hongyanhe plant is currently under construction. It is located in Liaoning Province, the first nuclear project in northern China. It includes four CPR-1000 units, with the first unit expected to begin operation in 2012. The Yangjiang plant located in western Guangdong Province comprises six CPR-1000 units, with the first two due to begin operation in 2013. Costs are projected to surpass 10 billion dollars (Zhou et al. 2011).

Zhejiang Sanmen and Shandong Haiyang are two generation III nuclear power stations using the AP1000 design—a two-loop PWR with a more powerful output. Construction on Sanmen-1 began in 2008, and operation is slated to begin in 2013. Construction on Sanmen-2 began in 2009, with operation due to start in 2014. Work on Haiyang units one and two began in 2009 and 2010, respectively, and these plants are expected to commence operation in 2014 and 2015 (CNNC 2011).

5.6.5 Construction, Operation, and Manufacturing

The nuclear power plants in China are operated by two corporations: China National Nuclear Corporation (CNNC) and China Guangdong Nuclear Power Group Holding Co. Ltd. (CGNPC). Although the country is also issuing permits to conventional thermal power plant construction parties to participate in nuclear power construction, these two corporations are expected to dominate China's nuclear power development for the foreseeable future (Zhou et al. 2011).

By 2009, the Chinese government had invested approximately ¥300 billion (US$46.9 billion) to support four main nuclear manufacturing companies: Shanghai Electric Company, Dongfang Electric Corporation, Harbin Power Equipment Company, and China First Heavy Industries Co. (Zhou et al. 2011). These four companies produce components for nuclear reactors and are critical to China's nuclear industry.

5.6.6 Policy and Regulations

China's *Medium- and Long-term Nuclear Power Development Plan (2005–2020)* emphasized the importance of nuclear power, and the country made plans to increase the share of nuclear power in its energy mix. Since then, it has taken an active role in developing nuclear power plants.

The significant progress and popular support has led the Chinese government to consider increasing the 2020 target to 70 GW. It is very likely that installed capacity in China will exceed the original 2020 target of 40 GW (Dynabond Powertech 2010; Lin et al. 2012).

The nuclear reactor planning, approval, and licensing process is very complicated in China, involving many government agencies, each with its own set of complex and confusing procedures. China underwent extensive efforts to ensure that safety measures are comprehensive and plants are reviewed often and thoroughly. The National Nuclear Safety Administration is the licensing and regulatory body in China in charge of nuclear safety (Lin et al. 2012).

5.6.7 Thorium-Based Fuel Cycles

In a thorium-based fuel cycle, thorium atoms absorb neutrons in the reactor and are, through a chain of reactions, transmuted to uranium, which is then used as the nuclear fuel. Natural thorium is three times more abundant than uranium, and China's thorium deposit level of 286,355 tons is the second-largest in the world (Yang and Huang 2010). Thorium dioxide has a higher melting point, higher thermal conductivity, and greater chemical stability than uranium dioxide. Thorium-based fuel cycles can also be more efficient. However, thorium has no fissile isotopes, which means that fissile materials have to be added to the reactor, making the process more complex. Current research on thorium reactors is limited mostly to small-scale units. India, with abundant thorium deposits of approximately 343,000 tons, has used this substance in its energy production. Some of the largest thorium-based reactors, India's Kakrapar Atomic Power Stations 1 and 2, are 220 MW pressurized heavy-water reactor (PHWR) reactors using thorium oxide pellets as fuel (ibid.).

5.6.8 Interim Conclusion

Nuclear power is a crucial part of China's future energy plan due to its low environmental effect and high energy efficiency. Furthermore, it is not as variable as some of the renewable energy sources, such as solar and wind. China has spent a significant amount of effort to research, develop, and construct nuclear power plants utilizing advanced technologies with the goal of diversifying its energy mix.

However, nuclear power expansion in China is limited by its low uranium reserves, and nuclear waste management will escalate as nuclear power production increases. There is also growing concern worldwide over the safety of nuclear power after recent incidents, as the potential effects of a nuclear plant meltdown are extremely detrimental.

After the Fukushima incident, China froze all nuclear power development and conducted thorough safety inspections of all existing nuclear plants. China has recently resumed nuclear power development and has not altered its goal of increasing domestic nuclear power capacity. Therefore, China will continue to be a key player in nuclear technology research and development in the foreseeable future.

5.7 CONCLUSION

In the past, the Chinese government held up its part of the social contract by promoting economic growth, albeit through the uninhibited exploitation of cheap, abundant labor and natural resources. Although this has resulted in impressive growth numbers, the patterns that these characteristics produced—large levels of investment with low productivity, overinvestment in export-oriented heavy industry and low-level manufacturing, reliance on cheaply priced energy, and inefficient energy use have proven unsustainable (Australian National University Crawford School of Public Policy: Centre for Climate Economics and Policy [Crawford School] 2012). In response to these systematic problems, the Chinese government is embracing alternative energy sources. This support has driven rapid growth in China's green energy industry, positioning the country as a major player in the global industry going forward. Serving as this movement's foundation, China's abundance of energy resources allows it considerable potential for continued development. There are, however, numerous problems underlying this heady story of green energy's explosion.

First, supporting infrastructure has not kept pace with the installation of new capacity. This is especially evident in the case of the wind industry, leading to valuable electricity going to waste despite China's ravenous hunger for energy. Deficient infrastructure also constrains the natural gas industry, as development was slowed for many years due to inadequate pipelines. The government has already shown signs that it plans to learn from these mistakes by controlling the expansion of solar power. Furthermore, the construction of a unified smart grid system by 2020, which includes accelerated construction of ultrahigh voltage transmission networks as the first step, is included in China's 12th Five-Year Plan.

Second, the status of the renewable energy industry as an industry driven almost exclusively by government support also causes unintentional consequences. Government-granted incentives have fueled the expansion but have not always aligned the interests of all the relevant parties. It is worth noting that the Chinese provinces have a certain degree of autonomy and can be somewhat decentralized (Kidd 2013). This has created problems in the hydropower industry because suboptimal dam placements stemming from a lack of coordination have exacerbated environmental harm. A lack of centralized planning for grid construction has also led to weak interregional connections because each grid company has had little incentive for intercompany cooperation. For a country with such a large geographical disparity between its supply of energy resources and its energy demand clusters, this problem is especially debilitating.

Third, rapid growth and imperfect government policy have resulted in a lack of quality in Chinese products. Technological innovation is still driven by development in other countries, despite China's rapidly growing share of global production. Fears of large-scale equipment failures in natural gas, wind, hydro, solar, and especially nuclear power, remain a concern going forward. Furthermore, to position itself as a technological and economic leader in alternative energy, China must continue to invest in and provide incentives for research and development.

Fourth, China's desire to promote its alternative energy industry and energy security has led to international tensions. As discussed earlier, this is reflected in downstream countries' complaints about China damming its rivers. Moreover, tensions between China and Europe and the United States escalated in 2012 as a result of complaints about the legality and fairness of China's aggressive solar industry subsidies. Finally, China's aggressive acquisitions of energy resources abroad have led to fear as well as criticism of the country's no-strings-attached, look-the-other-way approach when dealing with corrupt governments.

Fifth, China's abundant coal acts as a barrier to sustainable development, while increasing the energy and carbon intensities of Chinese industries. Currently, natural gas, hydro, and nuclear power represent cost-competitive sources of energy; but the low cost and abundance of coal means that the development of these industries still requires government support. Furthermore, as more variable power from solar and wind sources is fed into the grid, base-load power will become especially important. Coal-fired plants are extraordinarily inflexible; hence, a greater integration of hydropower, with its unique flexibility and storage capabilities, and natural gas, with its relatively greater flexibility, will be necessary to smooth out variability.

Despite all these imperfections, China remains a key player that will drive the energy industry for years to come. As long as the government's motivations do not change, the country's alternative energy industry is poised to grow tremendously as a result of abundant energy resources and the enormous potential for economies of scale.

REFERENCES

Australian National University Crawford School of Public Policy: Centre for Climate Economics and Policy (Crawford School). *China's Energy Reform and Climate Policy: the Ideas Motivating Change*, by Olivia T. Boyd. Acton, Australia: Australian National University, 2012. http://ideas.repec.org/p/een/ccepwp/1205.html (accessed April 15, 2013).

Bloomberg News. Wood Mackenzie lowers forecast China LNG demand for 2015. Last modified November 10, 2011 (accessed January 10, 2013). http://www.bloomberg.com/news/2011-11-10/wood-mackenzie-lowers-forecast-china-lng-demand-for-2015-on-price-concern.html.

BP p.l.c. *BP Statistical Review of World Energy: June 2011*. London: BP p.l.c., 2011. http://www.bp.com/sectionbodycopy.do?categoryId=7500&contentId=7068481 (accessed January 10, 2013).

Bradsher, K., and M. Wald. A measured rebuttal to China over solar panels. The *New York Times*, March 21, 2012. http://www.nytimes.com/2012/03/21/business/energy-environment/us-to-place-tariffs-on-chinese-solar-panels.html?pagewanted=all (accessed January 10, 2013).

Brennand, T.P. Natural gas, a fuel of choice for China. *Energy for Sustainable Development* 4: 81–83, 2001.

Brown, A., S. Müller, and Z. Dobrotková. *Renewable Energy Markets and Prospects by Technology*. Paris: International Energy Agency (IEA)/OECD, 2011.

Chen, J. Development of offshore wind power in China. *Renewable and Sustainable Energy Reviews* 9: 5013–5020, 2011.

Cherni, J.A., and J. Kentish. Renewable energy policy and electricity market reforms in China. *Energy Policy* 35: 3616–3629, 2007.

China Baike. China's oil and natural gas reserves report. Last modified April 6, 2011 (accessed January 10, 2013). http://www.chinabaike.com/z/yj/735878.html.

China Center for Energy and Development (CCED). *China's Natural Gas Reserves*, by W. Shi, N. Ma, and X. Li. Beijing: Peking University, 2012. http://www.nsd.edu.cn/cn/userfiles/Other/2012-07/2012072619140669246017.pdf (accessed April 24, 2013).

China Daily. China to increase coal-bed methane output. Last modified January 17, 2012 (accessed January 10, 2013). http://www.chinadaily.com.cn/bizchina/2012-01/17/content_14459876.htm.

China National Nuclear Corporation (CNNC). Nuclear plant. Last modified 2011 (accessed May 5, 2013). http://www.cnnc.com.cn/tabid/661/Default.aspx.

China National Petroleum Corporation (CNPC). 2010 Annual Report. http://www.cnpc.com.cn/resource/english/images1/pdf/10AnnualReportEn/CNPC_Annual_Report_2010.pdf?COLLCC=2019128045& (accessed January 10, 2013).

China United Coalbed Methane Corporation, Ltd (CUCBM). *Achievements of the International Cooperation in the Past 10 Years and the Progress of Science and Technology.* Beijing: CUCBM, 2009. http://www.uschinaogf.org/Forum9/pdfs/Benguang_English.pdf (accessed January 10, 2013).

China Wind Power Center (CWPC). Renewable energy law of the People's Republic of China. Last modified 2012 (accessed January 10, 2013). http://www.cwpc.cn/cwpc/en/node/6548.

ChinaMaps.org. Blank China Map. Last modified 2012 (accessed May 27, 2013). http://www.chinamaps.org/china/china-blank-map.html.

Chinese Renewable Energy Industries Association (CREIA), Global Wind Energy Council (GWEC), and Greenpeace. *2010 China Wind Power Outlook*, by Li Junfeng, Shi Pengfei, and Gao Hu. Beijing: Greenpeace, 2010. http://www.greenpeace.org/eastasia/Global/eastasia/publications/reports/climate-energy/2010/2010-china-wind-power-outlook.pdf (accessed January 10, 2013).

Cutler, R.M. China's gas imports jump. *Asia Times*. Last modified 2011 (accessed January 10, 2013). http://atimes.com/atimes/China_Business/MF24Cb01.html.

Dai, J., Y. Ni, Q. Zhou, C. Yang, and A. Hu. Significances of studies on natural gas geology and geochemistry for natural gas industry in China. *Petroleum Exploration and Development* 35(5): 513–525, 2008.

Deutsche Bank Group: DB Climate Change Advisors. *Hydropower in China: Opportunities and Risks*, by Michael Carboy, Camilla Sharples, Lucy Cotter, and Jane Cao. Frankfurt am Main, Germany: Deutsche Bank AG, 2011. http://www.banking-on-green.com/flash/Hydropower_in_China-Opportunities_and-Risks.pdf (accessed January 10, 2013).

Ding, Q., and Y. Liu. Nation plans to import more uranium. *China Daily*. Last modified March 13, 2012 (accessed January 10, 2013). http://www.chinadaily.com.cn/cndy/2012-03/13/content_14818316.htm.

Dynabond Powertech. China Medium and Long Term Development Planning for Nuclear Power (2005–2020). Last modified 2010 (accessed May 29, 2013).

Engelsiepen, J. Hydro power in China. *Ecology Global Network*. Last modified 2013 (accessed April 20, 2013). http://www.ecology.com/2013/03/28/hydro-power-in-china/.

The George Washington University Trachtenberg School of Public Policy and Public Administration (The Trachtenberg School). *China's Solar Policy: Subsidies, Manufacturing Overcapacity & Opportunities*, by Alim Bayaliyev, Julia Kalloz, and Matt Robinson. Washington, DC: George Washington University, 2011. solar.gwu.edu/Research/ChinaSolarPolicy_BayaKallozRobins.pdf (accessed April 20, 2013).

Global Wind Energy Council (GWEC). *Global Wind Report: Annual Market Update 2010*. Brussels: GWEC, 2010. http://gwec.net/wp-content/uploads/2012/06/GWEC_annual_market_update_2010_-_2nd_edition_April_2011.pdf (accessed January 10, 2013).

Greenpeace, European Photovoltaic Industry Association (EPIA), and World Wildlife Fund (WWF). *2007 China Solar PV Report*, by J. Li, S. Wang, M. Zhang, and M. Lingjuan. Beijing: China Environmental Science Press, 2007. http://www.wwfchina.

org/english/downloads/ClimateChange/china-pv-report-en.pdf (accessed January 10, 2013).

Haley, U.C.V., and D.A. Schuler. Government policy and firm strategy in the solar photovoltaic industry. *California Management Review*. HBS Case Study, 2011.

Hong Kong Nuclear Investment Company Limited (HKNIC). About Daya Bay. Last modified 2010 (accessed January 10, 2013). https://www.hknuclear.com/dayabay/about/pages/about.aspx.

Huang, H., and Y. Zheng. Present situation and future prospect of hydropower in China. *Renewable and Sustainable Energy Reviews* 13: 1652–1656, 2009.

IBISWorld. *IBISWorld Industry Report 4412: Hydroelectric Power in China*. Los Angeles: IBISWorld, 2011. Retrieved January 10, 2013 from IBISWorld database.

Institute of Defence and Strategic Studies Singapore (IDSS). *China in the Mekong River Basin: The Regional Security Implications of Resource Development on the Lancang Jiang*, by Evelyn Goh. Singapore: IDSS, 2004. http://www.rsis.edu.sg/publications/WorkingPapers/WP69.pdf (accessed January 10, 2013).

Institute of Energy Economics Japan (IEEJ). *Natural Gas in China*, by Kaoru Yamaguchi and Keii Cho. Tokyo: IEEJ, 2003. http://eneken.ieej.or.jp/en/data/pdf/221.pdf (accessed January 10, 2013).

International Energy Agency (IEA). *Natural Gas in China: Market Evolution and Strategy*, by Nobuyuki Higashi. Paris: IEA, 2009. www.iea.org/publications/freepublications/publication/nat_gas_china.pdf (accessed April 15, 2013).

International Energy Agency (IEA). *Integration of Renewables: Status and Challenges in China*, by Kat Cheung. Paris: OECD/IEA, 2011. http://www.iea.org/publications/freepublications/publication/Integration_of_Renewables.pdf (accessed November 19, 2012).

International Energy Agency (IEA), and Energy Research Institute of the National Development and Reform Commission of the People's Republic of China (ERI). *Technology Roadmap: China Wind Energy Development Roadmap 2050*, by Wang Zhongying, Shi Jingli, Zhao Yongqiang, Gao Hu, Tao Ye, Kat Cheung, Qin Haiyan, Liu Mingliang, Zhao Jinzhuo, Cao Boqian, Xie Hongwen, Wang Jixue, Guo Yanheng, Wang Hongfang, Chen Yong'an, Zhu Rong, Yang Zhenbin, Bai Jianhua, Jia Dexiang, Xin Songxu, Liu Jiandong, Yuan Jingting, and Zhu Shuquan. Paris: OECD/IEA, 2011. http://www.iea.org/publications/freepublications/publication/china_wind.pdf (accessed May 26, 2013).

International Hydropower Association (IHA). International Commission on Large Dams, Implementing Agreement on Hydropower Technologies and Programmes/International Energy Agency, and Canadian Hydropower Association. *Hydropower and the World's Energy Future*. Paris: IEA, 2000. http://www.ieahydro.org/reports/Hydrofut.pdf (accessed January 10, 2013).

International Renewable Energy Agency (IRENA). *Renewable Energy Technologies: Cost Analysis Series: Hydropower*. United Arab Emirates: IRENA, 2012a. http://www.irena.org/DocumentDownloads/Publications/RE_Technologies_Cost_Analysis-HYDROPOWER.pdf (accessed January 10, 2013).

International Renewable Energy Agency (IRENA). *Renewable Energy Technologies: Cost Analysis Series: Wind Power*. United Arab Emirates: IRENA, 2012b http://www.irena.org/DocumentDownloads/Publications/RE_Technologies_Cost_Analysis-WIND_POWER.pdf (accessed January 10, 2013).

Kidd, S. Comment—Nuclear in China—Now back on track? *Power Engineering*. Last modified 2013 (accessed April 20, 2013). http://www.power-eng.com/news/2013/04/15/comment-nuclear-in-china-now-back-on-track.html.

Lin, A., J. Li, J. Portner, and C. Xu. China moves to strengthen nuclear safety standards and moderate the pace of its nuclear power development. *Switchboard: Natural Resources Defense Council Staff Blog.* Last modified December 23, 2012 (accessed May 29, 2013).

Liu, C. China uses feed-in tariff to build domestic solar market. The *New York Times* 2011. http://www.nytimes.com/cwire/2011/09/14/14climatewire-china-uses-feed-in-tariff-to-build-domestic-25559.html?pagewanted=all (accessed April 20, 2013).

Loo, F. China's polysilicon prices slide on weak demand, rising supply. Last modified April 25, 2013 (accessed May 2, 2013). http://www.icis.com/Articles/2013/04/25/9662409/chinas-polysilicon-prices-slide-on-weak-demand-rising.html.

Marigo, N. The Chinese silicon photovoltaic industry and market: A critical review of trends and outlook. *Progress in Photovoltaics: Research and Applications* 15: 143–162, 2007.

Ming, Z., X. Song, M. Mingjuan, and Z. Xiaoli. New energy bases and sustainable development in China: A review. *Renewable and Sustainable Energy Reviews* 20: 169–185, 2013.

Ministry of Commerce of the People's Republic of China: Department of Foreign Trade (MOFCOM: DOFT). Summary of the economic development of domestic natural gas industry in 2010. Last modified January 31, 2011 (accessed April 23, 2013). http://wms.mofcom.gov.cn/article/zt_gypck/subjecto/201101/20110107386123.shtml.

Ministry of Commerce of the People's Republic of China, Department of Foreign Trade (MOFCOM: DOFT). Summary of the economic development of domestic natural gas industry in 2012. Last modified February 21, 2013 (accessed April 23, 2013). http://wms.mofcom.gov.cn/article/zt_gypck/subjecto/201302/20130200033304.shtml.

Mirae Asset Financial Group. *Emerging Markets Insight: Power Generation Sector China.* Seoul: Mirae Asset Financial Group, Ltd., 2012.

National Bureau of Statistics of China (NBSC). *China Energy Statistical Yearbook 2009.* Beijing: China Statistics Press, 2010.

National Bureau of Statistics of China (NBSC). *China Energy Statistical Yearbook 2010.* Beijing: China Statistics Press, 2011.

National Development and Reform Commission (NDRC). Policies on National Gas Use. Last modified August 2007 (accessed April 23, 2013). http://www.sdpc.gov.cn/zcfb/zcfbtz/2007tongzhi/W020070904363743792492.pdf.

Neiva, L., and L. Gama. The importance of natural gas reforming. In *Natural Gas*, edited by Primoz Potocnik, 71–86. New York: InTech, 2010. http://www.intechopen.com/books/natural-gas/the-importance-of-natural-gas-reforming (accessed April 23, 2013).

Ng, M. Economic impact of the photovoltaic industry in China after the financial crisis of 2009. *The Chinese Economy* 44(3): 22–44, 2011.

North West Shelf Venture. *North West Shelf Venture.* 2010. http://www.nwsg.com.au/download/NWSV%20Corporate%20Brochure%20Sep2009.pdf (accessed January 10, 2013).

OECD Nuclear Energy Agency (NEA) and International Atomic Energy Agency (IAEA). *Forty Years of Uranium Resources, Production and Demand in Perspective: "The Red Book Retrospective."* Paris: OECD Publishing, 2006.

OECD Nuclear Energy Agency (NEA) and International Atomic Energy Agency (IAEA). *Uranium 2005: Resources, Production and Demand.* Paris: OECD Publishing, 2006. http://www.mdcampbell.com/IAEARedBook2005.pdf (accessed January 10, 2013).

OECD Nuclear Energy Agency (NEA) and International Atomic Energy Agency (IAEA). *Uranium 2007: Resources, Production and Demand.* Paris: OECD Publishing, 2008.

OECD Nuclear Energy Agency (NEA) and International Atomic Energy Agency (IAEA). *Uranium 2009: Resources, Production and Demand.* Paris: OECD Publishing, 2010.

OECD Nuclear Energy Agency (NEA) and International Atomic Energy Agency (IAEA). *Uranium 2011: Resources, Production and Demand.* Paris: OECD Publishing, 2012.

Reuters. Chinamining.org. China's Sinopec to clinch huge Australia gas deal-Source. Last modified April 21, 2011 (accessed January 10, 2013). http://www.chinamining.org/Companies/2011-04-21/1303365240d44809.html.

Roberts, B.J. Solar Resource of China: Direct Normal Solar Radiation. Last modified July 26, 2012 (accessed May 26, 2013). http://www.nrel.gov/analysis/staff/billy_roberts.html.

Shi, G.H., Y.Y. Jing, S.L. Wang, and X.T. Zhang. Development status of liquefied natural gas industry in China. *Energy Policy* 38: 7457–7465, 2010.

Standard Chartered Bank. *Coalbed Methane in China*, by Han Pin Hsi and Evan Li. New York: Standard Chartered Bank, 2010. http://enviro-energy.com.hk/admin/uploads/files/1292913959Coalbed_combined%20%28Standard%20Chartered%29%28Dec%202010%29.pdf (accessed January 10, 2013).

UNESCO. Indicator name capability for hydropower generation. Last modified 2012 (accessed January 10, 2013). http://webworld.unesco.org/water/wwap/wwdr/indicators/pdf/H11_Capability_for_hydropower_generation.pdf.

U.S. Energy Information Administration (U.S. EIA). China Country Analysis Brief. Last updated February 12, 2013 (accessed March 10, 2013). http://www.eia.gov/countries/country-data.cfm?fips=CH#ng.

Wang, J., and S. Tseng. Small is Beautiful? Small Hydro Power and the Paradox of the Water-Energy Nexus in China. Paper presented at the *54th Annual Meeting of the American Association for Chinese Studies, Georgia Institute of Technology in Atlanta, GA, October 12–14.* Atlanta, GA: Georgia Institute of Technology, 2012.

Wang, Y., S. Guo, and W. Yu. Opportunities for and challenges facing China's oil and natural gas exploration and development in a low-carbon economy. *Energy Procedia* 5: 2048–2053, 2011.

Wang, Z., H. Qin, and J. Lewis. China's wind power industry: Policy support, technological achievements, and emerging challenges. *Energy Policy* 51: 80–88, 2012.

World Nuclear Association (WNA). Supply of Uranium. Last modified August 2012 (accessed January 10, 2013). http://www.world-nuclear.org/info/inf75.html.

World Nuclear Association (WNA). Nuclear Power in China. Last modified April 2013 (accessed April 10, 2013). http://www.world-nuclear.org/info/inf63.html.

Wu, Z., and T.A. Siddiqi. The role of nuclear energy in reducing the environmental impacts of China's energy use. *Energy* 20(8): 777–783, 1995.

Yamaguchi, K., and K. Cho. 2003. "Natural gas in China." Last modified 2003 (accessed January 10, 2013). http://eneken.ieej.or.jp/en/data/pdf/221.pdf.

Yang, C. Wind power being wasted. *Global Times*, 2012. http://www.globaltimes.cn/content/724922.shtml (accessed April 20, 2013).

Yang, G., and W. Huang. The status quo of China's nuclear power and the uranium gap solution. *Energy Policy* 38: 966–975, 2010.

Zhang, S., and J. Qi. Small wind power in China: Current status and future potentials. *Renewable and Sustainable Energy Reviews* 15(5): 2457–2460, 2011.

Zhang, D., Q. Chai, X. Zhang, J. He, L. Yue, X. Dong, and S. Wu. Economical assessment of large-scale photovoltaic power development in China. *Energy* 40: 370–375, 2012.

Zhou, S., and X. Zhang. Nuclear energy development in China: A study of opportunities and challenges. *Energy* 35: 4282–4288, 2010.

Zhou, Y. Why is China going nuclear? *Energy Policy* 38: 3744–3762, 2010.

Zhou, Y., C. Rengifo, P. Chen, and J. Hinze. Is China ready for its nuclear expansion? *Energy Policy* 39: 771–781, 2011.

Zhu, W. China Daily USA. Coal-bed gas production to be doubled by 2015. Last modified July 16, 2011 (accessed January 10, 2013). http://usa.chinadaily.com.cn/epaper/2011-07/16/content_12916283.htm.

6

Renewable Energy in Spain: A Quest for Energy Security

José Normando Bezerra, Jr.

CONTENTS

6.1 INTRODUCTION

Rapid economic growth and scarce fossil fuel reserves have pushed Spain to invest heavily in renewable energy over the past few decades. A successful feed-in tariff system has spurred private sector investment in renewable energy by offering competitive financial returns—helping the country become a world leader in both wind and solar energy technology. Spain still relies on foreign suppliers to provide three-quarters of its energy consumption, but investments in renewable sources have prevented an even greater dependence while reducing the country's carbon footprint. Major energy industry players in Spain have become internationalized, and some already derive most of their sales from overseas operations. This move has paid off in light of Spain's currently (in 2013) sluggish domestic economy and the sustained growth in markets outside Europe. This chapter will provide a brief overview of the renewable energy sector in Spain and touch on the sector's future prospects.

6.2 ENERGY CHALLENGES AND THE ONSET OF NUCLEAR POWER IN SPAIN

Recently rattled by the global financial crisis and suffering from a 25% unemployment rate, Spain saw its total energy consumption decline 12% between 2007 and 2011, which was equivalent to a decline of roughly half a ton of oil-equivalent per capita (Figures 6.1 and 6.2; Eurostat 2012). However, this recent economic stagnation stands in stark contrast to the unbridled growth the country enjoyed over the course of the past five decades, when it went from a predominantly rural society to one of the most successful economies in the world. Since the growth engine was set in motion in the early 1960s, the Spanish economy has grown in all but 4 years (1981, 1993, 2009, and 2010). In fact, Spain has increased its real gross domestic product (GDP) by more than 200% since 1970, 50% more than the average growth of its European peers (Figure 6.3; Eurostat 2012; The World Bank Group [WBG] 2011). The social and economic transformations during this time multiplied Spain's energy requirements sevenfold, whereas other major European countries' energy demands experienced less aggressive paces of growth (Eurostat 2012). The country's economic success, therefore, brought with it an unwanted reality of energy dependence.

In the 1960s, hydropower accounted for nearly all of the electricity produced in Spain. However, this energy source was unable to meet the demand of the following decades, as most of the country's hydroelectric potential had already been exploited (WBG 2011). In light of hydro's limited capacity and a lack of domestic energy-producing alternatives, the government was forced to rely on foreign fossil fuels, which were relatively

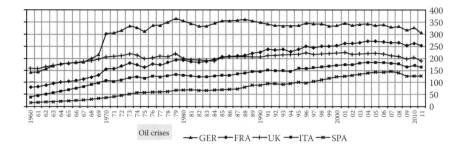

FIGURE 6.1
Historical primary energy consumption by country (in mtoe; million tons oil-equivalent). (Adapted from WBG. Data catalog. 2011. http://data.worldbank.org/data-catalog.)

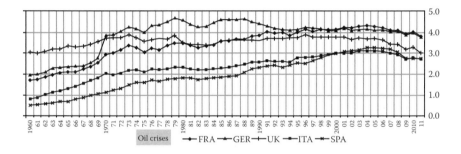

FIGURE 6.2
Historical energy consumption per capita by country (in toe). (Adapted from WBG. Data catalog. 2011. http://data.worldbank.org/data-catalog.)

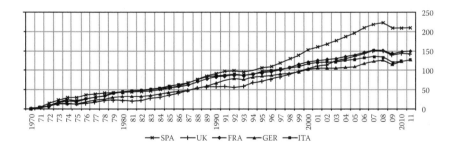

FIGURE 6.3
Historical cumulative real GDP growth by country (as a percentage, with 1970 as the base year). (Adapted from Eurostat, European Commission, Energy Statistics. http://epp.eurostat.ec.europa.eu/portal/page/portal/statistics/search_database; and WBG. Data catalog. 2011. http://data.worldbank.org/data-catalog.)

inexpensive and readily available at the time. Spain's growing appetite for energy showed no sign of abating, not even during the oil crises of the 1970s, when most of its European counterparts restricted their energy consumption (Figure 6.1). For example, although France and the United Kingdom each reduced consumption by 8.5% between 1973 and 1975, Spain increased its own consumption by 11.5% (ibid.).

But progress and its concomitant energy consumption growth were inevitably accompanied by greater dependence on foreign energy imports, which almost doubled between 1960 and 1970 to 74% of Spain's overall demand (Figure 6.4; ibid.). This growing dependence had been expected, however, because of Spain's limited reserves of coal and almost nonexistent reserves of oil and gas. To safeguard the country's economic growth, General Franco's regime began looking for alternatives as early as the late 1940s, finally settling on nuclear energy as the key to a future of greater energy security. Franco's government established the *Junta de Energía Nuclear* (Nuclear Energy Commission) in 1951, while the technology was still nascent; 15 years would pass before the first Spanish nuclear facility would be in full operation (Centro de Investigaciones Energéticas, Medioambientales y Tecnológicas [CIEMAT] 2013a).

As a viable but already controversial alternative at the time, nuclear energy was the Spanish government's chief choice to tackle the country's growing energy deficit. The regime announced the dawning of the "industrial nuclear era" in 1968 with the opening of the José Cabrera Nuclear Power Station in Almonacid de Zoritain the province of Guadalajara (ABC Hemeroteca 2011). This inauguration was marred, however, by Spain's young antinuclear movement, which had been inspired by the antinuclear movements in North America and Europe. The transition to

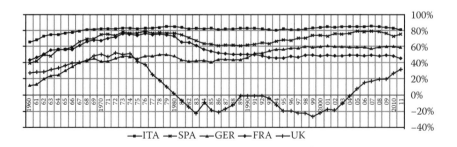

FIGURE 6.4
Net imported energy as a percentage of total consumption by country. (Adapted from WBG. Data catalog. 2011. http://data.worldbank.org/data-catalog.)

democracy in the late 1970s and early 1980s fueled this fledging movement, as massive activist protests in Spain's major cities culminated in a moratorium on new facilities, signed into law by the socialist government of Felipe González in 1984. The ban appeased the protesters but forced the country to look once again for new energy alternatives (Foro Nuclear 2009). It also left Spaniards with a financial liability felt to this day: utility companies that were building nuclear plants in 1984 were granted compensation packages that, even today, make energy bills more expensive. For example, in 2012, €53 million were paid to Spanish utilities, leaving a balance of €317 million at the end of the year in the nuclear liability owed by the government (Europa Press 2013).

Despite the ban on new facilities and the well-known potential hazards, nuclear power has proven to be a reliable source of energy since the Zorita reactor was first commissioned, reaching its peak share of the electricity matrix at 38% in 1989 (Figure 6.5; WBG 2011). Indeed, the construction of nuclear plants helped Spain reduce its energy import ratio to a historic low of 60% of its overall energy consumption in the late 1980s, down from 80% only a decade before. After the introduction of the construction moratorium, the existing reactors reached an annual output of 56 GWh in 1989. Production has not budged much since, fluctuating between 54 and 64 GWh/year (ibid.). In 2012, a total of eight active reactors in six different facilities provided approximately 20% of the total electricity produced in the country (Foro Nuclear 2013). That share may decline, however, if Spain follows in the footsteps of Germany and Switzerland, which both recently announced a phase-out of their nuclear facilities in the aftermath of the Fukushima-Daiichi disaster in Japan (Brackmann et al. 2011).

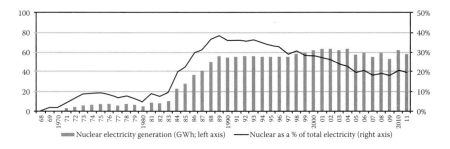

FIGURE 6.5
Historical development of nuclear power generation in Spain. (Adapted from WBG. Data catalog. 2011. http://data.worldbank.org/data-catalog.)

6.3 THE QUEST FOR RENEWABLE ENERGY

The 1984 ban on nuclear facilities did little to improve Spain's energy security. In fact, it has probably extended Spain's energy dependence. However, it may have spurred the government to establish the Instituto para el Ahorro y la Diversificación de la Energía (IDAE, Institute for Energy Savings and Diversification) that same year. The IDAE now operates under the auspices of the Secretary of Energy and has been responsible for stimulating and overseeing the renewable energy sector for the past three decades (IDAE 2011).

The IDAE's main contribution to Spain's energy security has been arguably the creation and implementation of a feed-in tariff system, or a special regime that guarantees above-market prices for producers who invest in renewable energy installations. This subsidy was necessary to overcome the high costs of nascent technologies (such as solar and wind energy) that would not have survived otherwise. The system was set up in the late 1990s amid rapid economic growth, a severe dependence on foreign energy, and high oil price volatility. One of its main goals was to establish well-defined targets for the share of renewables in Spain's energy mix (ibid.).

As will be described later in greater detail, the special regime has been extremely successful: between 1994 and 2011, renewables went from providing 6.1% to 11.3% of all the primary energy produced in Spain and from 17.7% to 33.1% of all the electricity generated (Section 6.3; Eurostat 2012). Despite the success of renewables in Spain thus far, the growth of the country's energy demand has far outstripped the additional supply provided by these sources. This growth has pushed Spain's energy import ratio back to the 70% to 80% range more recently. As explained previously, the country's increasing energy demand can be attributed to the rapid economic growth it experienced through 2008—221% in real terms since 1970, far greater than any other major European economy (WBG 2011). Its impressive performance has been even more pronounced since the mid-1990s: Spain averaged a strong 3.8% annual GDP growth rate between 1995 and 2008. This allowed the country to then claim the position of eighth largest economy in the world and its growth was considerably faster than that of other major European economies. The United Kingdom grew an average of 3.1% during the same period, France grew by 2.2%, Germany grew by 1.7%, and Italy grew by a less impressive 1.4%.

More investments are needed to improve Spain's foreign dependence ratio in energy, but efforts to encourage such investments have been undermined by the recent financial crisis. The country's economic woes forced the government of Mariano Rajoy to introduce austerity measures in early 2012, including a moratorium on new renewable energy facilities under the special regime (Thomson Reuters 2013).

These measures are meant to be temporary but will remain in effect until the "tariff deficit" is under control (Section 6.3.2), a goal that may take many years to achieve (ibid.). This could thwart the official energy plan for a greater share of renewables in the energy mix by 2020 (IDAE 2011). This could be compromised if subsidies are cut permanently or if technology does not advance quickly enough to reduce plant setup and maintenance costs.

The next section offers an overview of the regulation supporting the feed-in tariff system to provide a better understanding of Spain's special regime and to convey a sense of what the austerity moratorium really means for the renewable energy sector.

6.3.1 A Brief History of the Regulatory Framework and the Feed-In Tariff System

The Spanish renewables industry was born officially with the passing of the "Law for Energy Conservation," just after the second oil crisis of the 1970s (Law 82/1980). By encouraging investments in mini-hydraulic plants, this law set the stage for the development of a green energy sector that would potentially diversify Spain's energy supplies away from foreign fossil fuels.

The law was a major step toward the establishment of a renewable energy sector, but it did not provide enough incentives to entice inves- tors because long-term energy prices and other legal provisions were poorly defined (Mendonça 2007). It was not until 1994 that the current feed-in tariff system began to take shape: Royal Decree (RD) 2366/1994 set a fixed premium price for renewable energy for a period of at least 5 years (RD 2366/1994). Investors welcomed the news, but many thought that even longer-term guarantees were necessary to meet the minimum required returns for their investments.

Three years later, the Electric Power Act 54/1997 opened the doors for the establishment of a wholesale market for electricity, which was fully liberalized in 2003 (Law 54/1997). This liberalization meant that industrial and residential consumers would be able to choose their electricity providers. Most importantly, the act clarified the rights and obligations of renewable energy producers and established the idea of a special pricing regime for renewable energy. To attract investments to the nascent renewables sector, this act, together with RD 2818/1998, reinforced renewables' preferential access to the power grid, allowing plants of up to 50 MW of capacity to sell—at a premium—nearly all the power they could supply. Traditional sources of energy were now used only to the extent that renewables did not meet the electricity demand (RD 2818/1998).

The royal decrees that followed Act 54/1997 helped make the feed-in tariff system more predictable for investors by expanding the existing remuneration schedule to provide fixed energy prices for most of the expected operational life span of a renewable energy installation. There have been a few changes in the energy prices guaranteed to renewable energy producers over the years, but the general philosophy remains that producers who meet the requirements of the special regime should be paid prices above those observed in the wholesale market. This subsidization represents a sacrifice shared currently by all Spaniards to achieve more energy security and a healthier environment in the future.

The private sector responded vigorously to the aforementioned incentives by significantly increasing renewable energy capacity, first with wind power in the late 1990s and then with solar capacity more recently. According to Ángel González Palacios, Director of Special Projects with Gamesa, the world's second-largest wind turbine producer, Spain's feed-in tariff system was viewed favorably by investors because it eliminated most of the uncertainties surrounding their expected future revenues (it should be noted that comments and opinions not expressly attributed to interviewees in this chapter do not necessarily represent their points of view; the author takes sole responsibility for any inaccuracies and misrepresentations; González 2012). In the feed-in tariff model adopted by Spain, Germany, and Denmark, there is usually no government help upfront with respect to paying part of the initial capital costs of projects, but instead the government guarantees a legally set, minimum price per kilowatt-hour to investors over the long term (Fraunhofer Institute Systems and Innovation Research 2004).

As a result, in 2010, Spain was best in class among its European peers in terms of green energy. Renewables accounted for approximately 6% of all the primary energy produced in the country in 2005 (mostly hydro) and almost double that share in 2010 (Figure 6.6). The effect of the feed-in tariff system was even more pronounced in terms of electricity generation: Spain is ahead of second-best Italy by an 11% margin (Figure 6.7), as the share of electricity provided by renewables in Spain grew from 14.3% in 2005 to 33.1% in 2010 (Eurostat 2012).

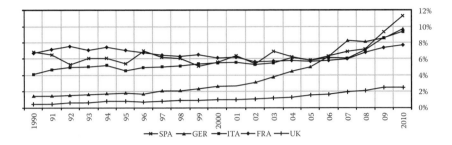

FIGURE 6.6
Development of renewables' share of gross primary energy production over time by country. (Adapted from Eurostat, European Commission. Energy Statistics. http://epp.eurostat.ec. europa.eu/portal/page/portal/statistics/search_database, accessed December 26, 2012.)

Until recently, the special regime was regulated by RD 1578/2008 for solar photovoltaic (PV) plants and RD 661/2007 for the other technologies. Both decrees have been superseded by RD-Law 1/2012—the first royal decree passed by the incoming government, which froze the addition of new plants to the special regime (RD 1/2012). Although additional capacity cannot be added to the feed-in tariff system for now, existing facilities do have guarantees for most of their financial incentives.

Finally, even though the special regime is best known for its support of solar, hydro, and wind energy, it also covers several other electricity-generating technologies, such as installations using biomass, geothermal power, marine energy (waves, currents, tides, thermal), waste power (from thermal heat), and cogeneration. In other words, other types of renewable technologies and even nonrenewable ones, such as cogeneration, have been strongly encouraged (RD 1578/2008). On a less positive note, this has also contributed to the huge tariff deficit (which is discussed next).

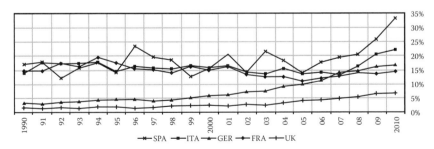

FIGURE 6.7
Development of renewables' share of electricity production over time by country. (Adapted from Eurostat, European Commission, Energy Statistics. http://epp.eurostat.ec.europa. eu/portal/page/portal/statistics/search_database, accessed December 26, 2012.)

6.3.2 The Tariff Deficit and the Critics of the Feed-In Tariff System

Although Spain has made remarkable progress in increasing the share of renewables in its energy mix, many argue that this success was achieved at a heavy cost—one of the highest energy prices in Europe and the "so-called tariff deficit" (Thomson Reuters 2013). Spanish households spent on average €0.1597 for every kilowatt-hour consumed in 2011, which is 60% more than the average price in nuclear-friendly France and 77% more than the €0.09/kWh found in Spain in 2005 (Figure 6.8). Industrial consumers, in turn, paid on average 50% more than their French counterparts and 58% more than they did in 2005 (Figure 6.9; Eurostat 2012).

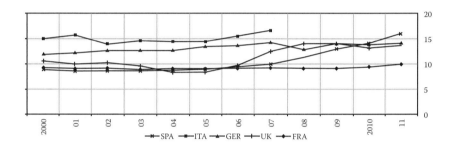

FIGURE 6.8
Average energy prices for households during the 2000s by country (€ cents per kWh). (Adapted from Eurostat, European Commission, Energy Statistics. http://epp.eurostat. ec.europa.eu/portal/page/portal/statistics/search_database, accessed December 26, 2012.)

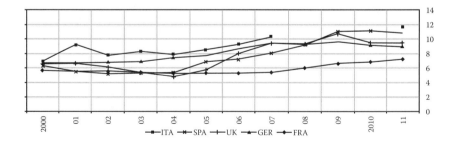

FIGURE 6.9
Average energy prices for industry by country (€ cents per kWh). (Adapted from Eurostat, European Commission, Energy Statistics. http://epp.eurostat.ec.europa.eu/portal/page/portal/statistics/search_database, accessed December 26, 2012.)

As with many other feed-in tariff systems, the Spanish model is based on a subsidy scheme that is ultimately paid for by electricity consumers according to their consumption. As noted in the previous section, renewable energy producers can supply all of their output to the power grid, and utility companies are mandated to pay premium prices for it (i.e., producers of biomass electricity receive higher prices than producers of coal or natural gas electricity). On the supply side, this system has created sufficient financial incentives for investors so that they willingly invest in new renewable capacity. On the demand side, however, the electricity prices charged to final consumers, which is regulated by the government, has not allowed utility companies to cover all their costs associated with purchasing renewable energy at premium prices. The costs and royalty rights payments that must be paid by utilities comprise their regular operating expenses, the premiums they pay to renewable energy producers, compensation for the nuclear plants closed down in the 1980s, and other miscellaneous costs (Thomson Reuters 2013).

The difference between the revenue per kilowatt-hour received (which is held artificially low by the law) and the sum of these costs has accumulated into a large debt owed to the utility companies, best known as the tariff deficit of the electric system (a large part of which has already been securitized by the utility companies). This liability is backed fully by the central government and has been paid back through extra charges on energy bills (La Moncloa 2012). Figures 6.8 and 6.9 show that electricity prices increased by more than 70% for both residential and industrial users between 2000 and 2011, reducing the disposable income of families and hurting the global competitiveness of Spanish businesses (Eurostat 2012). Despite these increasing energy costs, the system has run deficits every year since 2005, which suggests that there is still room for additional price hikes (La Moncloa 2012). Also, because the special regime offers long-term price guarantees (that raise utilities' costs) to existing green energy producers, there will be pressure on the cost side for at least the next several decades, even if advancements in technology allow renewables to become price-competitive in the short term (such advancements would result in reductions in the long-term prices guaranteed under the special regime, but only on a prospective basis).

The tariff deficit began to accumulate in 2000 and reached €30 billion in 2011 (Figure 6.10), with a net balance (accounting for past repayments) of €24 billion (ibid.). Although the premiums paid to green energy producers through the feed-in tariff system are not the only source of the deficit, they

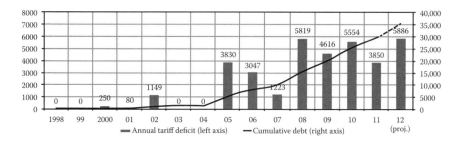

FIGURE 6.10
Annual and cumulative tariff deficits (in € millions); the cumulative debt does not account for past repayments or for past interest payments. (Adapted from La Moncloa, Spanish Government. *Estrategia Española de Política Económica: Balance y Reformas Estructurales Para El Próximo Semestre.* Madrid, Spain: Spanish Government, 2012.)

have contributed a great deal to the current debt level. Asociación Española de la Industria Eléctrica (UNESA—the Spanish Utilities' Association) estimates that the cost of the special regime comprises 20% of the average electric bill (UNESA 2012). To illustrate, approximately €14.7 billion in premiums were disbursed to producers of hydro, solar, and wind energy alone between January 2009 and December 2011 (Figure 6.11). Solar PV producers, for example, were remunerated approximately €0.40/kWh in 2012, or 2.5 times the final retail price charged to households (Comisión Nacional de Energía [CNE] 2012).

Faced with calls from the Eurozone to rein in spending because of international bond market pressures, the Spanish government passed legislation in January 2012 banning the addition of any new facilities to the special regime until the recurring tariff deficit is eliminated. Given the expected deficit of €5.9 billion for the 2012 calendar year—the highest

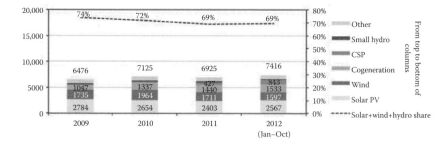

FIGURE 6.11
Premiums paid to participants in the special regime (in € millions). (Adapted from CNE. Información Estadística sobre las Ventas de Energía del Régimen Especial. http://www.cne.es/cne/Publicaciones?id_nodo=144&accion=1&sIdCat=10&keyword=&auditoria=F.)

ever—it may be quite some time before the doors of the feed-in tariff system reopen again, if at all (La Moncloa 2012).

In addition to the subsidies already discussed, another problem of the special regime relates to the preferential access to the grid enjoyed by renewables. This adds an extra layer of operational complexity to the power grid. A greater reliance on intermittent, weather-dependent energy sources reduces the overall predictability of the system, thereby making it more challenging and more costly to operate (RD 1578/2008).

Skeptics criticize the special regime because of this unpredictability problem and the tariff deficit for relatively straightforward cost reasons. They also argue that it is pointless to spend more to curb CO_2 emissions and to fight climate change in Europe without the firm commitment of polluting emerging economies (Bodansky 2011). Such economies typically prioritize industrial development and poverty alleviation much higher than environmental concerns. Thus, according to these critics of the special regime, any benefits obtained from a more environmentally friendly and costly energy mix, such as that of Spain, will simply be offset by the rapid growth of emissions in other countries. The skeptics conclude, therefore, that without binding, across-the-board commitments from all the major polluters to move toward green energy, the costs of the special regime are not worth bearing because Spain's efforts will simply be cancelled out by the emissions of other countries (ibid.).

Feed-in tariff systems are adopted to diversify the energy supply, lessen foreign energy dependence, and reduce a nation's carbon footprint. However, as seen previously, these noble pursuits come at a price, and more Spaniards will likely be critical of the special regime in this time of financial hardship.

6.4 THE RENEWABLE ENERGY PLAN OF 2011–2020

In addition to creating a more favorable investment landscape, the feed-in tariff system established specific targets for renewables in the energy mix. According to the 2005–2010 renewable energy plan, green energy was expected to cover 12.0% of the primary energy needs of Spain and provide 29.4% of the electricity consumed in the country (IDAE 2005). The first target was nearly attained, as renewables accounted for 11.3% of all primary energy in 2010 (or 13.2% of final energy), whereas the second goal

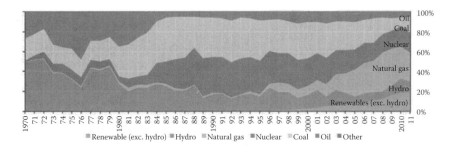

FIGURE 6.12

Electricity generation and its various sources over time. (Adapted from WBG. Data catalog. 2011. http://data.worldbank.org/data-catalog, accessed December 26, 2011.)

was achieved, as 32.3% of the electricity demand was satisfied by renewable resources in 2010, which was an especially good year for hydropower (Figure 6.12; IDAE 2011).

Following the European Union's (EU) broad directives, the IDAE issued a new renewable energy plan in 2011 laying out Spain's targets for 2020: Renewables should provide 19.5% of the primary energy consumed (or 20.8% of final energy), including 38.1% of all electricity generation and 11.3% of the energy used by the transportation sector (ibid.). These targets are expected to be reached through a combination of new installed capacity and improved energy efficiency. Tables 6.1, 6.2, and 6.3, and Figure 6.13 present the targets in greater detail.

The electricity produced in Spain became much greener between 2005 and 2010: Renewables increased their generation by 55 TWh in 2010, more than offsetting fossil fuels' decline of 50 TWh (ibid.). And if the current plan is implemented fully, renewables will add another 49 TWh by 2020,

TABLE 6.1

Renewables' Percentage Share of Final Energy Consumption by Sector, Historic and Projected

Sector	2005 (Actual) [%]	2010 (Actual) [%]	2015 (Target) [%]	2020 (Target) [%]
Heating and cooling	8.9	11.0	13.4	17.3
Transportation	0.8	5.0	8.3	11.3
Electricity	14.5	32.3	33.4	38.1
All sectors	8.2	13.2	17.0	20.8

Source: Adapted from IDAE. *Plan de Energías Renovables 2011–2020.* Madrid, Spain: IDAE, 2011.

TABLE 6.2

Primary Energy Sources, Historic and Projected (Kilotons of Oil-Equivalent), Based on Data Given by IDAE that Assumes Energy Efficiency Improvements over Time

(ktoe)	2005 (Actual)	2010 (Actual)	2015 (Target)	2020 (Target)	Total 2010 (%)	Total 2020 (%)	Growth from 2010 to 2020 (%)
Oil	71,765	62,358	56,606	51,980	47.3	36.4	−16.6
Natural gas	29,116	31,003	36,660	39,237	23.5	27.5	26.6
Renewables	8371	14,910	20,593	27,878	11.3	19.5	87.0
Nuclear	14,995	16,102	14,490	14,490	12.2	10.2	−10.0
Coal	21,183	8271	10,548	10,058	6.3	7.1	21.6
Net electricity import	−116	−717	−966	−1032	−0.5	−0.7	43.9
Total	145,314	131,927	137,931	142,611	100.0	100.0	8.1

Source: Adapted from IDAE. *Plan de Energías Renovables 2011–2020.* Madrid, Spain: IDAE, 2011.

the largest increase among all energy sources. Natural gas will contribute an additional 37 TWh and account for more than one-third of the total supply. Coal has lost considerable ground since 2005 when it represented 28% of all power generation. This decline will persist, as the new plan puts coal's share at 8.2% in 2020, about the same as in 2010. Oil's share, in turn, will become almost negligible, providing just 2.2% of Spain's electricity (ibid.).

Most of the growth in renewables observed until 2010 came from wind power, and this energy source is also expected to account for most of the increase in renewables until 2020 (ibid.). Solar energy has become a major source since 2007, when stronger economic incentives were created. However, these incentives, due to the moratorium, are no longer available. Before the freeze on long-term price guarantees for new projects, these incentives were provided for the two main types of solar technology: PV and concentrated solar power (CSP). PV is expected to cover 3.2% of the electricity generated in 2020. CSP, a technology in which Spain is the uncontested global leader, is expected to surpass PV and produce 3.7% of Spain's electricity needs (ibid.).

As already mentioned, Spain has almost no reserves of fossil fuels with the exception of coal, which is considered to be of low quality and highly polluting (Libre Mercado 2012). In 2010, the country saw net imports of 86.0% for coal, 99.9% for oil, and 99.2% for natural gas. The depletion of fossil fuel reserves and the advent of renewables increased the latter's share

TABLE 6.3

Electricity Generation, Historic and Projected (GWh), Based on Data Given by IDAE that Assumes Energy Efficiency Improvements over Time

	(GWh)	2005 (Actual)	2010 (Actual)	2015 (Target)	2020 (Target)	Total 2005 (%)	Total 2010 (%)	Total 2020 (%)	Growth from 2010 to 2020 (%)
	Hydro (renewable)	18,573	42,215	32,538	33,140	6.3	14.1	8.6	−21
+	Wind	21,175	43,708	55,769	73,485	7.2	14.6	19.2	68
+	Solar PV	41	6279	9060	12,356	0.0	2.1	3.2	97
+	CSP	0	691	8287	14,379	0.0	0.2	3.7	1981
+	Biomass, biogas, geothermal, and marine	2652	4228	7142	12,720	0.9	1.4	3.3	201
=	Renewables	42,441	97,121	112,797	146,080	14.5	32.3	38.1	50
+	Hydro (pumped)	4452	3106	6592	8457	1.5	1.0	2.2	172
+	Nuclear	57,539	61,788	55,600	55,600	19.6	20.6	14.5	−10
+	Natural gas	82,819	96,216	120,647	133,293	28.3	32.0	34.7	39
+	Coal	81,458	25,493	33,230	31,579	27.8	8.5	8.2	24
+	Oil	24,261	16,517	9149	8624	8.3	5.5	2.2	−48
=	Gross production	292,970	300,241	338,015	383,633	100.0	100.0	100.0	28
−	Used in generation	18,308	14,393	18,315	21,050	6.2	4.8	5.5	46
=	Net production	274,662	285,848	319,700	362,583	93.8	95.2	94.5	27
+	Net electricity import	−1344	−8338	−11,231	−12,000	−0.5	−2.8	−3.1	44
=	Gross demand	273,318	277,510	308,469	350,583	93.3	92.4	91.4	26
−	Used by transformers	5804	4100	5800	5800	2.0	1.4	1.5	41
−	Lost in distribution	25,965	24,456	26,894	29,839	8.9	8.1	7.8	22
=	Net demand	241,549	248,954	275,775	314,944	82.4	82.9	82.1	27

Source: Adapted from IDAE. *Plan de Energías Renovables 2011–2020.* Madrid, Spain: IDAE, 2011.

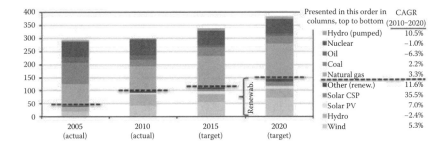

FIGURE 6.13
Electricity generation mix, historic and projected (TWh). (Adapted from IDAE, *Plan de Energías Renovables 2011–2020*. Madrid, Spain: IDAE, 2011.)

of domestically sourced energy from 30% in 1990, which was almost all hydropower, to more than 80% in 2010 (Figure 6.14; Eurostat 2012).

Although Spain's energy mix has become greener year after year, the energy produced with domestic resources has remained almost unchanged over the past two decades, fluctuating between 15,000 and 20,000 kilotons of oil-equivalent per year (Eurostat 2012; WBG 2011). Although renewables filled the gap left by dwindling fossil fuel reserves, they have not allowed Spain to increase production from autochthonous sources. As a result, the country's foreign dependence ratio has not improved.

This will change if the targets of the 2020 plan are met. These targets will also allow Spain to consolidate its green leadership further among its European peers. For these goals to materialize, at least one of two scenarios must take place. First, the government must resume the feed-in tariff system. That is probably out of the question until Spain's economy regains

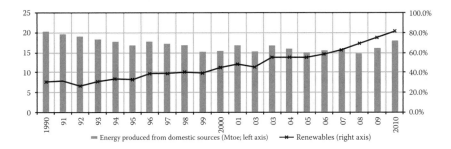

FIGURE 6.14
Energy produced from domestic sources over time (mtoe) and renewables' share of domestic production. (Adapted from Eurostat, European Commission, Energy Statistics. http://epp.eurostat.ec.europa.eu/portal/page/portal/statistics/search_database; and WBG. Data catalog. 2011. http://data.worldbank.org/data-catalog.)

momentum and the tariff deficit is under control. Second, technology must continue to advance, especially in the case of solar, to make renewable energy cost-competitive without subsidies. Although there have been strong signs that the second option is very likely, costs may not decline quickly enough to entice investors before the end of the decade.

This lack of purely economic competitiveness has greatly influenced the debate about which renewable technologies should be favored and whether Spain should even promote renewables. The wind energy sector, for example, installed 22 GW of capacity up until 2011 and is expected to reach 35 GW in 2020 if the current energy plan is fully implemented (IDAE 2011). But it is by no means certain that the government and the general public, who will eventually pay the bill, will support a larger expansion of renewables, which may put the success of the current energy plan at risk. This is a reality despite growing concerns about energy security and climate change in Spain, as expensive electric bills weigh especially heavily in times of high unemployment and low economic growth (Cartea et al. 2009).

Spain's past economic success provided a favorable political and economic setting for the creation of the special regime. This regime's financial incentives then allowed the renewables sector to flourish and become world-class in all parts of its supply chain. There is still potential for this sector to grow at home, but the limited size of the domestic market and the current state of economic stagnation have intensified the search for growth overseas. Spanish firms and Spanish professionals are already involved in the design, development, operation, and maintenance of renewable energy facilities all over the world—from onshore/offshore wind farms in Europe to PV and CSP plants in Latin America and Southeast Asia (POSHARP 2012). This is exactly what these players should do: try to replicate their success at home in countries where renewable energy is now gaining momentum.

6.5 THE MAIN SECTORS BEHIND THE GROWTH OF RENEWABLES IN SPAIN: WIND AND SOLAR

Owing to a combination of attractive policies and a well-known culture of entrepreneurship, Spain had built the second-largest solar park and the fourth-largest wind park in the world at the end of 2011 (Eurostat 2012; IDAE 2011). An impressive 27 GW of wind and solar capacity were installed between 1997 and 2011 (Figure 6.15), with a combined output

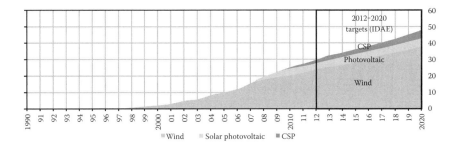

FIGURE 6.15
Cumulative renewable installed capacity over time (MW). (Adapted from Eurostat, European Commission, Energy Statistics. http://epp.eurostat.ec.europa.eu/portal/page/portal/statistics/search_database, accessed December 26, 2012; and IDAE, *Plan de Energías Renovables 2011–2020*. Madrid, Spain: IDAE, 2011.)

that supplied approximately one-sixth of the total electricity produced in the country (Table 6.3). That capacity is expected to reach 48 GW in 2020 and generate 26% of the country's total electricity output, unless the rules governing both sectors remain unstable (IDAE 2011). The following sections provide a brief overview of how the Spanish wind and solar sectors came into being, their current states, and what may lie ahead.

6.5.1 The Wind Energy Sector

Windmills have been embedded in the Spanish cultural consciousness ever since Miguel de Cervantes published his seminal *Don Quixote* in 1605. In this novel, the iconic character believes that windmills are ferocious giants controlled by wizards from afar. Apparently, modern wizards got the upper hand, as windmills are now much larger and can be found almost anywhere in the country.

Spain's wind power industry traces its origin to the 1981 establishment of the cooperative Ecotechnia. This organization was founded by a group of engineers who opposed the development of nuclear plants (Toke 2011). Inspired by Denmark's success in building wind turbines, it sought funding from the Spanish government to develop an alternative to nuclear energy. Ecotechnia gave Spain its first 20 to 30 wind turbines, but its most important contribution was perhaps motivating authorities to create the IDAE in 1984. As mentioned previously, the IDAE is the official face of the Spanish renewable energy program and has acted in cooperation with regional governments to foster the type of public–private partnerships necessary to attract and connect new producers to the power grid (ibid.).

Large-scale wind farms were first built in 1994 in the Autonomous Region of Navarra, and new developments surpassed the threshold of 1 GW/year in 2001 (Figure 6.16; ibid.). Ten years later, wind energy generated on average 16% of all the electricity consumed in Spain, at times covering more than 50% of the national demand (IDAE 2011).

Spain had the fourth largest amount of installed wind capacity worldwide in 2011 (Figure 6.17), representing approximately 9% of global capacity. However, it will most likely fall in terms of its global ranking if the level of new investments continues the downward trend of the past few years: 2011 was the worst year for the industry in a decade, with only 1050 MW of new installed capacity. This means that Spain contributed just 2.6% to global growth (Figure 6.18). At 2011's pace, India will overtake Spain in less than 3 years (ibid.).

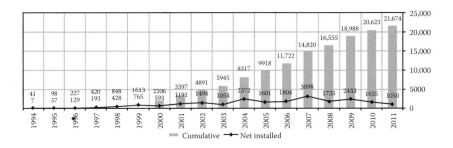

FIGURE 6.16
Annual installed and cumulative wind power capacity in Spain (MW). (Adapted from GWEC, *Global Wind Report 2011—Annual Market Update*, Brussels, Belgium: Bitter Grafik & Illustration, 2012; IDAE, *Plan de Energías Renovables 2011–2020*. Madrid, Spain: IDAE, 2011.)

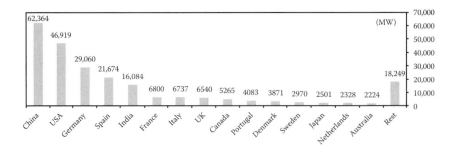

FIGURE 6.17
Global cumulative installed wind power capacity in 2011 (total worldwide capacity was 237,669 MW). (Adapted from GWEC, *Global Wind Report 2011—Annual Market Update*, Brussels, Belgium: Bitter Grafik & Illustration, 2012.)

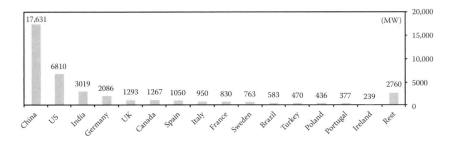

FIGURE 6.18

New wind energy capacity installed in 2011 (total worldwide installation was 40,564 MW). (Adapted from GWEC, *Global Wind Report 2011—Annual Market Update*. Brussels, Belgium: Bitter Grafik & Illustration, 2012.)

The renewable energy plan has set a wind power target of 35 GW of installed capacity by 2020, but that equates to less than 11% of Spain's wind potential, according to a study carried out by Meteosim Truewind (IDAE and Meteosim Truewind 2011). Considering all locations where the average wind speed is at least 6 m/s at 80 m above the surface, and where it would be feasible to install and connect a wind turbine to the grid, the potential capacity currently stands at 332 GW. In terms of power generation, that capacity could produce 700 to 750 TWh/year, or approximately three times Spain's total electricity demand in 2010 (ca. 250 TWh; ibid.). This is certainly great news for a country that relies on foreign supplies for three-quarters of its energy requirements. However, according to the Spanish Wind Energy Association (Asociación Empresarial Eólica, AEE), a combination of more stable policies, new technologies, and lower operating costs are necessary for Spain to tap into this vast potential (AEE 2011).

Spain's wind industry has become one of the most competitive in the world, as illustrated by the relatively large number of firms in the country. AEE had more than 200 members in 2011, including 43 parts manufacturers, 11 wind turbine producers, 53 wind farm developers, 99 service providers, and 5 associations (ibid.). Because of the recent domestic industry slowdown coupled with the cutting-edge technological and service-oriented offerings of its members, AEE has established partnerships with chambers of commerce and provincial governments to encourage international expansion to countries where wind power is experiencing considerable growth, such as Romania, Poland, Brazil, and Mexico (ibid.).

This seems to have paid off for some firms because in 2011, Gamesa, an international leader in wind energy project development, reported that 92% of its 2802 MW sales were made outside of Spain, with 57% in emerging markets (GAMESA 2011). The company has invested heavily in R&D over the years, building ever more powerful and efficient turbines for onshore use. It is also exploring the nascent offshore wind energy market and was expected to launch the G11X turbine (5 MW of capacity) in the United States between 2012 and 2013, followed by the G14X turbine (6 MW) in early 2015 in the United Kingdom (González 2012).

In summary, the Spanish wind energy sector has grown spectacularly over the course of the past 15 years but its future remains uncertain without stable, dependable policies. With the current moratorium on the special regime, Spanish companies have to look elsewhere for growth. It is important to keep ahead of the competition on a global scale through a pipeline of differentiated products and with the aid of well-trained "wizards," who can spot opportunities beyond the Spanish shores.

6.5.2 The Solar Energy Sector

The tourism industry revels in the millions of visitors who flock to Spanish shores every year to enjoy the country's famous beaches and celebrated sunshine. This sunshine lends positive energy not only to tourists, but also to Spain's world-class solar energy sector. A mix of economic incentives and rapidly decreasing manufacturing costs has allowed Spain to capitalize on its natural endowment, making the country a leader in two prominent technologies: solar PV and CSP.

PV is the most well-known solar technology currently available. It uses PV cells that produce electricity when stimulated by sunlight's photons. This technology produces energy that can be fed immediately into the power grid or, alternatively, used by residential or industrial clients located in remote areas not yet served by the grid. At the end of 2011, Spain had the fourth-largest installed capacity of PV in the world, with 4.4 GW (European Photovoltaic Industry Association [EPIA] 2012).

CSP, a promising technology, also referred to as solar thermal, concentrates solar radiation onto small areas to generate power. All CSP varieties ultimately transfer solar-generated heat to steam turbines connected to generators. The most interesting feature of this technology is perhaps the fact that heat can be stored to generate electricity when the sun is not shining (Protermosolar 2011).

Although Spain was not the first country to set up a CSP plant, it was the indisputable leader of this technology at the end of 2011, with 65% of the world's CSP capacity. According to Eduardo Iglesias, Deputy Secretary of Protermosolar—the national CSP association—Spanish companies and professionals are already leading most of the solar thermal projects around the world (Iglesias 2012).

The greatest challenge for solar energy remains the development of a manageable and cost-effective technology that stacks up well against traditional fossil fuel sources. According to the IDAE, solar PV energy is expected to become competitive (without subsidies) by 2023, and CSP no sooner than 2026 (IDAE 2011). But industry professionals are more optimistic in light of recent technological breakthroughs and rapidly decreasing manufacturing costs: the formerly independent Asociación Empresarial Fotovoltaica, which is now part of the Unión Española Fotovoltaica, projects that competitiveness could be achieved as early as 2013, whereas Protermosolar expects CSP to reach its economic maturity by 2020 (Buetas 2012; Iglesias 2012).

6.5.2.1 Solar PV

Although Spain is endowed with more sunlight than any other European country (Figure 6.19), it was not until 2007 that this energy source was

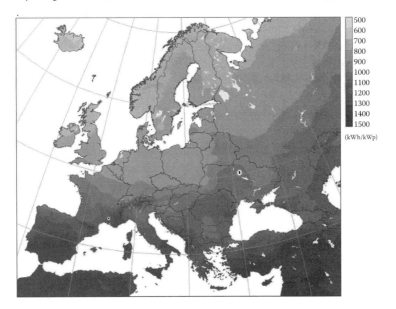

FIGURE 6.19
Solar radiation and solar photovoltaic potential across Europe. (Reprinted with permission from Šúri et al. Potential of solar electricity generation in the European Union member states and candidate countries. *Solar Energy* 81: 1295–1305, 2007.)

finally harnessed for commercial electricity generation. With the strong financial incentives and better guarantees to producers set forth by RD 661/2007, the national PV sector built, in just over a year, the second largest solar park in the world in 2008 (IEA 2009).

RD 661/2007 was published in May 2007 and provided the incentives that many investors had been waiting for. However, according to this decree, if the cumulative capacity for any renewable technology reached 85% of the target set for 2010, then new facilities for that renewable would be accepted in the special regime for only another 12 months (RD 661/2007). The 85% threshold for solar PV was surpassed in August 2007, just 3 months later, setting the 12-month countdown for new installations to be included in the special regime in motion. What followed was a rush of investors who installed more than 3 GW between September 2007 and September 2008, allowing Spain to reach a total of 3.5 GW of national capacity by the end of 2008 (Figure 6.20). That was approximately 10 times the 371-MW goal the government had set for the entire decade. Because solar energy is the most expensive energy source within the special regime (€0.40/kWh in contrast to about €0.09/kWh for wind power), the solar boom of 2007 to 2008 is frequently blamed for the current levels of the tariff deficit (CNE 2012).

Due to the boom in new capacity, the government issued RD 1578/2008 in September 2008, tightening the rules of the special regime for the PV sector. Annual caps of 500 MW for new additions were established, and regulated premium prices were cut by approximately 30%. At the end of 2010, yet another decree was issued, limiting the maximum number of annual hours a plant was allowed to operate under the special regime and affecting all PV producers. Given the price cuts and the retroactive

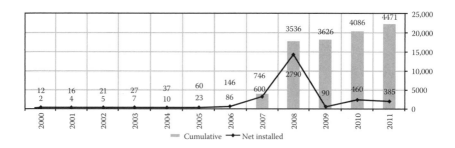

FIGURE 6.20

Annual solar PV capacity installed and cumulative installed PV capacity in Spain over time. (Adapted from REN21. *Renewables Global Status Report 2011*. Paris, France: REN21, 2012.)

measures against old producers, it was no surprise that new investments dwindled in the following years: the 500 MW annual cap was not even reached between 2009 and 2011. Spain then dropped from its second position worldwide in 2008 to fourth in 2011 (Figure 6.21) and will most likely fall further as both the United States and China ramp up their investments in solar energy (IDAE 2011).

However, experts believe the industry may take off again due to the pronounced decline in the costs of PV cells and modules observed in the past few years. According to Francisco Buetas, engineering director for the Spanish Photovoltaic Energy Association, the cost of new cells decreased by 70% between 2007 and 2011, and the technology was expected to become market-competitive (i.e., without subsidies) as early as 2013 (Buetas 2012). The decrease in manufacturing costs is attributable to the entry of relevant players in Germany and China who brought down costs by automating a previously labor-intensive process (Martinez 2012).

Spain is home to the well-known Solar Energy Institute of the Universidad Politécnica de Madrid and to the Institute of Concentration Photovoltaic Systems, two of the most advanced solar research centers in the world. Their scientists and engineers have forged partnerships with private sector players and research teams all around the world to create more efficient PV cells and modules. Developing a cost-effective solar technology represents perhaps the best remedy for the current stagnation surrounding the national PV industry. Although subsidies may have been necessary in the past, the key to a more financially sustainable sector may well lie in the laboratories of Madrid or of Puertollano (IDAE 2011).

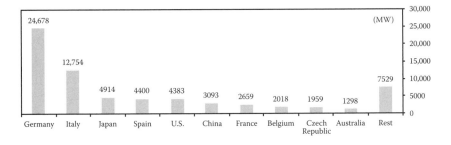

FIGURE 6.21
Global cumulative installed solar photovoltaic capacity in 2011 (total worldwide capacity was 69,685 MW). (Adapted from EPIA, *Global Market Outlook for Photovoltaics until 2016*, by Masson Gaetan, Marie Latour, and Daniele Biancardi, Brussels, Belgium: EPIA, 2012.)

6.5.2.2 Concentrated Solar Power

CSP technology has been around since the 1980s, with the commission of the first CSP plant having occurred in 1984 at Kramer Junction in California. Three years later, CSP was introduced in Spain as a joint effort between CIEMAT (Spain's Research Centre for Energy, Environment, and Technology) and the German Aerospace Research Centre (CIEMAT 2013a).

In 2011, Spain accounted for 65% (1149 MW) of global CSP capacity, with the United States in a distant second place with 29% (507 MW), and other countries just experimenting with the technology. By January 2013, the industry had almost doubled its capacity to 2 GW, which could generate approximately 5.1 GWh of electricity in a normal year (Protermosolar 2011). According to Eduardo Iglesias of Protermosolar, Spain was expected to reach approximately 2.5 GW of total capacity by the end of 2013 (Table 6.4 and Figure 6.22). These increases were not affected by the moratorium on the special regime since they were approved by the authorities in 2009 (Iglesias 2012).

CSP is unique among renewable energy alternatives in that it can generate power even when its primary source—direct solar radiation—is not available. CSP plants use molten salt systems that can store heat and generate electricity through the night or on cloudy days. Furthermore, these

TABLE 6.4

CSP Power Capacity in Spain (as of January 2013)

Technology	Operational			Total (Includes Operational, Under Construction, and Preauthorized)		
	Plants	Capacity (MW)	Electricity (GWh/year)	Plants	Capacity (MW)	Electricity (GWh/year)
Parabolic troughs	38	1873	4918	47	2323	6068
Central receiver tower	3	50	168	3	50	168
Fresnel linear	2	31	52	2	31	52
Sterling	—	—	—	3	32	73
Total	43	1954	5138	55	2436	6361

Source: Adapted from Protermosolar. Mapa de la Industria Solar Termoélectrica en España, 2012. http://www.protermosolar.com/mapa.html.

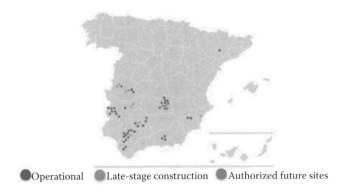

Operational   Late-stage construction   Authorized future sites

FIGURE 6.22

CSP plants in Spain and their current operating stages. (Reprinted with permission from Protermosolar, Mapa de la Industria Solar Termoélectrica en España. http://www.protermosolar.com/mapa.html.)

plants are hybrids, meaning that they can operate just like any other steam power plant when neither sunlight nor stored heat is available. For example, these plants can also burn natural gas or biomass to generate power. Thus, CSP is a much more predictable and stable technology than wind or PV from the standpoint of the power grid operator (Protermosolar 2011).

Currently, there are three major CSP technologies available in Spain: parabolic trough power (38 plants with 1873 MW), solar towers with heliostat fields (3 plants with 50 MW), and linear Fresnel reflectors (2 plants with 31 MW). A fourth type, using Sterling discs, has not yet been used in Spain, but the authorities have approved the construction of three plants using this technology in the city of Puertollano. All four technologies are pictured in Figure 6.23 (Protermosolar 2011).

The most common technology uses a series of parabolic trough reflectors placed in multiple rows, as shown in Figure 6.23. These mirrors concentrate solar radiation on thin tubes containing a fluid that will absorb heat while circulating through the plant. This fluid eventually reaches heat storage systems or steam engines, where it boils a working fluid, creating steam that can then generate power. The reflectors move according to the sun's movement in the sky, so as to collect as much radiation as possible. This technology accounted for 96% of all CSP capacity installed in Spain in 2012 (Protermosolar 2012).

Perhaps the most intriguing CSP technology is the central receiving tower variant shown in Figure 6.23, characterized by a system of thousands of flat mirrors—or heliostats—that concentrate solar radiation onto

FIGURE 6.23

CSP technologies commercially available in Spain: (top left) Parabolic troughs, the leading technology platform Almeria in Granada; (top right) central receiver towers—Platform PS10 in Andalusia; (bottom left) linear Fresnel—Platform Errado 1 in Mureia; (bottom right) Sterling dish–solar plant Casas de Los Pinos in Cuenea. (Reprinted with permission from [top left] CIEMAT [CIEMAT 2013b]. PSA-CIEMAT. http://www.psa.es/webesp/archivofoto.php [top right, bottom left, bottom right] Protermosolar, La Energía Termosolar. http://www.protermosolar.com/termosolar.html, accessed March 29, 2013.)

the top of a central receiver tower. Because of the much smaller concentration area, the temperatures achieved are much higher than those achieved using the parabolic trough technology. Furthermore, there is normally less energy loss because the heat transfer mechanism does not rely on a fluid that runs hundreds of meters before finally offloading its heat. With the central receiver, heat is transferred directly from the top to the base of the tower, where tanks of molten salts and steam turbines are located (Protermosolar 2011).

Spain currently has the largest CSP capacity in the world, but the government will not allow new plants beyond the 2.5 GW of capacity it has already approved into the special regime—at least until the tariff deficit problem is resolved (RD 1/2012). Nevertheless, as with the other renewable technologies, there is vast potential for development outside Spanish borders. According to a study conducted in 2008 by the Institute of Technical Thermodynamics at the German Aerospace Centre, large swaths of Africa, the Middle East, Australia, northern China, South America, and the United States are suitable for CSP plants (German Aerospace Center [DLR] 2009). Spanish companies, which are among the most sophisticated players in the world, were actively involved in the development of new projects in those same areas at the end of 2012 (Protermosolar 2012).

6.6 CONCLUSIONS

A combination of factors has led Spain to invest heavily in renewable energy over the past 30 years: scarce or nonexistent fossil fuel reserves, rapid economic growth, nationwide antinuclear sentiment, and the need to increase energy security through the creation of a diversified domestic base. The country has excelled in both wind and solar energy on a global scale. Its success has been based on three main pillars:

1. The entrepreneurial spirit of Spaniards, who have responded rapidly to market incentives and have established one of the most sophisticated and dynamic energy sectors in the world;
2. The leadership of the IDAE, which was instrumental in crafting the regulatory framework that allowed renewables to flourish; and
3. The support of energy consumers, even if tacit or nonexplicit, who have paid the premium prices demanded by renewables through more expensive electricity bills.

Although the Spanish story has largely been one of success for the past 30 years, a deficit of billions of Euros has led many to question the special regime's support for the renewable energy sector. Many would not necessarily advocate for cutting all subsidies permanently, but would rather they be targeted more wisely. Renewables hold the key to attaining lower foreign energy dependence, both in the short-term and in the long run. Thus, the government should not withdraw its support completely, lest investors decide to allocate their resources elsewhere and let the sector stagnate. Furthermore, if external environmental costs and renewables' overall contribution to GDP are taken into account, one may conclude that, even in the short-term, renewables' benefits outweigh the costs (AEE 2012; Protermosolar 2011).

Despite these arguments in favor of a continued expansion of Spain's renewable capacity, the domestic market may continue down its current path of austerity. This would leave at least two feasible alternative courses of action for Spain's firms to seek growth. The first is the installation and maintenance of solar and wind parks in foreign countries that are investing heavily in alternative energy. The second is helping other EU countries fulfill their commitments to renewables by selling green energy made in Spain today or by investing in the infrastructure necessary to transfer power from future solar plants in sun-drenched North Africa.

Just like Spain, other major economies are trying to diversify their energy base away from foreign fossil fuels. The world can benefit from all the green energy it can harness, and Spanish firms can now profit from their expertise gleaned over the past several decades. Many companies have already ventured outside Spain successfully, and others must do so if they wish to grow, stay relevant, and preserve Spain's reputation as a leader in the green power sector.

REFERENCES

ABC Hemeroteca. El Ministro de Industria pone en Marcha Oficialmente el Reactor Nuclear de Zorita (Guadalajara). Last modified 2011 (accessed December 26, 2011). http://hemeroteca.abc.es/nav/Navigate.exe/hemeroteca/madrid/abc/1968/07/18/049.html.

Asociación Empresarial Eólica (AEE, Spanish Wind Association). *Eólica '11: Asociación Empresarial Eólica la Referencia del Sector.* Madrid, Spain: AEE, 2011. http://www.aeeolica.org/es/new/eolica-11-toda-la-informacion-del-ano-2011-que-necesitas-conocer-sobre-el-sector/ (accessed March 29, 2013).

Asociación Empresarial Eólica (AEE), and EDP Renewables. *Impacto Macroeconómico del Sector Eólico en España,* by Deloitte. Madrid, Spain: Impression Artes Gráficas, 2012. http://www.aeeolica.org/uploads/documents/Estudio_macro_AEE_2012.pdf (accessed March 29, 2013).

Asociación Espanola de la Industria Eléctrica (UNESA). Calculadora de consumo. Last modified 2012 (accessed January 26, 2013). http://www.unesa.net/u/factura.html.

Bodansky, D. Belfer Center for Science and International Affairs. Whither the Kyoto protocol? Durban and beyond. Last modified August 2011 (accessed March 30, 2013). http://belfercenter.ksg.harvard.edu/publication/21314/whither_the_kyoto_protocol_durban_and_beyond.html.

Brackmann, M., M. Eberle, K. Stratmann, and G. Weishaupt. Die Welt kehrt der Atomkraft den Rücken. *Handelsblatt.* April 21, 2011. http://www.handelsblatt.com/politik/deutschland/ausstieg-aus-der-kernenergie-die-welt-kehrt-der-atomkraft-den-ruecken/4088010.html (accessed March 30, 2013).

Buetas, F.C. Interview with Francisco Campo Buetas, Engineering Director of AEF in Madrid, Spain, by José Normando Bezerra, Jr. January 11, 2012.

Cartea, P., M. Blanco, and P. Souto. *La Sociedad Ante el Cambio Climático: Conocimientos, valoraciones y comportamientos en la población española.* Santiago de Compostela, Spain: Fundación Mapfre and Universidade de Santiago de Compostela, 2009.

Centro de Investigaciones Energéticas, Medioambientales y Tecnológicas (CIEMAT 2013a). Historia. Last modified 2013 (accessed March 29, 2013). http://www.ciemat.es/portal.do?IDM=6&NM=2.

Centro de Investigaciones Energéticas, Medioambientales y Tecnológicas (CIEMAT 2013b). PSA-CIEMAT (accessed March 29, 2013). http://www.psa.es/webesp/archivofoto.php.

Comisión Nacional de Energía (CNE). Información Estadística sobre las Ventas de Energía del Régimen Especial: CUADROS_Octubre_2012. Madrid, Spain: CNE, 2012. http://www.cne.es/cne/Publicaciones?id_nodo=144&accion=1&sIdCat=10&keyword=&auditoria=F (accessed January 26, 2013).

Europa Press. La Tarifa de luz Asumirá 53 Millones de la Moratoria Nuclear en 2012, de la que quedan 317 Millones por Pagar. Last modified January 2013 (accessed March 29, 2013). http://www.europapress.es/economia/energia-00341/noticia-economia-tarifa-luz-asumira-53-millones-moratoria-nuclear-2012-quedan-317-millones-pagar-20130129113620.html.

European Photovoltaic Industry Association (EPIA). *Global Market Outlook for Photovoltaics until 2016*, by Masson Gaetan, Marie Latour, and Daniele Biancardi. Brussels, Belgium: EPIA, 2012. http://www.epia.org/fileadmin/user_upload/Publications/Global-Market-Outlook-2016.pdf (accessed March 30, 2013).

Eurostat. European Commission. Energy statistics. Last modified December 2012 (accessed December 26, 2012). http://epp.eurostat.ec.europa.eu/portal/page/portal/statistics/search_database.

Foro Nuclear. Moratoria nuclear en España. Last modified 2009 (accessed March 29, 2013). http://estaticos.soitu.es/documentos/2009/03/Moratoria_Nuclear.pdf.

Foro Nuclear. Energía nuclear en España. Last modified February 2013 (accessed March 29, 2013). http://www.foronuclear.org/es/energia-nuclear/energia-nuclear-en-espana.

Fraunhofer Institute Systems and Innovation Research. *Feed-in Systems in Germany and Spain and a Comparison*, by M. Ragwitz, and C. Huber. Karlsruhe, Germany: Fraunhofer ISI, 2004. http://www.worldfuturecouncil.org/fileadmin/user_upload/Miguel/feed-in_systems_spain_germany_long_en.pdf (March 29, 2013).

Gamesa Corporación Tecnológia, S.A. (GAMESA) 2011 Annual Report. December 31, 2011. http://www. gamesacorp.com/recursos/doc/accionistas-inversores/informacion-financiera/ memoria-anual/english/annual-report-2011.pdf (accessed March 30, 2013).

German Aerospace Center (DLR). *Global Potential of Concentrating Solar Power*, by Franz Trieb, Christoph Schillings, Marlene O'Sullivan, Thomas Pregger, and Carsten Hoyer-Klick. Stuttgart, Germany: DLR, 2009. http://www.dlr.de/tt/Portaldata/41/Resources/dokumente/institut/system/projects/reaccess/DNI-Atlas-SP-Berlin_20090915-04-Final-Colour.pdf (accessed March 30, 2013).

Global Wind Energy Council (GWEC). *Global Wind Report 2011—Annual Market Update*. Brussels, Belgium: Bitter Grafik & Illustration, 2012. http://gwec.net/wp-content/uploads/2012/06/Annual_report_2011_lowres.pdf (accessed March 29, 2013).

González, Á. Interview with Ángel González, Chief Engineer, Special Projects at the Gamesa Headquarters in Madrid, Spain, by José Normando Bezerra, Jr. January 13, 2012.

Iglesias, E.G. Interview with Eduardo García Iglesias, Deputy General Secretary of Protermosolar in Madrid, Spain, by José Normando Bezerra, Jr. January 12, 2012.

Instituto para la Diversificación y Ahorro de la Energía (IDAE). *Plan de Energías Renovables en España 2005–2010*. Madrid, Spain: IDAE, 2005. http://www.minetur.gob.es/energia/desarrollo/energiarenovable/plan/documents/resumenplanenergiasrenov.pdf (accessed March 29, 2013).

Instituto para la Diversificación y Ahorro de la Energía (IDAE). *Plan de Energías Renovables 2011–2020*. Madrid, Spain: IDAE, 2011.

Instituto para la Diversificación y Ahorro de la Energía (IDAE)–Meteosim Truewind. *Análisis del Recurso. Atlas Eólico de España—Estudio Técnico PER 2011–2020*. Madrid,

Spain: IDAE-Meteosim Truewind, 2011. http://www.idae.es/uploads/documentos/documentos_11227_e4_atlas_eolico_A_9b90ff10.pdf (accessed March 29, 2013).

International Energy Agency (IEA). *Trends in Photovoltaic Applications: Survey Report of Selected IEA Countries between 1992 and 2008.* By IEA PVPS Task 1. St. Ursen, Switzerland: IEA, 2009. http://www.iea-pvps.org/fileadmin/dam/public/report/statistics/tr_2008.pdf (accessed March 29, 2013).

La Moncloa, Spanish Government. *Estrategia Española de Política Económica: Balance y Reformas Estructurales Para El Próximo Semestre.* Madrid, Spain: Spanish Government, 2012. http://www.lamoncloa.gob.es/nr/rdonlyres/11bdf744-55af-4a0c-9bb8-57f9c538016c/0/120927_estrategiaespa%C3%B1olapoliticaeconomica3.pdf (accessed March 29, 2013).

Lancaster, T. *Policy Stability and Democratic Change: Energy in Spain's Transition.* University Park, PA: Pennsylvania State University, 1989.

Law 54/1997. Reference number: BOE-A-1997-25340. Cortes Generales, November 27, 1997. http://www.boe.es/buscar/doc.php?id=BOE-A-1997-25340.

Law 82/1980. Reference number: BOE-A-1981-1898. Cortes Generales, January 23, 1981. http://www.boe.es/buscar/doc.php?id=BOE-A-1981-1898.

Libre Mercado. El Carbón Español, un Negocio Ruinoso desde hace un Siglo. Last modified July 4, 2012 (accessed March 30, 2013). http://www.libremercado.com/2012-07-04/el-carbon-espanol-un-negocio-ruinoso-desde-hace-un-siglo-1276463068/.

Martinez, J.L. Interview with José Luis Martínez, former Executive Vice President of AEF and Director of Institutional Relations of UNEF in Madrid, Spain, by José Normando Bezerra, Jr., January 11, 2012.

Mendonça, M. *Feed-In Tariffs: Accelerating the Deployment of Renewable Energy.* London: Earthscan. 2007.

POSHARP. Renewable energy product exporters in Spain. Last modified 2012 (accessed March 30, 2013). http://www.posharp.com/renewable-energy-product-exporters-in-spain_renewable.aspx?btype=export>ype=country_ES.

Protermosolar (The Spanish Association of Solar Thermal Industry). *Macroeconomic Impact of the Solar Thermal Electricity Industry in Spain,* by Deloitte. Sevilla, Spain: Proyectos Editoriales, SA, 2011. http://www.estelasolar.eu/fileadmin/ESTELAdocs/documents/Publications/2011_Macroeconomic_impact_of_STE_in_Spain_Protermo_Solar_Deloitte.pdf (accessed March 29, 2013).

Protermosolar. Mapa de la Industria Solar Termoéléctrica en España. Last modified 2012 (accessed January 12, 2013). http://www.protermosolar.com/mapa.html.

Protermosolar. La energia termosolar. Last modified 2013 (accessed March 29, 2013). http://www.protermosolar.com/termosolar.html.

Renewable Energy Policy Network for the 21st Century (REN21). *Renewables Global Status Report 2011.* Paris, France: REN21, 2012. http://www.ren21.net/REN21Activities/GlobalStatusReport.aspx (accessed March 29, 2013).

Royal Decree 1/2012. Reference number: BOE-A-2012-1310. Cortes Generales, 24 January, 2012. http://www.boe.es/buscar/doc.php?id=BOE-A-2008-15595.

Royal Decree 1578/2008. Reference number: BOE-A-2008-15595. Cortes Generales, September 26, 2008. http://www.boe.es/buscar/doc.php?id=BOE-A-2008-15595.

Royal Decree 2366/1994. Reference number: BOE-A-1994-28980. Cortes Generales, December 31, 1994. http://www.boe.es/buscar/doc.php?id=BOE-A-1994-28980.

Royal Decree 2818/1998. Reference number: BOE-A-1998-30041. Cortes Generales, December 30, 1998. http://www.boe.es/buscar/doc.php?id=BOE-A-1998-30041.

Royal Decree 661/2007. Reference number: BOE-A-2007-10556. Cortes Generales, May 26, 2007. http://www.boe.es/buscar/doc.php?id=BOE-A-2008-15595.

Šúri, M., T.A. Huld, E.D. Dunlop, and H.A. Ossenbrink. Potential of solar electricity generation in the European Union member states and candidate countries. *Solar Energy* 81: 1295–1305, 2007. http://re.jrc.ec.europa.eu/pvgis/ (accessed December 26, 2012).

Thomson Reuters. UPDATE 2-Spain unveils fresh energy reforms that hit renewables. *Reuters*, February 1, 2013. http://www.reuters.com/article/2013/02/01/spain-energy-idUSL5N0B1B5J20130201 (accessed March 30, 2013).

Toke, D. *Ecological Modernization and Renewable Energy*. New York: Palgrave Macmillan, 2011.

The World Bank Group (WBG). Data catalog. Last modified December 2011 (accessed December 26, 2011). http://data.worldbank.org/data-catalog.

7

Renewable Energy in French Polynesia: From Unpredictable to Energy Independence?

Diana Townsend-Butterworth Mears

CONTENTS

7.1 INTRODUCTION

Energy influences all aspects of economic life for households, businesses, and communities, making it a top national policy issue around the world. What are the sources of energy, how secure are they, and what do they cost? For a country as isolated as French Polynesia—so far from international oil supplies and yet so dependent on oil—the effect of increasing oil prices or of a global oil shortage are even more significant. In 2008, French

Polynesia depended on foreign sources for 89% of its oil consumption by energy volume (Service de l'Energie et des Mines [SEM] 2009a). Forty-four percent of primary energy is used for electricity in French Polynesia, and electricity consumption has grown more than fivefold in the past 30 years (from 127 GWh in 1980 to 705 GWh in 2008; ibid.). The increase in consumption continues despite the frequently repeated claim that "French Polynesia has the most expensive electricity in the world" (Laurey 2009, p. 40). Transitioning electricity generation to renewable sources could significantly reduce the cost of electricity and increase energy security.

Recognizing the importance of reducing dependence on imported oil, in 2008, the government of French Polynesia set a goal to have electricity production from renewable sources reach 50% by 2020 and 100% by 2030 (SEM 2009b). According to articles in local Tahitian newspapers and confirmed by the author's conversations with citizens and government officials, French Polynesians support the government's plans. Not taking action today means the entire country, from households to businesses, will suffer the consequences of more expensive electricity or even no access to electricity. According to Nuihau Laurey, consultant for technical and strategic energy studies to the Minister of Economy and Energy, the incantation, "We must develop renewable energy. We will develop renewable energy resources," has been popular for at least a decade (Laurey 2011).

Given the importance of energy independence to the government and citizens alike, this chapter seeks to answer the following questions: How realistic are the government's renewable energy goals, and how much progress has actually been made? What is the future of renewable energy for this remote Pacific Island country?

The next section puts French Polynesia in historical, political, and economic contexts. The third section provides detailed energy background information for the country, looking first at primary energy consumption and then focusing on electricity. The fourth section presents the government's renewable energy goals, explores the existing situation, and evaluates the potential to meet the government's 2020 goal of 50% renewable energy. The final section moves the assessment of renewable energy's chances for success a step further by offering unique anecdotes and firsthand accounts of recent progress toward the 2020 goals, along with meaningful setbacks. This concluding section also places government mismanagement in historical and cultural contexts and analyzes France's role in its territory's renewable energy policy.

7.2 BACKGROUND

7.2.1 Geography

French Polynesia comprises 118 volcanic and coral islands and atolls in the middle of the South Pacific Ocean, between Australia and South America. Although the islands have a land mass of only 4000 km², they are spread across 2,500,000 km², approximately the size of Western Europe (Association of the Electricity Supply Industry of East Asia and Western Pacific [AESIEAP] 2011). The islands and atolls are divided roughly into four groups: the Society Islands, the Tuamotu Atolls, the Marquesas Islands, and the Austral Islands. The capital of French Polynesia, Papeete, is located on the island of Tahiti—the country's most densely populated island and is part of the Society Islands (Figure 7.1).

7.2.2 History

French Polynesia is one of France's five "overseas collectivities" (*collectivités d'outre-mer*). The other four are Saint Barthelemy in the Lesser

FIGURE 7.1
The geography of French Polynesia. (Reprinted with permission from Tahiti Tourism. The Islands of Tahiti. 2011. http://www.tahiti-tourisme.com/.)

Antilles, Saint Martin in the Lesser Antilles, Saint Pierre and Miquelon in the Atlantic Ocean off the coast of Newfoundland, Canada, and Wallis and Futuna in the Pacific Ocean. As a collectivity, French Polynesia is an administrative division of France, similar to France's own regions. It became an "overseas collectivity" in 2003, but gained the designation of "overseas country within the Republic" (*pays d'outre-mer*) in 2004. It operates with significant autonomy, which includes having its own president. The country has been a strategically valuable asset to France, particularly as a site for nuclear testing during the 1960s and 1970s. Although its use in this capacity (even though the tests ended in 1996) continues to color the relationship between local and French government officials, the country's main political agenda is independence (Giraud 2010).

The Pacific Islands (including what is now French Polynesia) are believed to have been settled 2000 to 3000 years ago, when Taiwanese and Melanesians crossed the Pacific Ocean in small boats in a series of great sea migrations (Encyclopaedia Britannica 2010). From the 1500s through the 1700s, Spanish, Portuguese, Dutch, and British traders discovered many of the islands, while also bringing the diseases that killed much of the indigenous population (ibid.).

Tahiti became a French colony in 1880. France then annexed other islands to form the French colony of Oceania. These islands became an "overseas territory" in 1946 and an "overseas collectivity" in 2003. They attained "overseas country" status in 2004. Pro-independence movements flourished in the 1970s and, over time, the islands took more control of internal affairs, culminating in a statute granting increased autonomy in 1996 (BBC News 2010).

France conducted 41 atmospheric tests on the Mururoa Atoll and neighboring Fangataufa between 1966 and 1996. In 1975, under international pressure, it switched to underground testing. After a 3-year moratorium, former French President Jacques Chirac announced that underground testing would resume in 1995. This move provoked international anger and violent protests in Papeete. Six of the eight planned tests were carried out, the last one in January 1996. At the end of the program, Paris agreed to a 10-year compensation package. In 1995, the United Nations' nuclear watchdog concluded that radiation levels around the atolls posed no threat. In 1999, Paris admitted that fractures had been discovered in the coral cone at the sites. The atolls continue to be monitored closely. In March 2009, the French government enacted legislation to provide compensation to former workers at France's nuclear weapons test sites (BBC News 2010).

7.2.3 Politics

The French Polynesian political system is characterized as a parliamentary representative democratic French "overseas collectivity" (Encyclopaedia Britannica 2010). Its head of state is the French president, but the local president (as of 2010) is Gaston Tong Sang. Over the past decade, the presidency has been an endless alternation of the same two leaders—one pro-independence (Oscar Temaru), the other anti-independence (Gaston Tong Sang). Sang became president of French Polynesia for the third time in November 2009, less than a year after he resigned from the post. He had resigned in February 2009, following a threatened no-confidence motion, and was succeeded by Temaru, who himself then became president for the fourth time in less than 5 years (BBC News 2010). The country has a 57-member assembly, elected every 5 years. The president is elected from within the assembly. France retains responsibility for foreign affairs, defense, justice, and security. French Polynesia is represented in the French parliament by two deputies and a senator, and is also represented at the European Parliament (ibid.).

7.2.4 Population

In 2010, of French Polynesia's 267,000 inhabitants, almost 70% lived in Tahiti and approximately 50% of that island's population lived in the capital city of Papeete. The territory's official language is French, which is spoken, read, and written by approximately 95% of the population as of the 2007 census.

French Polynesia enjoys a high standard of living. However, unemployment is high and wealth is distributed unevenly. The unemployment rate was nearly 12% in 2008 (versus 9.5% in France in 2009), and there is a large disparity between the average income level and the highest income brackets: gross national income (GNI) per capita for the high-income group (almost US$40,000 in 2008) was more than twice as high as the average GNI per capita for the country (~US$24,500 in 2010; World Bank Group [WBG] 2010).

7.2.5 Economy

To understand French Polynesia's economy today, it is important to look briefly at the country's economic history. France's nuclear tests transformed French Polynesia's economy from a quiet subsistence fishing and agricultural economy to one of civil servants. By 1968, 43% of the population was employed by the Ministry of Defense (L'Institut d'Emission d'Outre-mer [IEOM] 2010). With government salaries indexed to continental France, French Polynesia rapidly developed toward Western standards. When nuclear testing and France's military presence ended in the 1990s, money continued to flow from France and went toward developing a more independent economy, in which pearl-farming, fishing, and tourism became the dominant sectors (Ministère de l'Environnement [ME], and UC Berkeley Gump South Pacific Research Station [UCB GSPRS] 2009).

Despite the country's progress, French Polynesia continues to rely heavily on France's economic support (~US$2 billion in 2004), which represents approximately 50% of the country's gross domestic product (GDP; US$4.5 billion in 2010) (IEOM 2010). This makes French Polynesia the fifth largest economy in Oceania, after Australia, New Zealand, Hawaii, and New Caledonia (Energy Information Administration [EIA] 2010). The service sector and public administration are the two main employers of French Polynesians, engaging 47% and 32% of the workforce, respectively. In 2006, the service sector represented 78% of the country's GDP (ME and UCB GSPRS 2009).

7.2.5.1 Tourism

Tourism is the primary economic activity of French Polynesia, after the public sector. It represents 20% to 25% of GDP (varying according to the exact definition of "tourism"). It also represents the second-largest economic sector in terms of revenue, after transfers from France. However, tourism has been declining over the past decade. From 2006 to 2009, for example, the number of tourists visiting French Polynesia decreased 28%, from 221,549 to 160,447 (Figure 7.2; IEOM 2010).

To understand this decrease in the context of the global economy (the financial crisis and recession since 2008), we need to compare French Polynesia with other Pacific countries and the rest of the world. Figure 7.3 shows that, whereas tourism has decreased precipitously in French Polynesia in the last few years, it has remained constant or declined only slightly in other Asia Pacific countries and in the rest of the world. This trend can be

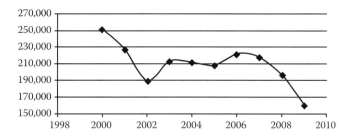

FIGURE 7.2
Number of visitors to French Polynesia, 2000 to 2009. (Adapted from IEOM. La Polynésie Française en 2009: Rapport Annuel. 2010. http://www.ieom.fr/IMG/pdf/ra2009_polynesie. pdf.)

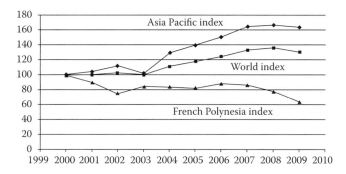

FIGURE 7.3
Index showing the number of tourists in the Asia Pacific, in the world, and in French Polynesia (index based on 100 in the year 2000). (Adapted from IEOM. La Polynésie Française en 2009: Rapport Annuel. http://www.ieom.fr/IMG/pdf/ra2009_polynesie.pdf; United Nations ESCAP. Statistical Yearbook for Asia and the Pacific 2011: Tourism. http://www.unescap. org/stat/data/syb2011/IV-Connectivity/Tourism.asp; and WBG. International tourism, number of arrivals. 2012. http://data.worldbank.org/indicator/ST.INT.ARVL?page=3.)

explained in part by the French Polynesian tourism industry's reliance on American tourists, whose international travel fell off the most because of the recession and the weak U.S. dollar. Other explanations include insufficient differentiation of French Polynesian tourism and fewer international flights (the number of seats offered decreased 19% from 2008 to 2009) as airlines have cut back their schedules due to higher oil prices (IEOM 2010).

7.2.5.2 *Pearl Farming*

Over the past 20 years, black pearls farmed from the local oyster species *Pinctada margaritifera* have become one of French Polynesia's primary exports, accounting for 60% of export revenues in 2009 (IEOM 2010). Yet the price of pearls—and thus the total value of exports—decreased dramatically in recent years as the total value of farmed pearl exports declined from more than 12 billion CFP francs in 2005 to less than eight billion CFP francs in 2009 (ibid.). This trend can be explained by the pullback in demand for luxury jewelry due to the global economic crisis, increasing external competition from countries such as Australia and Indonesia, and the fragmented nature of the market within French Polynesia (ME and UCB GSPRS 2009).

7.2.5.3 Trade Balance

In 2009, the trade deficit in French Polynesia was US$1.6 billion. The reserve ratio of 8.5% has decreased each year since as imports increase and exports decrease. Imports in 2009 totaled US$1.8 billion, primarily in food and raw materials (e.g., cement, natural gas, fuel), whereas exports totaled US$0.2 billion, primarily in pearls, noni fruit, and seafood (IEOM 2010).

7.3 ENERGY BACKGROUND

7.3.1 Primary Energy Sources and Uses

Total primary energy consumption in French Polynesia ranked 170th in the world in 2008, at 0.015 quadrillion Btu (EIA 2010). The top two uses of energy in the country are in the areas of electricity and transportation. In particular, 44% of primary energy is used for electricity generation, 44% for transportation, and 12% for other sectors (SEM 2009a). Figure 7.4 breaks down primary energy consumption in 2008 by source and shows which sectors were primarily responsible for consuming the energy produced. Percentages are calculated in terms of the French TEP or tons of oil-equivalent (similar to the U.S. system of measuring BOE, barrels of oil-equivalent), with total primary energy consumption totaling 336 kTEP that year. This total equates to primary energy consumption of 1.3 TEP per inhabitant (compared with 1.5 TEP per inhabitant in La Reunion and 3.1 TEP per inhabitant in New Caledonia, two other French overseas territories). The consumption of electricity per inhabitant was 2.4 MWh or 0.57 TEP (ibid.).

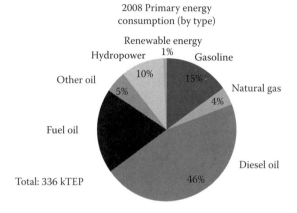

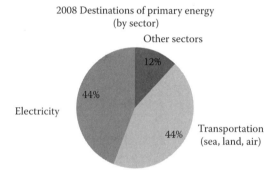

FIGURE 7.4

Total 2008 primary energy consumption, by sources and uses. (Adapted from SEM. 2008 Rapport Statistique Synthèse: Secteur de l'Energie, Electricité et Hydrocarbure. Luxembourg: Banque Européenne d'Investissement, 2009.)

Figure 7.4 illustrates that fossil fuel, in the form of imported oil and gas, comprises 89% of French Polynesia's primary energy consumption. In other words, the country was 89% energy-dependent in 2008. The degree of "energy dependence" is defined as the primary energy import divided by primary energy consumption. French Polynesia's energy dependence over time, as seen in Figure 7.5, is linked to fluctuations in hydroelectricity production over time (SEM 2009a). Despite a slight decrease from 2002 to 2008, energy dependence in the country remains high and is a primary driver of government initiatives to promote renewable energy (to be discussed in Section 7.4).

The destination of primary energy consumption can be broken down further within each sector, as seen in Table 7.1, with the bulk of energy

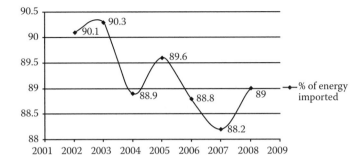

FIGURE 7.5
French Polynesia's energy dependence from 2002 to 2008. (Adapted from SEM. 2008 Rapport Statistique Synthèse: Secteur de l'Energie, Electricité et Hydrocarbure. Luxembourg: Banque Européenne d'Investissement, 2009.)

TABLE 7.1

2008 Destination of Primary Energy Consumption by Detailed Sector

Activity	Total Consumption (%)
Transportation	
Ground transportation	32
Public transportation	1
Flights between islands	5
Boat trips between islands	6
Subtotal transportation	44
Electricity	
Electricity–hotel/industry	20
Electricity–residential	17
Electricity–service/retail	6
Public lighting	1
Subtotal electricity	44
Other	
Natural gas	3
Other	10
Subtotal other	12
Total	100

Source: Adapted from SEM. 2008 Rapport Statistique Synthèse: Secteur de l'Energie, Electricité et Hydrocarbure. Luxembourg: Banque Européenne d'Investissement, 2009.

used for transportation going to ground transportation, and energy used for electricity split roughly between households and businesses.

The primary source of electricity in French Polynesia is oil (and its derivatives), with 75.3% coming from thermal power stations (SEM 2009a). Road infrastructure is extensive throughout the major cities of the Society Islands (such as capital city Papeete in Tahiti), and the country's boats and more than 50 airports also consume diesel fuel. According to the EIA, 20,000 tons of gas–diesel oil were used for transportation in 2007. Jet fuel consumption totaled 12,000 metric tons that same year. Oil consumption was more than 6000 bbl/day, and motor gasoline for road transport was 46,000 tons. The consumption of petroleum products and direct combustion of crude oil in 2009 was 7000 bbl/day (EIA 2010). This dependence on imported, foreign oil, common to all the Pacific Islands (Figure 7.6), has driven interest and investment in renewable energy whenever oil prices spike (first in the 1970s, and then again in recent years).

The share of electricity generated by diesel plants in different Pacific Island countries ranges from 50% to 100%, with French Polynesia falling toward the lower end of the range due to its natural hydropower resources.

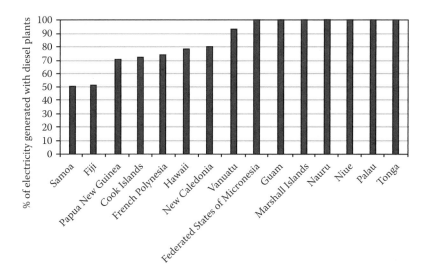

FIGURE 7.6

Electricity generation from diesel plants in Pacific Islands. (Reprinted with permission from M. Marconnet. Integrating Renewable Energy in Pacific Island Countries, 2007. http://researcharchive.vuw.ac.nz/bitstream/handle/10063/491/thesis.pdf?sequence=1.)

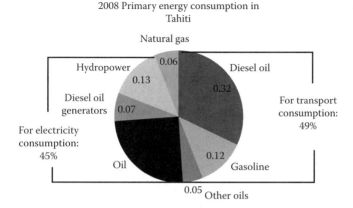

FIGURE 7.7
Energy consumption in Tahiti. (Adapted from SEM. 2008 Rapport Statistique Synthèse: Secteur de l'Energie, Electricité et Hydrocarbure. Luxembourg: Banque Européenne d'Investissement, 2009.)

Primary energy consumption in Tahiti is similar to consumption for French Polynesia as a whole, but slightly higher due to transportation needs. According to Figure 7.7, hydroelectricity, fuel oil, and approximately 7% of diesel oil generate electricity (~45% of total energy use in Tahiti). The remaining diesel oil, gasoline, and other oil derivatives are used for transportation (approximately 49% of total energy use in Tahiti). Natural gas is used primarily for cooking and hot water (~6% of total energy use in Tahiti). Renewable energy sources other than hydro are insignificant.

7.3.2 Electricity

7.3.2.1 Overview—French Polynesia

Ninety-seven percent of French Polynesians have access to electricity (AESIEAP 2011). The country had 227 MW of installed capacity as of 2008 and generated 705 GWh of power that same year. (According to the *AESIEA Gold Book,* French Polynesia's installed capacity was 283 MW in 2011. However, because the government could not verify this number, we have used the latest government data available, from 2008; AESIEAP 2011; ME and UCB GSPRS 2009; SEM 2009a). This puts French Polynesia in fifth place among the Pacific Islands in terms of electricity generated per year (see the chart comparison in Table 7.2, but note that the data are from a few years earlier). Power generation grew at a constant rate of 4% per year since 2000, but leveled off in 2008. Projections in 2009 for increases in electricity consumption were 2% per year for 2010 and 2011, and 25% (to 880 GWh) by 2020 (AESIEAP 2011).

Electricity use per capita is fairly high in French Polynesia, compared with other Pacific Island countries, as can be seen in Figure 7.8 (a statistic that is driven by population, electrification of the country, and electricity uses—with tourism and associated air-conditioning use being key drivers in French Polynesia).

TABLE 7.2

Power Capacity and Production in Pacific Island Countries (2005)

Country	Capacity, MW (year)	Electricity Generated, MWh (year)
American Samoa	93.75 (2005)	188,975 (2005)
Cook Islands	10.66 (2002)	29,758 (2002)
Federated States of Micronesia	34.49 (2005)	84,517 (1997)
Fiji	194.00 (2003)	6,989,000 (2003)
French Polynesia	215.49 (2005)	514,900 (2005)
Guam	552.20 (2004)	1,876,708 (2004)
Kiribati	7.75 (2003)	15,900 (2003)
Marshall Islands	31.30 (2005)	101,166 (2003)
Northern Mariana Islands	126.50 (2005)	432,000 (2005)
Nauru	18.90 (2004)	33,000 (2000)
New Caledonia	348.40 (2004)	1,677,527 (2004)
Niue	1.60 (2002)	3369 (2002)
Palau	28.00 (2005)	81,866 (2004)
Papua New Guinea	451.00 (2001)	3,178,000 (2003)
Samoa	36.02 (2003)	93,070 (2003)
Solomon Islands	22.40 (2002)	86,887 (2002)
Tonga	11.40 (2003)	34,000 (2003)
Funafuti (Tuvalu)	2.40 (2002)	4658 (2003)
Vanuatu	21.60 (2002)	47,148 (2002)
Total	**2207.86**	**15,472,449**

Source: Adapted from Marconnet, M. Integrating Renewable Energy in Pacific Island Countries, 2007.

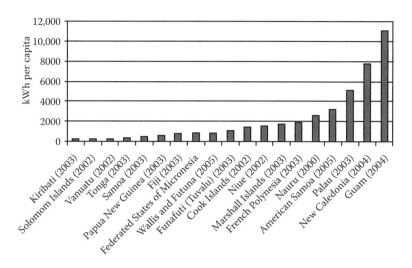

FIGURE 7.8
Electricity consumption (in kilowatt-hours) per capita in Pacific Island countries. (Reprinted with permission from M. Marconnet. Integrating Renewable Energy in Pacific Island Countries, 2007. http://researcharchive.vuw.ac.nz/bitstream/handle/10063/491/thesis. pdf?sequence=1.)

7.3.2.2 Overview—Tahiti

Power generation in Tahiti represented approximately 80% of the country's total production (or ~550 GWh) in 2008 (note that, together, Tahiti and the other Society Islands account for 96% of the country's power consumption; SEM 2009a). Thermal power stations produced approximately 72% of Tahiti's power (400 GWh) in 2008, with the rest supplied primarily by hydroelectric power (ME and UCB GSPRS 2009).

To forecast power consumption in Tahiti through 2020, the Public Works Ministry considered two scenarios. Under the first, the government does not pursue a comprehensive policy to promote energy efficiency and reduce consumption (such as tax rebates for energy-efficient appliances and air-conditioning systems). In this case, energy consumption is estimated to increase 2.7% per year from 2010 to 2020. Generation needs in 2020 would be 695 GWh, or 40% higher than in 2008. Under the second scenario, the government does pursue energy-efficiency plans, and energy consumption increases only 1.4% per year from 2010 to 2020. Generation needs would be 576 GWh in 2020, or 16% higher than in 2008.

In either case, the country will need to add more generating capacity to meet expected growth in consumption, in addition to adding renewable capacity, as it seeks to move away from energy dependence.

7.3.2.3 Electricity Sources

In 2009, 75.3% of French Polynesia's electricity was generated from thermal power stations, 22.4% from hydroelectric stations, and 2.3% from other sources (including solar photovoltaic [solar PV] and wind) (SEM 2009a). The country's primary natural sources of energy are hydropower (20% of total sources) and solar (<1%).

Comparing the sources of electricity generation in French Polynesia with the rest of the world in Figure 7.9, we find:

- Greater dependence on oil (>50%)
- No use of coal or natural gas (used only in the French overseas territories of La Reunion, Guadeloupe, and New Caledonia)
- Greater use of renewable energy (<30%) than the world average, but mostly undeveloped apart from hydroelectric power

Figure 7.9 shows that French Polynesia and other island nations have a much higher reliance on oil and hydro than the world average. To replace oil as a source of energy, the country could invest in other traditional sources of energy, such as coal or natural gas; but both require heavy infrastructure investment and are not natural resources for the country. Thus, a movement away from oil can happen only if French Polynesia makes a significant move into renewable energy, expanding its hydroelectric resources and developing solar and other technologies.

Total installed generation capacity and total production, broken down by fuel type at the end of 2009, included:

- *Thermal (fuel oil/diesel):* 162 MW installed capacity and 393 GWh production in Tahiti and approximately 48 MW installed capacity on the other islands of French Polynesia
- *Hydro:* 47 MW and 155 GWh production in Tahiti and approximately 1 MW installed capacity on the other islands of French Polynesia
- *Solar:* approximately 2.5 MW installed capacity in French Polynesia
- *Wind:* approximately 0.4 MW installed capacity in French Polynesia (SEM 2009a)

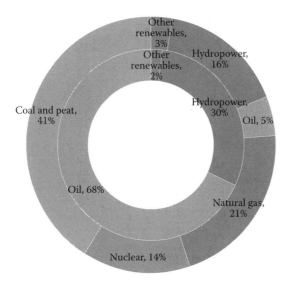

FIGURE 7.9
Sources of electricity generation in the world (outer circle, data for 2009) versus in French Polynesia (inner circle, data for 2008). (Adapted from OECD. OECD Factbook 2011: Economic, Environmental and Social Statistics. http://www.oecd-ilibrary.org/sites/factbook-2011-en/06/01/04/06-01-04-g1.html; SEM. 2008 Rapport Statistique Synthèse: Secteur de l'Energie, Electricité et Hydrocarbure. Luxembourg: Banque Européenne d'Investissement, 2009; U.S. EIA. International Energy Statistics: French Polynesia Electricity Net Generation by Type (billion kilowatt-hours), 2012. http://www.eia.gov/cfapps/ipdbproject/iedindex3.cfm?tid=2&pid=alltypes&aid=12&cid=FP,&syid=2010&eyid=2008&unit=BKWH.)

French Polynesia's largest thermal power station is in Tahiti in the Punaruu Valley. This plant runs on heavy fuel oil and has 122 MW of capacity and approximately 378 GWh of annual production (96% of Tahiti's total thermal production). The country's next largest thermal power plant is Vairaatoa, in the center of Papeete (the capital city, in Tahiti). This plant runs on diesel oil and has 40 MW of capacity and 15 GWh of annual production. The main sources of hydroelectricity in French Polynesia are in the Papenoo Valley. These plants have 28 MW of capacity and supply 50% of the hydroelectricity produced in Tahiti. Other hydro plants are spread across four additional valleys: Vaite (2 MW), Vaihiria (5 MW), Titaaviri (4 MW) in the west, and Faatautia (8 MW) in the east (Laurey 2009, p. 39; ME and UCB GSPRS 2009).

Among the rest of French Polynesia's atolls and archipelagos are a number of thermal and hydroelectric power plants (and distribution centers), as well as solar PV and wind installations. Thermal power is used throughout the country; hydroelectric is used primarily in Tahiti and in the Marquesas; and the Tuamotus and Gambier Islands have the highest number of solar PV installations.

7.3.2.4 Electricity Uses

Use of electricity can be divided into three categories, as detailed in Table 7.3.

- *Residential use:* 87% of customers consume 35% of the energy delivered through the grid. The average consumption is roughly 280 kWh/month. Average domestic consumption increased at a rate of 2% per year until 2008, after which it slowed due to the financial crisis.
- *Industrial use:* Less than 1% of customers consume more than 46% of the energy distributed. These consumers include hospitals and the tourism industry.
- *Commercial use:* This category represents 12% of consumers and 18% of the energy delivered. Commercial customers include government/public services, the retail sector (stores and restaurants), and public lighting. Commercial energy consumption had grown, on average, 3.6% for the last 10 years (prior to the financial and economic crises in late 2008; AESIEAP 2011).

7.3.2.5 Electricity Supply

There are three types of electricity supply arrangements in the Pacific Island countries: regulated monopolies, government-owned utilities, and public service concessions. In most countries, main grid electricity is provided by

TABLE 7.3

Number of Power Consumers by Sector

Sector	Total Consumers	Total Number of Consumers (%)
Industrial	688	0.80
Residential	70,276	87.50
Commercial	9352	11.70
Total	80,316	100

Source: Adapted from Association of the Electricity Supply Industry of East Asia and Western Pacific (AESIEAP), *AESIEAP Goldbook 2011*, Jakarta, Indonesia: AESIEAP, 2011.

a government-owned utility, which is often the sole provider. Most of these utilities were formerly government departments that have become corporate state-owned enterprises (Marconnet 2007). In the French territories of the Pacific (French Polynesia, New Caledonia, Wallis, and Futuna), private electricity companies have been awarded long-term public service concessions to provide electricity. These concessions allow the companies to operate within geographical areas specified by government contracts (ibid.).

In French Polynesia, the largest electric utility is Electricité de Tahiti (EDT), which is owned by energy affiliates of the French company GDF-Suez. EDT's contract for a public service concession to produce and distribute energy was initiated in 1960 and will run through December 31, 2030 (Laurey 2009, p. 38). This utility initially operated only thermal power stations. In 2000, it became the majority shareholder of another utility, Marama Nui, assuming responsibility for the latter's hydroelectric dams and power plants (ibid., p. 39).

The system for producing and distributing electricity in French Polynesia is broken down by geography and type of activity: Tahiti North's distribution concession with guaranteed service between the government and EDT; Tahiti South's distribution concession with guaranteed service between Secosud (a syndicate created to bring electricity to towns in South Tahiti) and EDT; a high-voltage transport concession between the government and the Société de Transport d'Energie en Polynésie (STEP); 19 islands with a distribution concession with guaranteed service between their towns and EDT; and 29 islands with state-controlled towns and six hydroelectric production concessions between the government and Marama Nui (now a subsidiary of EDT; Laurey 2009, pp. 38–39).

EDT has more than 80,000 customers, 60,000 of whom are in Tahiti (Laurey 2009, p. 40). It services the majority of grid-connected electricity consumers in French Polynesia, but several other public utilities and private independent power producers (IPPs) operate in more remote locations.

The SEM, a department of the French Polynesian government, is in charge of regulating the power sector and controlling electricity tariffs and power generation projects. An energy commission managed by SEM must validate all projects over 100 kW. This department is also responsible for verifying the compliance of EDT activities over the terms and conditions of its electricity public service contract. Each year, SEM conducts detailed due diligence for the government regarding EDT's activities, including its accounting practices. The French Polynesian government reviews electricity tariffs annually, evaluating them in light of shifts in key economic indicators (AESIEAP 2011).

7.3.2.6 Cost of Electricity

The tariff structure comprises 11 different rates, based on a reference rate, with five classifications: residential, industrial, commercial, street lighting, and prepayment (used for remote areas). The electricity reference rate, Pref = $E + T +$ ACE, includes E, the price per kilowatt-hour for primary energy used (fuel, hydroelectricity, renewable energy); T, the price per kilowatt-hour for transportation of energy between power plants to sub-stations; and ACE, EDT's internal accounting for all other expenses (see Table 7.4 for an example of the different electricity generation costs that might go into the E portion of the equation). The government and EDT negotiate the value of each term according to changes in cost-of-living indices, the technical yield of networks, total consumption, and growth in the past 5 years (AESIEAP 2011). The PETACE formula for calculating rates is complex and often difficult to interpret. According to Pierre Blanchard, a private consultant who has done extensive work on energy and power for the French Polynesian government, the formula has little relationship to economic reality. It is an "EDT formula" when it should be a "production-transportation-distribution formula" (Blanchard 2006, 2011).

Tariff rates are made public after they are approved by the local government, following a ministerial council meeting. Electricity rates have increased three times since early 2008, following changes in fuel prices and the US$/EUR exchange rates (AESIEAP 2011).

The price per kilowatt-hour charged for each *tranche* within the tariff structure is also based on energy consumption, with higher consumption corresponding to higher rates. Consumers are put into *tranches* based on their consumption (e.g., 0–100 kWh). In the last few years, rate increases have been significantly correlated with the price of oil—but only in one direction.

TABLE 7.4

Production Costs by Type of Energy in Tahiti

Type of Energy	Cost (US$)
Hydroelectric	$0.17–$0.22
Fuel oil-burning power station	$0.27
Gas power station	$0.30
Wind	$0.30
Solar (PV)	$0.44

Source: Adapted from MGT. *Renewable Energy Presentation.* Limited circulation Powerpoint presentation. Papeete, French Polynesia: MGT, 2008.

When prices have increased, rates have increased; but when prices have decreased, electricity rates have remained high. Crude oil's jump to almost US$140 a barrel in June 2008 brought with it a sharp increase in EDT's rates, from 5% to 20% higher, across all *tranches*. With the financial crisis in the fall of 2008, the price of oil decreased 75% at its low, but electricity rates in French Polynesia continued to increase across certain *tranches* of consumption (Table 7.5; Laurey 2009, p. 40). As discussed earlier in this section, 87% of customers are residential, and their average electricity consumption is 280 kWh/month. Thus, customers are paying, on average, US$0.44/kWh.

To put these rates in an international context, the first *tranche,* at US$0.18/ kWh, already puts French Polynesian tariffs higher than those found in most countries. The second *tranche,* at US$0.44/kWh, puts French Polynesia among a tiny number of countries with the most expensive electricity in the world. The third *tranche,* at US$0.54/kWh, puts French Polynesia at the absolute highest on an international scale. And finally, the fourth *tranche,* at US$0.60/kWh, "is an absolute record, unmatched in the world," according to Laurey (2009).

Table 7.6, from the International Energy Agency in 2008, shows the average electricity tariffs in other countries around the world. Although the list consists primarily of large, developed countries around the world, Hawaii is a somewhat comparable group of islands that EDT often cites as an example of a place where electricity rates are even higher than in French Polynesia—but the data do not support this claim. The US$0.47 rate listed for Hawaii is the highest rate paid, compared with French Polynesia's US$0.60 (Laurey 2009, p. 47).

Not only does French Polynesia have the highest electricity base rates in the world, but the French Polynesian consumer also pays taxes on these rates. In addition, the country utilizes a redistribution mechanism

TABLE 7.5

Electricity Tariffs (US$ per kilowatt-hour) in 2008 versus 2009

Tranche (kWh)	Price on July 1, 2008	Price on January 21, 2009	Difference (%)
0–99	0.13	0.18	36
100–150	0.38	0.18	−52
151–199	0.38	0.44	17
200–280	0.54	0.44	−18
281–400	0.54	0.54	0
401+	0.54	0.60	11

Source: Adapted from Laurey, N. *Energies Renouvelables: Plaidoyer pour une Véritable Politique de l'Energie en Polynésie Française.* Tahiti: Au Vent des Iles, 2009, 47.

TABLE 7.6

Electricity Tariffs (per kilowatt-hour) around the World in 2008

Country	Price (US$)
Venezuela	0.05
India	0.05
Mexico	0.09
Australia	0.10
Argentina	0.10
South Korea	0.10
US	0.11
France	0.16
New Zealand	0.16
Spain	0.17
Japan	0.18
Brazil	0.19
Germany	0.22
Italy	0.26
Netherlands	0.29
Hawaii*	0.47

Source: Adapted from Laurey, N. *Energies Renouvelables: Plaidoyer pour une Véritable Politique de l'Energie en Polynésie Française.* Tahiti: Au Vent des Iles, 2009, 47.

* The highest rate paid is in Hawaii.

to ensure that all consumers pay the same tariffs. To accomplish this, consumers in Tahiti are subject to "equalization" surcharges: They pay more for their electricity to subsidize consumers on the remote islands, who would pay much more if they were subject to market factors (e.g., distance and size of the market; Laurey 2009, pp. 48–49).

7.4 RENEWABLE ENERGY OPTIONS

7.4.1 Government Efforts to Promote Renewable Energy

7.4.1.1 Overview

Since the 2008 oil price spike and economic recession, the French Polynesian government has encouraged public utilities and private investors to invest in a wider range of technologies for generating electricity—with the goal of reducing dependence on imported oil. At the same time, the government has also pushed education efforts to manage the demand side of the equation, encouraging residents to manage consumption and modify energy-intensive behaviors.

In 2008, President Gaston Tang Song announced a goal whereby 50% of electricity would come from renewable energy sources by 2020 and up to 100% by 2030 (SEM 2009b). His announcement was accompanied by a technical investment plan for 2009 to 2020, establishing the desired mix of renewable energies needed to meet this goal (discussed below). Two new laws were also included, and the territory saw a dramatic increase in the number of solar PV installations between 2008 and 2010. These new installations should help the government reach its target for solar generation to cover 25% of peak demand through 2015, equivalent to 5% of the total electricity being generated from solar PV during this period, as detailed in Figure 7.10 (AESIEAP 2011).

Legislation passed in 2009 promotes renewable energy supplies from technologies such as solar PV and wind. These laws establish purchase tariffs (feed-in tariffs) for different kinds of renewable energy—including hydro, solar PV, and wind—with three different tariffs for solar PV installations, depending on the size of the project (<10 kW, 10–200 kW, and >200 kW). A "connection and purchase contract," established by the local government, requires EDT and all other local utilities to promote, facilitate, and connect every renewable energy generation project to the grid and to purchase the energy produced by these installations. Utilities

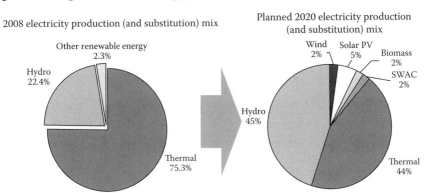

FIGURE 7.10

Electricity generation mix: planned changes from 2008 to 2020 to meet the 50% renewable goal. (Adapted from SEM. 2008 Rapport Statistique Synthèse: Secteur de l'Energie, Electricité et Hydrocarbure. Luxembourg: Banque Européenne d'Investissement, 2009; and SEM. L'Electricité a Tahiti. Bilan Prévisionnel et Programmation Pluriannuelle des Investissements. Papeete: Service de l'Energie et des Mines, 2009.)

(primarily EDT) are responsible for verifying that each project is in compliance with technical standards for the voltage and frequency of distributed energy on the grid (AESIEAP 2011).

In Tahiti, the construction of a new 60-MW thermal power station is planned for 2016 to 2018 to meet increases in demand and to replace older capacity. Depending on IPP development of hydro and solar, spurred by the 2009 legislation, new plant capacity may be reduced to 20 MW (AESIEAP 2011).

7.4.1.2 2009–2020 Government Plan to Meet the 50% Renewable Energy Goal

Based on the government's electricity consumption forecasts for 2020 and the goal of 50% of electricity to be generated from renewable sources by then, the government estimates it needs to add approximately 200 GWh of renewable power by that year. Table 7.7 helps to put this 200 GWh goal in context, showing how many plants of different types of power would have to be built to meet this goal. This table is indicative in order to provide context. Considering the feasibility of different renewable energy technologies and their costs, the government plans to meet the 200 GWh goal with a mix of sources. In Tahiti, for example, with 2020 electricity consumption expected to be approximately 700 GWh, the island will need 350 GWh of renewable capacity (50%). With just under 150 GWh of renewable energy already available from existing hydroelectric plants, the island still needs an additional 200 GWh of power. Table 7.8 shows the government's plan for adding renewable capacity in Tahiti to meet this goal.

TABLE 7.7

Equipment Needed for Different Types of Energy to Produce 200 GWh/year

	Structure	Equipment Required	Useful Life
Fuel oil/diesel	Conventional steam	2 units of 20 MW	30 years
Natural gas	Combustion turbine or combined cycle	4 units of 12 MW 1 unit of 40 MW	30 years
Wind	Turbines connected to grid	200 units of 500 kW 200 hectares	20 years
Solar PV	PV panels connected to grid	800,000 panels of 200 W each 240 hectares	20 years
Hydro	Dam/turbine/generator	16 units of 4 MW	40 years

Source: Adapted from MGT. *Renewable Energy Presentation.* Limited circulation Powerpoint presentation. Papeete, French Polynesia: MGT, 2008.

TABLE 7.8

Investments Needed to Reach Government Goal of 50% Renewable Energy by 2020 in Tahiti

Type of Energy	Hydro	Solar PV	Wind	SWAC	Biomass	Wave	Total
Capacity goal 2009–2020 (MW)	50	22	5	3	3	1	84
Power generation goal 2020 (GWh)	156	33	10	15	17	2	232
Cost of new capacity in million US$ (per MW)	5	9	3	9	7	12	
Total cost of capacity in million US$	273	191	15	28	20	6	532

Source: Adapted from SEM. *L'Electricité a Tahiti. Bilan Prévisionnel et Programmation Pluriannuelle des Investissements.* Papeete: Service de l'Energie et des Mines, 2009.

The costs to install renewable energy capacity, shown above, far exceed the two million dollar cost per megawatt to install thermal power (Ministère des Grands-Travaux [MGT] 2008). Taking into account all planned renewable energy capacity additions for French Polynesia, the mix of energy sources in 2020 would be 44% thermal, 45% hydroelectric, 5% solar PV, 2% wind, 2% seawater air conditioning (SWAC), and 2% biomass (detailed in Figure 7.10).

As seen in Figure 7.11, the government's plan involves doubling the percentage of electricity sourced from hydroelectric power and significantly expanding the production of electricity from solar PV, wind, SWAC, and biomass.

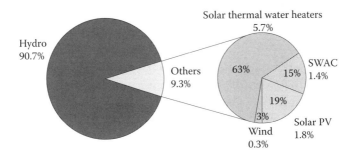

FIGURE 7.11

Components of renewable energy mix in 2008; percentages outside the smaller circle on the right add up to comprise the "9.3% others"; percentages inside the smaller circle on the right add up to 100%. (Adapted from SEM. 2008 Rapport Statistique Synthèse: Secteur de l'Energie, Electricité et Hydrocarbure. Luxembourg: Banque Européenne d'Investissement, 2009.)

7.4.1.3 Details of the 2009 Renewable Energy Legislation

Although the government may set goals for the mix of electricity to be sourced from renewable energy, actual investment in new capacity is, for the most part, in the hands of EDT and private investors. Thus, the government's primary tools for accomplishing its goals are policy and legislation that create investment incentives. The bylaws passed in 2009 were designed to provide appropriate incentives for investment in wind, solar PV, and hydroelectric power.

Details of a few of the most significant 2009 regulations include:

- Bylaw No. 0901 CM, June 25, 2009—feed-in tariffs (the guaranteed price for which an independent generator can sell power back to the grid) for renewable energy. Hydroelectric power, US$0.15/kWh; wind, US$0.18/kWh; and solar PV, US$0.42 to US$0.55 (depending on the power produced). Prices are fixed for 20 years for wind, 25 years for solar PV, and 35 years for hydroelectric power.
- Bylaw No. 760, May 29, 2009—tax incentives for renewable energy (a 45% tax credit for any investment in renewable energy, excluding hydroelectric power in valleys, which merits a 35% tax credit).
- Bylaw No. 0726 CM, May 29, 2009—a 5% tax credit to build housing with renewable energy installations (LEXPOL 2011).

7.4.1.4 TEP-VERTES: Regional Collaboration to Promote Renewable Energy in Isolated Sites

Although French Polynesia's goals for renewable energy generation encompass the entire country, Tahiti accounts for 80% of electricity consumption, and Tahiti plus the other Society Islands account for 96% of electricity consumption. Therefore, the author's conversations with government officials and the government documents they shared often focused on electricity in Tahiti alone. Here is a brief look at a regional collaboration also under way to address renewable energy use in more remote sites.

TEP VERTES—which stands for Tonne Equivalent Pétrole Valorisation des Energies Renouvelables Transfert d'Expérience et de Savoir-faire and translates as "Barrels of Oil-Equivalent: Promoting the Transfer of Experience and Knowledge in Renewable Energy"—is a project representing a regional collaboration cofinanced by the Ninth European Development Fund (le 9ème Fonds Européen de Développement, FED),

New Caledonia, French Polynesia, and Wallis and Futuna. The project's goal is to promote the use of sustainably built renewable energy sources to improve the quality of life for rural and isolated populations in the three French Pacific Ocean territories. The total cost of the project is US$14 million, of which the European Union (through FED) pays a little over 50%. French Polynesia and New Caledonia pay approximately 23% each, and Wallis and Futuna pays 4%. The project has three phases:

1. Phase one: Exchange experiences among the three countries and other overseas territories and define a plan of operation for each country
2. Phase two: Issue requests for proposals, inviting international investors to bid on executing the renewable energy plans for each country, sign contracts, and execute plans
3. Phase three: Present final analyses, expertise, and results of operations to all members of the Pacific Ocean countries with the goal of creating an even larger cooperation

As of this writing, the project was in phase two, with contracts signed and installation under way. In French Polynesia, the plan of operation includes the delivery, installation, and service of three solar–diesel hybrid systems on the remote atolls of Ahe, Fakahina, and Napuka in the Tuamotu Archipelago (SEM 2009c).

7.4.2 Existing Renewable Energy Capacity and Potential

Renewable energy potential in French Polynesia is significantly greater than the currently installed capacity and far surpasses the 2020 goals outlined above. Limitations to renewable energy to date include the high cost of installations and the lack of government incentives to subsidize investment (prior to 2009). A brief overview of current installed capacity and production versus potential for the six forms of renewable energy covered in the government's 2020 plan includes:

- *Hydro:* 47 MW installed capacity, 157 GWh generated in 2008. Potential production of electricity from hydropower plants not yet installed is 50 MW, with annual production estimated at 156 GWh.
- *Solar:* 2.5 MW installed capacity, 3 GWh generated in 2008. The 20,150 solar water heaters installed in 2008 (versus 15,500 in 2007) led to avoided consumption of 10 GWh. PV panels on roofs in

Tahiti could provide 78 MW, or more than 115 GWh/year. Adding 100 hectares of these panels could provide 100 MW of capacity, or around 150 GWh/year.

- *Wind:* 0.4 MW installed capacity, 0.5 MWh generated in 2008. A ridge approximately 70 km east of Tahiti offers a potential site for 70 MW of wind installation, with annual production estimated at 140 GWh.
- *SWAC:* the Hotel Intercontinental Resort and Thalasso Spa Bora Bora, which opened in May 2006, is fully air-conditioned with its own SWAC installation (1.6 MW cold). This system saves 80% to 90% of the energy used for cooling, compared with a conventional central air-conditioning compressor. Electricity consumption savings is estimated at 2.4 GWh/year.
- *Biomass:* The potential of biomass waste is limited, estimated at 9 MW, or up to 50 GWh/year. However, this type of electricity could help solve the problem of waste treatment and decrease the space needed for landfill in Tahiti.
- *Wave power:* This has been estimated at seven sites around Tahiti. The first wave power station, being built in Papara (on southern Tahiti), represents a potential of 500 kW, and 2 GWh/year. The four best sites represent a potential 8 MW, or 28 GWh/year, for Tahiti. However, in theory, every meter of coast represents 27 kW of potential power (Blanchard 2011; ME and UCB GSPRS 2009; SEM 2009a).

Table 7.9 displays the information detailed above for the primary sources of electricity production (thermal, hydro, solar PV, and wind) and substitution sources (solar thermal water heaters and SWAC) in 2008. In the context of electricity, a "substitute" is an alternate source of power. For example, a house might substitute a solar water heater for an electric water-heating system. This alternate source counts as electricity production, even though it is classified as a "substitution."

Although Figure 7.10 shows where French Polynesia was in 2008, it is also important to understand renewable energy sources in their historical context. Electricity generation from renewable energy grew 10% from 2000 to 2008, fueled, in great part, by the development of nonhydroelectric sources—most notably, solar thermal water heaters. Nonhydroelectric renewable energy in French Polynesia grew by 2000% during this period, but still constituted only 2.3% of the total electricity production in 2008. Percentage terms have little significance when we are discussing such small numbers. Table 7.10 shows changes in the capacity and production

TABLE 7.9

Installed Capacity and 2008 Production of Renewable Power Sources in French Polynesia

	Production					**Substitution**		
	Thermal*	Hydro	Solar PV	Wind	Total	Solar Thermal (Water Heaters)	SWAC	Total
Number of installations	25	13	**	3		20,150	1	
Installed capacity (MW)	177	47	2.5	0.4	227			
Production or substitution (GWh)	531	158	3	0.5	692	10	2	12

Source: Adapted from SEM. *2008 Rapport Statistique Synthèse: Secteur de l'Energie, Electricité et Hydrocarbure.* Luxembourg: Banque Européenne d'Investissement, 2009.

* Does not take into account EDT concessions.

** Among 1334 isolated installations, the exact number connected to the grid are unknown.

of the major sources of renewable energy in French Polynesia from 2000 to 2008. It is worth noting that SWAC was installed at the Intercontinental in Bora Bora in 2006, and increased incentives for installing solar water heaters were introduced in 2007 (Intercontinental Bora Bora Resort and Thalasso Spa 2007).

Table 7.10 also shows that in 2008, nonhydro renewable production comprised 9% of total renewable energy electricity production, of which 63% came from solar thermal water heaters (see Figure 7.11 for a graphic representation of the components of renewable energy power generation).

The rest of this section will consider the renewable energy technologies mentioned above in greater detail—first, the established technologies of hydroelectric, solar PV, and wind, and then the newer technologies of SWAC and ocean thermals.

7.4.3 Detailed Assessment of Existing Renewable Energy: Hydroelectric, Solar, and Wind

7.4.3.1 Hydroelectric Power

To generate hydroelectricity, a turbine converts hydraulic energy from a moving water source (e.g., a river, waterfall, or ocean current) into electrical energy. Unlike solar, hydroelectric energy is influenced significantly by economies of scale: 1 kWh produced by a large installation is much more expensive than 1 kWh produced by a small installation. Hydroelectricity is subject

TABLE 7.10

Changes in Renewable Energy Capacity and Electricity Production (2000–2008)

	2000	2001	2002	2003	2004	2005	2006	2007	2008
Hydroelectricity									
Capacity (MW)	47	47	47	47	47	47	47	47	47
Production (GWh)	157	154	119	124	156	147	160	174.0	158
Solar PV									
Capacity (MW)	0.5	0.7	0.8	1.0	1.2	1.5	1.8	2.1	2.5
Production (GWh)	0.7	0.9	1.1	1.3	1.6	2.0	2.3	2.7	3.1
Wind									
Capacity (MW)	0.1	0.1	0.1	0.1	0.1	0.1	0.1	0.2	0.4
Production (GWh)	0.1	0.1	0.1	0.1	0.1	0.0	0.1	0.1	0.5
Other renewable energy production									
Solar thermal (water heaters) (GWh)	–	–	4.0	5.0	5.0	6.0	6.0	8.0	10.0
SWAC Bora Bora (GWh)	–	–	–	–	–	–	2.0	2.0	2.0
Non-hydrorenewable production	1	1	6	6.0	7	8	11	13	16
Total production	**158**	**155**	**125**	**130**	**163**	**154.0**	**171**	**187**	**174**

Source: Adapted from SEM. *2008 Rapport Statistique Synthèse: Secteur de l'Energie, Electricité et Hydrocarbure.* Luxembourg: Banque Européenne d'Investissement, 2009.

to seasonal changes in water flow but is, nevertheless, fairly stable, as the locations of water sources rarely shift substantially. As discussed elsewhere in this chapter, hydroelectricity is the largest source of renewable energy found in French Polynesia—and is responsible for 22.4% of the electricity generation in 2008—due to the strong flows of Tahiti's rivers (EDT 2008).

Hydroelectricity has significant remaining potential for development in French Polynesia. As with other parts of the world, however, environmentalists point out that hydropower may be renewable, but it is usually not developed sustainably. When hydroelectric dams are built, they flood surrounding areas, resulting in deforestation and loss of animal and plant species (Meyer 2007). Dams in French Polynesia are built on a much smaller scale, however, than the dams that attract the most negative news headlines, such as Three Gorges Dam in China (22,500 MW) and Itaipu in Brazil (14,000 MW), because they are built into volcanic rock (Laurey 2009, p. 161). Thus, they represent a relatively environmentally sustainable option for the country, particularly when compared with the high greenhouse gas costs associated with importing fossil fuels.

French Polynesia currently has 16 hydroelectric power plants—13 in Tahiti and 3 in the Marquesas Islands. As noted in Section 7.3, installed capacity in Tahiti is just under 47 MW, and 2008 production was 155 GWh. The five valleys that hold Tahiti's power plants are Papenoo, Faatautia, Vaihiria, Vaite, and Titaaviri. In 2008, the government approved the construction and operation of hydroelectric plants in three new valleys: Taharuu, Vaitaara, and Papeiha. In the Papeiha Valley, EDT has plans for a 10-MW dam with 24 GWh of annual production, which will be completed in 2013. The project's expected cost is eight million dollars (EDT 2008). Looking at the entire island, five principal valleys, representing 44 MW or 123 GWh of capacity per year, as well as eight small sites in the southeast, representing less than 500 kW or 16 GWh of capacity per year, present the most promising areas for further hydroelectric development. Developing just the sites in the five valleys (44 MW) would almost double French Polynesia's existing hydroelectric capacity of 47 MW. Thus, this development is indispensable for reaching the government's goal of 50% renewable energy by 2020.

Outside of Tahiti, the Marquesas Islands are the only other group in French Polynesia that produce hydroelectricity. With less than 1 MW of installed capacity and 2008 production of 2.5 GWh, Marquesas production accounts for only 1.5% of the country's total hydroelectric production. Within the Marquesas, however, the effect is meaningful, with 23% of electricity production on Nuku Hiva (one of the islands) coming from hydroelectric power. Between 2009 and 2011, EDT had planned three projects for the Marquesas: optimization of the Taipivai power station, construction of a power station on Hakaui, and the addition of turbines to Aakapa (Giraud 2010). According to EDT, significant potential exists for additional hydroelectric development in the Marquesas and for microhydroelectric (installations up to 100 kW) development in the Society and Austral Islands. In addition, micro-hydro installations are being studied in Tahiti and in the Marquesas by the Polynesian Society of Electrotechnical Research and Development (la Société d'Etudes et de Développement Electrotechnique Polynésienne, SEDEP; ME and UCB GSPRS 2009).

7.4.3.2 Solar Power—PV Panels and Thermal Water Heaters

Solar energy is not yet a meaningful source of electricity in French Polynesia due primarily to the absence of the feed-in tariffs (prior to 2009) needed to promote investment. In 2008, the country's "solar park" comprised

1.7 MW of installations in isolated sites, 0.7 MW of installations connected to the grid (in the Society Islands), and 0.1 MW for EDT's solar station on Makatea in the Tuamotu Atolls. Table 7.11 divides French Polynesia's solar installations into three groups—(1) in isolated areas, (2) connected to the grid, and (3) EDT's Makatea installation—and shows how capacity and production have changed from 1997 to 2008. Seventy percent of the installed capacity and production was provided by PV installations in 1334 isolated sites (with the rest connected to the grid). Sixty-three percent of these isolated installations are located in the Tuamotu Atolls. Two government initiatives have been responsible for promoting the installation of much of the solar PV capacity detailed in Figure 7.11; the first was launched in 1997 and the second in 2005 (ME and UCB GSPRS 2009).

Hybrid solar systems, where solar panels are coupled with a generator and batteries for storage, are common in isolated areas that cannot be connected to the grid. Table 7.11 starts in 1997, the year the government launched its PHOTOM initiative, to encourage solar installations in remote sites. This program provides subsidies for purchasing and installing solar panels to inhabitants of islands that are outside the country's electrical grid. It benefits from a tax exemption, a customs exemption for importing equipment, and aid from the Agency for the Environment and Energy (Agence de l'Environnement et de la Maîtrise de l'Energie, ADEME). With the tax and financing subsidies, payments are only about US$68/month over the course of 15 years for the installation, about half the actual cost (ME and UCB GSPRS 2009; Laurey 2009, p. 124). The program has been renewed every year since its inception, benefitting, as of 2008, 1334 installations (Laurey 2009, p. 123).

The government launched its second program, Connectis DOM-TOM (a French acronym for overseas departments), in 2005 to connect PV installations in private homes or businesses to EDT's public grid to allow the users to sell electricity back to the grid. Approximately 30 installations have been connected to the grid each year under this program. Like PHOTOM, Connectis benefits from a tax exemption, a customs exemption for importing the equipment, and aid from ADEME (SEM 2009a).

Both of these solar programs are managed by ADEME and, by 2008, accounted for nearly all of French Polynesia's solar energy—1.7 MW for PHOTOM (just under 70% of the total) and 0.7 MW for Connectis (just under 30% of the total; ibid.). A large solar installation at the University of French Polynesia and a hybrid installation on Makatea Atoll in the

TABLE 7.11

Solar PV Capacity and Production (1997–2008)

	1997	1998	1999	2000	2001	2002	2003	2004	2005	2006	2007	2008
Installations in isolated sites												
Capacity (MW)	0.1	0.2	0.4	0.5	0.7	0.8	1.0	1.2	1.3	1.5	1.6	1.7
Production (GWh)	0.1	0.3	0.5	0.7	0.9	1.1	1.3	1.5	1.7	1.9	2.0	2.2
Installations connected to the grid												
Capacity (MW)	–	–	–	–	–	–	–	0.1	0.2	0.3	0.5	0.7
Production (GWh)	–	–	–	–	–	–	–	0.1	0.2	0.4	0.6	0.9
EDT's solar station in Makatea												
Capacity (MW)	–	–	–	–	–	–	–	–	–	0.05	0.05	0.05
Production (GWh)	–	–	–	–	–	–	–	–	–	0.09	0.04	0.05
Total capacity (MW)	0.1	0.2	0.4	0.5	0.7	0.8	1.0	1.2	1.5	1.8	2.1	2.5
Total production (GWh)	0.1	0.3	0.5	0.7	0.9	1.1	1.3	1.6	2.0	2.3	2.7	3.1

Source: Adapted from SEM. *2008 Rapport Statistique Synthèse: Secteur de l'Energie, Electricité et Hydrocarbure.* Luxembourg: Banque Européenne d'Investissement, 2009.

Tuamotu Atolls account for most of the remaining capacity (Laurey 2009, p. 124).

French Polynesia's main utility, EDT, has also experimented with solar energy, installing a hybrid solar–diesel system on Makatea, which supplies electricity to the 90 inhabitants there. The 2006 installation comprises 300 solar panels and a diesel generator that supplies additional energy if solar production is insufficient. EDT's cost of US$730,000 benefited from a 27% subsidy from ADEME. In 2007, approximately 63 MWh of energy were produced on Makatea, more than 60% of which came from solar energy. Thanks to EDT's hybrid solar system, consumption of diesel oil on the island was reduced by 70% in 2007. Sixty percent of electricity sourced from renewable energy is a record in French Polynesia (whereas it remains a goal for other atolls across the country; EDT 2008; Laurey 2009, pp. 119–120).

EDT is also conducting feasibility studies for similar installations on Tetiaroa and Maiao in the Society Islands and in several small valleys in the Marquesas Islands. Solar energy remains very expensive, however: the cost of generating 1 kWh on Makatea is more than three times higher than in Tahiti (US$0.97 compared with US$0.28). Because Makatea residents pay only US$0.34 for their solar electricity, the project is not profitable for EDT. Until the price of oil increases or the price of solar panels decreases to the point where the two technologies become cost-competitive, EDT believes that solar is best adapted for the most isolated areas (ME and UCB GSPRS 2009).

In addition to government-promoted remote installations, grid-connected installations, and EDT's project on Makatea, French Polynesia has seen some initiatives on the part of companies and educational institutions to meet their electricity needs through solar installations. In April 2009, the Bank of Polynesia opened an "eco branch" in Faa'a, powered completely by solar. The project, executed by independent power provider Moana Roa, cost US$360,000 for an installation of 250 m² of PV panels. A new project the bank is considering in Taravao will be powered by both solar PV and wind. Solar panels are being installed on the roof of the Carrefour Supermarket in Punaauia. The IPP ecoenergy will manage the electricity from this installation, which is expected to generate 1.5 GWh/year, approximately 25% of the store's electricity needs. In 2008, the University of Polynesia installed a PV park of 56 kW, with a potential production of 74 MWh per year (ibid.). A Total gas station in Papara installed 100 m² of solar panels in 2006, and the commercial center in

Punaauia, outside Papeete, installed solar panels in 2008 (Laurey 2009, p. 137).

Solar thermal water heaters replace traditional electric- or gas-powered heaters with a renewable energy source. Thus, they are considered a source of electricity "substitution" rather than production. These heaters are the only form of renewable energy in French Polynesia that is cost-competitive with traditional energy alternatives and does not rely on subsidies (Garnier 2008a). ADEME subsidized the installation of 15,000 solar water heaters prior to 2005. In 2006, concluding that the market was mature and the heaters cost-competitive, it discontinued this program. According to ADEME, the return on investment for a solar water heater is 2 years compared with an electric water heater and 5 years compared with a gas water heater, without taking any subsidies or tax benefits into consideration. It has estimated that electricity consumption per household decreases by 5% with an installed solar water heater. The number of such heaters in French Polynesia nearly doubled from 2004 to 2008, with 20,150 units providing 10 GWh of substituted electricity in 2008. In the same year, 30% of the hot-water supply in bathrooms in Tahiti came from solar energy (ME and UCB GSPRS 2009).

7.4.3.3 Wind Power

To generate electricity from wind, the blades of a turbine (which capture the wind's kinetic energy and turn it into mechanical energy) are connected to a drive shaft that turns an electric generator. Like hydroelectric power (and unlike solar power), wind power is highly sensitive to economies of scale. Large wind turbines can produce some of the cheapest renewable energy per kilowatt-hour of production. In areas where environmental conditions (such as the incidence of cyclones) prevent the installation of large 2-MW turbines, smaller turbines adapted to tropical wind conditions offer an alternative for generating wind power. The greatest problem with wind power systems is the wind's unpredictability (and instability) as a natural resource. When wind speeds are too low, the turbine does not rotate; and when they exceed a certain level (~90 km/hour), the turbine also stops to avoid mechanical failure. Such sudden halts in production can destabilize an entire electric grid if wind power is a large enough component of total power (Garnier 2008b).

The smaller turbines adapted to the tropical winds and cyclones described above are an important source of electricity in several French overseas territories. Reunion Island has 12 MW of installed capacity (40 GWh of annual production), New Caledonia has 25 MW of installed capacity (50 GWh of annual production), and Guadeloupe/Martinique have 15 MW of installed capacity (30 GWh of annual production; SEDEP, SEM, and Te Mau Ito Api 2008). For context, total world wind capacity was 197 GW in 2010 (World Wind Energy Association [WWEA] 2011). Yet French Polynesia had only 0.4 MW of installed capacity (0.5 GWh of annual production) in 2008 (SEM 2009a). During interviews conducted in Papeete, the author could find no satisfactory explanation, but reasons may include a lack of financial incentives for renewable energy investments and a lack of knowledge about the country's wind energy potential (see the discussion of wind power in Section 7.5.1.2).

The total potential for wind power by island or zone in French Polynesia is not well-established because very few studies have been conducted (Blanchard 2006). In 2008, EDT launched a study to measure wind potential across eight sites, using anemometers to measure wind speed. According to the initial results, EDT believes French Polynesia has low wind power potential, estimated at 25 to 300 kW on each island (EDT 2008). On the other hand, IPP Moana Roa has installed several wind turbines in the Tuamotu Atolls since 2004 and reports that the atolls do have wind consistent and strong enough to generate reliable energy (ME and UCB GSPRS 2009). Existing wind power capacity in French Polynesia consists of a hybrid wind–diesel farm in the Makemo Atoll in the Tuamotus, an EDT installation in Rurutu in the Austral Islands, and two isolated installations on Hao in the Tuamotus and Maupiti regions of the Society Islands. The Makemo installation contributes approximately 70% of annual wind energy production in the country. Details of capacity and production from 1999 to 2008 can be found in Table 7.12.

As seen in Table 7.12, the most significant wind initiative in French Polynesia began in January 2008, with a partnership to install a wind farm on the Makemo Atoll. This project began when the Ministry for Development of the Archipelagos (le Ministère du Développement des Archipels) created a Mixed Economy Society, called Te Mau Ito Api (la Société d'Economie Mixte Te Mau Ito Api), in January 2007 to expand electricity production from renewable energy sources in the Tuamotu Atolls (outside communities under EDT concession). This was the result of a plan begun in 2005 to promote energy independence in the Tuamotu Atolls, with a goal of 70% renewable energy by 2011 (Tahiti Presse 2007). The wind farm was installed by Te Mau Ito Api, in partnership with

TABLE 7.12

Wind Power Capacity and Production (1999–2008)

	1997	1998	1999	2000	2001	2002	2003	2004	2005	2006	2007	2008
Rurutu (EDT)												
Capacity (kW)	120	120	120	120	120	120	120	120	120	120	120	120
Production (MWh)*	65	65	65	65	65	65	65	65	41	65	63	65
Hao (isolated site)												
Capacity (kW)	–	–	–	–	–	–	–	–	–	–	22.5	22.5
Production (MWh)	–	–	–	–	–	–	–	–	–	–	45	45
Maupiti (isolated site)												
Capacity (kW)	–	–	–	–	–	–	–	–	–	–	15	15
Production (MWh)	–	–	–	–	–	–	–	–	–	–	30	30
Makemo (hybrid power station)												
Capacity (kW)	–	–	–	–	–	–	–	–	–	–	–	210
Production (MWh)	–	–	–	–	–	–	–	–	–	–	–	352
Total capacity (kW)	120	120	120	120	120	120	120	120	120	120	158	368
Total production (MWh)	65	65	65	65	65	65	65	65	41	65	138	492

Source: Adapted from SEM. *2008 Rapport Statistique Synthèse: Secteur de l'Energie, Electricité et Hydrocarbure.* Luxembourg: Banque Européenne d'Investissement, 2009.

* Exact numbers not available.

the Polynesian Society of Electrotechnical Research and Development (SEDEP), the Polynesian Society of Grid Research and Services (la Société Polynésienne de Réseaux d'Etudes et de Services, SPRES), and the French Polynesian government. The wind farm comprises a hybrid wind–diesel system: six 30- to 35-kW turbines, battery chargers, a 1.5-MWh battery park, a 100-kW inverter, two back-up generators, a new electric grid, and a group of prepayment meters. The system is designed to supply electricity to the community (~900 inhabitants) through 70% wind power (supplemented by battery power in case of short periods of low wind) and 30% back-up generators during longer periods without wind. According to Makemo's wind farm operator, SEDEP, the project has been a success, providing benefits that include consistent, stable electricity generation (due to moderate and constant winds throughout the year), reductions in noise pollution from the old diesel power station, reductions in diesel consumption (down to 8000 L/month from 24,000 L/month before the wind farm was built), and economic viability (producing wind power was found to be comparable or less expensive than the US$0.72 cost per kWh for thermal production on Makemo; SEDEP, SEM, and Te Mau Ito Api 2008).

The three other sources of wind power shown in Table 7.12 are EDT's installation on Rurutu and two isolated installations on Maupiti and Hao. On Rurutu, in the Austral Islands, an experimental program to test wind power began in 1999, with two 40-kWh turbines installed by EDT. These turbines generated 20% of the island's electricity needs, but the equipment is now obsolete and production is less reliable (Ministère du Développement et de l'Environnement de Polynésie Française [MDE] 2006). The number of wind turbines being used for private residences or small hotels in French Polynesia is small. A mixed 15-kW wind/2-kW solar installation was created on Maupiti in 2004, and a mixed 22.5-kW wind/3-kW solar installation was built for a private home on Hao in 2007. IPP Moana Roa also built the offices of the radio station Sailmail on Manihi in 2008, with a small hybrid 1.2 kW wind/1.5 kW solar/0.3 kW micro-hydro installation (ME and UCB GSPRS 2009).

The existing wind installations discussed in this section are all located on more remote islands in French Polynesia, primarily because winds have been found to be the most consistent (across seasons and across time of day) in the Tuamotu Atolls. To satisfy the government's goal of 5 MW of capacity and 10 GWh of production in Tahiti (Table 7.8), however, the government needs to promote large turbine (2 MW) installations on this island. The author could not obtain additional details about the study that found 70 MW of potential capacity on a ridge east of Papeete; but further feasibility studies clearly are needed to determine whether wind speeds are appropriate for wind farm installations (Blanchard 2011).

7.4.4 Newer Renewable Energy Technologies: SWAC and OTEC

The most cutting-edge development and deployment of new technologies in French Polynesia include SWAC technology and ocean thermal energy conversion (OTEC). SWAC uses cold deep-seawater, pumped from a depth of around 800 m, to run air conditioning. The process generates up to 80% savings on electricity used versus conventional air conditioning. OTEC uses the natural temperature differential between layers of ocean water to generate energy.

7.4.4.1 Seawater Air Conditioning

In French Polynesia, air conditioning can account for up to 50% of electricity consumption because of the country's year-round tropical climate and dependence on the tourism industry. Technology that uses deep ocean water rather than traditional sources of electricity to provide air conditioning offers an attractive way for the tourism industry (hotels and resorts) to save on the high costs of electricity and reduce carbon emissions (while also improving their "green" image for potential customers). SWAC technology was conceived by John Craven, the head scientist of the Common Heritage Corporation in Hawaii. Cold seawater from a depth of 800 m is pumped through a titanium thermal exchanger, where the cold air is transmitted to a closed freshwater circuit. From there, the freshwater carries the cold air throughout the hotel, air-conditioning the entire resort. The thermal exchanger functions like a large radiator, in that it comprises 200 titanium plates, which are noncorrodible thermal conductors. The SWAC system cuts electricity needs by more than 80% when compared with a normal air-conditioning system—equivalent to 2.5 million liters of fuel oil imported into French Polynesia per year. When finished, the water exits the heat exchanger by flowing back into the ocean through connecting pipes (Intercontinental Bora Bora Resort and Thalasso Spa 2007; Makai Ocean Engineering 2011). Figure 7.12 offers a rendering of a SWAC system.

The first SWAC system was implemented in 2006 at the Intercontinental Bora Bora Resort and Thalasso Spa, on Bora Bora in the Society Islands.

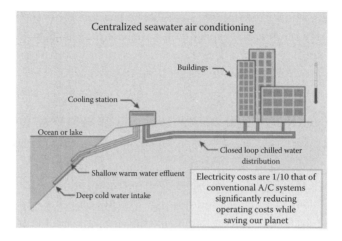

FIGURE 7.12

Artistic rendering of what a SWAC installation might look like. (Reprinted with permission from Hawaiian Electric Company. Hawaii's Energy Future. http://www.hawaiis energyfuture.com/articles/Sea_Water_Air_Conditionin.html.)

Since its installation, it has saved approximately 2500 tons of CO_2 per year. The resort's developer and owner, Pacific Beachcomber, also hopes to install SWAC at other resorts it owns, such as those on Moorea and Tahiti (Intercontinental Bora Bora Resort and Thalasso Spa 2007).

Pacific Beachcomber is also overseeing the late Marlon Brando's vision to turn his 13-island atoll, Tetiaroa, in the Society Islands, into an eco-resort. The Brando resort is under construction with a completion date set for 2012. In addition to using SWAC, Pacific Beachcomber hopes to achieve carbon neutrality so the resort will produce at least as much energy as it consumes. In addition, the company wishes to attain the Leadership in Energy and Environmental Design (LEED) Platinum certification by using solar panels and biofuel generators to provide electricity for the island. The PV panels were installed in May 2010. Although solar PV is already used in remote installations where central electricity generation is unfeasible, this will be the first major effort to commercialize solar power in the tourism industry. In addition, the biofuel generators are expected to emit 50% fewer greenhouse gases (hydrocarbons, carbon monoxide, and particulate matter) than the same energy generated with diesel. The Brando also plans to construct a new wastewater management system that will convert liquid wastewater into water that can be used for irrigation (Pacific Beachcomber 2010; Tetiaroa 2010).

7.4.4.2 Ocean Thermal Energy Conversion

Another promising renewable energy technology in French Polynesia is OTEC, which converts solar radiation to electric power. It uses the ocean's natural temperature difference between the warm surface layer and the cold deep layer to drive the power-producing cycle. If the thermal gradient is at least 20°C or 36°F, then this system can potentially generate 10^{13} W of base-load electric power. Because OTEC can supply power continuously throughout the year without being affected by climate conditions (unlike wind and solar), it is a much more stable option. An OTEC plant can be land-based, completely floating in water, or half-submerged. Furthermore, because the deep seawater is rich in nutrients, it can be used for other industries, such as cultivating fishing grounds, manufacturing mineral salt, producing pure water and mineral water, extracting rare metals such as lithium, and producing hydrogen. OTEC has the added benefit of low CO_2 emissions (0.014 kg of CO_2/kWh). Power generation cost is similar to a nuclear power plant of similar capacity and much cheaper than the current reliance on imported fuel (Solar Energy Research Institute [SERI] 1989).

Between 1978 and 1980, French companies conducted a preliminary environmental study for an onshore pilot OTEC plant of approximately 5 MW on the island of Tahiti, concentrating on thermal resource evaluation; detailed characteristics and timing of current, wave, and wind patterns; and biological and chemical issues. Unfortunately, the plant was never completed. Recently, however, research into OTEC's feasibility in French Polynesia has resumed. On March 10, 2010, Xenesys Inc., a Japanese specialist in thermal energy conversion technology, received an order from Pacific OTEC to conduct the power generation portion of the feasibility study for an offshore project near Tahiti. The plants would be connected to Tahiti's power grid and have the potential to generate 10% of the island's electricity consumption. During this 11-month-long study, with 50% financing from the French government, Xenesys worked with DCNS, a European leader in military shipbuilding. The project's goal is to complete a 10 MWh offshore OTEC platform, 80-feet-high and submerged 80 feet, to increase stability by mitigating strong currents and waves. Implementation of OTEC plants has already begun in other French overseas territories such as Reunion Island (Gauthier 2010; Xenesys, Inc. 2010).

7.5 ACHIEVING 2020 RENEWABLE GOALS—GOVERNMENT ITS OWN BARRIER TO SUCCESS?

Section 7.4 described the renewable energy landscape in French Polynesia and the government's goals for reducing energy dependence, given the country's significant untapped renewable energy potential. Yet, how realistic is the government's goal to source 50% of electricity from renewable energy by 2020? How effective have government efforts really been to promoting renewable energy? The majority of the data presented in Section 7.4 is from 2008, the most recent year for which the government has compiled and published comprehensive data. This section will use anecdotal evidence and primary source opinions gathered from the author's trip to French Polynesia in August 2010 and personal correspondence between August 2010 and April 2011 to push this chapter's analysis of the energy landscape in French Polynesia one step further. In particular, this section will first revisit hydro, wind, and solar energy, providing an inside look at these technologies in French Polynesia; an update on progress since 2008; and a view that the government is its own barrier to success. It will then consider the origins of government mismanagement and whether France might play a larger role in implementing French Polynesia's renewable energy plans.

7.5.1 Policy and Execution Failures for Existing Renewable Energy, Implications for 2020 Goals

7.5.1.1 Hydro

As seen in Section 7.4, the potential for increasing hydroelectric power in French Polynesia is clear, but any discussion of this potential must be tempered by an understanding of hydroelectric power's history in the country. Table 7.10 shows that hydroelectric capacity did not change from 2000 to 2008, and production was inconsistent. In Tahiti, hydroelectricity has generated 25% to 50% of electricity since 1997 (EDT 2008). In 1998, this percentage almost hit 50% and then decreased dramatically to less than 30% by 2008. Thus, the government's goal of "50% renewable energy by 2020" had actually been achieved over a decade ago. How did French Polynesia move backward in its use of hydroelectricity, and how can the country guarantee that this does not happen again?

Part of the decline can be attributed to EDT's lack of experience with hydroelectricity and lack of economic motivation to invest in additional capacity. Hydroelectricity in French Polynesia progressed steadily from 1985 to 1998, with an ambitious construction plan for small and medium-sized power stations directed by a tenacious entrepreneur named Dominique Auroy (now director of SEDEP) and the visionary mayor of Teva I Uta, Milou Ebb, under the auspices of the utility Marama Nui (Laurey 2009, p. 163). EDT acquired the hydroelectric plants it operates today when it took over Marama Nui in 2000. All the plants had been built by Marama Nui, and after the acquisition, according to Laurey, investments in maintaining and building new hydroelectric capacity dropped off significantly. Laurey notes that EDT did not invest in new capacity because thermal power was more profitable than hydroelectric power: "They were doing well with a tariff system that let them charge higher rates every year based on parameters including oil prices" (Laurey 2011). He adds that, "pursuing the boring tasks of managing hydro projects with a lot of opposition from people living near rivers made no sense for them" (ibid.). This investment freeze, coupled with an increase in electricity consumption over the past decade, explains the sudden decrease in the percentage of electricity sourced from hydro in French Polynesia (Blanchard 2011). However, when new regulations opened the hydro market to competition in 2008 and oil prices hit new highs, EDT was suddenly motivated to find lower-cost sources of electricity as quickly as possible.

This explanation indicates that EDT is at fault for hydroelectricity's lack of expansion in French Polynesia over the past decade. Yet, there was another motivating factor behind the utility's decision not to invest in new hydroelectric capacity when it acquired Marama Nui in 2000. The successful growth of hydroelectricity in French Polynesia ended abruptly when then-President Gaston Flosse passed the "French Polynesian Electrical Energy Charter No. 983668" on October 3, 1998. This charter stipulated that new hydroelectric plants should not be built until their production costs were compatible with the country's goal of reducing electricity tariffs—that is, until production costs were equal to or lower than those of thermal energy (Blanchard 2009). However, according to Blanchard, the charter made no sense at the time, as the per kilowatt-hour price of hydroelectricity being sold into the grid was lower than that of thermal electricity in 1998 (Blanchard 2011). Nevertheless, bound by government policy, first Marama Nui and then EDT ceased investing in new hydropower.

Thus, EDT's decision was, indeed, financial, but one made to comply with the government charter, not EDT's own analysis of thermal versus hydroelectric production costs.

Because the government reversed its prior point of view and is encouraging investment in hydroelectric capacity, EDT's 2009 to 2020 investment plan (written in 2008) now includes hydroelectricity development plans for the Marquesas Islands and Tahiti (as described in Section 7.4). However, based on interviews with government officials in the Ministry of the Environment during the summer of 2010, little progress seems to have been made (Giraud 2010). In particular, EDT was having trouble moving forward with its 10 MW dam in the Papeiha Valley because of protests from residents over the destruction of the environment and its effect on tourism. Laurey posits that the original management team at Marama Nui, which constructed French Polynesia's existing hydroelectric plants, was effective because it worked more closely with locals on project plans. For example, out of respect for residents' concerns about the environment, the team planted 10 new trees for each one it uprooted (Laurey 2009, p. 170). According to Blanchard, between problems with landowners and trouble with local politics, EDT has not been able to move forward with any of its projects, now blocked for over 10 years (Blanchard 2011).

The only positive update for hydroelectric power comes from Laurey, who was hired as a technical consultant to the Minister of Economy and Energy in 2008. He observes that his work and advice helped change government policy toward hydroelectricity, as shown by the introduction of a new regulatory framework to organize competitive bids for all hydroelectric projects. Such positive news might be short-lived, however, because Laurey, his superior (the Minister of Economy and Energy), and "half the government" (Laurey's words) were fired by the president in March 2011 for political reasons (Laurey 2011).

Although the government's plans for increasing hydroelectric power to 50 MW by 2020 seems to be achievable (given the potential discussed in Section 7.4), the government's past policies, EDT's lack of execution experience, and problems with political corruption may prevent them from becoming a reality.

7.5.1.2 *Wind*

Unlike hydropower, French Polynesia's limited investments in wind power cannot be explained by a single government regulation; but the inside story on wind power is also marked by government (or government agency) incompetence, casting doubts on the country's ability to achieve the 2020 goals for additional capacity. The "shining success" of wind power in French Polynesia, the Makemo installation, has actually been plagued by mismanagement and poor execution and is strongly disliked by residents.

To summarize, the Mixed Economy Company Te Mau Ito Api was created in 2006 to promote renewable energy on the Polynesian islands by developing electricity production from renewable energy sources adapted to each island, in a way that would be financially viable, clean, and sustainable. One of the company's first projects was to create a wind park on Makemo, in the Tuamotus Atolls, with the goal of sourcing 70% of the electricity needs of the atoll from renewable energy. The project, which began in 2007, installed six wind turbines with a capacity of 192 kW for an initial cost of three million dollars (40% of which was covered by a tax credit per Bylaw No. 1176 PR April 12, 2007; La Dépêche 2010; Laurey 2009, p. 64). However, according to an article in *La Dépêche* in 2008, nearly a year later, the results were disappointing. The project ultimately cost almost five million dollars and was plagued with electricity outages, which meant that installation of the wind turbines had not shifted any of the island's electricity consumption away from fossil fuel sources. An expert conducting an audit was quoted as suggesting that PV panels would have been a more appropriate choice of technology for the atoll, with fewer financial, maintenance, and installation problems (Prévost 2008). Results since then have been mixed. Another article in *La Dépêche* claimed that the six turbines did provide 50% of the island's electricity needs in 2009 (La Dépêche 2010). Yet, in May 2010, one of the turbines toppled over—an incredibly rare occurrence anywhere (Monnier 2010).

Interviews with residents on Makemo revealed that, although the cost of electricity did not change as a result of the wind turbines, the inclusion of private enterprise and the government in the venture significantly changed how electricity was paid for. Previously, residents of Makemo were allowed to run up large debts with the government provider but now are required to prepay for their electricity each month. Not surprisingly, they do not like this change, which conflicts with the flexibility typical of their culture and traditional island practices. Resident Dahlia A. was also critical of the turbines, explaining, "Makemo is just as reliant on the monthly cargo ship as before wind power came. The turbines break all the time, so instead of waiting for fuel, we wait for turbine parts." Although operator SEDEP claims wind power has been reliable, the school and several stores have installed their own generators to assure uninterrupted access to power (Dahlia 2010).

Based on the Makemo experience, it is difficult to determine whether wind is an appropriate renewable energy solution for French Polynesia or whether the fault lies solely in the execution. Do the turbines break down because they were installed incorrectly and are poorly maintained? Or do they break down because, despite the benefits of constant wind, overall weather conditions in French Polynesia are too problematic for existing wind technologies? The answer is a little bit of everything. Problems with the Makemo wind farm are linked to bad project management, equipment poorly adapted to Makemo's aggressive marine climate, and finally, politics in the form of personality conflicts (Blanchard 2011). Even the mayor of Makemo, according to Blanchard, "did not really seem to want the system to work properly. He 'won a toy,' but beyond that, he did not care about the economic advantages for inhabitants" (ibid.). In addition, Blanchard views the complaints as evidence that the atolls are generally so accustomed to benefitting from government help and subsidies that, when such benefits were taken away, the residents saw it as an injustice. That may not be an ideal base to work from, but it does need to be taken into account when promoting renewable energy on French Polynesia's more remote islands, in order to win the support and cooperation of the local populations.

As with hydroelectric power, little progress has been made toward the government's 2020 goals for wind power, despite the introduction of the 2009 feed-in tariffs. French Polynesia is waiting for the results of EDT's feasibility studies or for a more comprehensive government-funded study of capacity in Tahiti.

7.5.1.3 Solar

Although EDT may have limited plans for developing solar energy in French Polynesia beyond isolated installations (as discussed in Section 7.4), the potential to meet the government's goal of adding 22 MW of capacity by 2020 certainly exists. To understand what the country can accomplish in terms of solar generation, consider its solar capacity in the context of other island nations: In 2007, French Polynesia installed 0.3 MW of solar power, whereas all of France's 11 overseas territories in total installed 15 MW (Laurey 2009, p. 124). French Polynesia is the third largest of these territories (after French Guyana and New Caledonia) and should be responsible for a much larger share of this total. The country's solar installations have been so much lower due, primarily to a lack of financial incentives for solar investment, compared with these other territories (and the rest of the world). In other words, the government is again responsible for obstructing its own renewable energy goals. The majority of French Polynesia's solar capacity in 2008 was financed by ADEME. For the country to access its full solar energy potential, ADEME's support will not be sufficient. In 2008, the government finally realized that achieving its 2020 goals for solar would require investment incentives. Instead of the hoped-for successful implementation of policy to support its renewable energy goals, the government found the 2008–2010 period characterized by naïve missteps in legislation for solar investment incentives (Blanchard 2011).

To understand what the French Polynesian government should be doing to promote investment in renewable energy, consider a brief overview of incentive systems in place in other countries (Table 7.13). The most important policy tool for promoting solar energy investment is the feed-in tariff, a fixed rate at which the government (or utility) guarantees it will purchase energy generated from solar installations (whether owned by individual citizens or by businesses). The rate must also be fixed for a long period (usually 20 years or more) to compensate for the long-term nature of the capital investment required to install solar panels. Many countries also offer additional subsidies, such as tax credits or rebates, to reduce the initial capital investment.

TABLE 7.13

Examples of Solar Energy Incentives Offered by Governments around the World in 2009

Country	Government Intervention Method	Energy Purchase Price (US\$)	Duration of Purchase Agreement
Austria	Feed-in tariff, investment subsidy, tax incentives	\$0.68–\$0.87	13 years
France	Feed-in tariff, investment subsidy, tax incentives	\$0.43–\$0.79	20 years
Germany	Feed-in tariff	\$0.62–\$0.78	20 years
Greece	Feed-in tariff, investment subsidy	\$0.12	10 years
Italy	Feed-in tariff, investment subsidy, tax incentives	\$0.72–\$0.88	20 years
The Netherlands	Feed-in tariff, investment subsidy, tax incentives	\$0.43–\$0.58	10 years
Spain	Feed-in tariff, investment subsidy	\$0.32–\$0.59	Forever

Source: Adapted from N. Laurey. *Energies Renouvelables: Plaidoyer pour une Véritable Politique de l'Energie en Polynésie Française.* Tahiti: Au Vent des Iles. 2009, 112.

Note: Austerity measures in Europe, decreasing solar equipment prices, etc., have caused these numbers to fluctuate since 2009.

French Polynesia needs a similar set of incentives to attract private capital, but it did not consider such incentives until 2008. Even though feed-in tariff structures are well-established and French Polynesia had many role models for writing its own legislation (including from its mother country, France), every solar bylaw the government passed through 2011 was seriously flawed. A bylaw introduced in December 2008 initially set a feed-in tariff for independent generators of solar power at US$0.42/kWh, guaranteed for only 6 months, from January 1 to June 30, 2009. This policy was intended to encourage private solar PV installations, but the length of time was far too short to allow any real investments to take place. A new bylaw was then passed on June 25, 2009, to guarantee the feed-in tariff for 25 years and to extend the range of the repurchase price to US$0.42/kWh to US$0.55/kWh, depending on the amount of power produced (ME and UCB GSPRS 2009). The bylaw of June 29, 2009, however, attempted to treat all types of installations at once (from domestic to industrial), which resulted in 42 pages written in factory-style language that was incomprehensible to the average citizen looking to install a solar panel for a house. In addition, this bylaw forced installers to choose either to sell all the energy produced back to EDT's grid without consuming a single kilowatt-hour or to consume some and lose the rest. According to Blanchard, such methods and structure are "idiotic and unheard of in any other country." This bylaw was, thus, again unsuitable but remained in place for 10 months until it was replaced on May 18, 2010, by a new bylaw that was just as complex but eliminated the resale versus consumption constraint. This bylaw, however, introduced another complication: EDT no longer paid private installers for the kilowatt-hours of solar energy they generated and sold back to the grid. Instead, the company that installed the solar panels would send a bill to EDT and wait to be paid before reimbursing the private installer. This new structure added an unexpected administrative burden to solar developers and installers that they were not prepared to execute. In essence, the feed-in tariff (created to guarantee that a private solar installation could sell power back to the grid at a fixed price) became a disincentive because its structure made it almost impossible for installers to be paid that guaranteed price (Blanchard 2011).

Even if solar installations were successful in receiving the feed-in tariff for selling electricity back to the grid, Laurey believes that the tariffs are not high enough to attract the investment required to meet the government's 2020 plans. He explains that "the feed-in tariffs are not high enough, and the subsidy system on the investment (up to 75% of the equipment publicly financed) was too high" (Laurey 2011). In a 2009 interview with *Les Nouvelles*, he compares the feed-in tariffs in French Polynesia with those in France, where he believes tariffs are greater than the average price of electricity by enough of a margin to promote investment in solar energy. In France, the price of electricity is US$0.16, and the feed-in tariff for solar PV is US$0.43. Solar energy is, thus, guaranteed a feed-in tariff of 2.7 times the average tariff for electricity. Applying this same ratio to French Polynesia, Laurey argues, with an average electricity price of US$0.44, the feed-in tariff for solar should be US$1.12 (Laurey 2009). Terii Vallaux, technical counselor to the Public Works Ministry (Ministère des Grands Travaux), responded in an article 5 days later: "Nuihau Laurey uses a ratio… I don't think that's something you can extrapolate. The goal is that the incentive functions, not that it creates excess supply." He went on to make a convincing argument that feed-in tariffs in French Polynesia do not need to be as high as they are in France because solar energy production is more efficient. French Polynesia's level of sunshine is much higher than in France, so the same solar PV panel produces 50% more kilowatt-hours per year in French Polynesia than in France (Florence 2009). Blanchard agrees with Vallaux, citing his own private solar installation as an example: "Even with (the low end of the feed-in tariff range) US$0.42/kWh, installing photovoltaic is a viable investment" (Blanchard 2011). Although he believes the feed-in tariffs are appropriate, he emphasizes that the government policies are ineffective and dissuade solar investment for other reasons, including their use of abstruse legal descriptions and restrictive clauses.

Despite itself and its missteps, the French Polynesian government has succeeded in making progress toward its 2020 solar capacity goals, even if it has not yet succeeded with hydro and wind. Teva Rohfritsch, the Minister of Economic Redevelopment for Green Technology (le Ministre de la Reconversion Economique en Charge des Technologies Vertes), announced in April 2010 that 27 solar PV projects had been accepted by the Council of Ministers, representing 10.5 MW of new solar capacity (Tahiti Presse 2010). If implemented, these projects would fulfill half of the government's new solar capacity goal for Tahiti (Table 7.8). Four of the approved projects are solar PV farms conceived by SEDEP, part of seven sites the society has planned, through which it hopes to produce 11.7 GWh/year (a third of the government's 2020 solar goal), representing 2.5% of Tahiti's electricity consumption (Rabréaud 2010). This progress seems to demonstrate an important development for the French Polynesian government and to give credibility to its 2020 renewable energy goals. Further research, however, has uncovered a major caveat. The four approved SEDEP projects are planned for Papenoo, Arue, Mahina, and Punaauia, all suburbs of Papeete in Tahiti. According to a report from Rohfritsch's office, the existing electrical grid in these areas would not be able to absorb the new solar electricity production because it was designed to support the electricity needs of only the local population (only 6% of Tahiti's consumption). Rohfritsch is seeking the expertise of EDT and STEP to solve this problem (Giraud 2010; Ministère de la Reconversion Économique 2010). At the very least, this hindrance will cause delays in construction and operation of the new solar farms; at worst, the plants will be installed, but the power will go unused (without connection to the grid, the new capacity cannot help reduce French Polynesia's dependence on thermally generated electricity).

7.5.2 The Origins of Government Mismanagement

The government web site LEXPOL lists all policies and regulations passed in French Polynesia. However, prior to 2009, it displayed none of the policies related to renewable energy, and only four of the policies listed included real actions (Laurey 2009, pp. 61–62). The implication is that, like many governments, French Polynesia's is characterized by "too many committees, not enough action" (ibid., p. 61). This tendency is also evident in the current debate as to whether the country needs a dedicated Renewable Energy Minister. In January 2010, President Gaston Tang Song proposed a plan to create a new Renewable Energy Agency. One response was, "we must, at all costs, avoid creating a new sinecure (position requiring little or no work) and a new bureaucracy that will accomplish nothing" (Tahiti Punu News 2010). Auroy, director of SEDEP explains, "we must make sure that the government does not just sit for years reading the reports." In other words, policies need to be enacted, not read (Laurey 2009, p. 62).

According to Blanchard, the government incompetence presumed in such anecdotes has become endemic. "There is not a single government domain being managed seriously, with realistic strategic choices and with the goal of managing and developing lasting resources. In energy, as in other areas, there is not a single credible strategy or serious project, a reality that is clearly seen in the absence of progress toward energy independence" (Blanchard 2011). This unfortunate management style cannot be blamed on Polynesian culture, Blanchard explains. The history of the Polynesians includes preparing boats and resources for an extraordinary

Pacific Ocean crossing. Blanchard believes that government incompetence is a recent trend, linked to the easy flow of resources into the country, available without effort, and to the political and economic influences of former president Flosse, who had been president off and on for about 18 of the past 27 years, most recently in April 2008. He has been convicted of political corruption, served a jail sentence, and been accused of other crimes, including murder (News Wires 2009).

7.5.3 A Larger Role for France in French Polynesia's Renewable Energy Plans?

At a September 2009 conference (Délégation aux Etats Généraux) organized by the French Polynesian Environment Ministry and France's High Commissioner to French Polynesia, the stakeholders agreed that developing renewable energy is a key factor in sustainable development and increasing energy independence. The commission proposed the following agenda for the French Polynesian government to follow with respect to renewable energy:

- Provide incentives for solar energy production
- Encourage technological innovation in the energy sector
- Make French Polynesia a laboratory for the development of renewable energy by using public and private partnerships (ME 2009)

In this case, France's role is to help make renewable energy policy recommendations. French Polynesia has the policies in place, but the question is whether France might play a larger role in helping the latter country execute those policies and overcome its mismanagement issues.

In 2007, France created a comprehensive policy, "Toward a Model Overseas Territory," that lays out an environmental action plan for its overseas territories. This policy, written by France's Secretary of State for Overseas Territories, proposes an energy action plan, with the goal of energy independence in all overseas territories by reducing consumption and increasing the use of renewable energy to 50% by 2020. This plan lays out France's role in five areas:

1. Put governance resources in place
2. Support R&D
3. Develop regulations

4. Adapt monetary and economic mechanisms to incentivize virtuous behavior

5. Encourage reduced energy consumption and increased use of renewable energy (Ministère de l'Intérieure et Secrétariat de l'Etat a l'Outre-Mer [MISEO] 2007)

A more detailed description of what these five axes entail does not really clarify the role France intends to play. The details include expanded statements starting with verbs such as "help," "study," and "adopt," once again suggesting that France may recommend certain best practices, but it is leaving the actual funding, subsidies, and execution for renewable energy to the French Polynesian government.

Blanchard notes that the French Polynesian economy (including energy, taxes, industry, land management, trade, and agriculture) is managed exclusively by Polynesians. The French government cannot intervene in any way because of French Polynesia's legal level of autonomy. This does not apply to other French overseas territories, which may explain their greater success in adding solar capacity, for example, as they benefit more from France's experience and financial resources (Blanchard 2011). Unfortunately, then, French Polynesia cannot look to France to solve its policy and execution problems.

7.6 CONCLUSION

A review of existing renewable energy technologies in French Polynesia, informed by insiders' opinions and firsthand experience, calls into question the government's ability to meet its 2020 goal for 50% of the country's electricity production to come from renewable energy sources. Comparing the country's specific goals to increase the installed capacity of hydroelectricity, solar, and wind power with progress made up to 2011, the evidence indicates that the government's goals are overly ambitious. Yet, as the previous section has shown, the answer is more nuanced than that. Although most French Polynesians doubt the country can overcome its bureaucratic ways and political upheavals to achieve any of its goals, positive progress toward the solar capacity goals—despite the government's stumbling through incentive legislation—must be considered. The real answer is that the government's success—and, indeed, the future of renewable energy in French Polynesia—are *unpredictable*.

In 1998 French Polynesia achieved 50% energy independence through the successful exploitation of natural hydroelectric power resources. This renewable energy leadership dramatically reversed itself after the government passed the charter that prevented hydroelectricity investment. If French Polynesia succeeds in reclaiming success in this area, achieving the government's goal of 50% hydroelectricity by 2020, what is to prevent another short-sighted government policy from damaging that success once again? On the other hand, consider that the past 3 years of solar incentive policies read like a comedy of errors (e.g., misconstruing the need for a long-term guarantee, forcing consumers to decide between selling all electricity generated or consuming all of it, or writing policies in complex technical language). Yet the government is apparently on track to install half of its 2020 solar capacity goal only a few years into its 10-year renewable energy plan. What is to prevent continued renewable energy success despite the government getting in its way? Data gathered in the Tuamotu Atolls show that wind patterns are ideal for wind power (winds 365 days a year, 24 hours a day). Yet, if the country continues to build wind capacity in isolated areas, what is to prevent another turbine, such as the one on the Makemo wind farm, from toppling over "unprovoked" due to poor maintenance? Since 2008, Laurey and the Minister of Economy and Energy have created new legislation to increase the competitive market for hydroelectricity. What is to prevent another president from creating a new energy ministry and then changing his mind and firing that group too?

The answer to all of these questions can be summed up in a word: *unpredictable*. In all likelihood, this balance of steps forward and backward, good intentions balanced by poor execution, flowery objectives superseded by economic realities—which has characterized the development of renewable energy in French Polynesia thus far—will continue through the government's 2020 goals, as well as its 2030 goals.

With an appreciation for French Polynesia's political context of constant turnover and the lasting effects of Flosse's corruption, as well as the knowledge that France cannot step in and fix the government's mistakes, what prospects remain for the long-term future of renewable energy in French Polynesia?

Although high-level government officials continue to voice their support for renewable energy, some express a realistic understanding that the shift will not happen overnight and that no single form of renewable energy can offer a magic solution. In a 2008 interview, the CEO of EDT explained, "it would take 600 hectares of PV panels—that is, three-quarters of the entire surface of Papeete—to replace the two thermal power plants that are essential to the electricity supply of Tahiti" (La Dépêche 2008). French Polynesia has the right words and the right goals in place, but it needs a few realistic energy leaders who are able to sidestep the political turnover and focus on effective execution. Such leaders, followed by a more stable government at some point, might have the power to turn French Polynesia's *unpredictable* into a predictable and growing energy independence. Until then, the country with the most expensive energy in the world may find that, when the rising price of oil finally tips the economic balance fully in favor of renewable energy, no amount of government incompetence will keep the country from becoming 100% renewable.

REFERENCES

Association of the Electricity Supply Industry of East Asia and Western Pacific (AESIEAP). *AESIEAP Goldbook 2011*. Jakarta, Indonesia: AESIEAP, 2011.

BBC News. French Polynesia profile. Last modified September 15, 2010 (accessed September 30, 2010). http://www.bbc.co.uk/news/world-asia-16492623.

Blanchard, P. Production et Distribution de l'Energie Electrique en Polynésie Française: Présentation à M. le Ministre du Développement Durable. Limited circulation Powerpoint presentation. November 13, 2006.

Blanchard, P. Energies renouvelables, Partie C: Caractéristiques Principales des Energies Renouvelables à Etudier et Perspectives de Développement. Limited circulation Powerpoint presentation. January 31, 2009.

Blanchard, P. Interview with P. Blanchard, by D. Townsend-Butterworth Mears. 2011.

Buquet, B. Nuihau Laurey. Last modified June 25, 2009 (accessed September 30, 2010). http://www.lesnouvelles.pf/article/a-laffiche/nuihau-laurey.

Dahlia, A. Interview with Dahlia A., Resident of the Tuamotu Islands, by D. Townsend-Butterworth Mears. August, 2010.

Electricite de Tahiti (EDT). L'Energie Electrique dans les Archipels de Polynésie Française. Enjeux et Options de Développement (2009–2020). Powerpoint Presentation. 2008.

Encyclopaedia Britannica. French Polynesia. Last modified 2010 (accessed September 30, 2010). http://www.britannica.com/EBchecked/topic/219285/French-Polynesia/54080/History.

Energy Information Administration (EIA). French Polynesia. Last modified September, 2010 (accessed September 30, 2010). http://www.eia.gov/countries/country-data.cfm?fips=FP.

Florence, R. Du Soleil à Revendre: Entretien Avec Terii Vallaux. *Les Nouvelles*, June 30, 2009. http://www.lesnouvelles.pf/article/leconomie-du-fenua/du-soleil-a-revendre.

Garnier, B. Energies renouvelables: Quelles Perspectives pour Tahiti? French Polynesian Sustainable Development Association. Last modified 2008a (accessed September 30, 2010). http://www.2dattitude.org/dossiers/energie/72-energies-renouvelables-quelles-perspectives-pour-tahiti.

Garnier, B. L'Outre-mer uni, Force de Propositions et d'Actions. Paper presented at the *Congress Xviie Congrès de l'ACCD'OM (Association des Communes et Collectivités d'Outre-mer)*, French Guiana, November 2008b. http://www.france-accdom.org/OUTREMAG/outremagnovembre2008.pdf.

Gauthier, M. The French OTEC project in Tahiti: Preliminary results of the site environment study. Paper presented at the *Oceans '84 Conference, Washington, D.C., September 10–12, 1984, 359–363*. Piscataway, NJ: IEEE. 2010.

Giraud, C. Interview with C. Giraud, Technical Counselor to the Ministry of the Environment, by D. Townsend-Butterworth Mears. August, 2010.

Hawaiian Electric Company. Hawaii's Energy Future (accessed September 30, 2010). http://www.hawaiisenergyfuture.com/articles/Sea_Water_Air_Conditionin.html.

Intercontinental Bora Bora Resort and Thalasso Spa. Intercontinental Bora Bora—Deep Sea Water Air Conditioning. Last modified 2007 (accessed November 5, 2010).

L'Institut d'Emission d'Outre-mer (IEOM). La Polynésie Française en 2009: Rapport Annuel. Last modified 2010 (accessed September 30, 2010). http://www.ieom.fr/IMG/pdf/ra2009_polynesie.pdf.

La Dépêche. L'énergie renouvelable. *La Dépêche*, July 8, 2008.

La Dépêche. Makemo: De l'Electricité pas cher… Grâce au Vent. *La Dépêche*, May 18, 2010. http://www.ladepeche.pf/article/la-vie-des-communes/makemo-de-l'electricite-pas-chere-grace-au-vent (accessed November 20, 2010).

Laurey, N. *Energies Renouvelables: Plaidoyer pour une Véritable Politique de l'Energie en Polynésie Française*. Tahiti: Au Vent des Iles, 2009.

Laurey, N. Interview with Nuihau Laurey, Financial Consultant, by D. Townsend-Butterworth Mears. 2011.

LEXPOL, La Service Public de la Diffusion du Droit en Polynésie Française. *LEXPOL* (accessed April 2, 2011). www.lexpol.pf.

Makai Ocean Engineering. Seawater Air Conditioning (SWAC) System Basics. Last modified 2011 (accessed November 11, 2012). http://www.makai.com/p-swac.htm.

Marconnet, M. Integrating Renewable Energy in Pacific Island Countries. Last modified 2007 (accessed September 30, 2010). http://researcharchive.vuw.ac.nz/bitstream/handle/10063/491/thesis.pdf?sequence=1.

Meyer, J. Conservation des Forêts Naturelles et Gestion des Aires Protégées en Polynésie Française. *Bois et Forêts des Tropiques, 291*(1), 2007.

Ministère de l'Environnement (ME). Délégation aux Etats Généraux (DEG). *Etats Généraux de l'Outre mer en Polynesie Française: Document de Synthèse*. Papeete, Tahiti: ME, 2009.

Ministère de l'Environnement (ME), and UC Berkeley Gump South Pacific Research Station (UCB GSPRS). *Etat des Lieux sur les Enjeux du Changement Climatique en Polynésie Française* by A. Avagliano, and J. Petit. Moorea, French Polynesia, 2009.

Ministère de l'Intérieure et Secrétariat de l'Etat a l'Outre-Mer (MISEO). *Vers un Outre-mer Exemplaire: Grenelle de l'Environnement, Plan d'Action Outre-mer*. Paris: Secrétariat d'Etat à l'Outre-Mer, 2007. http://www.serge-letchimy.fr/wp-content/uploads/2011/05/vers-un-outre-mer-exemplaire.pdf (accessed November 5, 2010).

Ministère de la Reconversion Économique. du Commerce Extérieur, de l'Industrie et de l'Entreprise, en Charge de l'Economie Numérique et du Développement des

Technologies Vertes. Projets de Central d'Energie Photovoltaïque. Papeete, Tahiti: Ministère de la Reconversion Économique, 2010.

Ministère des Grands-Travaux (MGT). Renewable Energy Presentation. Limited circulation Powerpoint presentation. Papeete, French Polynesia: MGT, 2008.

Ministère du Développement et de l'Environnement de Polynésie Française (MDE). *Etat de l'Environnement en Polynésie Française 2006* by C. Gabrié, H. You, and P. Farget. 2006. http://www.sprep.org/att/irc/ecopies/countries/french_polynesia/30.pdf (accessed November 10, 2010).

Monnier, J. Une Eolienne Tombe à Makemo. *La Dépêche*, May 29, 2010. http://www.ladepeche.pf/article/la-vie-des-communes/une-eolienne-tombe-a-makemo (accessed November 5, 2010).

News Wires. Former President Flosse in detention facing new corruption charges. Last modified March 12, 2009 (accessed October 28, 2012). http://www.france24.com/en/20091203-former-president-flosse-detention-facing-new-corruption-charges-french-polynesia.

OECD. OECD Factbook 2011: Economic, Environmental and Social Statistics. Last modified 2011 (accessed March 11, 2012). http://www.oecd-ilibrary.org/sites/factbook-2011-en/06/01/04/06-01-04-g1.html.

Pacific Beachcomber. Hotel Maitai Polynesia Bora Bora upgrades to earthcheck. Last modified April 27, 2010 (accessed November 10, 2010). http://www.pacificbeachcomber.com/about-us/news-articles/823/.

Prévost, B. Vent de Colère Contre les Eoliennes. *La Dépêche*, November 15, 2008. http://www.ladepeche.pf/article/la-vie-des-communes/vent-de-colere-contre-les-eoliennes (accessed November 20, 2010).

Rabréaud, L. Tous au Solaire. *Les Nouvelles*, May 3, 2010. http://www.lesnouvelles.pf/article/la-vie-au-fenua/tous-au-solaire (accessed November 5, 2010).

Service de l'Energie et des Mines (SEM). 2008 Rapport Statistique Synthèse: Secteur de l'Energie, Electricité et Hydrocarbure. Luxembourg: Banque Européenne d'Investissement, 2009a.

Service de l'Energie et des Mines (SEM). L'Electricité a Tahiti. Bilan Prévisionnel et Programmation Pluriannuelle des Investissements. Papeete: Service de l'Energie et des Mines, 2009b. http://www.service-energie.pf/telechargement/BILAN%20PREVISIONNEL%20ELECTRICITE%20OFFRE%20DEMANDE%20TAHITI.pdf (accessed November 20, 2010).

Service de l'Energie et des Mines (SEM). Projets d'Investissement de la Polynésie Française. 2009c.

Société d'Etudes et de Développement Polynésienne (SEDEP), Service de l'Energie et des Mines (SEM), and Te Mau Ito Api. L'Energie Eolienne dans l'Archipel des Tuamotu. Presentation delivered at *Séminaire de l'Energie*, August, 2008.

Solar Energy Research Institute (SERI). *Ocean Thermal Energy Conversion: An Overview*. Golden, CO: Solar Energy Research Institute, 1989.

Tahiti Presse. Polynésie: Création de la Sem "Te Mau ito Api" Pour une Electricité Produite à Partir d'Energies Renouvelables aux Tuamotu. *Tahiti Presse*, January 23, 2007. tahitipresse.pf (accessed November 5, 2010).

Tahiti Presse. Teva Rohfritsch: Conforter le Développement Progressif et Harmonieux de la Filière Photovoltaïque en Polynésie. *Tahiti Presse*, April 30, 2010. tahitipresse.pf (accessed November 5, 2010).

Tahiti Punu News. La Polynésie a-t-elle Besoin d'une Agence pour les Energies Renouvelables? Last modified January 19, 2010 (accessed November 5, 2010). http://

hirofarepote.wordpress.com/2010/01/19/la-polynesie-a-t-elle-besoin-dune-agence-pour-les-energies-renouvelables/.

Tahiti Tourism. The Islands of Tahiti. Last modified 2011 (accessed September 20, 2010). http://www.tahiti-tourisme.com/.

Tetiaroa. The President of French Polynesia, Gaston Tong Sang, visits Tetiaroa. Last modified August 19, 2010 (accessed November 10, 2010).

United Nations ESCAP (UNESCAP). Statistical yearbook for Asia and the pacific 2011: Tourism. Last modified 2011 (accessed December 30, 2012). http://www.unescap.org/stat/data/syb2011/IV-Connectivity/Tourism.asp.

United Nations Statistics Division. French Polynesia. Last modified 2010 (accessed September 20, 2010). http://data.un.org/CountryProfile.aspx?crName=French Polynesia.

U.S. EIA. International energy statistics: French Polynesia electricity net generation by type (billion kilowatthours). Last modified 2012 (accessed May 19, 2013). http://www.eia.gov/cfapps/ipdbproject/iedindex3.cfm?tid=2&pid=alltypes&aid=12&cid=FP,&syid=2010&eyid=2008&unit=BKWH.

World Bank Group (WBG). Private sector at a glance: French Polynesia. Last modified April 26, 2010 (accessed September 30, 2010). http://data.worldbank.org/topic/private-sector.

World Bank Group (WBG). International tourism, number of arrivals. Last modified 2012 (accessed December 30, 2012). http://data.worldbank.org/indicator/ST.INT.ARVL?page=3.

World Wind Energy Association (WWEA). World Wind Energy Report 2010. Bonn, Germany: World Wind Energy Association WWEA, 2011. http://www.wwindea.org/home/images/stories/pdfs/worldwindenergyreport2010_s.pdf (accessed on: December 30, 2012).

Xenesys, Inc. Xenesys received the order to conduct the power generation portion of the feasibility study on the offshore OTEC project in the vicinity of Tahiti Island. Last modified March 10, 2010 (accessed November 8, 2010). http://www.xenesys.com/english/press_news/2010/20100310.html.

8

The Global Biopower Industry: Time for a Growth Phase?

Hannah Tucker González and José Luis González Pastor

CONTENTS

8.1 INTRODUCTION

Efforts to meet global demand for electricity (which increased 3.2% per year from 1993 to 2008 and is expected to grow 2.5% per year until 2030) while simultaneously tackling climate change are resulting in heavy investments in renewable energy (OECD 2010, p. 17). Although wind and solar power have received the most attention in recent years, biomass-based electricity—or "biopower"—also forms a significant part of the renewable energy mix. Once considered the world's most primitive source of heat, biopower today has the potential to diversify the field of renewable electricity production. In light of the growing global demand for accessible, clean sources of electricity, we pose the question: Will biopower soon experience a growth phase akin to those of wind and solar?

Although biopower is an efficient and sustainable source of energy, the pace of its development relative to other sources of renewable energy has been slow. As a frame of reference, in 2010, solar power was the renewable energy source that saw the most growth, with the installed base growing by 70% to 40 GW worldwide. Wind power also grew considerably, adding 24% of generating capacity to a global installed base of 197 GW. Biopower's growth rate was considerably smaller (The Economist Online 2011). Under current technological and regulatory circumstances, it is rarely profitable in competitive electricity markets worldwide and is, thus, a preferred source of electricity only under certain circumstances, as we will demonstrate in this chapter. This limits its potential for scalability. Although we argue, therefore, that biopower will not enter a growth phase over the next 10 years, we do not discount the possibility of significant development over a 10- to 30-year horizon, should the demand for electricity and the urgency to reduce carbon emissions increase as anticipated.

In support of this argument, we examine biopower's development, its current use worldwide, and its potential in the long term, particularly in relation to climate change. We also examine the case of Brazil, a country rich in feedstock sources, and pose the question: Why has biopower not played a greater role in providing electricity? We conclude by evaluating how far biopower is from competing with mainstream electricity sources and identify catalysts for broader commercialization.

8.2 DEFINITIONS AND SCOPE

Biomass refers to any organic material of plant or animal origin that can be used as feedstock for producing various forms of energy. The scope of this chapter is limited to dry-plant-material biomass (i.e., excluding animal origin and "wet biomass"). Plant biomass can produce energy in the forms of biopower, bioheat, biogas, and biofuels. To avoid any confusion, we define these terms in Table 8.1 and specify their applicability to this study.

Although many underdeveloped nations rely heavily on biomass for their residential energy needs at the individual consumer level, this chapter will focus on its commercial and industrial uses as a replacement for fossil fuel electricity. Before discussing electricity production, it is useful to review different orders of magnitude of energy (e.g., kilowatt-hours,

TABLE 8.1

Description and Applicability of Biomass Terms

Term	Description	Applicability
Biopower	Electricity produced from biomass, generally through the combustion of feedstock or of a derived product (such as biogas) to drive power-generating turbines	Focus of analysis herein
Biogas	Gas produced by the biological breakdown of organic matter in pressurized combustion (gasification) or in the absence of oxygen (anaerobic digestion)	The form relevant to this chapter is (1) derived from plant material and (2) used as an input for power generation
Bioheat	Thermal energy produced from biomass	Waste heat is relevant to this study to the extent that it is cogenerated with biopower
Biofuel	Fluid used to power engines, such as bioethanol, biodiesel, and vegetable oil	Biofuel will only be discussed in the context of the use of biofuel by-products for power generation (e.g., sugarcane bagasse from ethanol production)

TABLE 8.2

Description of Units of Power Terms

Unit of Power	Magnitude	Application (Order of Magnitude)	Energy Comparison
Watt (W)	1–1000 W	Lightbulb (50–150 W)	1 Wh can power a 100-W lightbulb for 1% of an hour
Kilowatt (kW)	10^3 W	Microwave (1.1 kW)	1 kWh can power a 100-W lightbulb for 10 hours
Megawatt (MW)	10^6 W	100 automobile engines (1 MW); standard wind turbine (1.5 MW)	1 MWh can power ~330 homes for 1 hour (CleanEnergyAuthority 2010)
Gigawatt (GW)	10^9 W	Peak capacity of Hoover Dam (2 GW); capacity of world's largest coal-fired power plant (4.1 GW)	1 GWh can power ~1 million homes for 1 hour (California Energy Commission 2006)
Terawatt (TW)	10^{12} W	Total electric capacity in the United States, 1 TW (U.S. EIA 2011)	1 TWh of electric energy can power the entire United States for 2 hours (OER 2011b)

joules, calories) and power (i.e., the rate at which energy is consumed or generated, typically given in watts, kilowatts, megawatts, etc.). Table 8.2 provides an overview of these magnitudes.

8.3 HOW BIOPOWER WORKS: FEEDSTOCK AND TECHNOLOGY

The first step in producing electricity from biomass is securing a reliable supply of feedstock. One of the advantages of biopower is that there are numerous feedstock choices, which means that producers have flexibility. The choice depends on the plant species that are native or abundant in the region, the supply's proximity to the power plant, and the feedstock's energy density (Bracmort 2010, p. 2). When evaluating feedstock, it is helpful to consider three classifications (1) "primary" refers to "energy crops" grown exclusively for use as biomass feedstock; (2) "secondary" includes byproducts from processing the primary feedstock for other purposes, also known as agricultural or forestry "waste or residue"; and (3) "tertiary" refers to postconsumer residues and wastes that are not derived directly from plant material, such as cardboard (ibid., p. 4). Our case study on Brazil discusses both primary (grass, wood chips from eucalyptus) and secondary (sugarcane bagasse and eucalyptus waste) sources.

Biopower is "available on demand," or "dispatchable," thanks to its origin in feedstock. In other words, the energy within biomass can be stored until it is needed. This characteristic allows biomass to be used in power plants that run continuously ("baseload power plants"), provided the feedstock supply is sufficient to meet the power plant's minimum level of demand ("the baseload"). This is in contrast with other renewable sources, such as solar and wind, that provide intermittent power due to variable environmental conditions, such as sunlight intensity or wind speed (Bracmort 2010, p. 1; U.S. Department of Energy [DOE] Biomass Program 2010). That being said, even though biomass energy is available on demand, the total energy that can be generated is dependent on the feedstock supply, which in turn is restricted by prohibitive transportation costs over significant distances. Sun and wind may not be constant, but they travel for free. The added cost of transporting biomass to the generator site, therefore, puts biopower at a disadvantage when compared with some of its renewable energy peers. Turning now to biopower technology, two main types of power plants are capable of converting biomass into electricity: (1) combustion and (2) gasification.

8.3.1 Combustion Biopower Plants

Direct combustion—the most common form of electricity generation—refers to burning biomass in dedicated power or cogeneration plants to power steam turbines and generate electricity (cogeneration, also known as combined heat and power [CHP], involves the use of a heat engine or a power station to generate both electricity and useful heat simultaneously; OECD 2010, p. 17). Biomass plants range in size from 1 to 100 MW, an order of magnitude smaller than typical coal-fired plants. Although this small scale is necessary, given the geographic constraint of feedstock availability, it prevents biopower plants from achieving an optimal level of thermal efficiency of conversion. The investment cost per kilowatt of capacity for the smaller biomass plants is roughly twice that of the larger plants (U.S. Energy Information Administration [U.S. EIA] 2011). To compensate for the absence of economies of scale and, consequently, higher investment costs, many contractors in the United States opted to use cheaper technology (such as lower-quality boiler material and lower-efficiency steam turbines) when building the first generation of biomass plants. This limited their thermal efficiency to 14% to 18%, meaning that only 14% to 18% of the feedstock's primary energy content is converted into final useful energy (electricity). When higher-efficiency boilers are used, thermal efficiency can surpass 30% (Bajay 2000, p. 218).

Another form of direct combustion is the combined use of biomass and fossil fuels in cofired plants (Bajay 2000, pp. 218–219). Conventional fossil fuel plants can source up to 10% of their energy output from biomass without having to make significant investments in new equipment and technology. Cofired plants generating more than 10% energy from biomass typically invest in separate equipment to burn the biomass in a separate chamber and then transport the hot gases to the coal boiler (ibid.). The advantages of cofired over dedicated biomass plants are (1) an approximately 10% increase in the thermal efficiency of utilizing feedstock, reaching 35% to 45%, and (2) cheaper costs for adapting a fossil fuel plant, as opposed to building a dedicated biomass plant (U.S. EIA 2011). However, the clear disadvantage is that the reduction in carbon emissions is minimal (approximately 90% less than in dedicated biomass plants; U.S. EIA 2011).

8.3.2 Gasification

Technological progress in the area of biomass gasification could offer new avenues for developing biomass electricity. Biomass gasification refers to the combustion of biomass-based gas to drive a turbine to produce electricity. The gasifier does not incinerate the biomass, but rather heats the feedstock in a high-temperature process (thermochemical) without sufficient oxygen for full combustion, which causes a gas to form (Kartha and Larson 2000, pp. 83–84). The gas can then be used to heat water, producing steam to drive the turbine and generate electricity. The advantages of gasification over direct incineration lie in enhanced combustion cleanliness (no ash or tar waste products) and the possibility of achieving higher yields through higher combustion temperatures (Observatoire des Energies Renouvelables [OER] 2010b).

The leading gasification technologies are (1) biomass integrated gasifier/gas turbine (BIG/GT) plants, also known as biomass integrated gasification combined cycle (BIGCC), and (2) cogeneration or CHP generation plants. The first refers to combined cycle plants in which the feedstock is processed in a hot, oxygen-starved environment to produce gas, which is then run through a steam turbine. Although this is still a developing technology, BIG/GT plants have reached a thermal efficiency of approximately 50% to 60%, which is around as high as modern natural gas combined cycle plants (Bajay 2000, p. 219). One example in the United States is the McNeil Generating Station's Vermont Gasifier, which generates 50 MW of electric power for Burlington residents using wood from nearby forestry operations, forest thinning, and discarded wood pallets. It heats the wood in a chamber filled with hot sand until the wood breaks into the basic chemical components. The gases and solids separate, resulting in a clean-burning gas fuel suitable for use in combined-cycle gas turbines. Because the gas is cleaned before combustion, the sulfur, nitrogen, and carbon emissions are very low (National Renewable Energy Laboratory [NREL] 1999).

The second gasification technology, cogeneration or CHP, has the additional feature of generating both electricity and useful heat (i.e., the heat is captured and distributed as opposed to being released). One example of such a plant is the Skive demonstration CHP project in Denmark, which began in 2007. The plant's owner, a local district heating company, I/S Skive Fjernvarme, served as the main contractor on the project, with the European Union (EU), Danish, and U.S. governments also providing subsidies. Carbona, a subsidiary of a pulp-and-paper technology firm, is the main supplier for the wood chips. The plant now produces 12 MW of district heat and 6 MW of electric power and operates at an overall efficiency of approximately 90% (Gas Technology Institute [GTI] 2010). Although ultimately a success, the project met with several roadblocks along the way, including the technical planning required for a new technology, the need for financing to cover the high investment and operating costs, the difficulty in managing the availability of the feedstock supply, and the uncertainty regarding future renewable energy prices and government subsidies (Salo and Horvath 2009).

Although demonstration projects around the world have proven the superior efficiency of gasification plants over combustion plants, as of late 2011, the biopower industry was still awaiting a large-scale build-out. In addition to the hurdles mentioned in relation to the Skive plant, an overarching problem is the supply constraint of biomass feedstock. Given that it is uneconomical to transport high-density waste materials across long distances, improved biomass densification methods—such as compression into pellets or dense acid oil—would be necessary to increase biopower's commercial use significantly. Furthermore, the development of a best-in-class gasifier technology is essential. Although the demonstration projects are a step in the right direction, the current technology is still far from mature (Bracmort 2010; Instituto de Pesquisas Tecnologicas [IPT] 2010, p. 130; OECD 2010, p. 17). Therefore, if biopower is to enter a growth phase, it will first be necessary for a leading player to (1) create industry-wide standards for gasifiers; (2) enable suppliers of densified biomass to improve the quantity and quality of production, through either increased demand or direct partnerships; and (3) produce biopower on a large enough scale to prove its commercial viability beyond single-plant use.

8.4 BIOPOWER AROUND THE WORLD

As of 2009, biopower supplied approximately 1.2% of the world's electricity supply (241.2 TWh/year). The bulk of bioelectricity is produced from direct combustion (71%) and, to a lesser extent, from gasification (15%), with the remaining 14% produced from liquid biomass and municipal waste. As illustrated in Table 8.3, biopower production has increased nearly twofold over the past 10 years and now accounts for 6.3% of electricity from renewable sources (OER 2010b).

Biopower use varies significantly by region. Western Europe—due to its natural endowments and supportive environmental policies—accounts for 41.7% of the world's biomass electricity and has steadily increased its investment by 12.5% annually over the past decade. North America follows close behind, with 26.3% of the global production. South America contributes 13.9%, and East and Southeast Asia are responsible for 9.8%. On a country level, the top four producers are the United States (55.6 TWh), Germany (33.8 TWh), Brazil (27.1 TWh), and Japan (11.8 TWh; OER 2010b). In the following sections, we will examine Western Europe, North America, and Asia in greater detail. Biopower in South America will be addressed in the case study on Brazil.

8.4.1 Western Europe

Western Europe's current biopower supply originates primarily from forestry—either directly from fuel wood or forest residues or indirectly from

TABLE 8.3

Growth in Global Renewable Electricity Production

Electricity Source (TWh)	1999	2009
Total Production	14,711.7	19,958.5
Total Renewable	2809.5	3810.2
Geothermal	49.8	65.0
Wind	21.2	268.2
Biopower	126.2	241.2
Solar	1.0	21.4
Hydro	2610.7	3213.9
Marine Energy	0.6	0.5
Total Conventional	11,902.2	16,148.3
Nonrenewable Waste	32.5	39.3
Nuclear	2531.1	2695.7
Fossil	9338.6	13,413.3
Renewable Total Share	19.1%	19.1%
Biopower Total Share	0.9%	1.2%
Biopower Share of Renewables	4.5%	6.3%

Source: OER. Electricity production from renewable sources: Details per region and per country. In *Worldwide Electricity Production from Renewable Energy Sources, 12th Inventory—Edition 2010.* Paris: Observ'ER, 2011.

industry byproducts such as sawdust and black liquor. Black liquor, according to the American Forest & Paper Association (2011), "is a thick, dark liquid that is a byproduct of the process that transforms wood into pulp. One of the main ingredients in black liquor is lignin, which is the material in trees that binds wood fibers together and makes them rigid, which must be removed from wood fibers to create paper." The remaining supply comes largely from waste and, to a much lesser extent, from agricultural residues and energy crops (European Climate Foundation [ECF] et al. 2010). It is, therefore, not surprising that the leading countries in biopower are those with the largest forestry operations, namely, Germany, Denmark, Finland, and the Netherlands. Overall, Western Europe accounts for 41.7% of the worldwide biopower production. In many cases, generous legislation supports biopower development. For example, Germany offers "biomass and cogeneration bonuses" and guaranteed feed-in tariffs, whereas Finland provides investors with subsidies of up to 30% of the cost of biopower plant construction (OER 2011b). Table 8.4 shows the role biopower plays in these leading countries' energy mixes.

TABLE 8.4

Biopower Production of Leading European Countries

Country (TWh in 2009)	Biopower Electricity Production	Total Electricity Production	Production (%)
Germany	33.8	596.7	5.7
Denmark	3.2	36.2	8.8
Finland	8.7	71.1	12.2
The Netherlands	6.1	113.2	5.4

Source: OER. Electricity production from renewable sources: Details per region and per country. In *Worldwide Electricity Production from Renewable Energy Sources, 12th Inventory—Edition 2010*. Paris: Observ'ER, 2011.

At the regional level, the European Commission also supports the development of biopower and anticipates that it will play an important role in meeting Europe's "202020" targets—that is, by 2020, reducing greenhouse gas emissions by 20%, relying on renewables for 20% of the energy supply, and increasing energy efficiency by 20% (ECF et al. 2010, p. 1). Biopower and bioheat production in Europe is estimated to reach 1110 TWh by 2020, 540 TWh below the target of doubling the 2010 production level of 800 TWh to 1650 TWh (ibid., p. 8). In pursuit of this goal, the European Commission (2010) has recommended the following initiatives:

- *Energy crops (primary source):* Increase the amount of crop land dedicated to growing feedstock for biopower from 50,000 to five million hectares. This corresponds to 2% to 3% of all European agricultural land and could come from the available six to seven million hectares of idle agricultural land. The primary barrier to this growth is the future market price for energy crops, which introduces uncertainty into farmers' business plans.
- *Agri-residue (secondary source):* Collect 30% to 50% of agricultural residue (i.e., field residues, such as the stalks, leaves, and crop stubble left in fields, and process residues, such as the materials left after processing crops, including husks, seeds, and bagasse) from at least half of Europe's farmland, following Denmark's model, providing an additional 250 TWh/year of primary energy for heat and power generation. This may not be as feasible because the current utilization rate of these residues is close to full, given that a certain amount needs to be left in the fields as nutrients or for livestock.
- *Forestry (secondary source):* Collect and use 30% of all forestry residues (an increase from the 3% collected at the time of this writing).

- *Biomass densification:* Import 500 kton/year from pellet mills outside Europe—the equivalent of 300 TWh (ECF et al. 2010, pp. 21–23).

8.4.2 North America

Canada is similar to the Northern European countries in that it has a large forestry industry that drives biopower production of 7.9 TWh/year (1.3% of its total electricity production). However, biopower as a commercial source of renewable energy (as opposed to simply an "in-house" energy source for stand-alone forestry or agricultural production plants) must compete with relatively inexpensive hydroelectricity—the country's primary source of electricity at 367 TWh/year. In this respect, Canada is comparable to Brazil (OER 2011b).

The United States is the leading country in biopower, with a production of 60.2 TWh and an installed capacity of 11 GW (OER 2011b). This large capacity allows North America to generate 26.3% of the world's biomass energy, the majority of which is located at pulp and paper mills, where byproducts are available for use as biomass feedstock (OER 2011c). Consistent with the world average, biopower plants in the United States supply approximately 1.4% of the country's electricity (Bracmort 2010, p. 1).

Biomass power–generating capacity increased rapidly in the United States in the 1980s, largely as the result of the Public Utilities Regulatory Policies Act of 1978 (PURPA). This act required utilities to purchase electricity from cogenerators and other qualifying independent power producers at a price equal to the utilities' avoided costs (Kartha and Larson 2000, p. 1). However, as previously mentioned, this ultimately led to the construction of biomass plants with low-cost technologies to compensate for a lack of economies of scale, resulting in plants with low thermal efficiencies.

The main biopower facilities in the United States are located in California (1217 MW) and Florida (1158 MW). In both states, a substantial number of biomass power plants use agricultural processing waste as fuel. For example, Florida is home to the largest direct combustion biopower plant in the country—the New Hope Power Partnership plant—which converts sugarcane bagasse and recycled urban wood waste to supply 140 MW, enough to power approximately 60,000 homes with electricity (OER 2011c). Also, as already mentioned, Vermont has one of the most advanced gasifier plants.

Prospects for biopower in the United States are unclear. On one hand, the EIA projects that biopower will grow to 5.5% of total generation by 2035, based on increased usage in cofired plants (Bracmort 2010, p. 1). On the other hand, biopower also competes with biofuels such as syngas, ethanol, and biodiesel for funding and sometimes feedstock. Indeed, in 2007, the U.S. DOE essentially stopped investing in biopower programs in favor of biofuels for transportation (Holmes and Papay 2011). That said, resources may increase in the future as a federal renewable energy standard is currently under consideration, and the 2011 DOE budget request included funding for a new initiative in biopower (U.S. DOE 2010, p. 3).

8.4.3 East and Southeast Asia

Dominant users of biomass in East and Southeast Asia include Japan, South Korea, China, Indonesia, the Philippines, Thailand, and Bhutan. These countries generate most of East and Southeast Asia's 10.6% share of worldwide biopower production. The demand is for both rural and commercial use (Nersesian 2010, p. 45). In the case of Japan, the country produces approximately 14 TWh of biopower per year, making it the fourth largest producer in the world. Biopower generation is associated primarily with the timber industry, which supplies waste wood for cofired plants. South Korea also sources a considerable amount of its electricity from biomass (5%) and is a global leader in biogas conversion into biopower (OER 2011b).

Although China is one of the world's largest electricity producers, it generates only approximately 2.4 TWh of bioenergy per year (OER 2011b). This represents much less than 1% of its electricity supply—the global biopower production proportion average. Nevertheless, China is committed to advancing renewable energy, and biopower seems to be no exception. In 2010, China attained its biopower target of 5.5 MW of installed capacity, as established in the Medium and Long-term Development Plan for Renewable Energy and the 11th Five-Year Renewable Energy Development Plan. Biomass was again featured in the 12th Five-Year Plan, with a target for power generation from agro-forestry of 24 GW by 2020 (Biomass Hub 2011). Furthermore, the Secretary General of the China Renewable Energy Society's Biomass Energy Committee, Yuan Zhenhong, has expressed interest in cultivating more primary sources of biomass to make use of the country's vast deserts, barren hills, and wastelands. The Chinese government is paying particular attention to breeding extreme climate-resistant plant species that can be combined with forestation (BJX News 2011).

8.5 BIOPOWER'S ROLE IN SLOWING CLIMATE CHANGE

The amount of carbon in the atmosphere is increasing each year due to the burning of fossil fuels. Atmospheric carbon acts like a blanket around the earth, trapping heat and warming the planet. The "thicker" that "blanket" becomes, the greater the effects and costs of climate change will be. As of December 2011, the atmosphere's carbon composition was 391 parts per million (ppm), well above the 350 ppm limit that many scientists deem necessary to avoid the detrimental effects of climate change (350.org 2010). By the end of this century, many expect the increase in atmospheric carbon to trap enough heat to warm the planet by 2°C to 5°C (or more). At a minimum, this would cause more frequent extreme weather events, disruptions to ecosystems, and threats to human health (Carbon Mitigation Initiative [CMI] 2010).

As a low-carbon source of electricity, biopower could play a significant role in reducing climate change. However, what would this take? And perhaps more importantly, is this course of action likely? Regarding the first question, biopower plants would need to replace a substantial number of coal-based electricity plants. Although biopower and fossil fuel–based electricity both come from organic material, the former is substantially better for the environment because plants used for feedstock absorb CO_2 from the atmosphere during their growing phase and then release that CO_2 during their decomposition phase (or when burned for energy). Although it is true that the cultivation, harvest, and transportation of the feedstock produce additional CO_2, these emissions should be taken into account only in the case of primary feedstock (i.e., those grown as a source of biomass), not secondary feedstock (i.e., byproducts with carbon emissions allocated to the primary product). The overall effect of biopower on the atmosphere therefore ranges from neutral to slightly positive. In contrast, coal-fired electricity plants release large quantities of CO_2 trapped in fossil fuels that otherwise would not have had an effect on the atmosphere.

The answer to the second question—whether substantially replacing coal-fired plants with biomass-fired plants is *likely*—will depend on the future plans of governments and investors worldwide. In the following section, we attempt to quantify how many coal-fired plants would have to be replaced with biopower plants to significantly reduce the amount of carbon released each year. We will then examine this number in the context of biopower's current development trajectory.

8.5.1 How Much Biopower Capacity Would it Take to Slow Climate Change?

To calculate how many biopower plants would be needed to slow climate change, it is important to first analyze how much carbon emissions need to be reduced. One can then compare carbon emissions from coal-fired plants to those from biopower plants with the same capacity and estimate the number of plants that would have to be replaced to have an effect.

Experts at Princeton University's CMI have spent the past decade researching the probable path of carbon emissions from fossil fuels and the changes that would be necessary to turn the increasing rate of emissions per year to a flat or decreasing rate. Results indicate that, to prevent dramatic climate change, the world must avoid emitting 200 billion tons of carbon (GtC) over the next 50 years. The CMI refers to this as the "stabilization triangle," illustrated as the light gray triangle in Figure 8.1a (CMI 2010).

To achieve this degree of emissions abatement, the CMI recommends taking action in eight different areas: production efficiency, wind, fuel switching, solar, carbon capture and storage, biomass fuels, nuclear, and natural sinks. It refers to each of these areas as "wedges" because, if each one were to achieve emission reductions of 25 GtC over a 50-year period, total abatement would reach the "stabilization triangle" target of 200 GtC (CMI 2010).

The CMI estimates that the emissions savings contained in each wedge are equivalent to decommissioning 683 individual 1-GW capacity–rated coal-fired plants over the next 50 years. Figure 8.1b illustrates the methodology applied to arrive at this number.

If we assume that coal-fired plants will increase in thermal efficiency over the next 50 years to 50% on average (from approximately a best-practices efficiency of 40% today), then the expected emissions rate per plant will be 186 tons of carbon per gigawatt hour (tC/GWh). This implies that in 50 years, a 1-GW plant operating at an average capacity factor of 90% (or for 7884 of the 8760 hours in a year) will emit 1.5 MtC over the course of a year (or 0.00146 GtC/year; see the box on the right side of Figure 8.1b). Because the stabilization triangle target is to reach a reduction rate of 1 GtC of emissions per year by the end of the 50-year period, it would require decommissioning 683 plants in total (1 GtC/year divided by 0.00146 GtC/year; see the graph on the left side of the

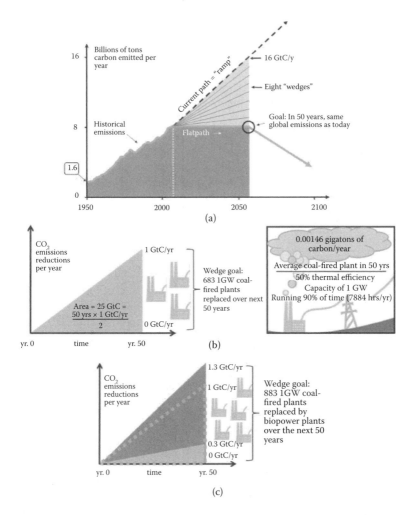

FIGURE 8.1
(a) CMI stabilization triangle and wedges; (b) CMI wedge methodology as applied to coal-fired plants; and (c) CMI wedge methodology showing the effect on emissions of replacing coal plants with biopower plants. (Adapted from C2ES. Climate Techbook: Biopower, 2011. http://www.c2es.org/technology/factsheet/Biopower; CMI. Stabilization Wedge Introduction, 2010. http://cmi.princeton.edu/wedges/intro.php; Oak Ridge National Laboratory Environmental Sciences Division, *Biopower—Overview and Context* by Anthony Turhollow. Oak Ridge, TN: Oak Ridge National Laboratory, 2006; OER, "Conclusion" in *Worldwide Electricity Production from Renewable Energy Sources 12th Inventory, Edition 2010* edited by Observ'ER. Paris: Observ'ER, 2011; Peter Behr. Recession Slows Electricity Demand and Renewable Energy Growth, NERC Finds. *New York Times*, October 29, 2009. http://www.nytimes.com/cwire/2009/10/29/29climatewire-recession-slows-electricity-demand-and-renew-37906.html; U.S. EIA. Biomass for Electricity Generation, 2002. http://www.eia.gov/oiaf/analysispaper/biomass; U.S. EIA. Electricity, 2011. http://www.eia.gov/electricity/; and WNA. World Nuclear Association Report, 2011. www.world-nuclear.org.)

figure). The idea is to replace these plants in a linear fashion over time— replacing 14 each year until year 50, when all will have been replaced. Emissions reductions will be small at first and then steadily grow to 1 GtC/year.

It is important to note that these 683 coal-fired plants must be replaced with another source of energy. Depending on the replacement's level of carbon emissions, even more coal plants may have to be decommissioned to achieve total net reductions of 25 GtC.

Although biopower was not included as a potential stabilization wedge, the CMI's methodology can be applied to determine how many biopower plants would be required to replace at least 683 coal-fired plants over the next 50 years. The first step in this process is to compare carbon emissions from biopower to those of fossil fuel–based electricity. A literature review conducted by the World Nuclear Association (WNA) concluded that the consensus estimate for average CO_2 emissions from biopower is 45 tC/ GWh. This compares to emissions from coal ranging from 186 tC/GWh (the CMI's estimate for a coal-fired plant with 50% efficiency) to 888 tC/ GWh (the WNA's estimate for an average coal-fired plant; WNA 2011). Although the WNA's research suggests that the carbon emissions savings from biopower are much more substantial than what the CMI assumes (95% vs. 76% reduction), we utilize the CMI estimate to keep the methodology consistent.

Given that biopower production is not completely carbon-neutral, simply replacing 683 coal-fired plants with biopower plants would not achieve the goal of 25 GtC. Rather, we would need to replace more than 683 plants so the emissions abated from coal-fired plants minus the emissions generated from biopower plants equal 25 GtC. By modeling the carbon emissions for each plant type according to the CMI triangle model (i.e., assuming a linear ramp-up of emissions reductions over 50 years), we found that if we were to replace a total of 883 1-GW coal-fired plants with 994 1-GW biopower plants (our model assumes that biopower plants have a lower capacity factor [80%] than coal-fired plants [90%]), the net emissions reduction would reach 1 GtC in year 50, equivalent to a 25 GtC total reduction (33 GtC abated from coal-fired plants minus 8 GtC emitted from biopower plants). The gray dotted line in Figure 8.1c outlines the original stabilization wedge, the light gray triangle represents carbon emissions from biopower plants, and the dark gray triangle represents net carbon emissions reductions from replacing coal-fired plants.

8.5.2 How Likely Is it that Biopower Will Replace Coal-Fired Plants?

Now that we have a meaningful target—adding 994 GW of biopower capacity over the next 50 years—we can assess the likelihood of reaching this goal. Achieving this increase incapacity would mean adding 20 GW/year (assuming a linear build-out for simplicity) to the existing installed base of approximately 35 GW. The EIA estimates the investment needed for biomass-integrated gasification-combined cycle plants to be US$1809/kW for an 80 MW facility (Oak Ridge National Laboratory Environmental Sciences Division [ORNLESD] 2006). If we ignore the potential economies of scale and cost reduction measures that could take place over 50 years, this implies an investment of US$36 billion per year, which amounts to US$656 billion in present value terms over 50 years. This number was obtained by calculating the present value of an annuity of paying US$36 billion per year for 50 years, discounted at 5%. To put this amount in context, in 2011, investment in clean energy worldwide (wind, solar, biofuels and other) was US$166.4 billion (McCrone and Aspinall 2012). Although US$36 billion is on the same order of magnitude as current investments in the clean energy sector, biopower would require a significant jump from today's minimal investments to 22% of the total investment in clean energy. Are national players anywhere close to committing to this aggregate level of investment?

One way of approaching this question is to review the existing goals of the leading producer countries. The previous section on global production detailed the commitments that China, the United States, and the EU have made. Table 8.5 summarizes these data and, in some cases, extrapolates them to compare commitments made through 2020.

This table illustrates that, whereas current goals for biopower production are not 100% on track to create a stabilization wedge, they are not far off. In fact, if these three regions achieve their biopower goals and continue along this development path, they could achieve emissions reductions of 12.5 GtC (half of a stabilization wedge). There is also the possibility that the development path could be exponential, as opposed to linear, meaning that biopower capacity could catch up to the wedge model's goals of 994 GW additional capacity by 2062. This possibility is in line with the forecasts of the International Energy Agency, which predicts a 10-fold increase in biopower by 2050 (Center for Climate and Energy Solutions [C2ES] 2011). Despite these positive indicators, it is important to note that

TABLE 8.5

Anticipated Biopower Capacity Increase by 2020

Country/Region	Biopower Capacity Increase Expected by 2020	Biopower Production Increase Expected by 2020
European Union[a]	44.2 GW	310,000.0 GWh
China[b]	18.5 GW	129,648.0 GWh
US[c]	11.9 GW	83,707.0 GWh
Total	**74.7 GW**	**523,355.0 GWh**
Wedge model forecast for 2020	**179 GW**	**1,253,698.4 GWh**
Comparison	**42%**	**42%**
Per year increase (8 years)	**9.3 GW**	**65,419.4 GWh**
Wedge model per year increase	**19.9 GW**	**139,300.0 GWh**
Comparison	**47%**	**47%**

Source: OER. Electricity production from renewable sources: details per region and per country. In *Worldwide Electricity Production from Renewable Energy Sources, 12th Inventory—Edition 2010* edited by Observ'ER. Paris: Observ'ER, 2011; and U.S. EIA. 2011. Electric Power Annual. http://www.eia.gov/electricity/annual/pdf/tablees1.pdf.

[a] The EU is on track to increase production from 800 to 1110 TWh by 2020, for a total increase of 310 TWh. Assuming a capacity factor of 80%, this is equivalent to an additional 44.2 GW of capacity.

[b] The 18.5 GW total is equal to China's goal of 24 GW by 2020, less current capacity of 5.5 GW.

[c] The U.S. aims to increase biopower production from 1.4% of total electricity generation to 5.5% of total electricity generation by 2035. Estimating that electricity production in the U.S. in 2035 will be 5470.1 TWh (1.5% growth rate), then total biopower production will be 300.9 TWh, representing a total increase of 240.7 TWh. The implied per year increase is 10.5 TWh. Over the 8 years from 2012 to 2020, the total increase will be 83.7 TWh, equivalent to 11.9 GW of additional capacity. It is worth noting that this increase in capacity may not achieve the same carbon emissions reductions that the model predicts, given that much of it corresponds to cofiring plants (OER 2011b; U.S. EIA 2011).

political commitments to a goal are not the same as actual investments toward that goal. Therefore, it remains to be seen whether the 2020 targets will be met.

8.6 EVOLUTION OF THE ELECTRIC GRID IN BRAZIL AND THE ROLE OF BIOMASS IN RECENT YEARS

The Brazilian electric grid has relied heavily on hydroelectricity to meet its increasing energy demand, thanks to the country's vast hydro resources. In 2001, the total hydro installed capacity (large hydro plants and mini-hydro facilities) represented 83% of the total installed electricity capacity. In that same year, due mainly to the droughts and the low level of the water reserves, the government of President Fernando Henrique Cardoso declared the national rationing of electricity for 1 year (U.S. EIA 2012). All industries faced a mandatory reduction of up to 25% of their energy consumption, hindering their production capacity and leading government and regulatory authorities to rethink the country's energy mix. With energy demand forecasted to increase every year and with a goal of reducing the country's dependence on hydroelectricity, the Brazilian government initiated a strategy of energy diversification, liberalization of prices (previously controlled by the government), and creation of a new regulatory system based on auctions per technology to add new installed capacity (Cunha et al. 2012).

Energy generated from biomass sources seemed to have support among industry players and regulators to become a viable alternative to hydro generation, as they lessened Brazil's dependence on the oil and gas markets while contributing to the development of the country's sugarcane and ethanol industries. With oil and gas prices escalating to all-time highs in 2008 and some political instability in the supply (oil comes from Arab countries or Venezuela, whereas gas comes mainly from Bolivia, the main supplier of gas to Brazil, which cut its supply in 2005 by 20% due to protests in the Chaco region), the regulatory body set ambitious goals for biopower (Blount and Smith 2005; Business News Americas Staff Reporters 2006; Schneyer 2006). For example, in 2006, the Mines and Energy Ministry predicted that biopower would represent 7% of the Brazilian installed capacity by 2015 (Empresa de Pesquisa Energética [EPE] 2006).

However, did biopower sources take advantage of the opportunity and increase their share in the energy matrix? How did biopower compete with other emerging sources, such as natural gas and wind, in adding new installed capacity to the grid through the auction system?

8.6.1 The "Battle" for New Installed Capacity (MW): The Auction System

To increase the grid's installed capacity (MW), the Brazilian government, through its energy regulatory agency, Agência Nacional de Energia Elétrica (ANEEL), designed an auction modeled on a free market system (after several decades of regulated prices) in which each year, based on the estimated growth of energy demand, new capacity to be installed is auctioned, granting long-term purchase contracts at fixed prices to the winning bidders. Depending on the auction, the new energy plants need to start production within 1, 3, or 5 years of license granting. A key consideration of the auction is the price offered—in other words, the fixed sale price of energy produced and sold to the system (in Brazilian reals per kilowatt-hour). This is the price the system will have to pay to the producer. In the government's view, the lower the price, the better because it reduces costs for all electricity buyers. This metric (price per kilowatt-hour) is a key factor for comparing and understanding the evolution and perspectives of biopower in Brazil (Cunha et al. 2012).

In the initial auctions, all the technologies competed together for the new installed capacity. This scheme resulted in a greater participation of nonrenewable sources of energy, such as natural gas and oil, in the grid due to lower operating costs and greater scalability than renewable sources (with the exception of hydro and mini-hydro). Natural gas and oil-based plants grew in both absolute and relative terms between 2003 and 2010 (representing 15% of the 85,857 MW total electricity capacity in 2003 and 17% of the 113,327 MW capacity in 2010; ANEEL 2011). Natural gas and oil-based energy plants added 5244 and 7763 MW, respectively, in that period. The reduction of renewable energy in the grid was not a desired result for the government. Because of the strong weight given to pricing in the auction system, fewer than expected renewable projects were able to compete (in terms of price) with fossil-based sources of energy, which resulted in a decrease of 2% in the participation of renewable sources at the end of 2010 compared with 2003. Without the addition of 12,907 MW in the same period from hydro sources, the decrease would have been very significant (16% assuming all new electricity was from conventional sources of energy).

Despite the slow growth of renewable sources, biopower experienced a successful period of growth, rising from 3% in 2003 to a significant 7% of the total installed capacity in 2010, increasing by 4253 MW in installed capacity due to the implementation of changes in the legal and competitive frameworks (see the next section; EPE 2011a). Biopower was developing quickly, as expected. Nevertheless, the continuation of this growth trend was uncertain due to structural changes in the natural gas market and the wind power industry, which will be addressed later in this chapter.

In 2005, the Brazilian government approved new types of auctions to increase the quota of clean sources in the energy grid, hoping to correct the unbalanced auction competition between fossil and renewable energies (because conventional sources are mature sources of energy, which have already driven down marginal costs and, therefore, can offer low prices—in terms of Brazilian reals per kilowatt-hour—in the auctions). In these specific auctions, clean energy sources—such as mini-hydro, different types of biopower, and wind—competed for new levels of installed capacity (in megawatts).

Between 2005 and 2010, 37 auctions resulted in the provision of 47,000 MW of capacity for the coming years (see Table 8.6 for the average sale prices of these capacity auctions). Prices ranged between R$84/kWh and R$162/kWh, and the highest average prices were the renewable energy auctions (which are referred to as LFA and LEN auctions). In addition, as an example of hydro's competitiveness, the Belo Monte hydro plant—the third largest in the world with 11.2 GW—won an auction by offering energy at R$78/kWh (EPE 2011b, p. 57).

As a result of the different auction systems, between 2001 and 2011, the Brazilian electric grid experienced several changes, as detailed in Tables 8.7 and 8.8:

- The system reduced its dependence on hydro sources from 83% to 70% and increased its reliability due to investments in transmission infrastructure
- Capacity diversification was achieved mainly through the increasing role of gas and other renewable sources

TABLE 8.6

Details of Capacity Auctions (2005–2010)

Auctions	Number of Auctions	Average Capacity (MW)	Energy Auctioned (TWh)	Average Sale Price (R$/MWh)
Of existing energy (LEE)	9	19,987	1391	88.4
Of Adjustment (LAJ)	9	2265	11	162
New Energy Auctions	19	24,751	4835	127.6
New energy (LEN)	10	14,799	2569	145.6
Structural projects (LPE)	3	6135	1524	84.2
Renewable auctions (LFA)	2	900	160	145.1
Reserve energy (LER)[a]	4	2917	582	157.3
Total/Average	**37**	**47,003**	**6237**	**119.0**

Source: ANEEL. BIG—Banco de Informações de Geração, 2011. http://www.aneel.gov.br/aplicacoes/ capacidadebrasil/GeracaoTipoFase.asp?tipo=2&fase=3; and EPE, Ministério de Energia do Brasil, *Relatório Final Balanço Energético Nacional (BEN) Anho Base 2010.* Rio de Janeiro: EPE, 2011.

[a] Includes Nuclear Angra III.

TABLE 8.7

Installed Capacity in Brazil (in MW, 2001–2011)

Year	2001	2002	2003	2004	2005	2006	2007	2008	2009	2010	2011
Hydro Power	62,409	64,397	67,611	68,998	70,961	73,571	76,757	77,391	78,437	80,518	82,030
Thermal	10,481	13,813	16,130	19,556	19,770	20,372	21,229	22,999	25,350	29,689	31,011
Marine	0	77	87	90	99	107	112	134	173	185	208
Nuclear	1966	2007	2007	2007	2007	2007	2007	2007	2007	2007	2007
Wind	21	22	22	29	29	237	247	398	602	927	1324
Solar	0	0	0	0	0	0	0	0	0	1	1

Source: ANEEL. BIG—Banco de Informações de Geração, 2011. http://www.aneel.gov.br/aplicacoes/capacidadebrasil/GeracaoTipoFase.asp?tipo=2&fase=3.

Note: Does not include imported electricity mainly from Paraguay, Uruguay, and Argentina. Thermal includes electricity from gas, oil, biomass, and coal sources. Hydro includes hydro and mini-hydro (PCH).

TABLE 8.8

Participation in the Electricity Matrix by Source (2001–2011)

Year	2001 (%)	2002 (%)	2003 (%)	2004 (%)	2005 (%)	2006 (%)	2007 (%)	2008 (%)	2009 (%)	2010 (%)	2011 (%)
Hydro	83.3	80.2	78.7	76.1	76.4	76.4	76.4	75.2	73.6	71.0	70.4
Thermal	14.0	17.2	18.8	21.6	21.3	21.2	21.2	22.3	23.8	26.2	26.6
Marine	0.1	0.1	0.1	0.1	0.1	0.1	0.1	0.1	0.2	0.2	0.2
Nuclear	2.6	2.5	2.3	2.2	2.2	2.1	2.0	1.9	1.9	1.8	1.7
Wind	0.0	0.0	0.0	0.0	0.0	0.2	0.2	0.4	0.6	0.8	1.1
Solar	0.0	0.0	0.0	0.0	0.0	0.0	0.0	0.0	0.0	0.0	0.0

Source: ANEEL. BIG—Banco de Informações de Geração, 2011. http://www.aneel.gov.br/aplicacoes/capacidadebrasil/GeracaoTipoFase.asp?tipo=2&fase=3.

Note: Does not include imported electricity mainly from Paraguay, Uruguay, and Argentina; thermal includes electricity from gas, oil, biomass, and coal; and hydro includes hydro and mini-hydro (PCH).

- The auction system helped channel needed private investments to the energy sector and helped create a steady stream of projects to meet future energy demand (Cunha et al. 2012)

By the end of 2011, natural gas–fired generation had doubled its capacity compared with 2003, supplying more than 10% of Brazil's electricity. Biopower also more than doubled its capacity and increased its total contribution to the grid to 7.6% (more than seven times the global average; Figure 8.2). Most of the new biopower capacity added to the grid came from projects associated with the sugarcane industry, as a result of favorable legislation and natural competitive advantages, which will be discussed later in this chapter (Granville et al. 2007). Hence, the installation of bagasse energy plants was the main driver of biopower growth. But why did other biopower sources, such as black liquor or rice, not take off in the same way, given the same legal framework? This issue will be addressed in the next section. As previously pointed out, sugarcane associations and some regulators were very optimistic about biopower's increasing importance or share in the grid. Were the advantages of biopower so strong as to compete against new players such as wind? The answer will follow after examining the different business models of the biopower industry and how they compare with other technologies, such as natural gas or wind.

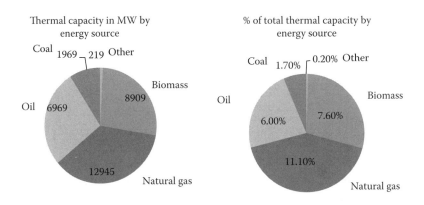

FIGURE 8.2

Thermal installed capacity by source (December 2011). (Adapted from ANEEL. BIG—Banco de Informações de Geração, 2011. http://www.aneel.gov.br/aplicacoes/capacidadebrasil/GeracaoTipoFase.asp?tipo=2&fase=3; EPE, Ministério de Energia do Brasil, *Relatório Final Balanço Energético Nacional (BEN) Anho Base 2010*. Rio de Janeiro: EPE, 2011.)

8.7 BIOPOWER AS AN ALTERNATIVE SOURCE OF ENERGY

From 2003 to 2010, biopower's contribution to the electric grid grew from 3% to 7% (ANEEL 2011). Most of this capacity in 2010 came from the sugarcane industry (344 plants, 5.74% of installed capacity). Other sources included black liquor (14 plants), several types of wood (42 plants), biogas (15 plants), and energy from rice (8 plants; ANEEL 2011; EPE 2011a). As in other cases, many of these plants were constructed to utilize residues from primary activities, generate additional revenue, and secure a source of energy for self-consumption. In addition to sugarcane, other alternative biopower sources—such as waste from black liquor, corn, soy, rice, and oranges—have not been price-competitive in the auctions. As a result, a limited number of new (non–bagasse type) facilities were built between 2005 and 2010. As a point of comparison, although all these sources combined accounted for 108 MW of new capacity sales in the *leilões de energia nova* ("auctions of new energy") during that period, electricity from the sugarcane industry (bagasse, essentially) increased by 4145 MW (EPE 2011a). Despite the minimal contribution to the grid of non–sugar-related biopower sources, according to a study by the EPE, Brazil today has the potential to build/utilize 10,000 MW of energy from different agricultural residues, especially from corn, soy, and rice operations (EPE 2011b). However, to build this capacity and make it profitable, the biopower industry will require strong investments and a favorable regulatory framework, in addition to a different price structure per kilowatt-hour. Expectations of biopower as an alternative to hydro generation in Brazil were focused on the sugarcane industry and, to a lesser extent, on other feedstocks such as eucalyptus and elephant grass. We turn now to a discussion of each of these sources.

8.7.1 Biopower from Bagasse

Before sugar mills began selling biopower to the grid in Brazil, the main residues of sugar/ethanol production—bagasse, *vinhaça,* and straw—were considered "waste" and producers had to find a way to eliminate them. These various waste products were either used as natural fertilizers or incinerated in mill boilers. A decade ago, bagasse not used as fertilizer was burned in the boilers to provide power to the mills, thereby eliminating waste in a profitable manner (Granville et al. 2007, pp. 4–6; Lora and Birth 2004). To facilitate this, the mills used low-pressure boilers (~21 bars), which are less efficient in terms of energy output but very effective at burning large amounts of "waste" (Granville et al. 2007, pp. 4–6; Nivaldo et al. 2008, p. 140).

In the past, due to the lack of a clear legal framework, selling excess energy to the grid was not an attractive option for sugar/ethanol producers and hardly justified investments in new boilers and other required equipment (Granville et al. 2007, pp. 4–6). As a result, few plants actually sold or considered selling electricity to the grid. However, in 2004, the regulatory framework and incentive system changed, making the sale possible and attractive for the sugar/ethanol mills. Consequently, many of the existing mills increased their boiler efficiencies (by increasing pressure to 65 bars), and almost all newly built mills incorporated infrastructure and new boilers, enabling them to sell electricity to the grid. Two other factors influenced the expansion of biopower production in the grid. The first was the gradual consolidation as independent sugarcane producers were acquired by larger players. These large groups were better organized to extract value across the value chain, had more resources to invest in new boilers and infrastructure, and could achieve economies of scale in the supply of feedstock for electricity production. The second factor was the continuous expansion of the sugarcane industry, increasing the amount of feedstock available (bagasse) and plants with biopower turbines.

The production of biopower from bagasse presents several advantages that make it price-competitive in auctions. The first is its competitive cost structure and low investment levels. Because the primary activity of the sugar/ethanol plants absorbs most of the costs (i.e., sowing, harvesting, transporting, separation from the sugar content), the production of electricity in the mills involves no additional fixed or variable costs, apart from the depreciation of new boilers and the infrastructure costs to connect to the grid.

The second advantage is the complementary timing between the sugarcane harvest (and, hence, production of biopower) and the cycle of lower rainfalls. As a result, the sugarcane industry provides most of its electric production in the drier months, when water reserves tend to be depleted, complementing the low hydro supply and helping to preserve water levels. From the regulator's point of view, biopower from bagasse improves the supply to the grid during the dry months and thus increases the reliability of the electricity supply (avoiding problems derived from droughts; Nivaldo et al. 2008, p. 140).

A third advantage stems from the Brazilian government's current strong interest in environmental sustainability. As noted previously, biopower from bagasse results in low carbon emissions. However, this assessment can be supported only if the development of sugarcane plantations has not incurred a "carbon debt." This term refers to CO_2 emissions that result from burning croplands or deforestation to prepare the land for sugarcane exploitation.

A fourth advantage of biopower from bagasse comes from the very low transportation/storage costs because harvest and transportation costs to the mill are absorbed by the primary activity of producing sugar or ethanol. These costs must be incurred to drive these more profitable activities. After extracting the sugar content from the sugarcane, the residues are transported to a boiler, where they are burned to produce energy. The bagasse is not stored for future utilization but rather is used immediately to produce energy. In the infrequent case of an excessive supply of bagasse, some companies have arranged transportation to other plants or industries. As a result, and in opposition to other crops that have to develop their own supply chain and incur logistical costs, the cost of transporting the bagasse is close to zero, presenting an important advantage over most other crops used for biopower.

These increased efficiencies and economies of scale have allowed bagasse-based biopower plants to offer attractive prices in auctions, decreasing from R$200, in the early years of the last century, to close to R$140 in the 2011 auctions. However, scale is probably the Achilles' heel of these projects (and possibly of the entire biopower industry) with respect to their ability to compete for a role in the Brazilian electric grid, as the installed capacity of any one plant rarely exceeds 50 MW of power. Natural gas plants, on the other hand, can easily reach 400 MW, and wind farms can reach 100 to 200 MW.

The advantages of biopower from sugarcane bagasse are difficult to replicate in other crops. Therefore, other sources of biopower have not been able to offer competitive prices in auctions. This does not mean they do not have a place in the Brazilian electric grid; rather, their role is going to be limited and conditioned on specific factors. Two additional business models are now introduced: biopower from eucalyptus and biopower from elephant grass.

8.7.2 Eucalyptus in Brazil

Eucalyptus is the most valuable and widely planted hardwood in the world (18 million hectares in 90 countries; Rockwood et al. 2008). Native to Australia, eucalyptus trees are utilized worldwide for a wide variety of products, including pulp for high-quality paper, lumber, plywood, veneer, solid and engineered flooring, fiberboard, wood cement, composites, mine props, poles, firewood, charcoal, essential oils, honey, tannin, and landscape mulch, in addition to shade, windbreaks, and phytoremediation. These trees are grown extensively as exotic plantation species in tropical and subtropical regions. In 2000, India had eight million mostly low-productivity hectares, followed by Brazil with three million mostly intensively cultivated hectares reaching average productivities of 45 to 60 m³ per hectare per year (Food and Agricultural Organization [FAO] 2006). Eucalyptus can be used as both primary and secondary feedstock (for electricity production). In other words, the entire tree can be cultivated and used in the biopower generation process (as a primary source) or, alternatively, only the residues from another industrial process may be collected and used for biopower (as a secondary source). Due to the high opportunity cost relative to the pulp and paper, furniture, and other forestry industries, eucalyptus is cost-effective for biopower production only when used as a secondary source.

Typically, forestry production plants are located in remote areas where connection to the high-tension grid is not always possible. Many producers aim to be energy self-sufficient by burning wood residues in conventional combustion boilers (Barros and Vasconcelos 2001). However, the distance to the grid means that, even if they had surplus electricity production, they would not be able to sell excess energy to the grid. In addition, the feedstock supply for biopower production does not have a sufficiently constant flow to make the production "efficient" in economic terms.

There are several other problems with the primary use of eucalyptus for biopower:

- The eucalyptus tree takes 3 to 5 years to grow before being cut, whereas sugarcane, rice, and the agricultural residues from other crops—such as oranges, corn, and soy—have up to three harvests per year. Therefore, massive plantations are needed, and returns are collected only in the long term.
- The eucalyptus tree has a lower energy density than other biomass feedstocks such as sugarcane and elephant grass, meaning that more hectares of feedstock are required to achieve the same energy output.

- Environmental concerns remain about deforestation due to the expansion of plantations because plantations have the potential to clear even more land than what is currently cleared for crops such as sugarcane.

In summary, the opportunity costs of eucalyptus, understood as alternative uses of the wood, render its use as a primary feedstock for biopower inefficient. Although eucalyptus residues are still valuable as a source of secondary feedstock, they are usually not collected in areas that are close to the main electricity grid, and the supply of residues is based on the 3- to 5-year cutting cycle, as opposed to a process that happens several times a year. This is not the case for all forestry residues, as exemplified by biopower in Northern Europe, but unfortunately it does limit the use of this source in Brazil. As a result, eucalyptus' potential energy production (outside self-generation–related industries) would be the conversion of leftover residues into pellets (with high calorific density and low volume) by industry players such as pulp and papers plants. Overall, in Brazil, a country blessed with a variety of natural resources, including rich biomass, the production of electricity from eucalyptus (also applicable to other woods) offers no advantage over other clean energies such as hydro, biopower from sugarcane, or wind. But it does allow some industries and plants located in isolated places to be self-sufficient.

8.7.3 Elephant Grass

Elephant grass, also known as Napier grass, gigante (Costa Rica), and mfufu (Africa) is consumed mostly by elephants and reaches significant heights. From a botanic point of view, it is a robust perennial Gramineae with a vigorous root system. It usually grows 180 to 360 cm tall, is branched upwards, and has leafblades that are 20 to 40 mm wide. It is native to subtropical Africa (Zimbabwe) and has been introduced into most tropical and subtropical countries. It grows best in areas with high rainfalls (>1500 mm precipitation per year), but its deep root system enables it to survive in dry times (FAO).

The generation of electricity from elephant grass is based mostly on steam-cycle plants, through the combustion of feedstock in conventional boilers (Mazzarella and Urquiaga 2006, p. 10). The steam turbine is very standardized, and only the boiler requires some adaptations for regular models to mix different feedstock as needed. With respect to this adaptation, Luis Felipe D'Ávila Sykué, one of the founders of the first elephant grass plant in Brazil, stated "We had to sit down with the suppliers that provided the machinery to the sugarcane industry and have them develop the specific machinery for our plant" (VIGtech Biotechnology 2010). Unlike eucalyptus, the utilization of elephant grass for electricity production is primary; that is, the feedstock is grown exclusively to generate electricity. Similar to sugarcane, the harvesting and transportation processes are highly mechanized and standardized. The feedstock is sometimes mixed with other grasses and burned in the boiler to generate electricity. Residual material from burning generally takes the form of ash, which can be used as fertilizers for croplands.

Elephant grass feedstock has three main virtues:

1. *Rapid growth yielding a large volume of dry biomass per hectare*: In 1959, Vicente-Chandler, Silva, and Figarella established a world production record of 84,800 kg of dry matter per year when they fertilized elephant grass with 897 kg of nitrogen (N) per hectare per year and cut it every 90 days under natural rainfall conditions of some 2000 mm per year (FAO 2004). Vicente Mazzarela, a well-respected researcher who has been studying elephant grass at the São Paulo state government's Institute for Technological Research (IPT) since 1991, states that, on average, the dry biomass generated in a year per hectare of elephant grass is 30 to 40 tons (compared with 7.5 tons and approximately 27 tons yielded by eucalyptus and sugarcane, respectively; Osava 2007). The crop yields two harvests per year.

2. *High energy productivity*: According to experts from Embrapa and IPT, the elephant grass energy balance of up to 24:1 (output energy yield per unit of input energy) is one of the highest energy productivities possible in nature. In comparison, Brazilian ethanol delivers an energy balance of 8:1 (some studies indicate up to 10.2:1), and ethanol from corn in the United States is either around energy neutral (between 2:1 and 1:1), or in some cases negative with an energy balance as low as 1:1.3 (Mazzarella and Urquiaga 2008, p. 6). The outcome is a higher productivity of electricity per ton burned, thus lowering relative costs.

3. *Low maintenance costs*: The elephant grass crop requires no special or expensive maintenance. Most important to its care is the use of natural (bacterial) or synthetic fertilizers, which enhance the crop's yield.

At the end of 2011, there was only one elephant grass–burning plant in Brazil, accounting for 30 MW of installed capacity and was sponsored by

Sykué Bioenergia. This company won one of the auctions (*leilão* A-3/2008) in 2008 with a price of R$138.2/kWh and is currently generating electricity in the town of São Desidério in the northeastern state of Bahia (ANEEL 2011; EPE 2011a).

Transporting elephant grass from croplands to biopower plants is problematic because the feedstock is relatively wet and has a high volume at harvest, thus necessitating compaction to reduce transportation costs. In addition, because the use of this grass for biopower is a primary activity, it needs to be dried and stored next to the plant for consumption which, again, requires innovative techniques and machinery for compaction. The logistical complexity is currently one of the most important challenges in making this grass a competitive and reliable source of biopower.

Nevertheless, elephant grass can be used to generate energy for several industries:

- *The ceramics industry*: According to Embrapa studies, the ceramics industry is likely to be the first to utilize elephant grass as an energy source. Medium-sized ceramics plants require less than 100 hectares of elephant grass grown nearby to satisfy their power requirements, so transportation and compaction costs would be manageable. The dried elephant grass can be used directly in furnaces, in place of wood or natural gas (Osava 2007). Urquiaga notes that, "there are more than 6000 ceramics plants in Brazil. Almost 98% of them use wood in the production process, and most of it comes from uncontrolled deforestation. The benefits of replacing the wood for sustainable elephant grass are evident. Embrapa is performing an experimental study where wood is substituted by elephant grass" (Mazzarella and Urquiaga 2008).

- *The pellet and mining industries*: The pellet and mining industries could also benefit from the energy potential of elephant grass. The mining industry, which imports coal to process iron ore into iron and steel for export, could use the grass compressed into pellets, similar to wood pellets, in its blast furnaces as an economical and environmentally friendly solution.

At the same time, according to Mazzarella, "in Europe, the use of dry, compacted biomass pellets for heating is growing rapidly, and elephant grass could open up export markets for Brazil similar to those for ethanol" (Osava 2007). Other energy industry experts agree that the potential for "pelletized" biomass in energy-intensive industries, such as mining or ceramics, is substantial (Granville 2011). It seems as though the pelletization of feedstock is the greatest opportunity cost facing biopower from elephant grass. The potential of this grass for agricultural uses is not significant enough to pose a threat to the economic viability of the biopower it supplies. Nevertheless, elephant grass feedstock faces a number of challenges. As noted earlier, the marginal costs are close to not being competitive compared with other sources of energy. In addition to price and scale, logistics are another crucial point for the development of this technology. "Drying and compacting the biomass is also a challenge," Mazzarella acknowledged. "Green elephant grass is 80% water, and it does not dry out in the sun, as eucalyptus does, but rots if left in piles. To dry, it must be cut up into small pieces, and some heat energy applied. Compacting is necessary for storage and transport because of the great bulk of the dry grass" (Osava 2007). From an investor's point of view, the fact that the model has not been replicated on a larger scale is an obvious source of discomfort.

According to Paulo Puterman, founder and leader of the Sykué project, "Brazil has the potential to double the electricity production if 60 million of degraded hectares were used to grow elephant grass for electricity production" (Yargas 2010, p. 116). Despite this promising potential, experts at the EPE are very skeptical that this potential will be realized (Basto and Ferreira da Cunha 2011). The main challenges for the rapid growth of biomass are, once again, price, scale, and competition from other proven technologies such as wind, hydro, and natural gas. In fact, Sykué's original plan, announced by the founders in 2010, to build up to 10 elephant grass plants across Brazil has been reconsidered, and Puterman has been looking for opportunities in gasification from different sources (Puterman 2011). Biopower faces competition as the choice power supply option to complement the hydro sources in the Brazilian electric grid. This competition comes mainly from natural gas–fired plants and windmills.

8.8 COMPETITION FROM NATURAL GAS AND WIND

8.8.1 Natural Gas

Natural gas is one of the beneficiaries of Brazil's diversification plan and has become an important source of electric energy due to its competitive price and proven and reliable technology. As opposed to hydro-based electricity, natural gas–based electricity does not depend on the level of water reserves or weather cycles. However, it still has a volatile price due to global market fluctuations even though transporting gas, except through pipelines, is expensive. Gas technology has three important advantages over other alternatives such as biopower and wind.

The first is its cost advantage. The cost per kilowatt-hour is more competitive than most biopower and wind projects, based on plant efficiencies and the low cost of natural gas following the 2008 financial crisis. The second advantage is its scalability. Gas-fired plants of 400 MW are feasible, which is not the case for biopower plants and is unusual for windmills. As a result, natural gas projects can offer more competitive prices than most biopower projects and can absorb a larger proportion of the new capacity installed countrywide (to the detriment of other sources). The third advantage relates to supply reliability: Gas, once a pipeline infrastructure is in place, is easy and inexpensive to transport and store. Moreover, it does not depend on harvest quality or wind intensity, that is, its production is predictable.

Because the key input to win auctions has thus far been price, natural gas, with the lowest variable price among the sources considered here, has had a predominant role in new installed capacity in recent years. It is worth noting that the substantial cost advantage is quite recent, as prices halved after 2008. Stability at current price levels is likely because of new extraction methods, including those that focus on shale gas deposits.

In Brazil, the price decrease in the natural gas market accelerated the consumption of electricity and stimulated the construction of extensive infrastructure. In 2010, the average consumption of natural gas in the electricity sector reached 22.1 million m^3 per day, which was significantly higher than the 8.0 million m^3 per day observed during the previous year and represented an increase of more than 180% (EPE 2011b, p. 18). The influence of natural gas in the electricity grid is likely to continue to increase.

Two recent facts support this prediction. First, in 2011, natural gas–fired plants accounted for 20% of the new installed capacity auctioned that year, even though natural gas could participate in only one out of the four auctions due to its nonrenewable nature. Second, in 2012, as can be seen in Table 8.9, 40% of the 25,850 MW preregistered for the A-3-2012 auction were natural gas projects. As illustrated by Table 8.10, natural gas has significant participation, but wind energy accounts for more than 50% of the new capacity added. Is wind power even more competitive than natural

TABLE 8.9

A-3 2012 Auction Preregistry Information

Source	No. of Projects	Offer (MW)	Percentage of Total MW Preregistered
Wind	524	13,180	51.0
Natural Gas	26	10,344	40.0
Biomass	23	1042	4.0
Hydro	25	1284	5.0
Total	**598**	**25,850**	**100%**

Source: EPE, Ministério de Energia do Brasil, *Relatório Final Balanço Energético Nacional (BEN) Anho Base 2010.* Rio de Janeiro: EPE, 2011.

gas? What might this mean in terms of biopower's future development in Brazil?

8.8.2 Wind Energy

The wind industry did not become a relevant actor in Brazil until late 2008, after an initial mapping of the best regions for wind farms was completed. The study revealed that most of Brazil's potential lies in its northeast. Since then, wind generation has experienced a boom, as evidenced by the number of new developments filed in recent auctions (Tables 8.9 and 8.10). The prices offered by wind operators, many of them private investors not related to the energy world, were highly competitive—close to the R$100/kWh barrier—enabling them to win 50% of the capacity in those auctions. As of December 2011, wind energy represented nearly 1% of Brazil's installed capacity (ANEEL 2011; Basto and Ferreira da Cunha 2011; EPE 2011a).

8.9 THE PERSPECTIVE OF THE REGULATOR AND THE OFFICE OF ENERGY PLANNING (EPE)

Experts working at EPE recognized that the potential of biopower in Brazil was enormous, but thought it faced fierce competition, not only from other clean energies, such as hydro and wind, but also from fossil fuels, especially natural gas (Basto and Ferreira da Cunha 2011). The cost per kilowatt-hour remained one of the key factors standing in the way of

TABLE 8.10

2011 Brazilian Capacity Auction Results

| | Year 2011 Auction Results | | | | | | | | | | |
| | Reserve Auction | | | A-3 Auction | | | A-5 Auction | | | Total | |
Energy Source	Number of Projects	Installed Power (MW)	Average Price (R$)	Number of Projects	Installed Power (MW)	Average Price (R$)	Number of Projects	Installed Power (MW)	Average Price (R$)	Number of Projects	Installed Power (MW)
Wind	34	861	99.5	44	1068	99.6	39	977	105.1	117	2906
Biomass	7	357	100.4	4	198	102.4	2	100	103.1	13	655
Hydro				1	450	102	1	135	91.2	2	585
Natural gas				2	1029	103.3				2	1029
Total	**41**	**1218**	**100.0**	**51**	**2745**	**101.8**	**42**	**1212**	**99.8**	**134**	**5175**

Source: EPE, Ministério de Energia do Brasil. *Relatório Final Balanço Energético Nacional (BEN) Anho Base 2010.* Rio de Janeiro: EPE, 2011.

biopower's greater participation in the energy grid—leading to the figures displayed in Table 8.11, which clearly shows that renewables other than hydropower are expected to play a relatively small role in Brazil's energy expansion plans.

Two examples illustrate the lack of traction in the biopower industry. The first is found in the figures published at the end of 2010 in EPE's *Plano de Desenvolvimento Energético 2020* (Ten-year Electric Development Plan). Of the 61 GW of new electricity capacity contracted to enter into operation between 2011 and 2020, as seen in Table 8.12, approximately 29% (17,814 MW) is expected to come from renewable energies such as wind, biopower, and mini-hydro. Wind will account for 59% of this 17,814 MW, whereas biopower represents only 26% of the new renewable capacity (EPE 2010). Figure 8.3 further supports the assertion that biomass is expected to play a relatively minor role in Brazil's future energy expansion in comparison with wind energy.

The overall effect is that wind energy is taking over the majority of the slice reserved for renewable energy sources (excluding small and large hydro plants) by offering low energy purchase prices of approximately

TABLE 8.11

Projected New Capacity Installations (in MW, 2011–2020)

	2011	2012	2013	2014	2015	2016	2017	2018	2019	2020
Hydro	1797	2005	2225	890	4197	4893	5469	4997	2212	3499
Fossil Sources	2297	2922	4743							
Nuclear						1405				
Renewables	1795	2603	2603	1529	1474	1290	1520	1460	1680	1860

Source: EPE, Ministerio de Minas E Energia, *Plano Decenal de Expansão de Energia 2020—Sumário.* Rio de Janeiro: EPE, 2010.

TABLE 8.12

Projected New Capacity Proportions (2011–2020)

	Projected MWs to be Installed	Total (%)
Total Hydro	32,184	52.4
Total Fossil	9962	16.2
Total Nuclear	1405	2.3
Total Renewable	17,814	29.0
Total	**61,365**	

Source: EPE. Ministerio De Minas E Energia, *Plano Decenal de Expansão de Energia 2020—Sumário.* Rio de Janeiro: EPE, 2010.

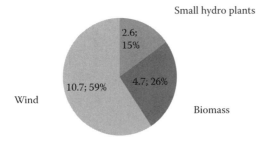

Small hydro plants

FIGURE 8.3
Breakdown of projected renewables installations (in GW, 2011–2020). (Adapted from EPE, Ministerio De Minas E Energia, *Plano Decenal de Expansão de Energia 2020— Sumário*. Rio de Janeiro: EPE, 2010.)

R$100/kWh in the auctions. This poses a significant challenge for biopower projects, which have traditionally been closer to the R$130/kWh to R$160/kWh range. As a consequence of this competition from hydro, gas, and wind generation (Figure 8.4), and with no major breakthroughs expected in the biopower sector, it is easy to see why experts and investors have become more skeptical about the future development of biopower in Brazil. This attitude contrasts with reports published in 2007 and 2008— when natural gas and wind had yet to take center stage—that had a much more optimistic outlook (Nivaldo et al. 2008).

In examining Figure 8.4, it would be erroneous to think that biopower has been able to increase efficiencies to offer competitive prices. The reality is that, with lower auction prices, fewer biopower projects make the grade. Only 13 biopower projects won auctions in 2011, with an average size of 50 MW. On the other hand, windmills won 119 projects, with an average

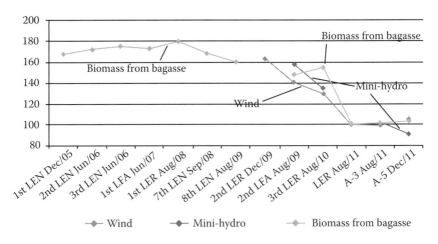

FIGURE 8.4
Prices in Brazilian reals per kilowatt-hour by energy type in auctions from 2005 through 2011. (Adapted from ANEEL. BIG—Banco de Informações de Geração, 2011. http://www.aneel.gov.br/aplicacoes/capacidadebrasil/GeracaoTipoFase.asp?tipo=2&fase=3.)

size of 25 MW per project. Natural gas–fired plants won only two projects, but added 1000 MW to the system, equivalent to 36% more capacity than biopower (ANEEL).

As a result, in the current competitive landscape, biopower seems to be at a crossroads where it cannot compete in price or in capacity with other renewable sources such as wind and hydro/mini-hydro due to feedstock supply constraints. To compete at least in price, biopower projects need to achieve high efficiencies in combustion. One way to achieve these efficiencies could be the development and implementation of gasification plants, which can drive down marginal costs substantially. If biopower does not find a way to compete, its role in the Brazilian electric grid will tend to decline, eroding the growth achieved in the last decade, and less investment will be directed to the sector. The role of all biopower plants may be pushed back to the same competitive status as the elephant grass or eucalyptus models, limited to remote locations, or forced to find nearby industrial players to supply electricity to on an exclusive basis.

8.10 CONCLUSION

If the global demand for energy and the urgency regarding climate change solutions continue to increase as anticipated, biopower will experience a significant growth phase. However, given current roadblocks, it is unlikely that this will take place in the next 5 years. A much more likely scenario is an exponential growth path to respond to urgent demand for clean sources of electricity, which could occur anytime over the next 10 to 30 years, depending on the nature and intensity of catalytic events.

Regardless of when a growth phase ultimately takes place, the industry needs to tackle two significant obstacles: (1) technological developments in densification and gasification processes and (2) pricing that includes the external costs of carbon. The first point relates to the high cost of transporting feedstock, which ultimately limits the scale of biopower plants. Although this does not pose a problem for rural consumption, widespread commercialization will require a large, secure, and cost-effective feedstock supply. Densification of feedstock into easily transportable pellets is the best solution to date; however, it presents a chicken-and-egg–like problem. Only when the demand for feedstock pellets is strong enough will the industry develop to support scale.

Gasification technology is still in its early commercialization phase. Moving from demonstration projects to widespread adoption will take a first-mover willing to bear the risk of building large-scale biopower operations. Carbon pricing, in the second point, remains a nascent concept subject to significant political resistance, even in Europe, home to the largest carbon trading system in the world. As long as fossil fuel power plants face little-to-no financial consequences for emitting carbon, uncertainty will persist about the future price of electricity. Although stronger regulations over greenhouse gas emissions seem inevitable, the current lack of clarity prevents many investors from making long-term commitments to biopower. As a result, biopower is much more likely to experience an exponential (rather than linear) growth path once new legislation is in place, as politicians continue to debate the issue.

In the specific case of Brazil, although there is tremendous potential, biopower has not been able to compete successfully in the price auction system. Most of the capacity recently auctioned off has gone to other energy sources, such as hydro, wind, and natural gas. The only business model that had some relevant contribution to the electricity grid was bagasse biopower from the sugarcane industry. Based on the substantial synergies with its primary activity, this form of biopower experienced notable growth from 2004 until the financial crisis, when two other sources—natural gas and wind—began to draw prices down to levels that became barely achievable for a biopower project. As a result, it seems that biopower's future growth in the Brazilian grid will be constrained by the low prices offered in the auctions by hydro/mini-hydro, wind, and natural gas sources. For this reason, biopower's share in the Brazilian grid will most likely decrease over time in favor of wind and natural gas. As a member of the EPE pointed out at the end of an interview, "Biopower is a very interesting source of energy, but it seems that it is not meant for Brazil," or at least not at the moment (Basto and Ferreira da Cunha 2011).

Although Brazil is unlikely to develop its biopower capacity anytime soon, it can help develop and export different business models to countries with different auction systems or legal frameworks or to countries where less access to other sources of energy makes biopower a viable alternative. The consulting company PSR points out that a niche with potential for energy crops can be the transformation of the crop (i.e., elephant grass) into high-density pellets. These pellets could then be shipped to other areas where energy resources are not as plentiful. One such destination could be the metal and mining industries in the north of Chile. Another contribution would be to advance gasification technology for sugarcane plants, which could be adopted in diverse industries worldwide.

Such niche strategies could prove essential because compounding the problem of biopower's current lack of competitiveness, the Brazilian regulating agency has little incentive to develop a system that creates a market for energy crops, as it has more important priorities, one of them being reducing the cost of the system (in Brazilian reals per kilowatt-hour) by as much as possible. Additionally, from the EPE experts' point of view, it makes more sense to invest in projects that use the enormous amount of waste generated by the main Brazilian cities, rather than investing in projects that use relatively benign agricultural or forestry wastes, as these urban wastes are currently overwhelming the capacity of landfills and recycling facilities. There is, according to the EPE, an opportunity for biogas or incineration to receive regulatory support in the form of higher prices or favorable legislation (Basto and Ferreira da Cunha 2011).

Overall, biopower offers a unique solution to the interrelated problems of increasing energy demand and imminent climate change. For the moment, however, this technology is waiting for positive catalytic events, such as the development of global, well-functioning carbon markets, or negative catalyst events, such as a global energy supply shock. Until then, it will continue to be an attractive source of electricity for specific end users, such as self-sufficient industrial plants and rural communities.

REFERENCES

350.org. 350 Science. Last modified 2010 (accessed November 30, 2011). http://www.350.org/en/about/science.

Agência Nacional de Energia Elétrica (ANEEL). BIG—Banco de Informações de Geração (accessed November 30, 2011a). http://www.aneel.gov.br/aplicacoes/capacidadebrasil/GeracaoTipoFase.asp?tipo=2&fase=3.

Agência Nacional de Energia Elétrica (ANEEL). Matriz de energia eletrica. Last modified November 25, 2011 (accessed November 30, 2011b). http://www.aneel.gov.br/aplicacoes/capacidadebrasil/OperacaoCapacidadeBrasil.asp.

American Forest & Paper Association. What is black liquor? (accessed November 30, 2011). http://www.afandpa.org/WorkArea/linkit.aspx?LinkIdentifier=id&ItemID=918&ei=KT-sUJrhKq-n0AHAiIDACQ&usg=AFQjCNGO5AOLCtDYo58pKU9PJGVFyoe0nw.

Bajay, S. *Industrial Uses of Biomass Energy: The Example of Brazil.* London: Taylor &Francis, 2000.

Barros, D., and E. Vasconcelos. Termelétricas a Lenha. In *Biomassa: Energia dos Trópicos em Minas Gerais* edited by M. Mello, 221–241. Belo Horizonte: LabMidia/FAFICH, 2001.

Basto, L., and S.H. Ferreira da Cunha. Interview with Luciano Basto and Sergio Henrique Ferreira da Cunha, Experts at the EPE, by J.L.G. Pastor. August 2011.

Behr, P. Recession slows electricity demand and renewable energy growth, NERC finds. *New York Times*, October 29, 2009. http://www.nytimes.com/cwire/2009/10/29/29climatewire-recession-slows-electricity-demand-and-renew-37906.html (accessed November 30, 2011).

The Biomass Assessment Handbook: Case Studies. Earthscan, 2007.

Biomass Hub. China's biomass industry falls short of goals, but support remains. Last modified April 26, 2011 (accessed December 10, 2011). http://biomasshub.com/china-biomass-industry-disappoints-despite-strong-policy/.

BJX News. The future of bio energy in China. Last modified April 24, 2011 (accessed November 30, 2011). http://deblockconsulting.com/blog/china-news/the-future-of-bio-energy-in-china/.

Blount, J., and M. Smith. Bloomberg L.P. Bolivia unrest may cut natural gas to Brazil in days (Update 1). Last modified June 9, 2005 (accessed November 30, 2011). http://www.bloomberg.com/apps/news?pid=newsarchive&sid=aQmCDD22O._g&refer=latin_america.

Bracmort, K. Congressional Research Service. Biomass feedstocks for biopower: background and selected issues. Last modified 2010 (accessed December 10, 2011). http://www.fas.org/sgp/crs/misc/R41440.pdf.

Business News Americas Staff Reporters. Pipeline accident, Chaco protests cut gas exports. *Business News Americas*. Last modified April 7, 2006 (accessed November 30, 2011). http://www.bnamericas.com/news/oilandgas/Pipeline_accident,_Chaco_protests_cut_gas_exports.

California Energy Commission. Consumer energy center: Glossary for the letter G. Last modified March 1, 2006 (accessed December 10, 2011). http://www.consumerenergycenter.org/glossary/g.html.

Carbon Mitigation Initiative (CMI). Stabilization wedge introduction. Last modified 2010 (accessed November 30, 2011). http://cmi.princeton.edu/wedges/intro.php.

de Castro, N., R. Brandão, and G. de A. Dantas. Sugar–ethanol bioelectricity in the electricity matrix. In *Ethanol and Bioelectricity: Sugarcane in the Future of the Energy Mix* edited by E. Leão de Sousa, I. de Carvalho Macedo, and O. Pilagallo, 136–153. Sao Paulo, Brazil: Brazilian Sugarcane Industry Association, 2010. http://sugarcane.org/resource-library/books/Ethanol and Bioelectricity book.pdf (accessed November 30, 2011).

Center for Climate and Energy Solutions (C2ES). Climate Techbook: Biopower. Last modified 2011 (accessed December 10, 2011). http://www.c2es.org/technology/factsheet/Biopower.

CleanEnergyAuthority. What is a megawatt and a megawatt-hour? Last modified May 4, 2010 (accessed December 10, 2011). http://www.cleanenergyauthority.com/solar-energy-resources/what-is-a-megawatt-and-a-megawatt-hour/.

Cunha, G., L.A. Barroso, F. Porrua, and B. Bezerra. Fostering wind power through auctions: the Brazilian experience. *The Quarterly Journal of the IAEE's Energy Economics Education Foundation*, no. 2, 25–28, 2012.

The Economist Online. An energised industry. *The Economist*. Last modified November 14, 2011 (accessed November 30, 2011). www.economist.com/blogs/dailychart/2011/11/renewable-energy.

Empresa de Pesquisa Energética (EPE). Ministerio De Minas E Energia. *Plano Decenal de Expansão de Energia Electrica 2006–2015*. Rio de Janeiro: EPE, 2006. http://portal2.tcu.gov.br/portal/pls/portal/docs/2062402.PDF (accessed November 30, 2011).

Empresa de Pesquisa Energética (EPE). Ministerio De Minas E Energia. *Plano Decenal de Expansão de Energia 2020—Sumário.* Rio de Janeiro: EPE, 2010. http://www.mme. gov.br/mme/galerias/arquivos/noticias/2011/SUMARIO-PDE2020.pdf (accessed November 30, 2011).

Empresa de Pesquisa Energética (EPE). Ministério de Energia do Brasil. *Relatório Final Balanço Energético Nacional (BEN) Anho Base 2010.* Rio de Janeiro: EPE, 2011a. https://ben.epe.gov.br/downloads/Relatorio_Final_BEN_2011.pdf (accessed November 30, 2011).

Empresa de Pesquisa Energética (EPE). *Panorama da Expansão da Geração de Energia Elétrica no Brasil.* 2011b.

European Climate Foundation (ECF), Södra, Sveaskog, and Vattenfall. *Biomass for Heat and Power: Opportunity and Economics.* European Climate Foundation, 2010. http://www.europeanclimate.org/documents/Biomass_report_-_Final.pdf (accessed November 30, 2011).

European Commission. *Report from the Commission to the Council and the European Parliament on Sustainability Requirements for the Use of Solid and Gaseous Biomass Sources in Electricity, Heating and Cooling.* Brussels: European Commission, 2010. http://ec.europa.eu/energy/renewables/transparency_platform/doc/2010_report/ com_2010_0011_3_report.pdf (accessed November 30, 2011).

Food and Agricultural Organization (FAO). *Pennisetum Purpureum Schumach.* Last modified 2004 (accessed December 10, 2011). http://www.fao.org/ag/AGP/AGPC/doc/ Gbase/data/pf000301.htm.

Food and Agricultural Organization (FAO). *Forest Resources Assessment 2005—Main Report.* Rome: FAO, 2006. ftp://ftp.fao.org/docrep/fao/008/A0400E/A0400E00.pdf (accessed November 30, 2011).

Gas Technology Institute (GTI). Combined heat and power from biomass gasification in Skive, Denmark. Last modified 2010 (accessed December 10, 2011). www.gastech nology.org/webroot/app/xn/xd.aspx?it=enweb&xd=1ResearchCap/1_8Gasification andGasProcessing/SignifResults/CHP_Biomass_Gasification.xml.

Granville, S. Interview with Sergio Granville, Technical Director and Head of Energy Risk Management at PSR, in Rio de Janeiro, by J.L.G. Pastor. August 2011.

Granville, S., P. Lino, L. Soares, L. Barroso, and M. Pereira. Sweet dreams are made of this: Bioelectricity in Brazil. *IEEE.* Last modified 2007 (accessed November 30, 2011). http:// www.ieee-pes.org/images/pes-ws/gm2007/html/SLIDES/PESGM2007P-001372.PDF.

Holmes, J., and L. Papay. Prospects for electricity from renewable resources in the United States. *Journal of Renewable and Sustainable Energy* 3(4): 0427011–0427015, 2011. http://sanat-danesh.com/danesh/pdf-m/for electricity.pdf (accessed November 30, 2011).

Instituto de Pesquisas Tecnologicas (IPT). Brazilian Sugarcane Yearbook 2010. Last modified 2010 (accessed December 10, 2011). http://www.ipt.br/en/technology_centers/ CETAE/projects/4-biomass_gasification.htm.

Kartha, S., and E. Larson. United Nations Development Program. Bioenergy Primer Modernised Biomass Energy for Sustainable Development. Last modified 2000 (accessed November 30, 2011). http://www.undp.org/content/dam/aplaws/publi cation/en/publications/environment-energy/www-ee-library/sustainable-energy/ bioenergy-primer-modernised-biomass-energy-for-sustainable-development/ Bioenergy Primer_2000.pdf.

Lemar, P. EPA. CHP and Biopower: Market Drivers and Outlook. Last modified 2008 (accessed November 30, 2011). http://www.epa.gov/chp/documents/meet ing_52508_lemar.pdf.

Lora, E., and M. Birth. *Geração Termelétrica: Planejamento, Projeto e Operação.* Rio de Janeiro: Editora Interciência, 2004.

Mazzarella, V., and S. Urquiaga. Workshop Sobre Produção Sustentável de Ferrogusa—Capim Elefante como fonte de Biomassa para a Siderurgia. Last modified September, 2006 (accessed December 10, 2011). http://www.capimelefante.org/trabalhos-apresentados/t13.

Mazzarella, V., and S. Urquiaga. 1 Seminário Madeira Energética—Maden 2008. Last modified August 21, 2008 (accessed November 30, 2011). http://www.revistafator.com.br/ver_noticia.php?not=50238.

McCrone, A., and N. Aspinall. Global trends in clean energy investment: Fact pack as at Q4 2011. *Bloomberg New Energy Finance.* Last modified 2012 (accessed February 15, 2012). www.bnef.com/Presentations/download/84.

National Renewable Energy Laboratory (NREL). Gasifier kindles biopower potential. Last modified September 1998 (accessed December 10, 2011). http://www.nrel.gov/docs/legosti/fy98/25372.pdf.

National Renewable Energy Laboratory (NREL). Project update: Vermont gasifier. Last modified August 1999 (accessed December 10, 2011). http://www.nrel.gov/docs/fy99osti/27236.pdf.

Nersesian, R. *Energy for the 21st Century: A Comprehensive Guide to Conventional and Alternative Sources.* Washington, DC: Library of Congress, 2010.

Nivaldo, J., R. Brandão, and G. de A. Dantas. *Sugar-Ethanol Bioelectricity in the Electric Matrix.* Sao Paulo: União da Indústria de Cana-de-açúcar, 2008. http://sugarcane.org/resource-library/books/SugarEthanolBioelectricityintheElectricityMatrix.pdf (accessed November 30, 2011).

Nogueira, L., E. Lora, M. Trossero, and T. Frisk. *Dendroenergia: Fundamentos e Aplicações.* Brasília: ANEEL, 2000.

Oak Ridge National Laboratory Environmental Sciences Division (ORNLESD). *Biopower— Overview and Context* by A. Turhollow. Oak Ridge, TN: Oak Ridge National Laboratory, 2006. http://bioweb.sungrant.org/NR/rdonlyres/F2270DFC-562E-404A-9815-8FE9F1D260EA/0/BiopowerTechnologyManuscript.pdf (accessed November 30, 2011).

Observatoire des Energies Renouvelables (OER). Conclusion. In *Worldwide Electricity Production from Renewable Energy Sources, 12th Inventory—Edition 2010* edited by Observ'ER. Paris: Observ'ER, 2011a. http://www.energies-renouvelables.org/observer/html/inventaire/Eng/conclusion.asp (accessed November 30, 2011).

Observatoire des Energies Renouvelables (OER). Electricity production from renewable sources: Details per region and per country. In *Worldwide Electricity Production from Renewable Energy Sources, 12th Inventory—Edition 2010* edited by Observ'ER. Paris: Observ'ER, 2011b. www.energies-renouvelables.org/observer/html/inventaire/Eng/preface.asp (accessed November 30, 2011).

Observatoire des Energies Renouvelables (OER). Survey of regional dynamics by sector. In *Worldwide Electricity Production from Renewable Energy Sources, 12th Inventory— Edition 2010* edited by Observ'ER. Paris: Observ'ER, 2011c. www.energies-renouvelables.org/observ-er/html/inventaire/Eng/preface.asp (accessed November 30, 2011).

OECD. *Bioheat, Biopower and Biogas Developments and Implications for Agriculture: Developments and Implications for Agriculture.* Paris: OECD Publishing, 2010.

Osava, M. Energy-Brazil: Elephant Grass for Biomass. Last modified October 10, 2007 (accessed November 30, 2011). http://www.ipsnews.net/2007/10/energy-brazil-elephant-grass-for-biomass/.

Puterman, P. Interview with Paulo Puterman, Founder of the Sykué Project, by J.L.G. Pastor. September, 2011.

Rockwood, D., A. Rudie, S. Ralph, J. Zhu, and J. Winandy. Energy product options for eucalyptus species grown as short rotation woody crops. *International Journal of Molecular Sciences* 9: 1361–1378, 2008. doi: 10.3390/ijms9081361 (accessed November 30, 2011).

Salo, K., and A. Horvath. Biomass Gasification in Skive: Opening Doors in Denmark. Last modified January 13, 2009 (accessed December 10, 2011). http://www.renewable energyworld.com/rea/news/article/2009/01/biomass-gasification-in-skive-opening-doors-in-denmark-54341.

Schneyer, J. Brazil, Chile, Argentina face gas shortage on Bolivia supply snag. Last Modified April 10, 2006 (accessed November 30, 2011). http://www.arizonaenergy.org/News_06/News%20Apr%2006/Brazil,%20Chile,%20Argentina%20face%20gas%20shortage%20on%20Bolivia%20supply%20snag.htm.

U.S. Department of Energy (U.S. DOE) Biomass Program. The ABC's of Biopower. Last modified 2005 (accessed November 30, 2011). http://www1.eere.energy.gov/biomass/abcs_biopower.html.

U.S. Department of Energy (U.S. DOE) Biomass Program. Biopower. Last modified March 2010 (accessed November 30, 2011). http://www1.eere.energy.gov/biomass/pdfs/biopower_factsheet.pdf.

U.S. Department of Energy (U.S. DOE). FY 2011 Congressional Budget Request: Budget Highlights. Last modified 2010 (accessed December 10, 2011). U.S. Department of Energy web site: http://www.cfo.doe.gov/budget/11budget/Content/FY2011Highlights.pdf.

U.S. Energy Information Administration (U.S. EIA). Biomass for Electricity Generation. Last modified 2002 (accessed November 30, 2011). http://www.eia.gov/oiaf/analysispaper/biomass/.

U.S. Energy Information Administration (U.S. EIA). Electricity. Last modified 2011 (accessed November 30, 2011). http://www.eia.gov/electricity/.

U.S. Energy Information Administration (U.S. EIA). Electric Power Annual. Last modified December 2011 (accessed February 15, 2012). http://www.eia.gov/electricity/annual/pdf/tablees1.pdf.

U.S. Energy Information Administration (U.S. EIA). Countries: Brazil. Last modified February 28, 2012 (accessed March 17, 2013). http://www.eia.gov/countries/cab.cfm?fips=BR.

VIGtech Biotechnology. O Capim que gera Energia. Last modified April, 2010 (accessed November 30, 2011). http://www.vigtech.com.br/index_002.html.

Waldheim, L., and E. Carpentieri. Update on the progress of the Brazilian wood BIG-GT demonstration project. *Journal of Engineering for Gas Turbines and Power* 123: 525–537, 2001. doi: 10.1115/1.1335482.

World Nuclear Association (WNA). World Nuclear Association Report. Last modified July, 2011 (accessed November 30, 2011). www.world-nuclear.org.

Yargas, A. A Força do Capim Talismo. *Veja*, August 11, 2010, 112–116.

9

The Business Case for Power Generation in Space for Terrestrial Applications

Andre Luiz Soresini

CONTENTS

9.1 WHY SPACE ENERGY?

This chapter sheds light on a sustainable power source that has been studied by the National Aeronautics and Space Administration (NASA) since the late 1970s: solar power satellites (SPS), a system that taps solar energy in outer space and transmits it to reception stations on Earth (National Space Society [NSS] 2010). The SPS concept was first conceived by Dr. Peter Glaser in 1968, in the middle of the Cold War and the Space Race. Its main goal is to take advantage of the readily available solar power that strikes Earth around the clock and, independent of seasonality and other intra-atmospheric issues, generate energy uninterruptedly in space to be beamed to this planet for terrestrial applications. Its primary benefit is its ability to generate electricity without causing endogenous problems for Earth's surface or atmosphere. Thus, it is truly an exogenous power source.

The context for this study is presented in the next section, which focuses on putting energy security issues in perspective. It also analyzes the current flawed global model of energy generation and the fact that economic growth is highly dependent on a country's ability to expand its energy generation capacity. The endogeneity of currently available power-generating technologies is discussed. Evidence is presented to support the main argument, that is, that all the existing power generation options—whether fossil-based or renewable—are inherently endogenous to Earth. Therefore, all available energy sources depend on and, in certain cases, aggravate rather uncertain long-term environmental conditions for their power generation capability to perform adequately in steady-state.

Although achieving the short-term goal of avoiding, in part, a worsening in environmental conditions, renewable energy sources still face a great deal of skepticism. Critics have argued that, with rapid worldwide population growth, real estate will not be as easily available in the future, which might threaten the sustainability of solar power plants—so-called solar farms (NSS 2013). Scholars have also expressed skepticism about wind energy, arguing that some consider onshore wind farms unpopular, and citing bird mortality as an undesirable side effect of this source of sustainable energy (Bryce 2009). Some economists have also argued that biofuels have been inducing agriculturists to produce fuel instead of food in what is, in effect, a natural cannibalization of agricultural goods destined for consumption, mankind's major source of sustenance. These economists have said that, if this trend continues, the current number of undernourished people, which amounted to an estimated 925 million people in 2010, will continue to increase (World Hunger Organization [WHO] 2010).

In addition, perhaps further exacerbating endogenous problems on the surface of the planet, all these renewable energy sources have one weakness in common: they depend on local environmental conditions. For example, windmills rely on the seasonality of wind gusts; solar power, on the availability of sunshine; and biomass, on a combination of the frequency of sunshine and the frequency of rain. This means that there is an inherent uncertainty regarding the power production capability of these resources, which may further increase the risk of blackouts, thus increasing the need for coal-, oil-, or gas-fired power plants to supply any resulting shortfall. This also results in a circular, illogical solution for a complex equation, that is, the improvement of environmental conditions, which is one goal of the CleanTech or Green Products industry, ultimately depends on one basic variable: those very environmental conditions in order to achieve the desired results. A characterization of today's energy landscape, which motivated NASA's study of the SPS as a means of overcoming obstacles currently faced by other green energy systems, is presented in the next section.

After a discussion of this main motivation for SPS and an illustration of the logical flaws in the current energy generation model, my focus will shift to the technological possibilities for space-based solar power. The third section provides information on architectural concepts and methods for capturing solar energy in space and beaming it to Earth, to shed light on how some of the main technological concerns for developing such an innovative energy-generating concept could be addressed. An economic viability assessment model follows this technological characterization of the SPS system. I then present several sensitivity analyses, highlighting possible combinations of the most influential variables and how they affect the financial base-case model. Possible future scenarios then demonstrate the potential of SPS concepts, depending on different stages of technological advancement. Political viability is discussed in the fifth section, which focuses on the need for joint action among countries around the world to make space-based solar power a reality. Finally, concluding remarks will focus particularly on the SPS system's unique dual characteristics of being the only existing power-generating concept that is completely sustainable due to its exogenous nature and of the ready availability of the virtually infinite solar energy that strikes Earth every day.

9.2 CONTEXT

In the mid-1990s, the American Physical Society issued a statement regarding its long-term view of the energy situation in the United States. From the society's point of view, the United States should have adopted a very aggressive plan to cut its consumption of oil then, even though oil costs at that time were very low (in 1996, for example, one barrel of crude oil cost less than US$30). The American Physical Society's assessment showed that low-cost oil resources outside the Persian Gulf were being depleted very rapidly, increasing the likelihood of sudden disruptions in supply (American Physical Society 1996). Its recommendations to improve the energy consumption scenario in the United States included the diversification of investments in energy research and development, as well as policies to promote efficiency and innovation throughout the energy system. The next subsection will detail the energy problem with a focus on the United States.

9.2.1 Current Issues

9.2.1.1 Energy Security

Access to increasing amounts of low-cost energy has become fundamental to the functioning of modern economies. However, the uneven distribution of natural energy resources among countries has led to formidable vulnerabilities for both suppliers and consumers. Threats to energy security include the political instability of several of the major global oil suppliers, cartelization of energy supplies, competition over energy sources, attacks on production facilities, and accidents and natural disasters (Gordes and Mylrea 2009).

Since the American Physical Society issued its statement on the energy situation in the United States in 1996, there has been a noticeable shift in the pattern of U.S. oil imports. Non-Organization of the Petroleum Exporting Countries (OPEC) countries are now responsible for a larger amount. Figure 9.1 shows the evolution of the United States' gross petroleum imports by country of origin as a percentage of total oil consumed. It is interesting to note that, whereas OPEC countries (the darkest-colored countries in the figure) continue to supply roughly 30% of U.S. oil consumption, non-OPEC countries (the upper, lighter-colored countries) are now responsible for supplying roughly 32% of the oil. This is a sign that the United States is actually moving away from extreme oil dependence on OPEC countries—an issue raised

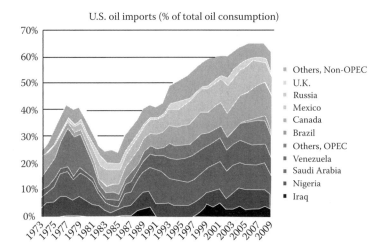

FIGURE 9.1
United States' gross petroleum imports by country of origin. (Adapted from U.S. EIA. Petroleum and Other Liquids: U.S. Imports by Country of Origin, 2010. http://www.eia. gov/dnav/pet/PET_MOVE_IMPCUS_A2_NUS_EPC0_IM0_MBBLPD_A.htm.)

by the American Physical Society in the 1990s (American Physical Society 1996). This clear move from dependence on OPEC-sourced oil has led to a major shift toward long-term, reliable neighbor countries, with Canada now being the United States' largest supplier. Since the 1980s, Mexico, the United Kingdom, and Brazil—also U.S.-friendly, stable nations—have increased their share immensely in supplying oil to the United States, as shown in Figure 9.1.

9.2.1.2 Climate Change

Data collected by various international agencies has shown that climate change is occurring. There is a scientific consensus regarding global warming because, regardless of geographic location, temperatures have been increasing over the past century (Sato 2010). Nevertheless, potential consequences are debated widely, both publicly and politically (Yale Project on Climate Change Communication [YPCCC], and George Mason University Center for Climate Change Communication [GMUCCCC] 2012).

Effects attributed to global warming include physical, ecological, social, and economic changes. Although physical and ecological consequences range from extreme weather changes and glacier retreat to ocean acidification and the rise of sea levels, social issues include effects on the food supply (due to the elevated concentration of CO_2 in the atmosphere, higher temperatures, altered precipitation, and the increased frequency of extreme events) and on health (changes in water, air, food quality, and changes in weather patterns). Scholars have argued that political risks and conflicts may arise due to heightened competition over natural resources with increased scarcity (Wilbanks et al. 2007; Zhang et al. 2007).

Most of the green sources of energy explored around the world could, however, also intensify political risks and the possibility of conflicts. There is inherent interest in specific windy or sunny locations because they can yield greater energy productivity for renewable energy sources. However, these locations are also optimal for food production or the settlement of populations. As a result, efforts to reduce CO_2 emissions from human-related activities have become a call to action in the international community.

9.2.1.3 Increasing Cost and Consumption of Energy

Figure 9.2 presents both nominal and real (consumer price index [CPI]-adjusted) crude oil prices for the past 40 years. Real prices of crude oil in late 2010 were comparable only to those observed during the Iranian Revolution from the late 1970s to the early 1980s. Although it is well known that supply and demand alone are not the only predictors of oil prices, the cartelization of the industry prevents countries from protecting themselves completely against unexpected price peaks, despite the short-term and medium-term palliative solution of hedging through financial derivatives.

Even though the increasing cost of energy is an unpredictable variable, it strongly influences the economic growth of a country whose GDP may be highly correlated to its energy consumption. As its economic output grows, the country will necessarily require greater energy consumption. To

FIGURE 9.2
Nominal and real annual imported crude oil prices. (Adapted from ChartsBin. Historical Crude Oil prices, 1861 to Present, 2009. http://chartsbin.com/view/oau.)

illustrate this point, Figure 9.3 shows the regression for 2007, the last year for which data were fully available. The correlation between total gross domestic product (GDP) (in purchasing power parity [PPP] terms) and total primary energy consumption for that year is roughly 0.95.

It is important to highlight the fact that the presence of the United States within the data set greatly influences the regression's fit curve, its statistical significance, and the resulting predicted curve. This chapter attempts to understand the United States' effect in the global context and how its

Summary output				
Regression statistics				
Multiple R	0.973258128			
R Square	0.947231383			
Adjusted R square	0.946742785			
Standard error	372.8124704			
Observations	110			
ANOVA				
	df	*SS*	*MS*	*F*
Regression	1	269454217.8	269454217.8	1938.671047
Residual	108	15010826.91	138989.1381	
Total	109	284465044.7		
	Coefficients	*Standard error*	*t Stat*	*P-value*
Intercept	71.54623917	37.50015532	1.907891809	0.059060831
Energy-2007	126.8836224	2.881731453	44.03034234	8.01795E-71

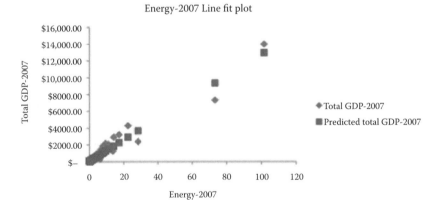

Energy-2007 Line fit plot

FIGURE 9.3

Total GDP (PPP) versus total energy consumed (quadrillion BTUs) in 2007. The following data for GDP in PPP terms is given in USD billion, whereas the total energy consumption is given in quadrillion BTUs. (Adapted from U.S. EIA. International Total Primary Energy Consumption and Energy Intensity, 2010. http://www.eia.doe.gov/emeu/international/energyconsumption.html; and World Bank Group. GDP, PPP [Constant 2005 International $], 2010. http://data.worldbank.org/indicator/NY.GDP.MKTP.PP.KD.)

energy patterns affect the energy consumption trend of emerging markets. Including the United States in the regression acknowledges the country's very important role in the current analysis because its presence greatly influences future outcomes for emerging economies. This country has been emerging economies' major role model in terms of economic success and growth and, therefore, might be a good representation of what lies ahead for these emerging countries if energy consumption patterns do not change.

As shown in Figure 9.3, the direct correlation between a country's economic growth and its consumption of energy is statistically highly significant. However, a country's ability to increase its primary energy production (and consequently its consumption) depends on the cost of adding capacity to its portfolio of power generation plants. If the fuel of choice is not readily available, a country's dependence on the international cartelized costs of coal, oil, and gas might lead to roadblocks in its development efforts. One of the fundamental issues for a country willing to grow economically is its capacity to generate more power. Therefore, the internationally observed cost of energy-related commodities plays a major role in determining the likelihood of a country's success in achieving lasting GDP growth.

Finally, with the rapid rise of emerging economies and the even faster growth of the worldwide middle class, the organic energy consumption growth rate is expected to increase, thus greatly affecting power generation capacity requirements. Studies show that, even recently, roughly two billion people were not yet connected to electric power grids (Mankins 1997). As countries grow economically, it is expected that the electric infrastructure will have more capillarity with deeper penetration. This imminent increase in the penetration of electricity among lower classes in developing nations will boost demand for electricity in the coming years to an unprecedented level and require substantial improvement in the quality and quantity of power output all over the world.

9.2.1.4 Vicious Cycle of the Current Energy Generation Model

There is a direct correlation between GDP growth and growth in power consumption, as already established. Therefore, the quick rise of emerging economies has been seen as a real threat in terms of emissions of pollutants and the inherent consequences, as several of these economies rely on fossil fuels. Issues such as the unavoidable exhaustion of natural reserves, the implicit increasing costs of power-related commodities such as oil and coal, and energy security needs have only been aggravated by the enormous increase in global power consumption as emerging economies have grown in recent decades. The current energy generation model is also causing an ever-increasing amount of stress on the environment. The soaring demand for fossil fuels in the developing world indicates that this problem will only increase in magnitude. This could lead to the need for massive expenditures to mitigate climate change's consequences in the future.

Earth is the only viable habitat for human beings (thus far), so any effect on the surface of the planet or in its atmosphere can be considered an endogenous problem—endogenous to our habitat and our environment—with potential consequences for humanity's survival. There is only one way to reduce the worsening conditions that arise from the increase in the generation and consumption of energy (however it is produced); that is, to somehow transform this inherently endogenous issue into an exogenous one. This is one of the main goals of the SPS system, that is, to generate power in such a way that it does not depend on endogenous variables (for example, the availability of fossil fuels, wind, sunlight, and water), therefore posing no potential harm to the planet.

The following subsection describes some of the best known options available for energy generation and briefly discusses the advantages and drawbacks of each.

9.2.2 Endogeneity of Current Energy Generation Options

9.2.2.1 Nuclear Power

Nuclear power plants are a great solution for the CO_2 issue because they are able to emit up to 98% less CO_2 per kilowatt-hour of electricity generated compared with coal power plants (Spadaro et al. 2000; World Nuclear Association [WNA] 2011). Nevertheless, these plants are probably the one source of energy that faces the greatest resistance from society. From the negative repercussions resulting from the accident at Chernobyl in 1986 (and the related popular manifestations such as the NIMBY—not in my backyard—movement) to the Fukushima-Daiichi nuclear disaster of 2011, and despite improved security measures, enhanced technological systems, and plant efficiencies, popular acceptance has continued to waver. In 1987, only a year after Chernobyl, Italy held a referendum in which the citizens decided to phase out the country's nuclear plants. It took almost two decades for Italy to pursue the development of a new generation of nuclear power plants, amid much public criticism. The March 11, 2011 earthquake off the coast of Japan and the resulting damage sustained by the Fukushima nuclear reactors recently contributed to public fears. At the time of this writing, there is still much uncertainty, but it is clear that one of the consequences of this disaster would be a complete reassessment of the viability and safety of nuclear power.

This is, however, not obvious from very recent surveys. For example, according to a public acceptance survey carried out by EDF Energy, although 11% of the U.K. population opposed any use of nuclear energy in June 2012, 63% believed that nuclear energy should play a role in the country's energy mix (World Nuclear News [WNN] 2012b). A strong trend in favor of atomic power was also noticed in the United States, with roughly 29% of the American population opposing power plants in September 2012 versus 65% in favor of its careful development (WNN 2012a).

Although such recent surveys had indicated increased public acceptance of nuclear power plants, the earthquake in Japan and its subsequent tsunami cooled this acceptance somewhat by showing the world how powerless mankind is in the face of environmental disasters (Pappas 2011). With the worsening of environmental conditions and the increased likelihood of extreme weather events, nuclear power plants might come to be seen as potential threats, capable of inflicting catastrophic effects on their surroundings.

Another major issue concerning nuclear power plants is the large generation of nuclear waste, whose decay process into acceptable radioactive levels can take thousands of years, thus necessitating safe, enclosed environments. These environments could be targets for terrorism or could be exposed to extreme weather conditions, cataclysms, or potentially catastrophic leakage.

9.2.2.2 Coal Power Plants

Due to its ready availability and the relatively low complexity of its plants, coal is still seen as an easy fix for growing energy needs, as seen in the recent industrialization process in China. For example, in 2010, China added a new coal-fired plant every 10 days (Bradsher and Barboza 2006). This incredible growth in electricity generation capacity, mainly through coal-fired power plants, has made that country the world's greatest emitter of greenhouse gases.

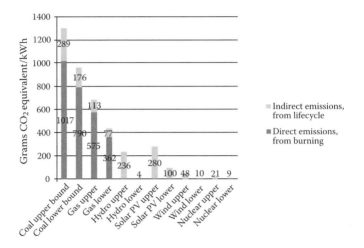

FIGURE 9.4

Greenhouse gas emissions from various means of electricity production. (Adapted from World Nuclear Association. Greenhouse Gas Emissions from Power Generation, 2011. http://www.world-nuclear.org/why/greenhouse_gas_from_generation.html.)

Coal also has a low cost. The direct correlation between economic growth and the ability to produce more energy leads emerging economies to seek resources that are economically viable to support in the short term without compromising their economic growth. Figure 9.4 presents a range of CO_2 emissions per kilowatt-hour of electricity generated for each type of power plant presented in this section. Although technologies have been developed that lower the emissions of coal power plants, coal still causes the greatest environmental effect among all energy sources in terms of greenhouse gases. In addition to the large emissions of CO_2 per kilowatt-hour of electricity generated, coal plants face resistance in coal-producing locations. Local populations have long opposed the environmental effects of mining—from human worker conditions to mountaintop removal to the destruction of vegetation and regional landscapes.

9.2.2.3 Hydroelectric Plants

Hydropower is the most widely used form of renewable energy. In 2008, it accounted for 85% of electricity generated from renewable sources (Renewable Energy Policy Network for the 21st Century [REN21] 2010). One of its biggest advantages is its virtual independence from fuel-related commodity cost increases, enabling a country to reach independence in terms of imported energy-related commodities, such as oil, coal, or natural gas. In addition, if well-maintained, hydroelectric power plants enjoy a long life cycle (ranging anywhere from 50 to 100 years). CO_2 emissions levels are also very low and originate almost entirely from life cycle causes (i.e., from construction and maintenance rather than from operations, unlike coal and gas plants).

Although hydropower enjoys a very high level of acceptance among populations in most parts of the world, it presents several major drawbacks, such as environmental effects, loss of land, population relocation needs, and the constant risk of a catastrophe in the event of a major failure in the dam structure. The most common causes of resistance from nongovernmental environmental organizations are the environmental effects and the loss of productive lands that result from the construction of these plants. A hydropower plant usually requires a large reservoir, resulting in a large area being inundated—virtually always destroying biologically rich and productive lowlands. The second greatest concern is the need to relocate people from their original habitats. It is estimated that 40 to 80 million people have been displaced physically as a direct result of dam constructions worldwide (International Rivers [IR] 2008). Finally, a dam collapse can cause very large areas to be submerged, affecting local populations. Furthermore, dams have been targets of sabotage, terrorism, and war efforts—for example, during World War II's Operation Chastise, the British Air Force destroyed German dams that provided not only electricity to major industrial towns but also potable water to the neighboring populations (TheDambusters.org.uk 2007).

9.2.2.4 Solar Power

One of the main advantages of utilizing solar power is a lower dependence on fossil fuel supplies, their cartelized players, and resultantly uncertain costs. Another advantage is that solar installations can increase the value of underutilized real estate and infrastructure, especially in single-family houses, where it is usually installed on rooftops, eliminating the problem of finding the required space for solar panels. It is also considered a very clean technology because it produces virtually no CO_2 while generating electricity (although it does have greater emissions than wind, hydro, and nuclear throughout its life) and no noise. Its economic viability is also a crucial decision factor for adopters, who usually see their initial investment recouped in just a few years.

At the same time, solar panels have one major drawback—their dependence on local weather conditions. Solar power efficiency relies on the availability of sunshine and seasonal factors (winters have fewer daily hours available for energy production). Weather predictability also enhances the efficiency of this system, and lower greenhouse emissions can help reduce extreme weather patterns. The success of solar panels requires renewable sources of energy. In other words, solar panels depend on local environmental conditions—although they are adopted mostly as a means to reduce the environmental effects created by electricity generation. This can be viewed as a circular reference, that is, the variable in question is dependent on itself, as the amount of greenhouse emissions that solar panels can avert depends on the effect of existing greenhouse gas emissions.

Several variables influence the efficiency of solar power, including pollution, the assumption that sunlight will always be available in the same location (i.e., nothing can block the sunlight to be absorbed by the solar panels to generate electricity), and the degradation of solar panels over their life cycle (on average, 2%–3% per year; Renewable Energy Sources [RES] 2009). This means that solar panels are expected to halve their electricity-generating capacity in 24 to 36 years, which is very significant for investors, as additional capacity will be needed to cover these generation losses to meet consumption that will likely continue to grow over time. This also means that reinvestments are needed for certain locations to continue to produce the same amount of electricity.

Finally, a major issue faced by large investors concerns real estate availability where electricity is needed most. Although several stakeholders—such as governments, societies, industrial firms, and others—have quickly accepted the concept of solar farms, the availability of reasonably low-cost real estate is a major hurdle for a solar farm to be economically and financially viable. This issue raises skepticism among specialists and especially nongovernmental organizations focused on fighting hunger and malnutrition. These organizations believe that large areas of land that receive an abundance of sunlight could be better used for agricultural production to help reduce the large numbers of undernourished people. This is especially relevant in tropical areas, where poverty levels are higher, as exemplified by the DESERTEC project, which intends to produce a vast amount of electricity in North Africa to cope with the increase in European demand (DESERTEC Foundation 2010). Even though the Sahara Desert covers most of the territories of the North African nations targeted for the DESERTEC Project, farms could be established on these lands with the proper availability of energy, as water could be pumped inland for agricultural purposes or for needed improvements to the quality of life there.

9.2.2.5 Wind Power

Wind and solar power share two similar advantages: (1) the capacity to generate electricity in remote areas that are not covered by national electricity grids, and (2) the fact that these sources of energy are readily available in nature at no cost. Also, because it takes up only a relatively small plot of land, wind power generation has quickly spread, especially across Europe. In addition to its considerable generation capability, wind energy has seen an immense growth fueled by government tax rebates and other incentives. This is one of the greatest motivations for investors to develop and install this technology. In general, it takes, on average, 10 years for investors to break even on their initial investments in windmills. However, this will continue only as long as governments maintain the tax incentives and continue to protect the indigenous players, as European nations have been doing thus far.

Although the wind power industry has seen a 112% growth worldwide from 2006 to 2010, and is expected to grow roughly 800% through 2030, it faces some skepticism from critics and specialists in the energy industry (Global Wind Energy Council [GWEC] and Greenpeace 2012). First, the 10-year break-even period depends both on local wind speeds averaging 10 miles per hour for long periods of time and on the continuation of existing subsidies. Locations where wind speeds sustain this average are usually in areas where there is a high likelihood of strong storms—in which case windmills could attract lightning due to their height. Once struck, both their structure and their capacity for generating power are in danger of being compromised. Therefore, the expected break-even time can actually vary widely, and maintenance costs can pose major threats to a project's economic viability, even with government tax benefits. Another frequent source of complaints from people who live near windmills is the noise, which is inversely proportional to the wind speed—that is, the lower the speed, the noisier the windmill. This invariably leads to neighboring communities having poor perceptions of wind turbines. In addition to these common complaints, the turbines are also perceived as bird-killers because they are located exactly in the optimal wind paths used by birds on their migratory routes (Bryce 2009). Finally, windmill technology can also be seen as having a major logical drawback in its functioning, as it depends on the environmental conditions of certain locations, leading to issues such as those described for solar energy technology.

9.2.3 Historical Background of SPS

The previous section shed light on the currently most popular and widely utilized energy sources. It becomes clear from the descriptions and the pros and cons listed that they are all endogenously intertwined, either aggravating environmental conditions (coal, hydro, nuclear) or depending on them (wind, solar) to operate appropriately.

Because none of the available options present an optimal solution for the energy issues currently faced by most nations around the world, the emergence of new technologies is crucial. This section explains the historical and strategic reasons that led to the development of the SPS system and how this purely exogenous power source can help reshape energy generation. It also summarizes current studies being carried out in this field and identifies some critical developmental breakthroughs that are expected in the years ahead.

9.2.3.1 The Cold War, Space Exploration Programs, and the Oil Crises of the 1970s

The unique possibility of generating large amounts of energy in space, to be beamed by satellites for terrestrial applications, was first suggested during the Cold War (Glaser 1968). At that time, both the Soviet Union and the United States were engaged in a fierce competition for space exploration. Major technological breakthroughs were achieved in the late 1960s, which enabled the Soviet Union to send the first man into space and the United States to land the first man on the surface of the moon. Advanced space and satellite technologies enabled scientists to envision concepts of power generation in space that would beam electricity down to Earth for various terrestrial applications (Glaser 1973; Glaser et al. 1974). The energy shortages of 1973 revitalized conceptual discussions that had begun in the late 1960s, as the United States needed a higher level of energy security to drive economic growth. In early 1976, the Department of Energy (DOE) and NASA initiated an SPS concept development and evaluation program, which was completed in 1980 (NSS 2010).

This joint study led to a preliminary concept and several recommendations, with suggestions regarding the next steps, so the concept could be refined and developed further (NSS 2010). The preliminary concept resembled one that had been suggested earlier by Dr. Peter E. Glaser (1968). It comprised a device, placed in a geosynchronous orbit, that generated electrical power from solar energy, transmitting the power to Earth via focused microwave beams and collecting and converting the beams into useful electricity on the planet's surface with receiving antennas (rectennas). Even though additional modular features were considered and analyzed, the technological breakthroughs required for such a system to operate optimally had not yet been developed. Projecting that the concept could become a reality by the year 2000, it was set aside until the anticipated breakthroughs became technologically feasible (NSS 2010).

9.2.3.2 9/11 as an Opportunity to Move away from Oil/Coal Dependency

Strategically speaking, there could have been no better moment to further develop such a complex energy-generating source than the post-9/11 period. Assuming extremely popular public support and the full commitment of taxpayers and industrialists, this point in time could have marked the beginning of a new era in space exploration to take advantage of the virtually infinite amount of energy that is independent of seasonal factors and local weather conditions. The intensity of radiation intercepted in near-Earth space is approximately 1.35 kW/m^2, on average roughly 580% higher than the radiation that strikes Earth's surface (Lior 2001). Figure 9.5 shows the seasonality of the solar radiation that strikes Earth due to the planet's inherent solstice–equinox positioning along its orbit around the sun.

As a result of the economic downturn that followed the terrorist attacks on the World Trade Center, oil prices plummeted. Figure 9.6 presents a set of historical data on oil prices for 1970 to 2009 (in 2009 US$). Although public support and investments could have been directed toward the development of an SPS system to reduce U.S. dependence on foreign energy–related commodities (especially from the highly cartelized OPEC countries), the lower prices of fossil fuel commodities, due to much-reduced demand, caused most decision-makers (including SPS stakeholders) to maintain reliance on oil.

Despite the lack of significant efforts to further develop the SPS technology, important developments took place in the 1990s, even though the investments involved were several orders of magnitude lower than those directed

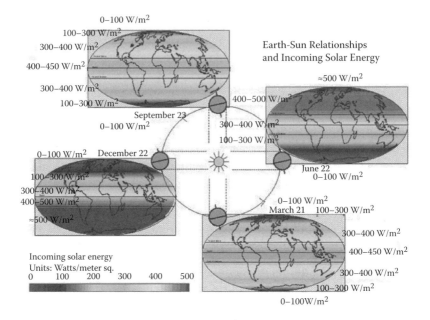

FIGURE 9.5

Solar radiation on the surface of the Earth. (Reprinted with permission from Sharing of Free Intellectual Assets. Earth–Sun Relationships and Incoming Solar Energy. http://sofia.fhda.edu/gallery/geography/images/seasons_insolation.gif, accessed September 19, 2010.)

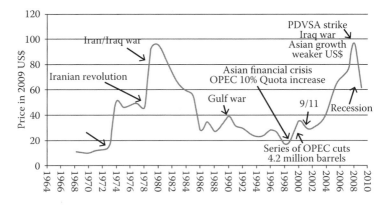

FIGURE 9.6

Timeline of historic oil prices and influential events. (Adapted from ChartsBin. Historical Crude Oil prices, 1861 to Present, 2009. http://chartsbin.com/view/oau; and J. Williams. Oil Price History and Analysis, 2011. http://www.wtrg.com/prices.htm.)

toward other renewable energy sources. That decade marked a series of studies (by NASA, Boeing, and other government-sponsored agencies) that began to renew the popularity of the SPS concept (Mankins 1997).

Funds have not yet been made available to develop space-based solar power concepts. In the last three decades, NASA and the U.S. DOE have collectively spent US$80 million on SPS studies, whereas nuclear fusion development has received US$21 billion (NSS 2008). Although initial studies labeled financial and economic factors as the major difficulties in the development of SPS systems, they relied on the simple extrapolation of oil prices. This meant that they drew conclusions based on a curve with only a slightly positive slope, which assumed small incremental increases in prices over time (NSS 2008). In the decade since 9/11, this scenario has changed considerably. Oil prices skyrocketed from US$15 per barrel to more than US$145 per barrel (price at the close of the New York Mercantile Exchange trading session on July 3, 2008) in less than a decade. SPS systems now seem like a much more attractive proposition from an economic standpoint, as this competing energy source has seen a 10-fold increase in costs.

9.2.3.3 Current Studies

A detailed study, conducted in 2007 by the National Security Space Office (NSSO), on the technical, strategic, economic, and financial feasibilities of space-based solar power reached several conclusions. First, the strategic opportunity offered by SPS could significantly advance the United States' and its partners' security, capability, and freedom of action, which in turn could draw further attention from both public and private sector investments. Second, although some technical issues remain partially unanswered, such an energy source is now executable more than ever before due to recent technological advancements. Third, a proof-of-concept prototype could further increase the chances of private investments as well as international cooperation, therefore mitigating some of the technological risks identified thus far.

Several other studies followed, and a number of private firms began to invest time, expertise, and resources heavily into this technology (Boyle 2009; David 2012; Mitsubishi Electric 2010). The high visibility brought about by recent discussions and developments spurred interest from U.S. state governments, with the goal of minimizing dependence on oil, coal, and gas while mitigating the risks of further environmental effects. For example, the government of California has signed a contract for procuring of renewable energy resources with Solaren Corporation, which, by 2016, will provide the state with power generated in space by SPS and beamed to Earth for terrestrial applications (Public Utilities Commission of the State of California [PUCSC] 2009). This agreement, the first of its kind, can pave the way for a new wave of investments from the private sector in this field, providing the United States with a head start in this cutting-edge era of space exploration, with sustainable and renewable energy-generating objectives.

Other private investments can be found in different countries, from Japan to the United Kingdom, and studies have been converging in terms of technology choices and even economic and financial viability (Mitsubishi Electric 2010; Powersat Corporation 2012). These developments signal that the 40-year-old concept of space-based solar power can become a reality in the current decade if, among other prerequisites, countries align efforts, appropriate policies are put in place, and incentives similar to those given to nuclear fusion, wind, and solar power are provided.

9.2.4 Advantages and Drawbacks of SPS

Although most of the issues raised as potential challenges for the development of SPS are related directly to costs (development, production, launching, and maintenance, to name but a few), other drawbacks are worth mentioning. For example, power collected by solar arrays needs to be beamed electromagnetically to Earth for terrestrial use. However, this technology has not yet been extensively tested (at least not at this order of magnitude) and could be a potential hurdle to the development of SPS. In addition, because the SPS system is expected to orbit Earth, it faces potential damage from meteorites and cosmic dust. Finally, international agreements are needed for space use and power distribution.

Although some of these risks deserve careful and thorough assessment, so as to mitigate the potential drawbacks of space-based solar power systems, the SPS system presents many advantages over any other energy source. First, the energy generation potential is enormous, in that a single kilometer-wide band of geosynchronous Earth orbit experiences as much solar flux in 1 year as the amount of energy contained within all known remaining oil reserves on the planet today (NSS 2008). In addition, this energy could be tapped without emitting greenhouse gases or generating any other environmental, social, or political side effects, that is, in a purely exogenous way, unlike any other currently existing energy source. Second, SPS can generate electricity around the clock, and its performance does not depend on seasonality, weather conditions, or atmospheric pollution, unlike solar and wind power sources. Third, with the absence of gravitational force, an SPS's structure could be much lighter than any other infrastructure associated with terrestrial energy sources, resulting in much less complex structures. Fourth, there are reasons to believe that an SPS system can last much longer than any other power plant on Earth because of the absence of oxidizing agents, rain, dust, hail, and vandalism (Lior 2001). Fifth, unlike the scenarios for hydro or nuclear power plants, accidents would cause no harm to people or nature. Sixth, due to the inherent flexibility of the operation and deployment of the SPS system, power could be beamed directly to isolated areas that have been affected either by natural catastrophes or by armed conflicts, unlike any other existing power-generating method. Finally, energy can be beamed easily to the direct consumer as needed, which enables this system to operate in a unique manner: capturing the entire upside of the energy where it is needed at peak times and selling electricity at peak prices on the spot market in different countries, according to local demand and seasonality issues.

9.3 TECHNOLOGY OF POWER GENERATION IN SPACE

The previous sections of this chapter established the foundations for understanding why a reliable, secure, expandable, nonpolluting, and truly exogenous source of energy is a global necessity. Space-based solar power is well-positioned to capture this opportunity and become the leading source of electricity around the world in the future. This section sheds light on its technological concept, which involves one very basic idea: place large solar arrays into orbit, collect large amounts of electrical energy, beam this energy to Earth electromagnetically, and receive it on the surface for terrestrial uses (NSS 2008).

9.3.1 Architectures

Two SPS architectures are described in this chapter: the Sun Tower and the Solar Disc. NASA and Boeing have studied both concepts extensively over the past decades and have published a vast number of articles on these emerging technologies (Mankins 2002; Oda 2004; Spice 1999). The descriptions are based largely on the ideas within Mankins' *A Fresh Look at Space Solar Power*, published in 1997, which was further analyzed and detailed in the NSSO study of 2007 (Mankins 1997).

9.3.1.1 Sun Tower

The Sun Tower concept comprises four major components: the backbone, the solar power arrays, the transmitting phased array, and other miscellaneous elements. A gravity gradient tether-backbone is deployed as part of the tension-stabilized structure design. The backbone is 15 km long and is connected to 340 pairs of solar collectors and a circular transmission array facing Earth. Each solar collector is 50 to 60 m in diameter, with 1 MW of net electrical output. The transmitter is 200 to 300 m in diameter, with a transmitting frequency of 5.8 GHz. Electronic beam steering, instead of mechanical pointing, is used to guide the radio frequency (RF) beam. The system would initially be deployed in low Earth orbit (LEO) and later moved to a middle Earth orbit (MEO) or geostationary Earth orbit (GEO). It would be located in a sun-synchronous orbit, so a solar pointing system would not be required. Figure 9.7 presents an artistic rendering of the Sun Tower system.

FIGURE 9.7
Artistic rendering of the sun tower. (Reprinted under NASA's media resources usage policy: NASA. NASA Looks for New Ways to Harness Sun's Energy for Earth and Space, 1999. http://www.msfc.nasa.gov/news/news/photos/1999/photos99-096.htm.)

Due to the extensive modularity of the design, very small individual systems can be built and tested with existing facilities, and the actual Sun Tower can be mass-produced, benefiting from the economies of scale provided by the large number of solar panels and other components needed. Heavy lift launch vehicles (HLLV), transporting the four components separately, are suggested as a means to minimize the costs associated with launching the Sun Tower into space. However, combining and stabilizing the four parts in space would require assembly by either personnel or equipment, which could imply high costs, or by very elaborate kinetically deployed mechanical systems, which could be exceptionally complex and carry high risks (Mankins 2002).

The general design of the ground receiver is a rectifying antenna—a 4-km diameter site with direct electrical feed into a commercial power utilities interface (Figure 9.8). Once the beam is intercepted by the rectenna, it is converted back into electricity, which is then rectified to alternating current (AC) and fed into the power grid (Combs 2010). Although rectennas require large parcels of land, it is important to note that the land can be utilized simultaneously for other purposes. Because the intercepting array absorbs the microwaves while allowing sunlight and rainfall to come through, the land could be used for farming or ranching without causing any harm to plants or people thereon. Alternatively, the rectenna could be built as a vast set of greenhouses, thus serving multiple purposes: intercepting and

FIGURE 9.8
SPS power is beamed directly to a rectenna. (Reprinted with permission from T. Prohov, A. Globus, and J. Sercel. Space Studies Institute. Solar Power Satellite—Rectenna. http://ssi.org/space-art/ssi-sample-slides/, accessed September 19, 2010.)

converting the power beamed from space, while enabling greenhouses to be installed underneath, and creating a potentially optimal environment for food to be grown (ibid.). This is an important advantage of the SPS system over any other green energy source. As mentioned previously in this chapter, other sources of green energy present inherent endogenous issues.

For example, consider the use of a potentially lucrative piece of land to produce energy to be consumed by richer countries. This can be seen as cannibalization of the land, where a silent auction takes place behind the scenes between hungry, impoverished countries and richer countries in need of additional sources of energy. Developed nations, with aligned needs and similar goals, could outbid local agricultural or real estate players in an unethical, but potentially realistic, scenario for specific pieces of the land that are especially suitable for the installation of solar, wind, or hydropower plants. This effect of globalization is rather negative for impoverished nations, where the main focus is on the survival of the population rather than on protecting their assets. In other words, sometimes economic incentives are misaligned with environmental goals.

9.3.1.2 Solar Disc

The Solar Disc is a single, rotationally stabilized, spirally constructed, GEO-based concept utilizing differentially spinning elements. The disc would have a diameter of approximately 3 to 6 km. Power would be collected from groups of solar panels connected to provide the required voltage. The center of the disc would integrate the power collected and conveyed to the transmitter array. The array would be approximately 1 km in diameter and approximately 1.5 to 3.0 m in thickness. A single transmitting element would be a hexagonal surface about 5 cm in diameter. One Solar Disc system would be expected to generate power anywhere within a 1 to 12 GW range. The rotation would ensure that the transmitter array is continuously pointing to Earth and the solar panels are continuously pointing to the sun. Figure 9.9 presents an artistic rendering of the Solar Disc system. A unique and cost-effective LEO-to-GEO in-space transportation system is critical for the technical viability of this system.

Similar to the Sun Tower system, the Solar Disc would also have a highly modularized design that would minimize potential transportation and maintenance issues. The ground receiver would be a 5 to 6 km site and would connect directly to the local utilities' interface. This receiver would be very similar to that needed for the Sun Tower. The characteristics of

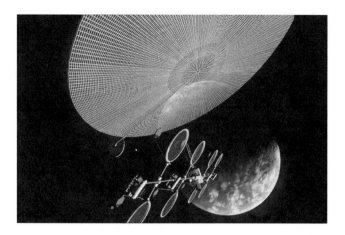

FIGURE 9.9
Artistic rendering of the solar disc. (Reprinted under NASA's media resources usage policy: NASA. NASA Looks for New Ways to Harness Sun's Energy for Earth and Space, 1999. http://www.msfc.nasa.gov/news/news/photos/1999/photos99-096.htm.)

the rectenna would also be similar, and its benefits would be identical. Another important shared similarity and benefit is that multiple ground sites could be served from one system, enabling the SPS to beam its produced energy to locations that lack energy the most or that present a more urgent need for electricity, due to damaged existing power sources, environmental issues, or insufficient electricity infrastructure. This would be extremely beneficial from an economic standpoint because it would enable the SPS operator to beam the power generated in space to places with the greatest willingness to pay, thereby capturing the peak of the surplus curves in each location and boosting its revenues from these places.

9.3.2 Space Satellite Positioning

Possible deployment locations for SPS systems include LEO, MEO, and GEO, as presented by Figure 9.10, and the moon. The following subsections will discuss the advantages and weaknesses of each option.

9.3.2.1 Low Earth Orbit

LEO is defined as an orbit between 160 and 1400 km above the Earth's surface. If deployed within LEO range, a satellite would be in a specific sun–synchronous orbit, in which it passes all latitudes at the same time as the sun every day. LEO is closest to the Earth among all the options and has

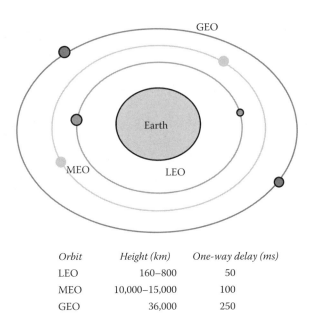

Orbit	Height (km)	One-way delay (ms)
LEO	160–800	50
MEO	10,000–15,000	100
GEO	36,000	250

FIGURE 9.10
Depiction of the three Earth orbits. (Adapted from InetDaemon Enterprises. Satellite Orbits, 2010. http://www.inetdaemon.com/tutorials/satellite/satellite-orbits.shtml.)

been identified as the lowest *cost to first power*, that is, the lowest total cost that would enable a space-based solar power source to become operational. The shallow distance minimizes the complexity of power transmission, the required size and weight of the transmitting antenna, and the energy cost of Earth-to-orbit (ETO) transportation. LEO conceivably could also become a primary assembly and transfer location for higher orbits.

However, satellites in LEO would spend a great deal of time in Earth's shadow and, therefore, have significantly lower productivity. One satellite could serve an individual receiving site on Earth for only a few minutes, thus requiring sophisticated beam steering and a large number of ground stations to maximize productivity. In addition, without an effective capability to remove space debris from orbit, an LEO-based object of such size and surface area would be exposed to a large amount of artificial debris and greater risks of damage.

9.3.2.2 Middle Earth Orbit

MEO is defined as an orbit between 10,000 and 15,000 km above Earth's surface. The so-called low MEO orbit sits at an altitude of 6000 km, whereas a high MEO sits at 12,000 km. The higher altitude substantially reduces the required beam steering. In addition, the satellite spends much less time in Earth's shadow; hence, its productivity is greater. A continuous power supply can thus be achieved by a constellation of satellites, either in the same orbit or in multiple orbits.

However, the higher altitude would also require larger apertures and greater system weights. The transportation cost from Earth to orbit would also be much higher. In addition, exposure to radiation may result in equipment degradation. Because power transmission would have to cross many other satellite orbits to reach Earth, the potential effect of interference also needs to be considered.

9.3.2.3 Geostationary Earth Orbit

GEO is an altitude of 36,000 km. The orbital period of a GEO-based satellite corresponds to the speed of Earth's rotation, which means that the satellite would always be above the same location on the planet. This characteristic allows very little beam steering but facilitates a continuous power supply from a single satellite to one receiving station on Earth. Due to the planet's axial tilt with respect to the sun, the satellite would spend less than 1% of the total time in Earth's shadow, thus optimizing productivity. Other benefits of GEO include fewer environmental effects because there is no beam slewing, less debris in the orbit, more available space in orbit, and more surface visibility on Earth.

On the other hand, because GEO is the highest orbit, satellites would have to be bigger and use more power to compensate for their distance from Earth. Their beams would need to pass through all other orbits; and, similar to MEO objects, they would be at risk of potential impacts. Transportation costs would also skyrocket. In addition, GEO is already used for many other satellites, such as communications and television satellites, as it allows them to maintain continuous communication with a specific point on Earth's surface. This poses more complications when deploying the large SPS system in that orbit.

9.3.2.4 Surface of the Moon

Another option is to construct power-collection and power-beaming equipment on the moon. Because lunar raw materials can be utilized, the tonnage to be transported from Earth would be minimized, along with its associated costs. However, because of the moon's 14-day lunar night, the collectors would have to be constructed on two equidistant sites to generate constant power. In addition, as the moon is not geostationary, this power transfer would require a long-distance beam with high-precision steering. It would also require reflectors or re-transmitters in Earth's orbit or a global distribution grid to facilitate a flow of continuous power to different locations on Earth. The initial launch and installation costs would be extremely high; however, the lunar infrastructure could be utilized for later space exploration purposes.

This option is recognized as being the least viable due to the additional technical constraints—from the need to install sets of power collectors on two sides of the moon to the need to install reflectors or re-transmitters. Furthermore, a global distribution system, integrating power grids worldwide, seems highly unfeasible.

9.3.3 Energy-Capture Methods

9.3.3.1 Photovoltaic

A photovoltaic (PV) cell facilitates the conversion of solar radiation into electricity. PV materials absorb photons upon exposure to light and release electrons. The free electrons move through the cell and fill in orifices, creating direct current (DC) electricity. One single PV cell produces only about 2 W of power. Thus, a number of PV cells must be connected electrically and physically in a support structure to form a PV solar module that can provide sufficient output to power a variety of applications. Multiple modules wired together form a PV solar array. Overall, PV solar array efficiencies range from 6% to 25%. Experimental high-efficiency solar cells have also been developed, with an efficiency of more than 40% (Fraunhofer-Institut 2009). Solar concentrators can be manufactured at low cost and could be used to augment the PV cell output. Crystalline materials have a uniform molecular structure and, therefore, have higher efficiency compared with noncrystalline materials. Among these materials, silicon and gallium arsenide (GaAs) are the current prime candidates for PV cells.

Silicon has an electricity output to weight ratio of 30 W/kg (this is a typical measurement of efficiency for these types of materials, representing the energy output per unit of mass of material employed; Lior 2001). There are two types of crystalline silicon, single-crystal and polycrystalline. A single-crystal solid is a material with a continuous crystal lattice over the entire sample, with no grain boundaries. The absence of defects from grain boundaries gives it special characteristics. In particular, single-crystal silicon has a high conversion efficiency of 15% to 20%, and it is very reliable in energy output and outdoor environments (Mah 1998). However, it incurs high manufacturing costs, due mainly to silicon loss in the wafering process. At the opposite end of the spectrum is an amorphous structure. Between the two extremes is polycrystalline, a material composed of many crystallites of varying sizes and orientations. The orientations can be either random or directed. Compared with single-crystal silicon, polycrystalline silicon is less efficient—ranging between 10% and 14% efficiency—but carries significantly lower manufacturing costs (ibid.).

GaAs is a compound semiconductor made of gallium (Ga) and arsenic (As). Some of its electronic properties are superior to those of silicon. For example, it creates higher saturated electron velocity and electron mobility. Its electricity output to weight ratio is approximately 48 W/kg, 60% higher than silicon on average (Lior 2001). It also has a direct band gap, which allows for much higher light absorption and conversion efficiency, roughly 25% to 30% (Mah 1998). Its high resistance to heat is ideal for concentrator systems, and it has strong resistance to radiation, generating less noise when operating at high frequency. These characteristics make it a strong candidate for space applications (ibid.). However, GaAs does carry higher production costs when compared with silicon because the latter is abundant in nature and is relatively cheap to process. Moreover, GaAs panels are estimated to degrade at 2% to 3% per year in space (Lior 2001). Figure 9.11 depicts GaAs solar panels.

Other crystalline materials include cadmium telluride (CdTe) and copper indium gallium selenide (CIGS). CdTe is the only thin film PV that is cheaper than crystalline silicon PV. However, it is not as efficient as polycrystalline silicon; the best cell efficiency achieved to date is 16.5% (see Figure 9.12 for a diagram of a CdTe PV cell). CIGS is used in PV cells primarily in the form of polycrystalline thin films; the highest efficiency achieved to date with these cells has been 19.5%.

Amorphous silicon (a-Si) is the noncrystalline form of silicon. It has inferior electronic performance and is less efficient than crystalline materials. One advantage, however, is that it is more flexible and can be made thinner than crystalline silicon, thus lowering material costs. It can also be deposited at very low temperatures, favoring its use in space, which represents an almost perfect heat sink location (Figure 9.13 presents a diagram of a-Si PV cell composition).

Due to its inherent mechanical and chemical properties and resistance to radiation, GaAs is generally the compound chosen when evaluating SPS

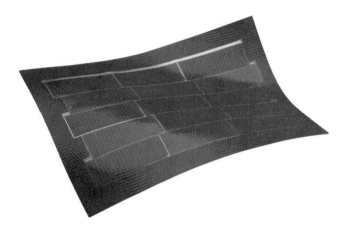

FIGURE 9.11
Gallium arsenide solar panels. (Reprinted with permission from Alta Devices, as furnished by Rich Kapusta, Vice President, Marketing. Alta Devices Unveils High Efficiency Gallium Arsenide Solar Cell, 2011. http://www.ecofriend.com/alta-devices-unveils-high-efficiency-gallium-arsenide-solar-cell.html.)

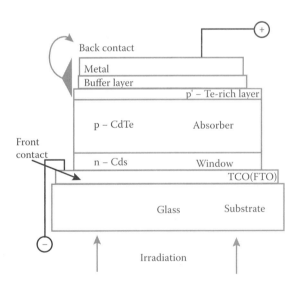

FIGURE 9.12
Cadmium telluride photovoltaic cell composition. (Reprinted with permission from M. Terheggen. Massachussetts Institute of Technology [MIT]. Cadmium Telluride Photovoltaic Film, 2011. http://ocw.mit.edu/courses/mechanical-engineering/2-626-fundamentals-of-photovoltaics-fall-2008/chp_CdTe.jpg.)

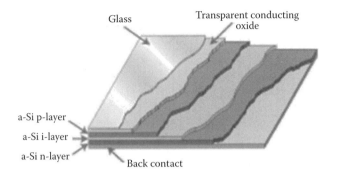

FIGURE 9.13
Amorphous silicon photovoltaic cell composition. (Reprinted under U.S. DOE's media resources usage policy: U.S. DOE EERE. Amorphous Silicon Photovoltaic Cell Composition, 2010. http://www.eere.energy.gov/basics/renewable_energy/images/illust_nip_structure.gif.)

systems. The economic and financial viability analysis in the next section utilizes this compound.

9.3.3.2 Solar Dynamic Systems

Solar dynamic systems, also referred to as concentrated solar power (CSP), concentrates sunlight into a receiver where solar heat is used to drive a heat engine connected to an electric generator. Each system includes a solar concentrator, which collects sunlight; a receiver, which receives, stores, and transfers the concentrated solar energy; a Brayton heat engine, which generates electric power; and a radiator, which emits waste heat (Farmer 1993). The receiver contains a phase-change salt storage material that absorbs energy and provides continuous heat throughout the orbit (Lior 2001). The Brayton heat engine comprises a turbine, a compressor, and a rotary alternator and uses helium-xenon as the working gas. The helium provides superior heat transfer characteristics for the heater, the cooler, and the recuperator. The xenon gas adjusts the molecular weight to provide superior aerodynamic performance for maximized turbine and compressor efficiency (Harty et al. 1994). A recovery heat exchanger between the turbine discharge and receiver inlet is often used to improve cycle efficiency. The gas expands through the turbine when heated; cools through the heat exchanger, where waste heat is transferred to a liquid coolant; and pressurizes within the compressor. Waste heat is dissipated into space via the radiator. The rotary alternator typically provides three-phase alternating

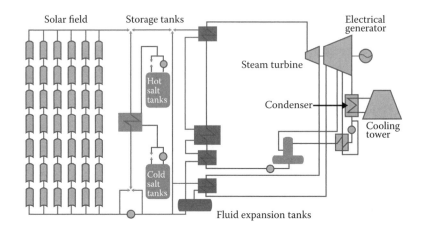

Solar field   Storage tanks   Electrical
                                                               generator

Steam turbine

Hot
salt
tanks

Condenser

Cooling
tower

Cold
salt
tanks

Fluid expansion tanks

FIGURE 9.14

Diagram of a CSP dynamic system. (Reprinted with permission from Abengoa Solar. Solana's Technology, 2013. http://www.aps.com/main/green/Solana/Technology.html.)

current at approximately 1 kHz (Mason 1999). Figure 9.14 presents a diagram of a solar dynamic system with water as the working fluid.

Compared with PV systems, this system can provide continuous power in LEO/MEO without electricity storage systems (Lior 2001). It also has the potential to provide a much higher efficiency of 20% to 30% and incurs lower maintenance costs (Farmer 1993; Lior 2001).

9.3.4 Power Transmission

9.3.4.1 Radio Frequency

RF is the transmission of electromagnetic waves with frequencies below those of visible light. It is generated by oscillations of electromagnetic fields. Current uses include audio, video, navigation, and heating. Wireless power transmission is an already validated technological breakthrough and has demonstrated great potential. In 2008, Dr. Neville Marzwell, of the NASA Jet Propulsion Laboratory, successfully transmitted energy via RF over a distance of 92 miles (148 km) and achieved greater than 90% conversion efficiency of energy to electricity (PUCSC 2009).

The DC-to-microwave conversion system could be performed by a microwave tube system, a semiconductor system, or a combination of both. The microwave tube generates high power with high voltage, whereas the semiconductor system generates low power with low voltage. Some commonly used methods of microwave tubing include the magnetron, the traveling wave tube

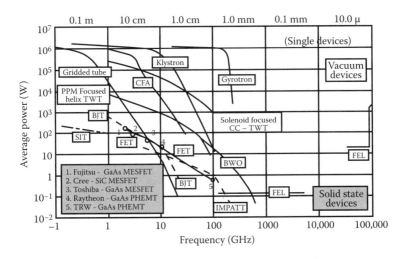

FIGURE 9.15

Average RF output power versus frequency for various electronic devices and semiconductor systems. (Reprinted with permission from N. Shinohara. Wireless Power Transmission for Solar Power Satellite [SPS], 2008. http://www.sspi.gatech.edu/wptshinohara.pdf.)

amplifier (TWTA), and the klystron (Figure 9.15 compares different types of systems). Microwave tubes are more economical and have a higher power/weight ratio than semiconductor amplifiers (Shinohara 2008). For the same amount of power generated, microwave tubes would be lighter and more feasible in an SPS system in which weight is an important consideration due to the major role of launching costs in the economic feasibility of SPS systems.

Transmission is accomplished through a phased array antenna, which generates the beam with specific shape and direction and can steer the beam if needed (Table 9.1 compares different types of antennas). Because the antenna array must be very large (>0.5 km in diameter), it requires great on-orbit weight. It is a key driver of whole system mass; as a result, it is not very plausible for small amounts of power (NSS 2008). The RF used in wireless power transmission would be 2.45 or 5.8 GHz (Shinohara 2008). The beam directed toward Earth would have an intensity about one-sixth that of noon sunlight, delivering 5 to 10 GW of electricity (PUCSC 2009). As mentioned earlier, a rectenna is the method of power reception.

9.3.4.2 Laser Beams

The use of lasers to transmit power was first proposed by Grant Logan in 1988, with technical details worked out in 1989 (Landis 2011).

TABLE 9.1

Typical Parameters of the Transmitting Antenna of a SPS

Model	Old JAXA Model	JAXA1 Model	JAXA2 Model	NASA/DOE Model
Frequency	5.8 GHz	5.8 GHz	5.8 GHz	2.45 GHz
Diameter of antenna	2.6 km	1 km	1.93 km	1 km
Amplitude taper	10 dB Gaussian	11 dB Gaussian	12 dB Gaussian	13 dB Gaussian
Output power (beamed to Earth)	1.3 GW	1.3 GW	1.3 GW	6.72 GW
Maximum power density at center	63 mW/cm²	420 mW/cm²	114 mW/cm²	2.2 W/cm²
Minimum power density at edge	6.3 mW/cm²	42 mW/cm²	11.4 mW/cm²	0.22 W/cm³
Antenna spacing	0.75 l	0.75 l	0.75 l	0.75 l
Power per one antenna (number of elements)	Max. 0.95 W (3.54 billion)	Max. 6.1 W (540 million)	Max. 1.7 W (1.95 billion)	Max. 185 W (97 million)
Rectenna diameter	2.0 km	3.4 km	2.45 km	1 km
Maximum power density	180 mW/cm²	26 mW/cm²	100 mW/cm²	23 mW/cm²
Collection efficiency	96.50%	86%	87%	89%

Source: N. Shinohara. Wireless Power Transmission for Solar Power Satellite (SPS), 2008. http://www.sspi.gatech.edu/wptshinohara.pdf.

Unfortunately, the high cost of taking the concept to operational status and the distant payoff horizon halted its development (ibid.). A large-scale space demonstration has yet to be executed. Figure 9.16 depicts what this demonstration would look like. Laser beams have much shorter wavelengths than RF transmissions. Thus, transmission can be achieved through much smaller apertures. This characteristic facilitates (1) reaching first power at much lower on-orbit weights and (2) the use of much smaller receiver sites on Earth. Tuned PV arrays or solar dynamic engines would serve as the ground receivers.

The main disadvantage is that laser beams have lower efficiencies in both generation and reception. They are less mature at high power levels, may pose eye-safety concerns, and may be unacceptable to the public (NSS 2008). The perceived risks of such exposure could create negative public perceptions about the technology, and NIMBY manifestations could occur, resulting in a complete end to the development of the power-receiving site.

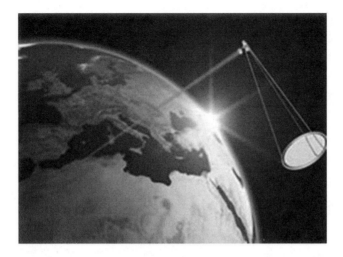

FIGURE 9.16
Concept demonstration of laser power beaming. (Reprinted with permission from Lawrence Livermore National Laboratory. Concept Demonstration of Laser Power Beaming, 2010. https://lasers.llnl.gov/multimedia/publications/images/solar.jpg.)

9.3.5 Earth-to-Orbit Transportation

It is argued that direct launch from Earth to orbit, using chemical propellants, may be possible, but highly undesirable due to the technology's low efficiency levels. A nodal architecture would be preferred, with an Earth to LEO as one segment, followed by an LEO to GEO segment. The HLLV would rely on a two-stage-to-orbit (TSTO) approach. Reusable launch vehicles (RLV) could be used from Earth to LEO. They could be a TSTO system, a rocket-powered vertical takeoff–horizontal landing (VTHL) system, or a Maglev-assisted air-breathing or airborne oxygen enrichment horizontal takeoff and landing (HOTOL) system (examples of VTHL and VTL are shown in Figure 9.17). Solar electric transfer vehicles (SETV) would then be used to transport from LEO to GEO. These infrastructures would require extensive operations and maintenance, but a set number of Earth-to-space launches would make these systems viable because the cost of the RLVs would be amortized over many launches, thus lowering the total cost of launching SPS systems into space.

Other suggested options include a gun launch utilizing a space elevator and a magnetic levitation (Maglev) launcher, as shown in Figure 9.18. The space elevator is the most commonly studied nonrocket means of space

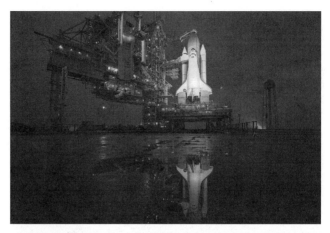

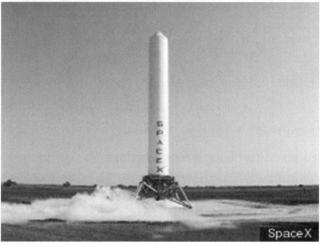

FIGURE 9.17

Examples of vertical takeoff and horizontal landing (top) and vertical takeoff and landing (bottom). (Reprinted under NASA's media resources usage policy [top]: B. Ingalls. NASA. Atlantis Reflection. http://apod.nasa.gov/apod/ap110709.html; and with permission [bottom] SPACE.com Staff. 2012. SpaceX 'Grasshopper' Reusable Rocket Prototype Makes Two-Story Test Flight. http://www.huffingtonpost.com/2012/11/16/spacex-grasshopper-test-flight_n_2145428.html.)

launch. First proposed by Konstantin Tsiolkovsky in 1895, its operating principle is similar to that of a regular elevator, with a physical structure providing guidance for launching objects (Price 2000). Although some concepts are now considered technologically feasible, Earth's gravitational field is too powerful, and there are no engineering materials that are strong and light enough to facilitate the development of these concepts on

FIGURE 9.18
Maglev launching system. (Reprinted under NASA's media resources usage policy: S. Siceloff. NASA. Emerging Technologies may Fuel Revolutionary Launcher, 2010a. http://www.nasa-usa.de/topics/technology/features/horizontallaunch.html.)

Earth (Price 2000). Maglev systems were first patented in 1907 and developed for railroad systems. For example, top train speeds of 581 km/h (361 mph) were attained in Japan in 2003 (Guinness World Records [GWR] 2011). The most well-known implementation of high-speed Maglev is the Shanghai Maglev Train, connecting the city's Pudong section with the International Airport. This train covers 30 km (19 mi) in roughly 7 minutes, with a top speed of 431 km/h (268 mph) and an average speed of 250 km/h (160 mph; ibid.). Extensive testing continues to identify additional uses for this technology.

9.4 ECONOMIC VIABILITY OF SPS SYSTEMS

Most of the issues in evaluating SPS's business viability are related to various costs, which are sometimes seen as impediments for the technology's acceptance by governments and by the private sector. The assumptions used to develop a financial model to assess the financial opportunities associated with building, launching, and operating an SPS system are presented in this section, together with model results and sensitivity analyses.

Other important issues currently faced by the energy industry, discussed previously, range from environmental effects (greenhouse gas emissions, pollution, and the need to handle nuclear waste) to social effects (moving populations to build hydro power plants or the cannibalization of highly productive land to produce biofuels instead of food, in a world where nearly a billion people are undernourished) to potential conflicts between neighboring nations for natural resources (both energy-related commodities and highly desirable real estate for hydro, solar, or wind power). The eventual exhaustion of fossil fuel reserves and the real estate issues already mentioned make the search for a definitive, reliable, and sustainable solution essential to mitigate security, social, political, and environmental issues.

Although the costs associated with developing, launching, and maintaining SPS systems have been labeled by some as impeditive, they may prove surmountable if the following potential scenarios materialize in the future:

- Electricity prices are expected to increase, due to greater demand by emerging economies and the potential exhaustion of readily and cost-effectively available natural fossil fuel reserves, which will automatically make other alternatives for power generation more attractive, especially the SPS systems.
- International agreements may provide further incentives for countries to reduce their carbon emissions, which could further affect the cost of available sources of electricity. An increase in current available power option costs would draw additional attention to a truly exogenous solution, the SPS systems.
- There are several possibilities for government subsidies to support the development of SPS systems, which include energy security funding sources or a similar treatment given to other sources of energy (such as nuclear, solar, or wind power)—in terms of either funds to be made available for major developments or tax holidays for firms that invest in these technologies.
- ETO transportation costs are expected to be reduced considerably as private firms enter this sector, which had begun at the time of this writing. The magnitude of this cost reduction remains unclear, but scientists and other specialists state that costs can be lowered by several orders of magnitude. The expected transportation cost reduction would signal the maturity of this technology and assure that SPS systems can be a profitable venture, as will be seen in subsequent sensitivity analyses.

- SPS systems could last longer than previously estimated. It is common practice in the aerospace industry to be conservative in choosing a set of assumptions satisfying the operating conditions of equipment in space. Because scientists want to have their tests carried out in space with a high degree of certainty, most space equipment has actually been overdesigned and has often provided more benefits than expected. In this sense, it is possible that the degradation of an SPS system, initially estimated to range from 2% to 3% per year (leading to an average power generation life of roughly 40 years) could be much lower, boosting the system's operational life and leading to longer-lasting streams of revenues from the sale of electricity and, therefore, a greater financial return than initially anticipated.

9.4.1 Assessment of Prices

Figure 9.19 presents the average retail price of electricity in the United States by state in 2009. Because the cost of generation and transmission of power generated by an SPS system does not depend on the location to which the electricity is beamed, the system's operator can take advantage of various locations' optimal retail prices. For example, based on 2009 prices, by beaming power to Hawaii, Connecticut, Alaska, New York State, and Massachusetts, the average price per kilowatt-hour for an SPS system operator was US$0.19+ (U.S. Energy Information Administration [U.S. EIA] 2009). It is important to note, however, that Figure 9.19 presents data for the United States only. The average retail price per kilowatt-hour of electricity in the United States was US$0.115. Worldwide prices vary even more. Figure 9.20 presents data on the worldwide average price of electricity. Although these are the prices that utility companies charge industries (and thus are lower than the average price), it is obvious that SPS systems can capture the upside of selling large amounts of electricity to heavy industrial users in Japan, Italy, or Nicaragua at US$0.15+ per kilowatt-hour, 30% more than the average retail price charged in the United States in 2009 (Maritime Electric 2011).

Figure 9.20 demonstrates that more isolated locations have a greater willingness to pay for readily available energy sources. The figure shows that two of the three most expensive locations are in the Federal Territory of Nunavut in Canada, where the price per kilowatt-hour averages US$0.65. The flexibility of SPS systems enables their operators to take advantage of

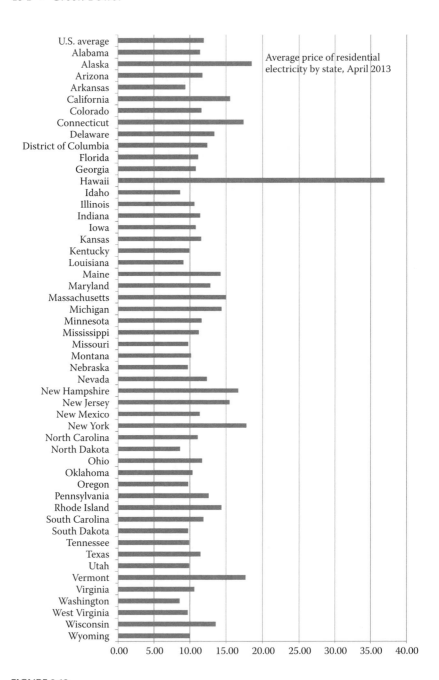

FIGURE 9.19

Average retail price of electricity in the United States in April 2013. (Adapted from U.S. EIA. Average Retail Price of Electricity to Ultimate Consumers by End-Use Sector, by State, April 2013 and 2012 [Cents per Kilowatt-hour]. http://www.eia.gov/electricity/monthly/epm_table_grapher.cfm?t=epmt_5_6_a.)

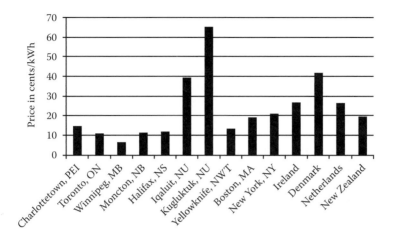

FIGURE 9.20
Average worldwide electricity prices. (Adapted from Maritime Electric. 2011. Electricity Prices Vary All Around the World. http://www.maritimeelectric.com/understanding_ electricity/el_comparing_prices.asp.)

the surpluses in these isolated areas. One can expect the demand at these sites to be low when compared with well-developed countries. According to the figure, however, it is clear that well-developed countries with a heavy demand for electricity also pay much higher average retail prices per kilowatt-hour. This is the case in Denmark (US$0.42), Ireland (US$0.27), and the Netherlands (US$0.26) (Maritime Electric 2011).

The possibility of capturing the consumer surplus or the greater willingness to pay for a cleaner, readily available source of energy around the globe is a major advantage of SPS systems versus other energy sources. The prices of electricity per kilowatt-hour, given previously, will be used to develop sensitivity analyses to evaluate the systems' economic and financial viability.

9.4.2 Assessment of Costs

This cost assessment is based largely on the assumptions adopted by Boeing's 1997 study entitled *Commercial Space Transportation Study* (Commercial Space Transportation Study Alliance [CSTSA] 1997). The assumptions used in the base-case of the financial model of SPS systems (Table 9.2) are based on a scenario in which technological breakthroughs have managed to lower launch costs to GEO to as little as US$400/kg

TABLE 9.2

Assumptions Adopted for the Initial Financial Viability Analysis of SPS Systems

Solar Power Satellite—Assumptions			
Conversion Efficiency		**Other Information**	
Grid conversion efficiency[2]	90%	Power price, $/kWh[1]	0.1151
Rectenna efficiency[2]	90%	*Power price growth, %/year*	0%
Transmission efficiency[2]	96%	Maintenance cost, $/kWh[2]	0.0200
Power purchased from SPS	100%	*Maintenance cost growth, %/year*	0%
Power availability year-round	100%	30-year capital cost, $/kWh[2]	0.0093
Total development cost, $M[2]	$8000	Upper stage prop mass, kg[2]	31,396,520
Total cost of SPS, $M[2]	$5044	Total SPS mass to be launched, kg[2]	54,331,613
Total cost of rectenna, $M[2]	$7243	Power generation capacity, MW[2]	11,433
Launch cost to GEO, $/kg[2]	$400	Lifetime, years[3]	40
LEO delivery cost, $/kg[2]	$110	Degradation/year[3]	2.50%

Source: (1) U.S. EIA. Average Retail Price of Electricity by State, 2009. http://www.eia.doe.gov/cneaf/electricity/epa/fig7p5.html; (2) CSTSA. *Commercial Space Transportation Study.* NASA HQ, 1997. http://www.hq.nasa.gov/webaccess/CommSpaceTrans/; and (3) J.C. Mankins, A Fresh Look at Space Solar Power: New Architectures, Concepts and Technologies. Paper presented at the *38th International Astronautical Congress, Brighton, U.K., October 10–17, 1997.*

(CSTSA 1997). The suborbital space tourism market, spearheaded by Virgin Galactic, has aided the initial development of the space transportation market. Following the success of Virgin Galactic's spaceships, Elon Musk, founder of PayPal and Tesla Motors, recognized the potential for the private sector to compete with government spacecraft in the commercial and scientific space transportation markets. He started Space Exploration Technologies, Corp. (SpaceX) to bring innovation to a market that has traditionally been dominated by governments and large aerospace companies. With the retirement of the U.S. space shuttle fleet, he expects to be able to make space transportation immensely cheaper than the current government costs for launching and operating spacecraft per kilogram of space cargo (SpaceX 2010).

Even with such an optimistic scenario, launching costs would still represent 77% of the total cost-to-first-power of an SPS system. Boeing estimates that one 11.4-GW SPS system would cost US$5.04 billion, whereas its rectenna would cost US$7.24 billion, totaling roughly US$12.3 billion in hardware costs. Meanwhile, the transportation costs would be about US$40.3 billion, given an SPS system's total structural weight of roughly 54,331 metric tons and a launch cost of US$400/kg of hardware to be transported into space (CSTSA 1997).

9.4.3 Financial Model

Given the assumptions presented in Table 9.2, a financial model was developed, assuming that all the power generated by an SPS system and received by the rectenna would be sold at an average retail price of US$0.1151, the average residential retail price in the United States in 2009, according to the U.S. Energy Information Administration (2011). Other assumptions include a U.S. corporate tax rate of 35%, a discount rate of 7%, and an SPS system lifetime of 40 years (i.e., 2.5% degradation of solar cells per year). Table 9.3 presents the financial model's base-case, along with its respective net present value (NPV) and internal rate of return (IRR). The resulting negative NPV of −US$21.8 billion might seem like a bad deal for such an innovative technology. However, nuclear fusion power, for example, has received US$21 billion in government funding (NSS 2008). If this amount of funding had been made available to SPS developers, the NPV of the systems would be at approximately its break-even point. However, this forecast scenario could be even better if certain favorable conditions materialize, such as

- Governmental subsidies, in the form of either technological funding or tax holidays
- Provision of power only to the spot market
- Increasing costs of energy-related commodities and, consequently, existing power generation methods
- Production of more than one satellite, amortizing development costs totaling US$8 billion over more units, and substantially reducing the development costs allocated to each satellite, and
- The potential establishment of price surcharges for polluting fossil fuel power generation technologies, so as to cover costs associated with pollution and other socioenvironmental side effects of these power sources

To further investigate the effect of each variable of the financial model of the SPS systems, the next subsection provides a series of sensitivity analyses aimed at presenting possible scenarios in which the space-based solar power concept could bear fruit, while generating a positive financial return, creating proprietary intellectual knowledge and transforming the energy industry into a purely exogenous, truly sustainable endeavor.

TABLE 9.3

Financial Model for the Base-Case Model with Assumptions Presented in Table 9.2

Figures in US$ M (Unless Otherwise Indicated)	2011 0	2012 1	2013 2	2014 3	2015 4	2016 5	2017 6	2018 7	2019 8	2020 9	2021 10
Power Generation Capacity											
Installed capacity (MW)	—	11,433	11,147	10,861	10,575	10,290	10,004	9718	9432	9146	8860
Cumulative Degradation	0.0%	0.0%	2.5%	5.0%	7.5%	10.0%	12.5%	15.0%	17.5%	20.0%	22.5%
Annual capacity (MWh)	—	100,152,029	97,648,228	95,144,427	92,640,627	90,136,826	87,633,025	85,129,224	82,625,424	80,121,623	77,617,822
Power Installation Costs ($M)											
a. Total development cost	(8000)										
b. Launch cost	(40,268)										
c. SPS + rectenna cost	(12,287)										
x. Government subsidies											
Salvage value											
Depreciation	—	307	307	307	307	307	307	307	307	307	307
d. Depreciation tax shield	—	108	108	108	108	108	108	108	108	108	108

Operations

Revenues from power		8964	8740	8516	8291	8067	7843	7619	7395	7171	6947
Power supply consumed		100%	100%	100%	100%	100%	100%	100%	100%	100%	100%
Price per kWh ($)		0.12	0.12	0.12	0.12	0.12	0.12	0.12	0.12	0.12	0.12
Annual price increase per kWh		0.0%	0.0%	0.0%	0.0%	0.0%	0.0%	0.0%	0.0%	0.0%	0.0%
Cost of power generation		(2003)	(1953)	(1903)	(1853)	(1803)	(1753)	(1703)	(1653)	(1602)	(1552)
Maintenance cost per kWh ($)		0.02	0.02	0.02	0.02	0.02	0.02	0.02	0.02	0.02	0.02
Annual cost increase per kWh		0.0%	0.0%	0.0%	0.0%	0.0%	0.0%	0.0%	0.0%	0.0%	0.0%
Capital cost 30-year amortization		(931)	(908)	(885)	(862)	(838)	(815)	(792)	(768)	(745)	(722)
30-year capital cost ($/kWh)		0.0093	0.0093	0.0093	0.0093	0.0093	0.0093	0.0093	0.0093	0.0093	0.0093
e. Before tax operating profit	—	6029	5879	5728	5577	5426	5276	5125	4974	4823	4673
f. Tax expense	—	(2110)	(2058)	(2005)	(1952)	(1899)	(1846)	(1794)	(1741)	(1688)	(1635)
After tax cash flows (a+b+c+d+e+f)	(60,555)	4027	3929	3831	3733	3635	3537	3439	3341	3243	3145
Terminal value											32,275
Discount Factor	1.00	1.07	1.14	1.23	1.31	1.40	1.50	1.61	1.72	1.84	1.97
NPV of cash flows	(60,555)	3763	3431	3127	2848	2591	2357	2141	1944	1764	1599
NPV of terminal value											16,407
Total NPV	(60,555)	3763	3431	3127	2848	2591	2357	2141	1944	1764	18,006
NPV of SPS	(21,802)		IRR of SPS		−4.1%						

9.4.4 Sensitivity Analyses

In each of the sensitivity analyses presented here, two of the base-case model assumptions were varied whereas the others remained constant. Each analysis was intended to investigate the conditions under which an SPS system could yield positive financial returns to a prospective investor in this technology. The following subsection will present potential scenarios that could favor the adoption of SPS systems.

9.4.4.1 Discount Rate and Corporate Tax Rate

In the first sensitivity analysis, both the discount and corporate tax rates were varied. The real discount rate was varied from 9% to negative 9% to capture both the potential downside of a higher-than-expected discount rate and the potential upside from lower discount rates. The negative real discount rates reflect the possibility that inflation could prove higher than the interest rate. In this case, a negative rate means an indirect subsidy is being employed by either the country's central bank or another potential government lender. This was the case during the South Korean boom from the early 1970s through the early 1990s (Woo 1991). Figure 9.21 presents the real interest rates in South Korea during this period. These subsidies were intended to boost investments and drive indigenous industries into an unprecedented growth cycle.

The other assumption that was varied simultaneously was the corporate tax rate, which started at the U.S. statutory rate of 35% and was varied down to 0%, indicating the maximum possible tax holiday given as an incentive to this clean energy source. It is important to note that the effective tax rates of major multinationals in the United States have been low due to enormous benefits, as exemplified by GE, which, from 2007 through 2009, had average effective tax rates of only 7% (O'Keefe 2011).

The output of this sensitivity analysis is presented in Table 9.4. It is clear from this analysis that, if governments were willing to adopt economic measures similar to those adopted by South Korea in the past to incentivize the space-based solar power industry, such a venture could become profitable, even at a full corporate tax rate of 35%. However, a compromise between low rates and tax holidays could be achieved to minimize the total effect of these industry incentives and subsidies from the government's perspective. For example, for an effective tax rate of 10% (slightly higher than the real rate for GE, for example) and a discount rate of 5%, an

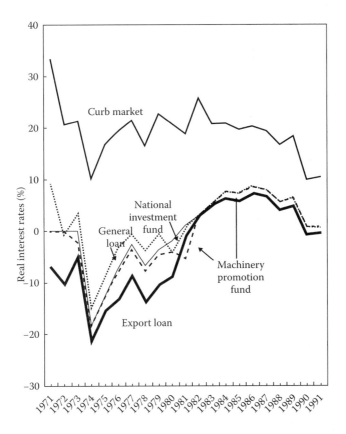

FIGURE 9.21
South Korean real interest rates, 1971 to 1991. (a) Real (i.e., inflation-adjusted) interest rates were calculated by subtracting the GNP deflator rate from a nominal loan rate. (b) The curb market rate is the informal rate at which small savers lend money to commercial and business borrowers. (Adapted from K. Fields, *Enterprise and the State in Korea and Taiwan.* Ithaca, NY: Cornell University Press, 1995; and J. Woo, *Race to the Swift: State and Finance in Korean Industrialization.* New York: Columbia University Press, 1991.)

SPS project becomes profitable if the other assumptions are held constant. These assumptions are certainly not extreme and could help make SPS systems economically and financially viable.

9.4.4.2 Retail Price of Electricity and Launch Costs to GEO per Kilogram of Mass Transported

SPS systems have the potential to sell energy on the spot market and the flexibility to deploy energy all over the world, depending solely on the

TABLE 9.4

Sensitivity Analysis Varying the Real Discount Rate and the Corporate Tax Rate

In US$ M		Discount Rate									
NPV	(21,802)	9%	7%	5%	3%	1%	−1%	−3%	−5%	−7%	−9%
Corporate Tax Rate	35%	(27,567)	(21,802)	(13,968)	(2957)	13,106	37,517	76,256	140,575	252,385	455,876
	30%	(25,283)	(19,134)	(10,786)	934	18,013	43,927	84,984	153,026	271,074	485,475
	25%	(22,999)	(16,467)	(7605)	4826	22,919	50,337	93,712	165,476	289,762	515,073
	20%	(20,715)	(13,799)	(4423)	8718	27,825	56,747	102,440	177,927	308,451	544,672
	15%	(18,431)	(11,132)	(1242)	12,609	32,732	63,157	111,168	190,378	327,139	574,271
	10%	(16,147)	(8464)	1940	16,501	37,638	69,568	119,896	202,828	345,827	603,869
	5%	(13,863)	(5797)	5121	20,392	42,544	75,978	128,624	215,279	364,516	633,468
	0%	(11,579)	(3129)	8303	24,284	47,450	82,388	137,352	227,729	383,204	663,067

installation of a rectenna. This would enable SPS electricity to be beamed where it is needed most or where its operator deems it to be most profitable. Some aforementioned countries have high electricity costs per kilowatt-hour (i.e., the potential revenues per kilowatt-hour sold in those countries), ranging from US$0.1151 (the average retail price in the United States) to US$0.6534 (the retail price in the Federal Territory of Nunavut in Canada; Maritime Electric 2011). Between these two figures, a spectrum of other possibilities were considered, including selling energy to selected states within the United States (New York, Connecticut, Massachusetts, Alaska, and Hawaii) and to countries with high demand and high prices in Europe (e.g., France, the Netherlands, Germany, Italy, and Denmark). Moreover, the use of energy for military applications is more expensive. Charania and colleagues (2010) calculated that the real costs per kilowatt-hour consumed by the U.S. military in its missions in Iraq and Afghanistan were US$0.92 and US$1.16, respectively. Revenues derived from such lucrative military applications have also been considered in the present sensitivity analysis.

The second assumption that was studied simultaneously was the launch cost to GEO. Using the premises set forth by SpaceX and assuming that NASA's agreement to use SpaceX spaceships for the next generation of space missions remains in place, it has been determined that the best current price per kilogram launched into GEO is US$4872 (SpaceX 2010). The price in the sensitivity analysis has been reduced to a small fraction of this, following expectations of studies carried out by Boeing and NASA that arrived at a target price per kilogram of US$400 (CSTSA 1997). With the other assumptions held constant, according to Table 9.2, the outcome of this sensitivity analysis is presented in Table 9.5.

9.4.4.3 Price of Electricity Growth and Longer-than-Anticipated SPS Lifespan

Two other important variables are those that answer the following hypothetical questions: "What if the retail price of electricity continues to increase?" and "What if the SPS system lasts longer than anticipated?" The former intends to address a phenomenon currently observed around the globe. For example, the average retail price per kilowatt-hour has nearly doubled in several European countries over the past 5 years (EuroStat 2012). Starting from the basic assumptions adopted for the initial financial model, retail price growth has been set at a minimum

TABLE 9.5

Sensitivity Analysis Varying Retail Price of Electricity and Transportation Cost to GEO

In US$ M		Retail Price of Electricity ($/kWh)									
		(1)	(2)	(3)	(4)	(5)	(6)	(7)	(8)	(9)	(10)
NPV	28,585	0.1151	0.1675	0.1921	0.2649	0.3147	0.3607	0.4181	0.6534	0.9200	1.1600
Launch to GEO ($/kg)	4872	(380,859)	(339,734)	(320,498)	(263,399)	(224,359)	(188,244)	(143,272)	41,230	250,275	438,463
	4000	(306,122)	(264,996)	(245,761)	(188,661)	(149,622)	(113,507)	(68,535)	115,968	325,013	513,201
	2000	(134,666)	(93,540)	(74,304)	(17,205)	21,834	57,950	102,921	287,424	496,469	684,657
	1000	(48,937)	(7812)	11,424	68,523	107,562	143,678	188,650	373,152	582,197	770,385
	800	(31,792)	9334	28,569	85,669	124,708	160,823	205,795	390,298	599,343	787,531
	600	(14,646)	26,479	45,715	102,814	141,854	177,969	222,941	407,443	616,488	804,676
	400	2499	43,625	62,861	119,960	158,999	195,115	240,086	424,589	633,634	821,822
	200	19,645	60,771	80,006	137,106	176,145	212,260	257,232	441,734	650,780	838,968

Source: (1), (3) U.S. EIA. Average Retail Price of Electricity by State, 2009. http://www.eia.doe.gov/cneaf/electricity/epa/fig7p5.html; (2), (5), (6) Europe's Energy Portal. "Europe's Energy Portal." http://www.energy.eu/; (4), (7), (8) Maritime Electric. Electricity Prices Vary All Around the World, 2011. http://www.maritimeelectric.com/understanding_electricity/el_comparing_prices.asp; and (9), (10) A.C. Charania, J. Jung, and J.R. Olds. A Rational Roadmap for Developing a First Revenue Space Solar Power Satellite. Paper presented at the *61st International Astronautical Congress, Prague, CZ*. Prague, CZ: The International Astronautical Federation, 2010. http://www.sei.aero/eng/papers/uploads/archive/IAC-10.E6.3.11.pdf.

of 2% annually (i.e., the average retail price of electricity in the United States would double in roughly 36 years) and a maximum of 18% annually (e.g., it would double in roughly 4 years) over the course of the 10 years considered in the financial model. The latter variable is used to determine the benefit if the SPS system lasted longer than its expected life of 40 years, that is, if the degradation rate of the solar cells was lower in space than on Earth (which is a very reasonable assumption, given the working conditions in space) .

Table 9.6 plots the output of this sensitivity analysis. It is obvious that the growth in the retail price of electricity plays a major role in determining the economic feasibility of the SPS concept. Although a longer system life span is desirable, it is not a major factor in determining its design because the longer it is designed to last, the heavier it will be. At the same time, the transportation costs into GEO have been shown to be, by far, the most important cost component for determining the realization of an SPS system. Also, the financial benefit of a longer system lifetime is not as relevant as other benefits—for example, lowering transportation costs or the growing worldwide retail price of electricity—because most of the cash flows from this longer life span will occur too far in the future and, for most realistic discount factors, they become practically negligible. Based on these three sensitivity analyses, the two variables with the greatest effect will be varied simultaneously in the next sensitivity analysis.

9.4.4.4 Price of Electricity Growth and Launch Costs to GEO

This sensitivity analysis considered current commercially available launch vehicles as the starting point for the launch cost variable. The maximum transportation cost to GEO considered here was for a Delta IV, an American launch vehicle dating from 2002. With a total cost of US$155 million and a payload to GEO totaling 10,843 kg, its cost per kilogram launched to GEO amounted to US$14,295 in 2012 (United Launch Alliance [ULA] 2013). Two other existing launch vehicles were also considered: SpaceX's Falcon 9 and Falcon 9-Heavy, whose costs per kilogram transported would amount to US$11,966 and US$4872, respectively (SpaceX 2010). It is important to note that the former is in the late stages of testing, whereas the latter is in its early development. This is a possible reason for such a great discrepancy between their costs. Other costs considered in the analysis are specialists' estimates of the cost reduction potential, taking into account not only

TABLE 9.6

Sensitivity Analysis Varying the Growth of Retail Price of Electricity and SPS Life Span

In US$ M				Growth of Retail Price of Electricity (% p.a.)				
	(1)	(2)	(3)	(4)	(5)	(6)	(7)	(8)
NPV 28,585	0.0%	2.0%	4.0%	6.0%	8.0%	10.0%	12.0%	18.0%
Lifetime of SPS (years) 40	−5251	16,048	44,803	84,510	140,571	221,414	340,301	1,204,951
45	−1456	22,711	56,347	104,543	175,628	283,415	451,080	1,864,531
50	1741	28,585	67,031	124,127	212,067	352,347	583,445	2,858,039
60	6775	38,395	86,094	161,920	289,150	514,114	931,326	6,647,922
70	10,512	46,179	102,484	197,872	372,007	714,010	1,431,473	15,485,142
80	13,365	52,440	116,623	232,011	460,905	961,336	2,154,376	36,384,576
90	15,601	57,543	128,865	264,391	556,183	1,267,910	3,204,802	86,391,867
100	17,393	61,757	139,507	295,079	658,239	1,648,724	4,738,853	207,219,215

Note: (1) Current assumption—prices are stable forever; (2) prices double every 36 years; (3) prices double every 18 years; (4) prices double every 12 years; (5) prices double every 9 years; (6) prices double every 7 years.

technological advancements but also the potential economies of scale of several launches taking place in 1 year. This assumption is based on the fact that SPS systems will require multiple launches for a single satellite to be transported into space.

Examining growth in the retail price of electricity involved a wide range of possibilities, from a more conservative approach of 6% annually to a more extreme price increase of 18% annually. Hence, the price per kilowatt-hour is assumed to double every 4 to 12 years, based not only on past data, but also on potential future trends of power consumption, which very likely will force energy-related commodity prices upward.

Based on the output obtained from this sensitivity analysis (Table 9.7), it is possible to conclude that, if the energy industry sees a huge increase in its costs—derived either from the increasing costs of energy-related commodities or from other factors (e.g., surcharges on polluting energy sources to cover associated socioenvironmental costs or the annulment of tax holidays and other government subsidies toward specific energy sources)— SPS systems can become a viable alternative, even at the launching cost

TABLE 9.7

Sensitivity Analysis Varying the Growth of Retail Price of Electricity and Transportation Costs to GEO

In US$ M			Growth of Retail Price of Electricity (% p.a.)					
			(1)	(2)	(3)	(4)	(5)	(6)
NPV		28,585	6.0%	8.0%	10.3%	12.0%	14.4%	18.0%
Launch to GEO ($/kg)	(a)	14,295	−1,032,775	−944,835	−778,300	−573,456	−74,441	1,701,138
	(b)	11,966	−833,098	−745,158	−578,623	−373,779	125,236	1,900,815
	(c)	4872	−224,941	−137,001	29,534	234,378	733,393	2,508,971
		3000	−64,475	23,465	189,999	394,844	893,859	2,669,437
		1500	64,117	152,057	318,591	523,436	1,022,451	2,798,029
		800	124,127	212,067	378,601	583,445	1,082,460	2,858,039
		400	158,418	246,358	412,892	617,737	1,116,752	2,892,330
		200	175,563	263,504	430,038	634,882	1,133,897	2,909,476

Source: (a) Mark Wade. Encyclopedia Astronautica, Encyclopedia Astronautica: Delta IV. http://www.astronautix.com/lvs/deltaiv.htm, November 2010; (b) SpaceX. Falcon 9 Overview. http://www.spacex.com/falcon9.php, November 2010; (c) SpaceX. Falcon Heavy Overview. http://www.spacex.com/falcon_heavy.php, November 2010.

Note: (1) Prices double every 12 years, (2) prices double every 9 years, (3) prices double every 7 years, (4) prices double every 6 years, (5) prices double every 5 years, (6) prices double every 4 years.

of US$4900/kg of space cargo utilizing SpaceX's Falcon 9-Heavy launch vehicle. I propose two potential scenarios based on these sensitivity analyses. I have classed them into two outcome categories: high probability and medium probability.

9.4.5 Potential Scenarios

Two potential scenarios will be subsequently outlined. I have also supplied the sets of assumptions upon which each scenario is based, as well as results of the financial analyses (e.g., the sets of cash flows obtained, the NPVs, and the IRRs). They are followed by a critical analysis of the assumptions used, as well as a discussion of other components that could have been improved, had a more optimistic scenario been assumed.

9.4.5.1 High Probability Outcome

The first scenario is based on the assumption that the lowest possible cost to launch an SPS system will be 20% lower than that derived from SpaceX's Falcon 9-Heavy launch vehicle, which corresponds to roughly US$4900/kg of hardware transported into GEO (SpaceX 2010). This is based on the fact that this price includes a healthy profit margin for SpaceX. Taking into account the need for multiple launch flights into GEO for an SPS system to operate in space, there might be economies of scale to be captured, which will be reflected here in the cost of launching a SPS. Compared with the base-case model, three other assumptions were changed: the initial power price (in US$ per kilowatt-hour), the power price growth (expressed as an annual percentage), and the system's life span. For the initial power price, it was assumed that the system's operator would beam its power to selected U.S. states (New York, Connecticut, Massachusetts, Alaska, and Hawaii), where the price of electricity per kilowatt-hour consumed is highest, thereby achieving the highest possible revenues from the U.S. market. The power price growth was assumed to be 4% per year in real terms. This corresponds to the assumption that prices will increase slowly, doubling within 18 years. Finally, the life span considered was 50 years instead of 40 years, that is, a 2% degradation of the solar cells per year instead of the previously assumed 2.5%. The set of assumptions is presented in (Table 9.8).

TABLE 9.8

Set of Assumptions Chosen for the High Probability Outcome

Solar Power Satellite—Assumptions			
Technical Assumptions Affecting Revenues		**Cost Assumptions**	
Grid conversion efficiency[2]	90%	Total development cost, $M[2]	8000
Rectenna efficiency[2]	90%	Total cost of SPS, $M[2]	5044
Transmission efficiency[2]	96%	Total cost of rectenna, $M[2]	7243
Power purchased from SPS	100%	Launch cost to GEO, $/kg[2]	3898
Power availability year-round	100%	LEO delivery cost, $/kg[2]	110
Power generation capacity, MW[2]	11,433	Maintenance cost, $/kWh[2]	0.02
Lifetime, years[3]	50	Maintenance cost growth, %/year	0%
Degradation/year[3]	2%	30-year capital cost, $/kWh[2]	0.0093
Technical Assumptions Affecting Costs		Price Assumptions	
Upper stage prop mass, kg[2]	31,396,520	Power price, $/kWh[1]	0.1921
Total SPS mass to be launched, kg[2]	54,331,613	Power price growth,%/year	4%

Source: (1) U.S. EIA. Electric Power Annual—Average Retail Price of Electricity, 2011. http://www.eia. doe.gov/cneaf/electricity/epa/epat7p4.html; (2) CSTSA. *Commercial Space Transportation Study.* NASA HQ, 1997. http://www.hq.nasa.gov/webaccess/CommSpaceTrans; (3) John C. Mankins. A Fresh Look at Space Solar Power: New Architectures, Concepts and Technologies. Paper presented at the *38th International Astronautical Congress.* International Astronautical Federation, 1997. http://spacefuture.com/archive/a_fresh_look_at_space_solar_power_new_ architectures_concepts_and_technologies.shtml.

In addition, it was assumed that governmental subsidies would take two distinct forms: tax holidays (in which the effective corporate tax rate would be as low as GE's, i.e., 7%) and a low discount rate (not as extreme as the negative rates in the South Korean model of the 1970s, but as low as 3% in real terms; O'Keefe 2011). The assumed corporate tax rate is realistic because it resembles the benefits provided to companies in similar industries. Given the state of the economy at the time of this writing and the U.S. federal rates that reflect this state, a 3% real discount rate is not unrealistic, as these types of projects are usually heavily supported by local governments. Nevertheless, no direct government funding was assumed, even though this could be expected, given the strategic role that power from space can play in terms of energy security, technological advancement, proprietary knowledge development, etc.

Table 9.9 presents the financial model output. Even with no direct governmental subsidies, the model yields a positive NPV, indicating that,

TABLE 9.9

Financial Model Output for the High Probability Outcome

Figures in US$ M, (Unless Otherwise Indicated)	2011	2012	2013	2014	2015	2016	2017	2018	2019	2020	2021
	0	1	2	3	4	5	6	7	8	9	10
Power Generation Capacity											
Installed capacity, MW	—	11,433	11,204	10,976	10,747	10,518	10,290	10,061	9832	9604	9375
Cumulative degradation	0.0%	0.0%	2.0%	4.0%	6.0%	8.0%	10.0%	12.0%	14.0%	16.0%	18.0%
Annual capacity, MWh	—	100,152,029	98,148,988	96,145,948	94,142,907	92,139,866	90,136,826	88,133,785	86,130,745	84,127,704	82,124,664
Power Installation Costs, $M											
a. Total development cost	(8000)										
b. Launch cost	(340,145)										
c. SPS + Rectenna cost	(12,287)										
x. Government subsidies											
Salvage value											
Depreciation		246	246	246	246	246	246	246	246	246	246
d. Depreciation tax shield		17	17	17	17	17	17	17	17	17	17
Operations											
Revenues from power		14,960	15,248	15,534	15,819	16,101	16,381	16,658	16,931	17,198	17,461

Power supply consumed		100%	100%	100%	100%	100%	100%	100%	100%	100%	100%
Price per kWh, $		0.19	0.20	0.21	0.22	0.22	0.23	0.24	0.25	0.26	0.27
Annual price increase per kWh		4.0%	4.0%	4.0%	4.0%	4.0%	4.0%	4.0%	4.0%	4.0%	4.0%
Cost of power generation		(2003)	(1963)	(1923)	(1883)	(1843)	(1803)	(1763)	(1723)	(1683)	(1642)
Maintenance cost per kWh, $		0.02	0.02	0.02	0.02	0.02	0.02	0.02	0.02	0.02	0.02
Annual cost increase per kWh		0.0%	0.0%	0.0%	0.0%	0.0%	0.0%	0.0%	0.0%	0.0%	0.0%
Capital cost 30-year amortization		(931)	(913)	(894)	(876)	(857)	(838)	(820)	(801)	(782)	(764)
30-year capital cost, $/kWh		0.0093	0.0093	0.0093	0.0093	0.0093	0.0093	0.0093	0.0093	0.0093	0.0093
e. Before tax operating profit	—	12,026	12,372	12,717	13,060	13,402	13,740	14,076	14,407	14,734	15,054
f. Tax expense	—	(842)	(866)	(890)	(914)	(938)	(962)	(985)	(1008)	(1031)	(1054)
After tax cash flows (a+b+c+d+e+f)	(360,397)	11,201	11,523	11,844	12,163	12,481	12,796	13,108	13,416	13,719	14,018
Terminal Value											285,961
Discount Factor	1.00	1.03	1.06	1.09	1.13	1.16	1.19	1.23	1.27	1.30	1.34
NPV of cash flows	(360,397)	10,875	10,862	10,839	10,807	10,766	10,716	10,658	10,591	10,515	10,430
NPV of terminal value											212,782
Total NPV	(360,397)	10,875	10,862	10,839	10,807	10,766	10,716	10,658	10,591	10,515	223,212
NPV of SPS	1950	IRR of SPS		0.03%							

although the launch cost to GEO is still a major concern (as it amounts to US$340 billion in prerevenue costs), the imminent increasing costs of electricity allied with a strategic choice of markets to which power will be supplied makes space-based solar power an interesting alternative. The IRR is approximately zero, because the initial cash outlays are extremely high, due mainly to the launch costs of the SPS system to its intended GEO. Although this scenario seems positive, the initial sum needed for the system to become operational is extremely high under the launch costs of the currently available technologies.

9.4.5.2 Medium Probability Outcome

The second scenario differs in three aspects. First, the launch cost is assumed to be US$800/kg of hardware to be transported to GEO. This represents a 100% higher cost than that estimated by Boeing and NASA in their 1997 study (CSTSA 1997). Nevertheless, it is still approximately 83% lower than the lowest cost estimated to soon be available for commercial hardware to be launched into space (SpaceX's Falcon 9-Heavy, US$4900/ kg). Second, the electricity price increase per year is assumed to be only 2% because the SPS systems are expected to lower these price increases by introducing a new method for power generation, which could lower the increase in prices of other energy sources by lowering demand for them. Finally, it is also assumed that costs associated with maintaining the SPS system are expected to grow at a 2% annual rate, similar to the average retail price in this scenario. This would reflect a natural tendency toward cost increases as the technology ages. The effective corporate tax rate was assumed to be 20%, that is, the SPS system operator still enjoys a tax break, but not as high as that expected under the previous scenario. The real discount rate chosen was 7%, also much higher than previously and the likely rate to be used for this type of project. Table 9.10 presents a full list of the assumptions used to generate the medium probability outcome.

As presented in Table 9.11, this scenario yields a much higher NPV. However, the expectation that launch costs will be lowered by as much as 83% remains unclear, as there are currently no available mature technologies capable of achieving such low cost levels. Unless a major breakthrough is brought to the market (such as those listed in the Earth-to-Orbit Transportation section), this scenario does not seem to be realistic in the short run.

TABLE 9.10

Set of Assumptions Chosen for the Medium Probability Outcome

Solar Power Satellite—Assumptions			
Technical Assumptions Affecting Revenues		**Cost Assumptions**	
Grid conversion efficiency[2]	90%	Total development cost, $M[2]	8000
Rectenna efficiency[2]	90%	Total cost of SPS, $M[2]	5044
Transmission efficiency[2]	96%	Total cost of rectenna, $M[2]	7243
Power purchased from SPS	100%	Launch cost to GEO, $/kg[2]	800
Power availability year-round	100%	LEO delivery cost, $/kg[2]	110
Power generation capacity, MW[2]	11,433	Maintenance cost, $/kWh[2]	0.02
Lifetime, years[3]	50	Maintenance cost growth, %/year	0%
Degradation/year[3]	2%	30-year capital cost, $/kWh[2]	0.0093
Technical Assumptions Affecting Costs		**Price Assumptions**	
Upper stage prop mass, kg[2]	31,396,520	Power price, $/kWh[1]	0.1921
Total SPS mass to be launched, kg[2]	54,331,613	Power price growth, %/year	2%

Source: (1) U.S. EIA. Electric Power Annual—Average Retail Price of Electricity, 2011. http://www.eia. doe.gov/cneaf/electricity/epa/epat7p4.html; (2) CSTSA. *Commercial Space Transportation Study.* NASA HQ, 1997. http://www.hq.nasa.gov/webaccess/CommSpaceTrans; (3) John C. Mankins. A Fresh Look at Space Solar Power: New Architectures, Concepts and Technologies. Paper presented at *38th International Astronautical Congress.* International Astronautical Federation, 1997. http://spacefuture.com/archive/a_fresh_look_at_space_solar_power_new_architectures_concepts_and_technologies.shtml.

9.5 POLITICAL FEASIBILITY

The previous sections have described the energy generation industry as inherently endogenous. Although some benefits of traditional energy sources are achieved locally (e.g., farmers leasing land for wind turbines, landowners receiving royalties for the exploitation of fossil fuels on their properties, etc.), on the whole, the game being played by institutions and organizations is zero-sum. The consumption of energy-related commodities—particularly in the case of fossil fuels—reduces the amount of these commodities available to others. Additionally, because most of the so-called green energy sources

TABLE 9.11

Financial Model Output for the Medium Probability Outcome

Figures in US$ M (Unless Otherwise Indicated)	2011	2012	2013	2014	2015	2016	2017	2018	2019	2020	2021
	0	1	2	3	4	5	6	7	8	9	10
Power Generation Capacity											
Installed capacity, MW	—	11,433	11,204	10,976	10,747	10,518	10,290	10,061	9832	9604	9375
Cumulative degradation	0.0%	0.0%	2.0%	4.0%	6.0%	8.0%	10.0%	12.0%	14.0%	16.0%	18.0%
Annual capacity, MWh	—	100,152,029	98,148,988	96,145,948	94,142,907	92,139,866	90,136,826	88,133,785	86,130,745	84,127,704	82,124,664
Power Installation Costs, $M											
a. Total development cost	(8000)										
b. Launch cost	(74,559)										
c. SPS + Rectenna cost	(12,287)										
x. Government subsidies											
Salvage value											
Depreciation		246	246	246	246	246	246	246	246	246	246
d. Depreciation tax shield	—	49	49	49	49	49	49	49	49	49	49

Operations

	Initial										
Revenues from power		14,960	14,954	14,942	14,924	14,898	14,866	14,826	14,779	14,724	14,661
Power supply consumed		100%	100%	100%	100%	100%	100%	100%	100%	100%	100%
Price per kWh, $		0.19	0.20	0.20	0.20	0.21	0.21	0.22	0.22	0.23	0.23
Annual price increase per kWh		2.0%	2.0%	2.0%	2.0%	2.0%	2.0%	2.0%	2.0%	2.0%	2.0%
Cost of power generation		(2003)	(2002)	(2001)	(1998)	(1995)	(1990)	(1985)	(1979)	(1971)	(1963)
Maintenance cost per kWh, $		0.02	0.02	0.02	0.02	0.02	0.02	0.02	0.02	0.02	0.02
Annual cost increase per kWh		2.0%	2.0%	2.0%	2.0%	2.0%	2.0%	2.0%	2.0%	2.0%	2.0%
Capital cost 30-year amortization		(931)	(913)	(894)	(876)	(857)	(838)	(820)	(801)	(782)	(764)
30-year capital cost, $/kWh		0.0093	0.0093	0.0093	0.0093	0.0093	0.0093	0.0093	0.0093	0.0093	0.0093
e. Before tax operating profit	—	12,026	12,039	12,047	12,050	12,047	12,037	12,021	11,999	11,970	11,934
f. Tax expense	—	(2405)	(2408)	(2409)	(2410)	(2409)	(2407)	(2404)	(2400)	(2394)	(2387)
After tax cash flows (a+b+c+d+e+f)	(94,846)	9670	9681	9687	9689	9686	9679	9666	9648	9625	9596
Terminal Value											106,627

(continued)

TABLE 9.11 (Continued)

Financial Model Output for the Medium Probability Outcome

Figures in US$ M (Unless Otherwise Indicated)	2011	2012	2013	2014	2015	2016	2017	2018	2019	2020	2021
	0	1	2	3	4	5	6	7	8	9	10
Discount Factor	1.00	1.07	1.14	1.23	1.31	1.40	1.50	1.61	1.72	1.84	1.97
NPV of cash flows	(94,846)	9037	8455	7908	7392	6906	6449	6020	5616	5236	4878
NPV of terminal value											54,204
Total NPV	(94,846)	9037	8455	7908	7392	6906	6449	6020	5616	5236	59,082
NPV of SPS (incl. TV)	27,255		IRR of SPS (incl. TV)		3.7%						

depend on the same variable—environmental conditions (either sun or wind)—they are susceptible to fluctuations caused by fossil energy sources, as they cause environmental conditions to change. Thus, the only way the energy industry can become a non–zero-sum game is for an exogenous alternative to come into play. By boasting no direct dependence on local environmental conditions, but rather by being dependent solely on a constant solar influx that strikes Earth's atmosphere around the clock, space-based solar power offers the hope of transforming energy generation.

Using existing energy sources, an illustrative conceptual Pareto frontier can be built (Figure 9.22), showing the trade-off between the investment needed for a 1% GDP increase in a given country (given the direct relationship between economic growth and energy consumption growth detailed earlier and the direct environmental effect per kilowatt-hour of electricity generated). In Figure 9.22, smaller values are preferred over larger ones because they present a lower need for investment or a lower environmental impact. Any of the solutions presented conceptually on the frontier can be described as Pareto-efficient outcomes. However, some of the energy sources presented in this way might, in fact, not be so because the direct environmental effect (or risk) plotted on the x axis is an ex ante estimate of the effect caused by these power plants and does not take into account potential catastrophic accidents that might take place. That said,

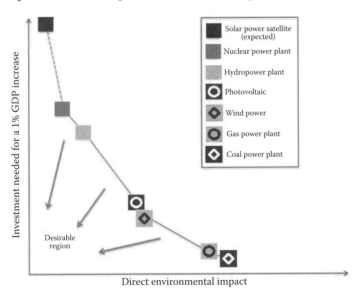

FIGURE 9.22
Conceptual Pareto frontier for currently existing energy sources.

it is important to reclassify some of the power sources as Pareto-suboptimal, to account for potential catastrophic failures of these plants, such as nuclear and hydropower. Ex post, therefore, nuclear and hydro power could be seen as not belonging on the Pareto frontier due to the risk of radioactive leaks or dam ruptures, for instance. SPS has been presented in Figure 9.22 for illustrative purposes only because, as noted, it has the potential of not being a zero-sum game player due to its inherent exogenous characteristics.

Government measures have deeply affected the shape of this Pareto frontier by enabling some power sources (through the removal of tariffs, tax holidays, or other financial subsidies) to move toward the frontier and broadening the portfolio of existing energy source options. The importance of government support, therefore, is paramount for the development of newer technologies that will facilitate a radical shift in the global energy landscape. This section presents recommendations that support the development of SPS systems and, simultaneously, move the current heavily endogenous power generation model into sustainable, exogenous, non–zero-sum game territory.

9.5.1 Political Support: Tax Benefits, Government Funding, and Public–Private Partnerships

A number of countries (most of Europe, the United States, India, and China, among others) have established a series of government support incentives to develop indigenous sources of renewable energy. The energy security issue is still regarded as the most important problem to be solved by the global energy industry, and solar, wind, nuclear, geothermal, biofuels, and other technologies are seen as promising in the movement toward energy sustainability. Worldwide tax benefits and government-sourced funding accounted for a total global investment in clean energy of US$155 billion in 2008 (Clean Edge, Inc. 2009). The Obama administration's American Recovery and Reinvestment Act of 2009 secured investments of up to US$70 billion in direct spending and tax credits for clean energy. Moreover, more than 70 countries have established some type of policy to promote renewable power generation (REN21 2010).

Although other sources of renewable energy, regardless of their endogenous influence on the environment, have been receiving enormous government support and funding, space-based solar power has received virtually nothing. More funding and support are crucial for the necessary technologies to be thoroughly developed and extensively tested, to ensure reliable applications of SPS systems in the near future. Due to a lack of funding, NASA has decided to launch the Commercial Crew and Cargo Program to stimulate efforts within the private sector to develop and demonstrate safe and cost-effective transportation capabilities (Turnbough 2010). This solution has the potential to develop cost-effective space transportation, given that this is the most important variable for the SPS concept's economic viability. SpaceX is one of NASA's partners in the development of new commercial launch vehicles, and successfully launched its Falcon 9 rocket in September 2010, when it flew unmanned into orbit and returned safely (Siceloff 2010b). This is a clear sign that the participation of private firms may be able to accelerate innovation toward new breakthroughs, driving costs down and making it more likely that SPS systems can become a reality.

Further government support is necessary to achieve this goal—in funding and tax credits for private firms, assets, or cash. The total costs involved to develop space-based solar power are extremely high. Therefore, countries have a strong incentive to cooperate in this process, given the global benefits this technology can provide, not only in terms of the huge amount of energy readily available in space but also in terms of reducing the environmental effects and other endogenous problems associated with the current model of power generation on Earth.

9.5.2 International Cooperation

As Dr. Bobby Braun, NASA chief technologist, noted, "When you are out in space, you don't see boundaries. You just see the beauties of the world" (Braun 2011). At first, space exploration was perceived as a matter of honor. Capitalism fought against socialism for space supremacy. Getting there first was a matter of political superiority and pride. The progress from the Soviets' launch of the first manned rocket into space to the Americans first setting foot on the moon represented a huge technological leap, a scientific treasure that later directly influenced the rapid development of commercial aviation, telecommunications, and the health care industry, while helping other industries indirectly. With the fall of the Berlin Wall and the consequent disintegration of the Soviet Union, space was no longer a neutral location with bipolar "owners"; the Space Race was over. The collapse of the Soviet Union gave way to cooperation on space exploration. In June 1992, American President George H.W. Bush and Russian President Boris Yeltsin signed a cooperation agreement. This resulted in the International Space

Station (ISS), primarily a combined venture by the United States and Russia, but later joined by a number of other countries (Japan, Canada, Belgium, Denmark, France, Germany, Italy, The Netherlands, Norway, Spain, Sweden, Switzerland, and the United Kingdom; Kauderer and Dunbar 2013).

Unofficial estimates indicate that more than US$100 billion has been invested in the ISS over the last two decades (Boyle 2006). Although some countries have invested more than others (and also enjoyed more crew time, electrical power rights, and rights to purchase the supporting services necessary to conduct scientific research in space, such as data upload and download), none would have been capable of carrying out these investments individually. Nor would some of the recent scientific research initiatives have been possible. This cooperation in space can be seen as a lesson for the future because it not only spreads risks and costs among different stakeholders but it also benefits the stakeholders through multicultural development.

> Fossil fuels were the energy source that shaped 19th and 20th century civilization. But burning coal, oil, and gas has proved highly damaging to our environment. Carbon dioxide emissions, greenhouse effect gases, and fumes all contribute to the disruption in the balance of our planet's climate. Global energy consumption is set to triple by the end of the century. And yet supplies of fossil fuels are depleting and the environmental consequences of their exploitation are serious. Two questions loom over humanity today: how will we supply all this new energy, and how can we do so without adding dangerously to atmospheric greenhouse gases? No single nation can face these challenges alone (ITER Organization 2011).

This statement introduces the International Thermonuclear Experimental Reactor (ITER) project on its web site. ITER is a large-scale fusion scientific project, which aims to yield 10 times the energy initially invested to start the reaction. Scientific experiments are expected to commence in 2019. Because this will be a large-scale project, no nation dared meet the challenge by itself. Rather, the European Union (EU), already a consortium of several nations per se, envisioned this project as a global collaboration effort and invited other nations to participate. The ITER location is now under construction with financial and technological support from the EU, the United States, China, India, Japan, Russia, and South Korea. Total costs are estimated to be approximately 10 billion Euros (or US$13.9 billion, as of March 13, 2011) for an estimated net output of 450 MW of energy (production of 500 MW with the simultaneous consumption of 50 MW; ITER Organization 2011).

Although the ITER project is still in the early stages of development, the greatest lesson learned to date is the rationale used in designing this joint effort: affording such an expensive, technically challenging power source through international collaboration among nations with aligned interests, that is, to discover new, reliable, and sustainable energy sources. Coincidentally (or not), both examples of some of the world's largest international cooperation projects (ISS and ITER, as previously mentioned) involve space exploration and power generation, the exact definition of what an SPS project should be based on.

9.6 CONCLUDING REMARKS

Space-based solar power has been studied only as a byproduct of the Space Race. In the 1960s and 1970s, innovation was fueled by the Cold War, when political motivation made it easier for the Soviet Union and the United States to fund their space agencies. Several technological breakthroughs originated from the Space Race of the late 1960s, such as cutting-edge materials and fabrication processes, satellites and their direct influence over telecommunications, and medical research. However, the technologies available in the 1960s did not facilitate the production or testing of SPS. The main technological breakthroughs needed for an SPS system have already been achieved in part—for example, wireless power transmission (or power beaming). Although there is definitely a need for extensive testing and technology risk mitigation, at the time of this writing, there are no major areas of concern that would prevent these satellites from becoming a reality. The major restriction facing their development is related to ETO transportation costs. Although private investors (such as Sir Richard Branson's Virgin Galactic and Elon Musk's SpaceX) have been exploring and innovating in this area—and thus lowering costs—there is currently no readily available technology for affordable transportation to space, especially for a huge satellite that is projected to weigh approximately 55,000 metric tons.

However, the desirability of a purely exogenous energy source is real, and no other power source—available or in development—is capable of claiming to be not purely endogenous. As a result, although new power sources might be less polluting and have less impact than the more primitive energy sources still in use in most countries (such as those derived from fossil fuels), the already delicate state of the environment continues to worsen as time goes by. Moreover, all of the currently existing power generation options can be characterized as players of a zero-sum-game, that is, they shift commodities and benefits from one side of the globe to the other, whereas potentially having negative environmental effects. Issues currently faced by the energy industry worldwide point to an increasing need for a non–zero-sum game to be developed, allowing mankind to benefit from a new power source, without jeopardizing the environment or individual nations.

This chapter has detailed several potential factors that could support SPS systems—for example, direct government subsidies (funding) or indirect support (tax holidays or lower interest rates on the necessary investments). Some possible scenarios have been presented, all of which illustrate the potential of the SPS concept and the potential for investors who choose to undertake projects in this promising energy field. Finally, international cooperation has been highlighted as being crucial for the success of any venture of this magnitude, given the sheer size of the investments needed and the global potential of the resulting benefits. The first Space Race, driven by two separate economies with much secrecy and no cooperation whatsoever, brought many global benefits for mankind. A second Space Race, aimed at discovering and exploring the virtually infinite source of solar energy available in space, is the only long-term feasible solution for the current global energy crisis. This solution causes no environmental harm, poses no threat to mankind on Earth's surface, and is absolutely unique in one aspect: it is truly exogenous.

REFERENCES

Abengoa Solar. Solana's Technology. Last modified 2013 (accessed April 5, 2013). http://www.aps.com/main/green/Solana/Technology.html.

Alta Devices. Alta Devices Unveils High Efficiency Gallium Arsenide Solar Cell. Last modified 2011 (accessed September 28, 2012). http://www.ecofriend.com/alta-devices-unveils-high-efficiency-gallium-arsenide-solar-cell.html.

American Physical Society. National Policy 96.2 Energy: The Forgotten Crisis. Last modified May 6, 1996 (accessed September 19, 2010). http://www.aps.org/policy/statements/96_2.cfm.

Boyle, A. NBCNews.com. What's the Cost of the Space Station? Last modified August 25, 2006 (accessed September 19, 2010). http://www.msnbc.msn.com/id/14505278/.

Boyle, A. NBCNews.com. PG&E Makes Deal for Space Solar Power. Last modified April 13, 2009 (accessed March 22, 2013). http://www.nbcnews.com/id/30198977/.

Bradsher, K., and D. Barboza. Pollution From Chinese Coal Casts a Global Shadow. *New York Times*, June 11, 2006. http://www.nytimes.com/2006/06/11/business/worldbusiness/11chinacoal.html?_r=0 (accessed October 3, 2010).

Braun, B. Keynote Speech at the 13th Annual Emerging Technology Update Day at the Wharton School of Business. Philadelphia, PA, February 25, 2011.

Bryce, R. Windmills Are Killing Our Birds. *The Wall Street Journal*, September 7, 2009. http://online.wsj.com/article/SB10001424052970203706604574376543308399048. html (accessed September 30, 2010).

Charania, A.C., J. Jung, and J.R. Olds. A Rational Roadmap for Developing a First Revenue Space Solar Power Satellite. Paper presented at the *61st International Astronautical Congress, Prague, CZ*. Prague, CZ: The International Astronautical Federation, 2010. http://www.sei.aero/eng/papers/uploads/archive/IAC-10.E6.3.11.pdf.

ChartsBin. Historical Crude Oil Prices, 1861 to Present. Last modified 2009 (accessed December 31, 2012). http://chartsbin.com/view/oau.

Clean Edge, Inc. *Clean Energy Trends 2009* by Joel Makower, Ron Pernick, and Clint Wilder. 2009. http://www.cleanedge.com/sites/default/files/Trends2009.pdf (accessed September 19, 2010).

Combs, M. The Space Settlement Art Gallery (accessed September 19, 2010). http://space. mike-combs.com/gallery.htm.

Commercial Space Transportation Study Alliance (CSTSA). Boeing Defense and Space Group, General Dynamics Space Systems Division, Lockheed Missiles and Space Company, Inc., Martin Marietta Astronautics, McDonnell Douglas Aerospace, and Rockwell Space Systems Division. *Commercial Space Transportation Study* by Commercial Space Transportation Study Alliance. NASA HQ, 1997. http://www. hq.nasa.gov/webaccess/CommSpaceTrans/ (accessed March 6, 2011).

TheDambusters.org.uk. Operation Chastise. Last modified June 9, 2007 (accessed September 19, 2010). http://www.thedambusters.org.uk/chastise_index.html.

David, L. SPACE.com. Proposed Satellite Would Beam Solar Power to Earth. Last modified April 6, 2012 (accessed March 22, 2013). http://www.space.com/15189-solar-power-beaming-satellite.html.

DESERTEC Foundation. DESERTEC Homepage (accessed September 19, 2010). http:// www.desertec.org/.

EuroStat. Energy Price Statistics. Last modified December 2012 (accessed March 22, 2013). http://epp.eurostat.ec.europa.eu/statistics_explained/index.php/Energy_price_statistics.

Farmer, J.T. *Solar Dynamic Power System Development for Space Station Freedom*. National Aeronautics and Space Administration, 1993.

Fields, K. *Enterprise and the State in Korea and Taiwan*. Ithaca, NY: Cornell University Press, 1995.

Fraunhofer-Institut. World Record: 41.1% Efficiency Reached for Multi-Junction Solar Cells at Fraunhofer ISE. Last modified January 14, 2009 (accessed October 4, 2010). http://www.ise.fraunhofer.de/press-and-media/press-releases/press-releases-2009/ world-record-41.1-efficiency-reached-for-multi-junction-solar-cells-at-fraunhofer-ise.

Glaser, P.E. Power from the sun: Its future. *Science*, November 1968, 857–861.

Glaser, P.E. Method and Apparatus for Converting Solar Radiation to Electrical Power. U.S. Patent 3,781,647, issued December 25, 1973.

Glaser, P.E., O.E. Maynard, J. Mackovciak, and E.L. Ralph. Arthur D. Little, Inc. Feasibility Study of a Satellite Solar Power Station. NASA CR-2357, NTIS N74-17784, February 1974.

Global Wind Energy Council (GWEC), and Greenpeace. *Global Wind Energy Outlook 2012* by GWEC, Greenpeace, DLR, Ecofys, and The University of Utrecht. Brussels, Belgium, 2012. http://www.gwec.net/wp-content/uploads/2012/11/GWEO_2012_ lowRes.pdf (accessed May 14, 2013).

Gordes, J.N., and M. Mylrea. A new security paradigm is needed to protect critical U.S. energy infrastructure from cyberwarfare. *Foreign Policy Journal*, September 14, 2009. http://www.foreignpolicyjournal.com/2009/09/14/a-new-security-paradigm-is-needed-to-protect-critical-us-energy-infrastructure-from-cyberwarfare/ (accessed September 19, 2010).

Guinness World Records (GWR). Fastest Maglev Train. Last modified 2011 (accessed March 7, 2011). http://www.guinnessworldrecords.com/records-1000/fastest-maglev-train/.

Harty, R.B., W.D. Otting, and C.T. Kudija. Applications of Brayton cycle technology to space power. *IEEE AES Systems Magazine*, January 1994. http://ieeexplore.ieee.org/stamp/stamp.jsp?tp=&arnumber=257140&userType=inst (accessed October 4, 2010).

InetDaemon Enterprises. Satellite Orbits. (accessed October 4, 2010). http://www.inetdaemon.com/tutorials/satellite/satellite-orbits.shtml.

Ingalls, B. NASA. Atlantis Reflection. Last modified July 9, 2011 (accessed April 5, 2013). http://apod.nasa.gov/apod/ap110709.html.

International Rivers (IR). The World Commission on Dams Framework—A Brief Introduction. Last modified February 29, 2008 (accessed October 4, 2010). http://www.internationalrivers.org/resources/the-world-commission-on-dams-framework-a-brief-introduction-2654.

ITER Organization. The ITER Story (accessed March 7, 2011). http://www.iter.org/proj/iterhistory.

Kauderer, A., and B. Dunbar. International Cooperation. Last modified February 26, 2013 (accessed May 19, 2013). http://www.nasa.gov/mission_pages/station/cooperation/index.html.

Landis, G.A. My Involvement with Laser Power Beaming. Last modified March 6, 2011 (accessed May 12, 2011). http://www.geoffreylandis.com/laser.htp.

Lawrence Livermore National Laboratory (LLNL). Concept Demonstration of Laser Power Beaming (accessed September 19, 2010). https://lasers.llnl.gov/multimedia/publications/images/solar.jpg.

Lior, N. Power from space. *Energy Conversion and Management* 42: 1769–1805, 2001.

Mah, O. Department of Energy, and National Solar Power Research Institute, Inc. Fundamentals of Photovoltaic Materials. Last modified 1998 (accessed September 19, 2010). http://userwww.sfsu.edu/ciotola/solar/pv.pdf.

Mankins, J.C. A Fresh Look at Space Solar Power: New Architectures, Concepts and Technologies. Paper presented at the *38th International Astronautical Congress, Brighton, U.K., October 10–17, 1997*. International Astronautical Federation. http://spacefuture.com/archive/a_fresh_look_at_space_solar_power_new_architectures_concepts_and_technologies.shtml (accessed September 19, 2010).

Mankins, J.C. Technical overview of the "suntower" solar power satellite concept. *Acta Astronautica* 50, 369–377, 2002.

Maritime Electric. Electricity Prices Vary All Around the World. Last modified March 8, 2011 (accessed September 19, 2010). http://www.maritimeelectric.com/understanding_electricity/el_comparing_prices.asp.

Mason, L.S. A Solar Dynamic Power Option for Space Solar Power. Paper presented at the *34th Intersociety Energy Conversion Engineering Conference Sponsored by the Society of Automotive Engineers, Vancouver, British Columbia, August, 1999*. Hanover, MD: NASA Center for Aerospace Information, 1999. http://ntrs.nasa.gov/archive/nasa/casi.ntrs.nasa.gov/19990062655_1999094443.pdf (accessed September 19, 2010).

Mitsubishi Electric. Space Systems: Solarbird (accessed October 4, 2010). http://www.mitsubishielectric.com/bu/space/rd/solarbird/index.html.

National Aeronautics and Space Administration (NASA). NASA Looks for New Ways to Harness Sun's Energy for Earth and Space. Last modified 1999 (accessed April 5, 2013). http://www.msfc.nasa.gov/news/news/photos/1999/photos99-096.htm.

National Space Society (NSS). Space-Based Solar Power as an Opportunity for Strategic Security. Last modified August 16, 2008 (accessed March 6, 2011). http://www.nss.org/settlement/ssp/library/nsso.htm.

National Space Society (NSS). Satellite Power System Concept Development and Evaluation Program. Last modified September 15, 2010 (accessed October 4, 2010). http://www.nss.org/settlement/ssp/library/doe.htm.

National Space Society (NSS). Space Solar Power: Limitless Clean Energy from Space. Last modified March 4, 2013 (accessed March 22, 2013). http://www.nss.org/settlement/ssp/.

Oda, M. Japan Aerospace Exploration Agency (JAXA). Solar Power Satellite—OOS Workshop 2004. Last modified October 2, 2004 (accessed March 22, 2013). http://www.on-orbit-servicing.com/pdf/OOS2004_presentations_pdf/OOSIssuesOverview_Oda.pdf.

O'Keefe, W. GE Tax Rate Shows Hypocrisy of Obama's Energy Tax Plan. Last modified February 14, 2011 (accessed March 6, 2011). http://fuelfix.com/blog/2011/02/14/ge-tax-rate-shows-hypocrisy-of-obamas-energy-tax-plan/.

Pappas, S. Japan's Disaster May Cool U.S. Acceptance of Nuclear Power. Last modified March 15, 2011 (accessed May 12, 2011). http://www.livescience.com/13240-japan-disaster-sway-perception-nuclear-power.html.

Powersat Corporation. Energy Market Drivers Behind Space Solar Power (SSP). Last modified 2012 (accessed November 6, 2012). http://www.powersat.com/.

Price, S. Audacious & Outrageous: Space Elevators. Last modified September 7, 2000 (accessed September 19, 2010). http://science.nasa.gov/science-news/science-at-nasa/2000/ast07sep_1/.

Prohov, T., A. Globus, and J. Sercel. Space Studies Institute. Solar Power Satellite—Rectenna (accessed September 19, 2010). http://ssi.org/space-art/ssi-sample-slides/.

Public Utilities Commission of the State of California (PUCSC). Energy Division Resolution E-4286. Last modified December 3, 2009 (accessed September 19, 2010). http://docs.cpuc.ca.gov/PUBLISHED/FINAL_RESOLUTION/110808.htm.

Renewable Energy Policy Network for the 21st Century (REN21). *Renewables 2010 Global Status Report* by Janet L. Sawin, Eric Martinot, Virginia Sonntag-O'Brien, Angus McCrone, Jodie Roussell, Douglas Barnes, and Christopher Flavin. 2010. http://www.ren21.net/Portals/97/documents/GSR/REN21_GSR_2010_full_revised Sept2010.pdf (accessed October 10, 2010).

Renewable Energy Sources (RES). PV Panels Efficiency Degradation Factors. Last modified July 30, 2009 (accessed October 4, 2010). http://www.renewable-energy-sources.com/2009/07/30/pv-panels-efficiency-degradation-factors/.

Rich Solar. Products: Monocrystalline Modules. Last modified 2012 (accessed January 1, 2013). http://richsolar.com/monocrystalline_modules.php.

Sato, M. Goddard Institute for Space Studies—Surface Temperature Analysis. Last modified September, 2010 (accessed October 10, 2010). http://data.giss.nasa.gov/gistemp/graphs_v3/.

Sharing of Free Intellectual Assets (SOFIA). Earth–Sun Relationships and Incoming Solar Energy (accessed September 19, 2010). http://sofia.fhda.edu/gallery/geography/images/seasons_insolation.gif.

Shinohara, N. Wireless Power Transmission for Solar Power Satellite (SPS). Last modified 2008 (accessed September 19, 2010). http://www.sspi.gatech.edu/wptshinohara.pdf.

Siceloff, S. NASA. Emerging Technologies May Fuel Revolutionary Launcher. Last modified September 10, 2010a (accessed April 6, 2013). http://www.nasa-usa.de/topics/technology/features/horizontallaunch.html.

Siceloff, S. NASA. SpaceX Launches Success with Falcon 9/Dragon Flight. Last modified December 9, 2010b (accessed September 19, 2010). http://www.nasa.gov/offices/c3po/home/spacexfeature.html.

SPACE.com Staff. SpaceX 'Grasshopper' Reusable Rocket Prototype Makes Two-Story Test Flight. Last modified November 16, 2012 (accessed December 19, 2012). http://www.huffingtonpost.com/2012/11/16/spacex-grasshopper-test-flight_n_2145428.html.

Space Exploration Technologies Corp (SpaceX). Falcon Heavy Overview (accessed September 19, 2010). http://www.spacex.com/falcon_heavy.php.

Spadaro, J., L. Langlois, and B. Hamilton. Greenhouse Gas Emissions of Electricity Generation Chains Assessing the Difference. Last modified February, 2000 (accessed September 19, 2010). http://www.energienucleaire.ch/upload/cms/user/IAEA_GreenhouseGasEmissions.pdf.

Spice, B. Post–Gazette.com. NASA Eyes Solar Energy Collectors in Space by 2015. Last modified July 12, 1999 (accessed March 22, 2013). http://old.post-gazette.com/healthscience/19990712solar1.asp.

Terheggen, M. Massachusetts Institute of Technology (MIT). Cadmium Telluride Photovoltaic Film 2011. (accessed March 6, 2011). http://ocw.mit.edu/courses/mechanical-engineering/2-626-fundamentals-of-photovoltaics-fall-2008/chp_CdTe.jpg.

Turnbough, L. Commercial Crew & Cargo. Last modified October 8, 2010 (accessed October 10, 2010). http://www.nasa.gov/offices/c3po/home/.

United Launch Alliance, LLC (ULA). Delta IV: The 21st Century Launch Solution. Last modified 2013 (accessed March 22, 2013). http://www.ulalaunch.com/site/pages/Products_DeltaIV.shtml.

U.S. Department of Energy, Energy Efficiency and Renewable Energy (EERE). Amorphous Silicon Photovoltaic Cell Composition (accessed October 10, 2010). http://www.eere.energy.gov/basics/renewable_energy/images/illust_nip_structure.gif.

U.S. Energy Information Administration (U.S. EIA). Average Retail Price of Electricity by State, 2009. Last modified 2009 (accessed October 10, 2010). http://www.eia.doe.gov/cneaf/electricity/epa/fig7p5.html.

U.S. Energy Information Administration (U.S. EIA). International Total Primary Energy Consumption and Energy Intensity. Last modified 2010a (accessed October 10, 2010). http://www.eia.doe.gov/emeu/international/energyconsumption.html.

U.S. Energy Information Administration (U.S. EIA). Petroleum and Other Liquids: U.S. Imports by Country of Origin. Last modified 2010b (accessed September 19, 2010). http://www.eia.gov/dnav/pet/PET_MOVE_IMPCUS_A2_NUS_EPC0_IM0_MBBLPD_A.htm.

U.S. Energy Information Administration (U.S. EIA). Short-term Energy Outlook: Real Prices Viewer. Last modified 2010c (accessed September 19, 2010). http://www.eia.gov/forecasts/steo/realprices/.

U.S. Energy Information Administration (U.S. EIA). Electric Power Annual—Average Retail Price of Electricity. Last modified 2011 (accessed March 7, 2011). http://www.eia.doe.gov/cneaf/electricity/epa/epat7p4.html.

Wade, M. Encyclopedia Astronautica. Encyclopedia Astronautica: Delta IV. http://www.astronautix.com/lvs/deltaiv.htm.

Wilbanks, T., J. Lankao, P. Romero, M. Bao, F. Berkhout, S. Cairncross, J.P. Ceron, M. Kapshe, and R. Muir-Wood. Industry, settlement and society. In *Climate Change 2007: Impacts, Adaptation and Vulnerability. Contribution of Working Group II to the*

Fourth Assessment Report of the Intergovernmental Panel on Climate Change, edited by M. Parry, O. Canziani, J. Palutikof, P. van der Linden and C. Hanson, 357–390. Cambridge, U.K.: Cambridge University Press, 2007. http://www.ipcc.ch/pdf/assessment-report/ar4/wg2/ar4-wg2-chapter7.pdf (accessed September 19, 2010).

Williams, J. Oil Price History and Analysis. Last modified 2011 (accessed March 13, 2011). http://www.wtrg.com/prices.htm.

Woo, J. *Race to the Swift: State and Finance in Korean Industrialization*. New York: Columbia University Press, 1991.

The World Bank Group (WBG). GDP, PPP (Constant 2005 International $). Accessed September 19, 2010. http://data.worldbank.org/indicator/NY.GDP.MKTP.PP.KD.

World Hunger Organization (WHO). 2010 World Hunger and Poverty Facts and Statistics. Last modified 2010 (accessed October 4, 2010). http://www.worldhunger.org/articles/Learn/world hunger facts 2002.htm.

World Nuclear Association (WNA). Greenhouse Gas Emissions from Power Generation. Last modified 2011 (accessed November 5, 2012). http://www.world-nuclear.org/why/greenhouse_gas_from_generation.html.

World Nuclear News (WNN). Gradual Growth in U.S. Support for Nuclear. Last modified September 24, 2012a (accessed November 5, 2012). http://www.world-nuclear-news.org/NP-Gradual_growth_in_US_support_for_nuclear-2409124.html.

World Nuclear News (WNN). U.K. Nuclear Support Bounces Back. Last modified July 3, 2012b (accessed November 5, 2012). http://www.world-nuclear-news.org/NP-UK_nuclear_support_bounces_back-0307127.html.

Yale Project on Climate Change Communication (YPCCC), and George Mason University Center for Climate Change Communication (GMUCCCC). *Public Support for Climate and Energy Policies in March 2012* by Anthony Leiserowitz, Edward Maibach, Connie Roser-Renouf, and Jay Hmielowski. New Haven, CT: YPCCC, 2012. http://environment.yale.edu/climate/files/Policy-Support-March-2012.pdf (accessed March 22, 2013).

Zhang, D., P. Brecke, H. Lee, Y. He, and J. Zhang. Global climate change, war, and population decline in recent human history. *Proceedings of the National Academy of Science* 104: 19214–19219, 2007. doi: 10.1073/pnas.0709955104.

10

Strategic Considerations for
Electric Vehicle Adoption

Luca Ratto and Natalie Volpe

CONTENTS

10.1 INTRODUCTION

The market for electric vehicles (EVs) has been gaining considerable momentum recently as a "clean" energy transportation alternative. Car manufacturers have presented battery-powered vehicles at the latest automotive industry shows, and policymakers worldwide have enacted incentives to foster the commercialization of EVs. According to a Boston Consulting Group study, the Chinese, French, and American governments, among others, have pledged up to US$15 billion over the next 5 years to promote the development and production of electric cars via tax incentives, subsidies, levies, and other stimulants (e.g., consumer bonuses; Moffet 2009).

Sources generally agree that the sector really gained traction more than 10 years ago with the launch of the hybrid vehicle, which combines the power of a traditional combustion engine with that of an electric engine (Universidad Carlos III de Madrid 2010). Since then, attention has shifted gradually toward fully electric vehicles as part of a worldwide effort to decarbonize the transportation sector. Experts argue that it is difficult to foresee a truly sustainable transport system without a shift from oil toward more sustainable energy sources. Supporting this is the fact that in the European Union (EU), transportation consumes two-thirds of all oil, leading to 28% of all CO_2 emissions (European Federation for Transport and Environment 2009, p. 4). EVs are expected to play an important role in the transition toward a more sustainable transport system, creating substantial reductions in CO_2 emissions while improving local air quality. If these vehicles run on renewable electricity—such as wind, solar, or hydro energy—CO_2 emissions can, in fact, be effectively eliminated. However, the effects—both positive and negative—that large numbers of EVs will have on electricity grids represent another key component in predicting the long-term success of this technology (Kampman 2010, p. 4).

This chapter outlines the main issues and challenges associated with the rise of EVs. Many skeptics have concerns about the functionality of these vehicles, as several structural issues have not yet been clearly defined and solved, including limitations of battery life, grid capacity, scarcity of charging sites, electromagnetic radiation, and safety. This chapter broadly addresses some of the issues through both technological and institutional lenses and examines their economic implications. We focus more specifically on four areas: battery limitations and their effect on driving range, EVs' energy and environmental efficiencies, the charging infrastructure challenge, and the business model selection problem that arises when deploying EV solutions.

The first key issue relates to battery technology and the need for infrastructure standardization. Current batteries cannot provide enough energy for long-distance trips. In addition, a certain degree of standardization among car manufacturers is necessary for battery-charged vehicles to become universal. Battery-switching stations provide a potential solution, as they provide an alternative to charging stations. To gain further insights about this matter, we will analyze the Johnson Controls case study, in which a leading manufacturer of hybrid and electric battery systems received a grant from the U.S. government to build domestic manufacturing capacity for advanced EV batteries.

Second, an increase in the number of EVs will lead to higher levels of electricity demand. The grid's capacity (or lack thereof) to support this increased demand is a major concern. Smart grids, which are expected solve the energy-sourcing problem, will recharge cars during off-peak periods, when demand is low and energy costs are cheaper. Furthermore, smart girds will also permit users to send power back into the grid when demand is high and energy more costly. This chapter includes an overview of the grid problem and the U.S. Department of Energy's (U.S. DOE) initiative to create a reliable national clean electric power system. We will also assess the feasibility that this new system will be able resolve EV-related and general power grid matters.

Third, the charging infrastructure is presently inadequate to support a fleet of EVs. These vehicles must be able to travel long distances without needing to *return home* to refuel or recharge. They are currently practical only for short-distance driving because of the scarcity of charging stations. Local governments across the globe have created incentivizing policies to improve charging technologies and to launch infrastructure partnerships, to increase the number of EVs in the coming decades. To illustrate one such localized effort, this chapter examines the Greater London Authority's attempt to alter London's infrastructure to better accommodate EVs.

Finally, it is not yet clear how best to market EV solutions to customers; in fact, a variety of business models may ultimately prove to be successful. The integrated model of the company Better Place did, however, stand out in many respects. Better Place, although founded in California, created a system of battery-switching stations in Israel and was expanding to other global markets to provide EV services in an integrated fashion (i.e., offering customers a combined package encompassing both an EV and access to an extensive network of convenient switching stations), comparable to the way telecom providers market their phone and data services. Better Place had made battery swapping possible using a standardized system and was characterized by significant technological and organizational innovations (Better Place 2012). However, diseconomies of scale in its two active markets of Israel and Denmark led to its filing for liquidation in an Israeli Court on May 26, 2013. Even so, Better Place's executives, along with industry commentators, have largely laid the blame for the company's failure on factors *other* than the company's unique strategy of separating EV batteries from the actual cars (Pearson and Stub 2013; Rabinovitch 2013). We will therefore discuss Better Place's history and business model to highlight the important lessons the company's short life continues to impart.

10.2 OVERVIEW

This section provides an overview of the EV space. We begin by presenting the historical context surrounding the evolution of EVs and then introduce the concept of sustainable mobility. This latter aspect has been embraced by twenty-first century policymakers, some of whom are increasingly focused on postponing and avoiding the depletion of the planet's resources. Finally, this section examines the situation faced by the EV market as of late 2010. This market has gained momentum worldwide, leading car makers to rush their prototypes to market in light of the optimistic forecasts of EV adoption rates. This momentum is supported globally by governments that are drafting legislation to provide incentives for the EV industry.

10.2.1 History

The first EV was reportedly invented around 1832, pre-dating the 1885 introduction in Germany of the first affordable internal combustion–powered car, Karl Benz's MotorWagen (European Federation for Transport and Environment 2009, pp. 8–10). As a result of design improvements in 1881, EV development continued into the late nineteenth century. France and Great Britain were the first countries to see the widespread use of electric cars, followed by the United States' first large-scale commercial application, which encompassed the entire New York City taxi fleet in 1897. Electric cars outsold gasoline- and steam-powered cars at the turn of the century. The first battery-powered vehicles were heavier (due to their batteries' weight) than their internal combustion engine counterparts, but featured benefits such as quieter motors, no gear changing, and shorter start-up times (European Federation for Transport and Environment 2009).

In 1912, EV production was reported to have peaked. By 1918, there were approximately 50,000 EVs in the United States (J.D. Power and Associates 2010, pp. 4–6). These vehicles, however, enjoyed success only until around 1920. By this time, a series of technological and logistical developments set in motion in the early 1900s had established internal combustion vehicles' dominance. Previously, some of the main obstacles that precluded the adoption of combustion engine cars in the United States were the lack of a stable oil supply and the burdensome hand crank essential to the car-starting process. Improvements in the road network, which made longer-range cars more attractive; the discovery of vast reserves of crude oil in Texas, which helped make gasoline more affordable; and Henry Ford's mass production of combustion engine vehicles (beginning in 1908), which made these cars much less expensive, were responsible for the decline of EVs (European Federation for Transport and Environment 2009, p. 8). Then, in 1911, Charles Kettering invented the electric engine starter, which eliminated the need for the hand crank. Due to these developments in combustion engine vehicles and the continued increase in the cost of battery-powered cars, EVs disappeared (Massachusetts Institute of Technology [MIT] 2000).

Gasoline rationing for citizens during World War II inspired a short-lived revival of the EV industry. This receded quickly, however, when the abundant supply of gasoline after the war interrupted new developments in battery-powered vehicles, ushering in a trend that would last for the next 30 years. In the late 1960s and early 1970s, interest in EVs among manufacturers and consumers was renewed, due primarily to the 1973 increase in oil prices and concerns over air pollution. The fuel shortages experienced by developed countries during the oil shocks of the 1970s led to increased public interest in seeking alternative fuels. In the United States, regulations were passed to encourage the development of new technologies, including EVs. Japan invested in technical research programs, and France attempted to create a market for the commercial use of EVs through its network of state-owned firms. However, most of the initiatives launched at that time relied on the assumption that battery technology would improve rapidly. This assumption proved misguided, leaving EVs at a competitive disadvantage vis-à-vis other types of vehicles (European Federation for Environment and Transport 2009, pp. 8–9).

Historically, environmental concerns have joined energy security concerns in spurring interest in EVs. Beginning in force in the 1970s, heightened environmental awareness in political and intellectual circles of the threats to collective well-being from emissions and resource scarcity led citizens to urge their respective governments to devote attention to developing alternatives to oil-based fuel. In the 1990s, California attempted to force car manufacturers to increase the supply of EVs. However, as regulatory pressure diminished, interest in and production of these vehicles stopped. After a resurgence following the Iraqi invasion, oil prices entered a period of steady decline, despite rising consumption levels driven by the fast-growing Asian economies. Several European countries—including France, Germany, Italy, and Sweden—invested in research and development of EVs through government- or industry-sponsored programs. However, the price of EVs remained prohibitively high, even after accounting for tax incentives. Over the past 3 years, rising oil prices and climate change concerns have ignited a series of energy efficiency and CO_2 emissions regulations and a renewed interest from the automobile industry. These new regulations, in particular in the EU, have increased the likelihood that car manufacturers' investments in low-emission technologies will benefit from a future payoff (Boston Consulting Group [BCG] 2009, p. 6).

10.2.2 Sustainable Mobility

The concept of sustainability encompasses a human desire to create a better world and guides decision-making that will result in a way of life in which humans do not deplete resources. A recent example of these efforts is the push for EVs to eventually replace petroleum- and fossil fuel–based vehicles. Sustainable mobility must be environmentally, socially, and economically advantageous. According to the European Council of Ministers of Transport, a sustainable transport system must allow basic access and developmental needs of individuals, companies, and societies to be met while promoting equity with successive generations. It must be affordable, operate fairly and efficiently, and support development and a competitive economy. An effort to develop EVs and other alternative-fueled vehicles will be key for creating global sustainable mobility in the future (European Conference of Ministers of Transport 2004). Eventually, these vehicles will be capable of leading the effort in sustainable mobility, but they currently face infrastructure issues and high initial production costs, among other drawbacks. Electrification has the potential to generate the greatest reduction in CO_2 when compared with advanced internal combustion engines and alternatively fueled vehicles (Book et al. 2009, p. 2). EVs are more energy efficient, and they promote other sources of green energy, including renewable energy (Kampman 2010, p. 56).

New research on the harmful effects of climate change, especially global warming, has renewed our interest in developing sustainable mobility. Increased concerns about mankind's negative effect on the planet have encouraged policy-makers to combat the potentially irreversible damage caused by internal combustion engines. A recent study has shown that as oil prices increase, consumer interest in EVs increases, and consumer interest follows a similar pattern when oil prices decline, as depicted in Figure 10.1 (European Federation for Transport and Environment 2009,

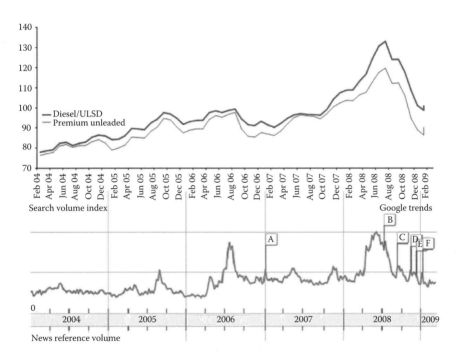

FIGURE 10.1

The top graph shows the development of fuel prices over time, whereas the bottom graph shows public interest in EVs, as determined by the frequency of various search terms in Google. As can be seen, public interest is highly correlated with movements in fuel prices. (Reprinted with permission from European Federation for Transport and Environment. *How To Avoid an Electric Shock—Electric Cars: From Hype to Reality.* Brussels, Belgium: European Federation for Transport and Environment, 2009.)

pp. 10–11). This inconsistent interest in the development of EVs prevented their widespread adoption in previous years, as oil prices remained stagnant and led skeptics to question EVs' feasibility. Increasing scientific knowledge and pressure from environmental groups are factors that have recently contributed to more consistent pushes for sustainable transport systems.

A transport system based on nonrenewable resources and harmful emissions is not sustainable for the future. The biggest issue with commercializing a sustainable transport system is making the use of clean energy as inexpensive, readily available, and socially acceptable as petroleum. A number of countries have been working to create systems of sustainable transport for the masses. Further improvements in technology, increased government support, and enhanced public awareness will be necessary to supplant our petroleum-based transport economy. The transition from internal combustion vehicles to EVs may take decades or possibly longer. This is especially true for less-developed countries that cannot absorb the costs of advanced technologies (World Energy Council 2007, p. 3). Uncertainty over the availability of future oil reserves is a key factor in the sustainable mobility movement. Some scientists claim we have already passed the maximum rate of global petroleum extraction, and that the rate of production is now declining (Energy Watch Group 2007). If these scientists are correct in their observations, alternative energy and fully electricity-based vehicle technology are now more important than ever. It is becoming too risky for car manufacturers to continue to utilize a nonrenewable resource because of the uncertain quantity of reserves.

As we wait to develop a zero–carbon emissions vehicle, internal combustion engines continue to inflict environmental damage. Atmospheric concentrations of CO_2 have increased by 100 parts per million above preindustrial levels and are expected to increase another 100 parts per million as the developing world becomes more industrialized (Intergovernmental Panel on Climate Change [IPCC] 2007). Increasingly high atmospheric concentrations of CO_2 will exacerbate the greenhouse effect, creating a possible 2.5°C to 6.0°C increase in temperature over the next century. Although the transportation sector is one of the largest contributors to the increase in CO_2, it is also one of the most innovative sectors in attempting to offset this environmental effect. By utilizing subsidies, incentives, regulations, and government loans, policymakers and automobile manufacturers are moving to facilitate a smooth transition toward sustainable mobility.

10.2.3 Current Situation

The first EVs produced for mass-market consumption in the current era reached car showrooms in the United States, Europe, and Japan in 2010 and 2011, as shown in Table 10.1 (*The Economist* 2010). Purely electric cars, such as the Nissan Leaf, can be used continuously for 150 km before needing to be recharged, a process that, under normal circumstances, takes 6 to 8 hours (in 2011). Vehicles commonly known as range extenders, such as GM's Volt (known as the Ampera in Europe), recharge their batteries through a gasoline-powered onboard combustion engine once the electric charge has been depleted. Figure 10.2 depicts a Volt driving on the highway. The third variant of EV is the hybrid vehicle, an example being Toyota's standard Prius, which can operate on gasoline, electricity, or on a simultaneous mixture of the two. Because of its relatively less powerful batteries, the hybrid is characterized by a lower electric-only range.

TABLE 10.1

Current Product Offerings in the EV Market

Electric Car	Release Date	Technology	Electric Only Range (km)	Market Price (£)
Mitsubishi i-MiEV	April 2010	Battery	160	23,990
Nissan Leaf	December 2010	Battery	160	23,990
GM Volt/Ampera	December 2010	Battery/range extender	64	25,000
Renault Fluence	Mid-2011	Battery	160	18,000
Toyota Prius Plug-in	2012	Hybrid	20	TBA

Source: A sparky new motor. *The Economist*, 2010. http://www.economist.com/node/17202405.

FIGURE 10.2
The GM Volt in action. (Reprinted with permission from Jeff Cobb, Editor-in-Chief at GM-Volt.com and HybridCars.com.)

The latest Prius plug-in Hybrid has swapped its nickel-metal batteries for a lithium-ion battery pack, which allows the vehicle to travel 13 miles further on an electric charge alone before the gasoline engine kicks in (*The Economist* 2010). These EVs are charged via a relatively conventional power cord, as can be seen in Figure 10.3. Governments seem to strongly support manufacturers of EVs, as these vehicles are considered essential for reducing carbon dioxide emissions and pollution. In fact, GM's Volt was a major factor that influenced the U.S. government's decision to support the company when it entered bankruptcy.

Several companies have decided to invest heavily in the development of EVs. Renault and its Japanese ally, Nissan, are jointly investing four billion Euros in an EV project and expect sales in this market segment to account for 10% of total car sales by 2020. PSA Peugeot Citroen will commercialize its partner Mitsubishi's iMiEV in Europe with a more conservative estimate in mind, as the company forecasts that only 5% of its total car sales will come from EVs in 2020.

Sector experts do not yet concur on the economic potential of the fully EV market. The variety and interconnection of the political, economic, and social interests in this sector make market forecasting very challenging. Car-sector analysts believe that the greatest obstacle for electric cars is drivers' skepticism. This stems from their fears of having to wait to recharge or to switch their batteries constantly (Massey-Beresford 2010). The Automotive Group of IHS Global Insight, a leading market intelligence company, published an optimistic study in which EVs were expected to account for 20% of the global market for light vehicles by 2020. The study claims that sufficient base-load electric power generation capacity exists in Europe and the United States to support vast fleets of fully electric vehicles (IHS Global Insight 2010). A less optimistic study, conducted by the global

FIGURE 10.3

Example of a power cord for a plug-in EV. (Reprinted under U.S. DOE's media resources usage policy: U.S. DOE EERE. Analysis Shows Significant Advances in Electric Vehicle Deployment, 2011. http://apps1. eere.energy.gov/news/daily.cfm/ hp_news_id = 290.)

marketing information company J.D. Power and Associates (2010), stated that future demand for battery-powered vehicles might be exaggerated. The research firm expects battery-powered vehicle sales to account for only 2% of global passenger vehicle sales, as seen in Figure 10.4. Its report justified its assertions by emphasizing consumer concerns about the reliability of new EV technologies and general dissatisfaction with the overall power and performance of electric cars. Another market study, performed by management consulting firm Boston Consulting Group, expects that 18% of city cars globally will be fully electric by 2020. Overall, this study forecasted that sales of electric cars will comprise 5% to 10% of new car sales in 2020 (BCG 2009, pp. 6–7).

In April 2010, the European Commission approved a new strategy aimed at promoting clean and energy efficient vehicles, focusing primarily on the role of electric cars. As part of a package to support the automotive industry, the EU devoted five billion Euros to the Green Car Initiative. This plan includes incentives to develop clean technologies for cars. According to the European Commission, "the aim of the European Green Car initiative is to protect jobs in the car sector and to ensure its long-term viability by encouraging a sustainable reform that embraces new environmentally friendly technology" (EurActiv 2008). However, a study prepared for the Danish Petroleum Industry Association showed that, over the next 15 years, as long as Europe's electricity supply remains based on fossil fuels, switching transportation from diesel to electric cars will not create an appreciable improvement in the European carbon footprint (ibid.). Thus, it is clear that simply switching over to fully electric vehicles is not

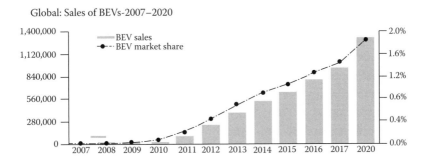

FIGURE 10.4

Global BEV sales. (Reprinted with permission from J.D. Power and Associates. *Drive Green 2020: More Hope than Reality?* Columbus, Ohio: McGraw Hill Companies, November 2010.)

enough to reduce carbon emissions. Rather, a more comprehensive effort, involving shifting countries' energy generation mixes over to renewables, must be undertaken.

The current momentum behind EVs is supported by several incentives provided by national governments to individual consumers and to businesses. Within the EU, the governments of Germany, France, Spain, Denmark, Portugal, Poland, Belgium, Norway, and the United Kingdom have introduced programs to foster the progress of electric mobility in their areas. Outside the EU, countries with government-funded EV programs include China, Japan, Australia, and the United States. In particular, China, Japan, and the United States have well-developed plans to encourage the rollout of EVs on a large scale (Ministry of Industry, Tourism and Commerce 2010). The examples of the United States and Belgium are described below to illustrate this momentum.

In 2009, the United States' federal income tax credit for electric cars acquired after December 31, 2009, was increased to US$7500 per car for individuals. This tax credit was limited to the first 200,000 qualifying cars purchased within the United States. Many states provided additional incentives to their citizenry (IRS 2009). The federal government underwrote US$2.6 billion in loans to Fisker, Nissan, Tesla, and other manufacturers under the Advanced Technology Vehicle Manufacturing Loan Program, which supports the establishment of manufacturing facilities for EVs in Tennessee, California, and Delaware, respectively. Moreover, the U.S. DOE provided a grant of US$5 billion to electrify the U.S. transportation sector, appropriating US$2.4 billion to build 30 battery and component manufacturing plants. Construction on 26 of these plants began in 2010 (U.S. DOE 2010).

Turning now to Belgium, as of January 1, 2010, the Belgian federal government introduced specific incentives for citizens who purchased new EVs. Individuals received tax benefits, allowing them to deduct 30% of the purchase price for electric cars (up to €9000) and 15% of the purchase price for electric quadricycles, tricycles, and motorcycles. These deductions directly reduced consumers' income tax bills. Companies could deduct 120% of the acquisition costs for zero-emission cars. In addition, they were able to deduct from their taxable income 100% of the costs of vehicles emitting less than 60 g of CO_2 per kilometer (Belgium Council of Ministers 2009).

Corporations such as FedEx, PG&E, and Nissan Motors have lobbied for U.S. tax credits aimed at increasing the number of EVs in corporate and government fleets to 200,000 by 2015. Executives from these companies created an activist group, called the Electrification Coalition, which promotes the deployment of EVs. The Electrification Coalition also advocates for tax credits for EVs and for the extension of government assistance for charging station construction until 2018 (Hughes 2010). In November 2010, General Electric (GE) announced one of the largest EV commitments to date, promising to convert half of its global fleet by 2015 and started by signing an agreement to purchase 12,000 GM Volt vehicles. GE expects the broader EV market to deliver up to US$500 million in near-term business, derived in part from the rapid deployment of the WattStation, GE's brand of commercial and residential charging options (GE 2010).

10.3 MAJOR ISSUES

Despite governmental support, the growth of the EV market still faces a number of challenges. These challenges are often believed to be intertwined with varying technology developments across different regions of the world. This section presents a review of the main issues that have heretofore been identified.

10.3.1 Technological Issues

The development of battery technology has been identified as the foremost technical challenge impeding full adoption of EVs. The average driving range of these vehicles currently equates to only a fifth to a quarter of that achieved by other vehicle types (Japan Automobile Manufacturers Association [JAMA] 2010). Plug-in hybrid vehicles, such as the Chevrolet Volt, can attain a noncombustion driving range of 30 to 50 miles, whereas fully electric vehicles, like BMW's Mini E and the Nissan Leaf, boast driving ranges of up to 100 miles on a full battery charge. As far as energy storage options for EVs are concerned, the natural choice is the use of lithium-ion batteries, due to their relatively high energy densities and long life spans. However, this type of battery costs approximately US$2000 per kilowatt-hour (as of 2010). Industry experts expect greater demand to decrease costs to approximately US$500 to 700 per kilowatt-hour due to economies of scale. The decreased cost notwithstanding, and assuming the need for a 20 kilowatt-hour battery to achieve an EV driving range of 80 miles, the battery would cost between US$10,000 and US$14,000 (BCG 2009, p. 3).

In the summer of 2009, the Obama administration announced a US$2.4 billion grant program to support the domestic production of EV batteries to decrease production costs. Battery makers have claimed that they could not ramp up production in the absence of a solid supply chain. The program was described as the largest auto-battery investment ever and was to be split among 48 projects (Rahim and Leber 2009). The high density and weight of electric batteries usually results in EVs weighing more than conventional vehicles, which has been associated with longer braking distances in EVs. However, the heavier weight has also been associated with a safety benefit. On average, occupants of heavier vehicles suffer fewer and less serious injuries in car accidents (Ehsani 2004).

This is important, as safety concerns have long influenced regulations related to EVs. In 1958, the United Nations adopted the first international regulation, Regulation 100, on the safety of EVs to ensure that high-voltage electric power train vehicles have the same safety standards as combustion engine vehicles. Typically, EVs operate at very high voltages of around 500 V. This regulation was based on the belief that it was essential to require protection from electric shocks (United Nations Economic Commission for Europe [UNECE] 1958). The European Commission indicated its willingness to incorporate the UN's regulation in its corresponding rules on technical standards for vehicles (European SmartGrids Technology Platform 2006).

Stepping away from the realm of EV-specific technology, we assert that electricity generation is central to the debate about EVs' potential to decrease carbon emissions and for determining the long-term viability of battery-based vehicles. Research provides conflicting results regarding the degree to which EVs can contribute to actual reductions of CO_2 emissions (even though electric systems in EVs have zero emissions, the electricity generation needed to recharge the batteries still emits CO_2; European Federation for Transport and Environment 2009, pp. 18–22). Some preliminary analyses of the global power generation market have shown that in countries where the power mix is very carbon-intensive, such as in India and China, replacing combustion engine vehicles with EVs would have a negligible effect on reducing overall carbon emissions. At the same time, because of the extensive use of renewable and nuclear energy, EVs in Europe are expected to generate 50% to 60% fewer CO_2 emissions (European Federation for Transport and Environment 2009, p. 21). Consequently, it is clear that EVs and renewable energy sources are two inseparable sides of the larger sustainable mobility coin.

Continuing along this thread, much of the debate about EVs' energy consumption is focused on their effect on the electric utility grid (European Federation for Transport and Environment 2009, pp. 31–33). The most effective way to manage EVs' additional energy requirements is to charge them during off-peak hours, when energy demand is lowest. The incentive that utilities envision to achieve this may take the form of time-of-use (TOU) rates. These rates operate on the premise that consumers will be able to lower their energy costs by charging their vehicles during off-peak hours. However, few TOU systems have been implemented to date. To understand how to manage the energy load successfully will require smart meters to provide internal data that allow utilities to implement the TOU rates. An important issue identified with TOU rates is that EV consumption needs to be metered separately. Otherwise, the lower TOU rates would apply to customers' total electricity consumption at the specified TOU times, incentivizing them to modify their lifestyles to run other appliances and home devices off-peak during these preferred rate periods (Mulder 2010). At a household level, experts note that installing an EV charger might change customers' coincident peak—the time during the day when the demand for electricity is highest. The energy used by the charger could also overload the customer's service if all the appliances are in use at the same time (European Federation for Transport and Environment 2009, p. 33). This speculation indicates that EVs present many challenges (and opportunities) outside of the realm of their range limitations and climactic effects.

Among the other various attractive features and advantages, owners may welcome the quieter sound of their EVs. However, as use of electric cars spreads, there is a growing concern that blind people, other pedestrians, and cyclists might not hear these vehicles approaching. Despite the lack of data related to accidents, recent studies have shown that reduced noise is a serious concern. When electric cars run at less than 20 miles per hour, they are especially difficult to hear. The U.S. Congress and the EU are currently exploring initiatives to establish a minimum sound level for EVs. A potential solution currently under consideration is incorporating external sound systems in the vehicles (The Economist 2009).

There are also concerns about the electromagnetic radiation emissions caused by EVs. There seems to be no broad agreement about what level constitutes a health hazard, and no standard has been set for allowable exposure levels. However, the U.S. National Institutes of Health published a study showing that EVs have lower electromagnetic fields than conventional combustion engine vehicles, which is in line with similar findings by the U.S. DOE (National Institute of Environmental Health Sciences 2002).

10.3.2 Institutional Issues

On the regulatory side, there has been no global consensus about the methods to shift energy from petroleum-based to greener alternative technologies. The UN has attempted to unite countries on several issues involving renewable energy and climate change. However, this entity lacks the authority to enforce its regulations and to monitor compliance. Some observers believe that an international regulatory mechanism should be authorized to penalize countries that are unable to achieve emission reduction targets (J.D. Power and Associates 2010, p. 14).

Lacking such a transnational power, the main challenge in the current situation lies in persuading a mix of developed and emerging countries to adopt a convergent plan aimed at promoting and advancing green technologies. The current lack of consistency in regulations across jurisdictions is leading car manufacturers to seek alliances and technology-sharing agreements due to the prohibitive costs of developing multiple power train options. Collaboration among auto companies is necessary to control costs and remain competitive at this stage of the industry cycle. Otherwise, it would cost much more for each major manufacturer to have its own isolated EV R&D projects. Sharing technological breakthroughs and capital costs, therefore, helps to mitigate risk and the high up-front costs of innovation (J.D. Power and Associates 2010, p. 14).

Taking a step into the future now, as EVs increase in popularity, public-charging infrastructure will become essential and may depend on the public sector for funding. Unless the duration of the charging process can be shortened substantially from the current 4 to 8 hours, these vehicles will require an extensive charging infrastructure in residential areas, hotels, cinemas, shopping centers, and other points of interest and public gathering. Although a home recharging unit is considered a viable way to satisfy the demands of most daily commuters, widespread infrastructure will be required for longer trips, such as vacations, in which users are less likely to be in locales where they have control over access to recharging units. According to the Boston Consulting Group, which discussed Germany as an example, if electric utilities were to bear the full cost of building a public charging infrastructure (and assuming an investment amortization period of 15 years), they would have to more than double the price of electricity (BCG 2009, p. 5). In addition, technology breakthroughs might facilitate faster charging mechanisms, resulting in the obsolescence of the initial infrastructure. Thus, it is unlikely that power companies will commit huge amounts of capital to the needed charging infrastructure without powerful government incentives or certainty about the future of the technology (J.D. Power and Associates 2010).

Furthermore, it seems that the public sector will need to educate and incentivize consumers because studies show that some car buyers are not willing to sacrifice performance in exchange for lower CO_2 emissions. A substantial segment of the market seems unwilling to forego the power of combustion engine cars, especially when driving on hilly roads, which put a strain on EV performance. Living in colder climates has also been found to negatively influence consumers' perception of battery-powered vehicle performance due to the greater amount of energy needed to power all of the vehicles functions (J.D. Power and Associates 2010, pp. 15–16).

Finally, research from the largest automotive markets in the world has shown that EV purchasers occupy a well-defined demographic niche. They are generally older, possess graduate degrees, have higher disposable income, and tend to be early adopters of new technologies (J.D. Power and Associates 2010, p. 20). Thus, it is still unclear to what extent EVs will appeal to the broader population.

10.3.3 Economic Implications

Although most of the attention has focused on how EVs will perform on roadways and in the public infrastructure, these vehicles are expected to spend a considerable amount of time at their owners' homes. This gives rise to an entirely different set of issues that must be considered when looking strategically at the future of EVs. For example, the high costs of installation and the additional energy demand stemming from home chargers play key roles in determining a consumer's decision. In addition, regulations must be factored into the analysis, as local authorities may require permits to install the chargers and their required circuit breakers.

Recent global surveys reveal that consumers may be unwilling to pay substantial premiums for EVs unless the initially high investment leads to a total-cost-of-vehicle-ownership advantage. The total cost of ownership is a comprehensive cost measurement calculated over a period of 5 years or longer. On a 5-year basis, assuming a battery cost of US\$700 per kilowatt-hour, combustion engine vehicles are currently cheaper than electric cars when oil prices are below US\$280 per barrel (BCG 2009, pp. 5–9).

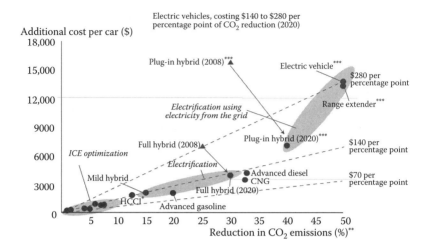

FIGURE 10.5

Forecasted total cost of ownership of an EV under varying assumptions. Each label in italic type represents a set of technologies. *HCCI, homogeneous charge compression ignition. **All CO_2, improvement numbers refer to a base gasoline engine. ***Values are calculated assuming 586 grams per kilowatt-hour (g/kWh) of carbon intensity from power generation. (Reprinted with permission from M. Book et al. *The Comeback of the Electric Car?* Boston Consulting Group [BCG] Report, 2009. http://www.bcg.com/documents/file15404.pdf.)

Figure 10.5 displays the costs of owning an EV based on carbon emission reductions.

However, consumer choices are not driven solely by cost analyses. Emotional factors, such as passions for new technologies or for the environment, can drive a certain proportion of EV purchases, despite their relatively high total cost of ownership. Consumers also consider society's perception of their individual contributions to creating a cleaner environment when deciding to purchase an EV. These factors' influences currently seem weak: although many consumers report having a strong interest in reducing their carbon footprint, that consideration does not yet substantially affect their purchase decisions (BCG 2009, p. 9).

10.4 BATTERY LIMITATIONS AND DRIVING RANGE

The limitations of electric car batteries are at the core of the challenges facing the full adoption of EVs. According to most consumers, driving range represents the major hurdle to adopting this technology. In a recent survey conducted by the research firm Market Strategies, 28% of respondents labeled range or battery life as the most meaningful deterrents to their potential purchase of an EV (Lloyd 2010). Therefore, this section will examine these limitations to analyze the current situation and try to gain insights into the future trajectory.

An EV with a 100-mile driving range incurs a manageable weight and cost penalty. However, such a limited range could not be considered competitive when compared with conventional vehicles with internal combustion engines, especially considering the difference between the aforementioned long recharge time for EVs and the quick, efficient refueling process for conventional vehicles. As shown in Table 10.2, increasing the driving range to 200 miles would make the EV substantially heavier and more costly (European Federation for Transport and Environment 2009).

Because of their deterministic role in the costs and weights of EVs, we will first focus on the characteristics and limitations of electric batteries. Batteries have the potential to overcome their long recharging time limitations through the concept of the battery swap, which has been a significant innovation in overcoming the driving range issue. This method allows drivers to exchange their car's depleted battery for a fully recharged one in a short period at convenient battery-switching stations. The venture capital–backed, California-based company Better Place planned to leverage battery exchanges by implementing a business model whereby customers enter into contracts to purchase driving distances, similar to contracts with mobile phone operators. Section 10.7.2 provides a case study of Better Place's innovative proposition, along with a review about how this plan may have advanced the adoption of EVs across a select group of countries, had the company not filed for insolvency.

TABLE 10.2

Vehicle Characteristics of EVs with Varying Ranges

	Units	100 Miles	200 Miles	400 Miles
Battery energy	kWh	25	48	112
Battery weight	kg	170	320	750
Vehicle weight	kg	1300	1620	2260
Battery cost	US$	6250	12,000	28,000

Source: European Federation for Transport and Environment. *How To Avoid an Electric Shock— Electric Cars: From Hype to Reality.* Brussels, Belgium: Transport and Environment, 2009.

10.4.1 The EV Battery

For several decades, the battery of choice for the automotive sector was based on lead acid chemistry. Due to its ability to provide high currents for a short duration, this battery became an optimal solution for use alongside the internal combustion engine. In addition, it is relatively cheap, with a cost of approximately US$100 to US$200 per kilowatt-hour. Due to the short duration of its available power and heavy weight, however, car manufacturers had to search for more suitable alternatives for powering EVs (Electrification Coalition 2009).

Nickel metal hydride (NiMH) was the first compound to replace lead acid in automotive batteries. This alternative offered a mix of features that was considered superior to its lead acid predecessors across most categories, except for cost. Despite being more expensive, these batteries provided better energy density, leading to lighter weights. The NiMH battery technology was employed in the first generation of EVs, which included the GM EV1 and Toyota RAV4-EV, which were sold in California under the state's Zero Emission Vehicle regulation. The latest advancement in EV battery technology, the lithium-ion (Li-ion) battery, has overtaken all other compounds because of its superior energy density, which results in even smaller and lighter batteries. The most basic component of these batteries is the cell, which can have different designs depending on the type of vehicle chosen and the performance required. Figure 10.6 compares the performance of lead acid, nickel metal hydride, and lithium-ion batteries (Electrification Coalition 2009, p. 59).

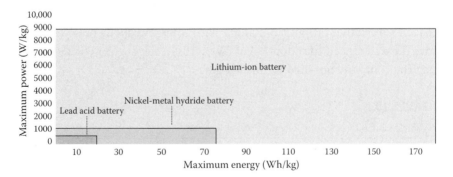

FIGURE 10.6

Battery performance by technology. (Reprinted with permission from Electrification Coalition. *Electrification Roadmap: Revolutionizing Transportation and Achieving Energy Security.* Washington, DC: Electrification Coalition, 2009. http://www.electrificationcoalition.org/sites/default/files/SAF_1213_EC-Roadmap_v12_Online.pdf.)

In electric cars, regardless of the compound used in the batteries, a balance needs to be reached between power density and energy density. The former is needed to provide rapid acceleration, and the latter influences the length of charge depletion. Lead acid batteries reportedly provide a maximum energy density of 25 watt-hours per kilogram (Wh/kg) and power densities of up to 200 to 300 watts per kilogram (W/kg). Nickel-metal-hydride batteries can reach energy densities of 50 to 75 Wh/kg and power densities from 10 to 1000 W/kg. Li-ion batteries provide the most attractive features because they can achieve maximum energy densities of 50 to 175 Wh/kg and power densities of 10 and −9000 W/kg (Electrification Coalition 2009, pp. 30, 77).

Most of the battery research initiatives among car producers currently involve the Li-ion category. Some laboratory results have suggested that these batteries may be able to reach energy densities up to 6 to 10 times greater than those mentioned above. The raw materials in these batteries consist of graphite for the negative electrode (the anode) and lithium compounds for the positive electrode (the cathode; see Figure 10.7 for a full description). The lithium compounds used are most often either derivatives of lithium carbonate (Li_2CO_2) or of lithium hydroxide (LiOH). Nickel and cobalt are often used together with lithium to form the cathodes. Due

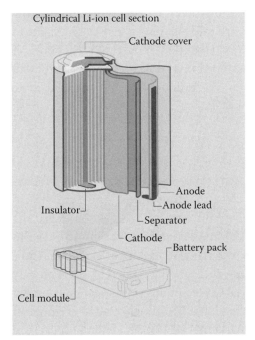

Cylindrical Li-ion cell section

Cathode cover

Anode
Anode lead
Separator
Insulator
Cathode
Battery pack
Cell module

FIGURE 10.7
Lithium-ion battery. (Reprinted with permission from Electrification Coalition. *Electrification Roadmap: Revolutionizing Transportation and Achieving Energy Security.* Washington, DC: Electrification Coalition, 2009. http://www.electrificationcoalition. org/sites/default/files/SAF_1213_ EC-Roadmap_v12_Online.pdf.)

to these three relatively expensive inputs, cathodes account for more than 40% of the final battery price (Electrification Coalition 2009, p. 80).

Of the current demand for lithium, only 20% comes from batteries, which are used mostly in laptops and mobile phones. However, EVs require approximately 100 times more lithium than laptops do. Thus, a large-scale deployment of EVs is expected to put considerable stress on the lithium supply. As the demand increases, both production and recycling capacity are expected to increase to satisfy the spiraling demand (Deutsche Bank 2010). It is also plausible that the increased demand will result in higher prices, which may counteract the cost reductions inherent to the large-scale production of EV batteries.

Lithium can be found naturally in minerals (spodumene), in salt form (brine pools), and in hard clays. Between 1950 and 1985, Australia and the United States dominated its international production. More recently, Chilean production expanded rapidly and currently satisfies 50% of the global demand. The world's largest known deposits are located in Bolivia, but they are not developed commercially because of a number of economic restrictions. A geological survey conducted by the U.S. government indicated that substantial lithium deposits can also be found in Tibet, Austria, Afghanistan, India, Spain, Sweden, Ireland, and Zaire (Electrification Coalition 2009). Most researchers agree that the world does not face a lithium shortage, the expected penetration rate of EVs notwithstanding. Any concerns about lithium shortages are mitigated by its recyclability, one of its key characteristics (Electrification Coalition 2009, p. 80). Therefore, at least in the long term, once sufficient mining capacity has been achieved, increased demand for lithium for EV production will not contribute greatly to increased raw material prices.

10.4.2 Battery Subsidies

The U.S. government has taken steps toward increasing battery production and improving technology by providing grants and subsidies to various corporations. The Obama administration has acknowledged the sluggish adoption of EVs in the U.S. market and has committed itself to increasing the country's presence by setting an ambitious goal of one million plug-in vehicles on the road by 2015. As the United States is already falling behind the latest transportation technology, this accomplishment should help America jumpstart the global EV vehicle era. To accomplish this goal, President Obama, in 2009, launched a US$2.4 billion program to spur the development of plug-in EVs. Of this sum, US$1.5 billion has been allocated to improving battery manufacturing (LaMonica 2009).

The U.S. government distributed the funds by soliciting grant proposals from EV battery makers. These proposals were sent to the DOE, which was then in charge of reviewing the proposals and apportioning the stimulus package supplied by the American Recovery and Reinvestment Act (ARRA). In addition to advancing battery technology, the grants are expected to promote long-term growth, new jobs, and American global leadership in EV technologies. With car manufacturers such as General Motors deciding to develop battery technology in-house for their EVs, these grants could prove essential to the adoption of new technologies. The Obama administration expects that the United States will account for 40% of the battery-manufacturing industry by 2015, compared with the negligible 2% it held in 2010. To achieve this goal, the U.S. DOE distributed grants to battery companies, such as A123 Systems, to improve battery technology and capacity, thereby addressing concerns about range anxiety. Unfortunately, only about 31,000 battery-powered and plug-in EVs were sold in 2012, accounting for 0.28% of all the vehicles sold that year and falling behind schedule on the DOE-funded efforts to get one million plug-in vehicles on the road by 2015 (Harder 2012).

10.4.3 Case Studies of Johnson Controls and A123 Systems

To gain a better understanding of the grant program under ARRA and its progress to date, we will analyze the stories of two grant recipients. In 2009, Johnson Controls, a leader in hybrid and electric battery systems, received a grant from the U.S. government to build domestic manufacturing capacity for advanced EV batteries. This grant, totaling US$299.2 million, was the single largest grant awarded from the US$2.4 billion EV program. Johnson Controls works primarily with various types of hybrid EVs, but the DOE grant specifically targeted the company's battery storage technology. Johnson Controls supplemented this grant with incentives received from the State of Michigan, totaling US$168.5 million, to build a U.S. manufacturing facility for Li-ion cells. The establishment of this battery plant is expected to reduce the in-house costs for many car manufacturers, which are importing batteries from other major battery-producing countries, such as China. This will ultimately reduce the costs and difficulties of producing EVs within the United States. In addition to building the first Li-ion manufacturing plant in the United States, Johnson Controls received US$8.2 million from the United States Advanced Battery Consortium in 2008 to develop Li-ion battery systems for plug-in hybrid EVs. In April 2012, the U.S. DOE awarded the company an additional US$5.48 million contract for Li-ion battery technology development to foster its goal of a "low-cost, long-life, high-power and high-energy vehicle system" (United States Council for Automotive Research 2012).

The support that companies such as Johnson Controls received from the federal and state levels was indicative of the serious push from the United States to ultimately encourage consumer-level demand. The Obama administration aimed to advance the technology of batteries and infrastructure to convince car manufacturers and consumers to invest in EVs. The administration also reached out to consumers by providing a US$7500 tax credit to those who purchased these vehicles. Some critics have noted that supply, rather than demand, is where the government faces its biggest hurdle. U.S. auto companies believe there will not be enough batteries to support the mass production of EVs in the coming years. The DOE has responded by supporting investments that will create a competitive advantage for the United States in advanced battery technology.

However, many of the Obama administration's EV and clean energy investments have faced strong criticism, despite successes such as Johnson Controls. A recent example of the backlash from the administration's green energy projects emerged from A123 Systems' bankruptcy in late 2012. The company, a Massachusetts-based Li-ion battery maker, received a US$249 million grant from the U.S. DOE (Hoffman 2012). It filed for bankruptcy due to low demand for EVs and a failed business deal worth US$465 million with Chinese auto parts supplier Wanxiang Group (Seetharaman and Rascoe 2012). The company had spent US$132 million of the allocated grant and created only 168 jobs, which, when compared with the expected 38,000, represented a major failure. Johnson Controls bought out A123 Systems for US$125 million and will acquire its automotive business assets, automotive technology, and cathode powder manufacturing facilities in China (Zacks Equity Research 2012).

The failure of A123 Systems followed the highly publicized collapse of another DOE investment recipient, solar panel manufacturer Solyndra, which had both politicians and investors doubting the promise of EV and clean energy investments in the United States. In addition, at least US$813 million of the total stimulus that has gone to clean energy programs has been awarded to firms that have since filed for bankruptcy, creating uncertainty as to whether a cleaner power grid will emerge to support increased numbers of EVs (Seetharaman and Rascoe 2012). The next section will address the importance of developing a smarter grid and renewable energy resources to harness the benefits of EVs.

10.5 ENERGY SOURCES AND ENVIRONMENTAL EFFICIENCY

EVs not only ideally release less harmful emissions but also use scarce fuel resources more efficiently than their conventional counterparts. Both of these benefits, when paired with an energy mix comprised mostly of renewables and a smart grid, will lead to increased energy security at a country level and greater environmental benefits on a global scale. EVs are referred to as "zero-emission vehicles" because they do not pollute through tailpipe emissions, fuel evaporation, or fuel transport to service. Hybrid EVs produce lower emissions by leveraging the combination of their electric power trains and internal combustion engines. Pollution from battery and hybrid EVs comes from power plants that supply electricity to the grid. EVs are 90% cleaner than conventional gasoline-powered vehicles when the electricity running them comes from green energy sources, such as wind or solar (U.S. DOE 2003).

However, the United States' main energy source for electrical power is coal, as seen in Figure 10.8. This is detrimental, as American coal plants typically have primary energy efficiencies of between 35% and 40% (after power transmission losses are factored in), that is, if 3 kWh of raw coal is fed into a plant, then only about 1 kWh of usable electricity is obtained. Then, when this energy is fed into a hybrid's or fully electric EV's battery, an additional 10% of the energy is lost, resulting in a final primary energy efficiency from the power plant to the wheels of the EV of approximately 30% to 35%, when the car is powered by electricity from coal. This is still better than internal combustion engines, which have primary energy efficiencies of between 20% and 25%. However, we must always keep in mind that gasoline's carbon emissions are only approximately 75% of coal's carbon emissions for an equal amount of energy generated. This indicates that, given the above primary energy efficiencies, that EVs powered by coal-derived electricity have similar carbon emission intensities as their internal combustion counterparts. On the other hand, if natural gas is used to generate the electricity for an EV in modern combined cycle power

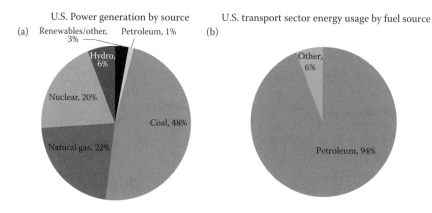

FIGURE 10.8

U.S. power (a) and transport (b) energy generation. (Adapted data from Electrification Coalition. *Electrification Roadmap: Revolutionizing Transportation and Achieving Energy Security.* Washington, DC: Electrification Coalition, 2009. http://www.electrificationcoalition.org/sites/default/files/SAF_1213_EC-Roadmap_v12_Online.pdf.)

plants, emissions reductions of approximately 60% are possible when compared with traditional combustion engines burning gasoline (Greenpeace, European Renewable Energy Council [EREC], and Global Wind Energy Council [GWEC] 2012; U.S. EIA. 2011).

Consequently, clean electricity needs to be generated, not only to reduce the harmful effects of climate change but also to ensure that EV adoption will be environmentally beneficial. Countries are facing increasing pressure to develop alternative energy sources, especially those without adequate reserves of fossil fuels. To address these pressures, new generation technologies are being improved or created to improve energy reliability and decrease environmental effects. Some observers suggest hydro and nuclear power plants—which have already been established as potential methods of generation with nearly zero greenhouse gas emissions—as possible energy alternatives (U.S. DOE 2008). Wind farms represent another recent generation technology that is in its early commercial stages.

Aside from reduced vehicle-related emissions, plug-in hybrids and all EVs use fuel more efficiently than gasoline-powered transportation. DOE researchers estimate that a shift to EVs, given America's current electricity generation matrix, could reduce foreign oil imports by 52% and, for every vehicle-mile traveled, reduce CO_2 emissions by 27% and energy consumption by 30% (Pacific Northwest National Laboratory 2010). However, because EVs are only as environmentally beneficial as the electricity that powers them, environmental advantages will be recognized only if power comes from cleaner energy. Consequently, the lack of a national renewable energy target in the United States and the fact that renewable energy accounts for only 7% of the United States' domestic electricity generation means that EVs are currently suboptimally beneficial (Lowe et al. 2011, p. 15). Conversely, 20% of the electricity in Organization for Economic Co-operation and Development (OECD) Europe is generated from renewable sources (International Energy Agency [IEA] 2010). Based on the EU's electricity mix, EVs consume three times less energy than fossil fuel vehicles (due to more efficient sources of electricity) and emit less than half the CO_2 of internal combustion engines (European Association for Battery Electric Vehicles 2009).

These are only average reduction numbers, however, as the exact global warming emissions and environmental benefits of EVs depend on where they are being charged. This is because of the variety of power plants providing electricity to the grid both across the country and around the world. Germany has a more progressive renewable energy policy, with renewable sources accounting for 23% of the country's electricity needs in 2012 (Radowitz 2012). Thus, EVs charged from the German electricity grid will likely produce lower global warming emissions than average compact gasoline-powered vehicles. In the United States, emissions from charging EVs vary by region. Regions that supply energy to California, parts of New York, and the Northwest have the lowest emission intensities, whereas regions known for their coal production, such as the Rockies, the Upper Plains, and the Midwest (with the exceptions of Dakota and Iowa, which now have disproportionately large shares of wind in their matrices), have the highest emission intensities (U.S. DOE 2003).

10.5.1 Smart Grid Technology

Changing the way electricity is delivered to the grid is essential for increasing access to clean electricity. The "smart grid" refers to the application of information technology that amplifies visibility and control within the grid's infrastructure. The term also refers to the expanded use of communications, sensing, and control systems across all levels of the electricity grid (Pacific Northwest National Laboratory 2010). The European Technology Platform defines a smart grid as an electricity network that integrates the actions of all users connected to generators, including consumers and those who generate and conserve in order to efficiently deliver sustainable, economic, and secure electricity. These grids utilize energy generated at all levels—from large power plants all the way down to individual consumers—allowing all economic agents to play a part in optimizing the operation of the system, whereas providing consumers with increased access to information and power (European SmartGrids Technology Platform 2010, p. 6). The smart grid is characterized not by a specific technology but rather by a distributed system that provides more reliable, secure, economic, and environmentally friendly electricity to the grid.

The goals of smart grids are to achieve a more efficient generation and transmission system and a more liberalized energy market. The objectives of increasing participation for energy consumers, investing in grid renewal in preparation for the next 50 years of operation, and handling grid congestion while reducing uncertainties have propelled their advancement. Another important driver behind the technology is the emergence of national policies, as seen in Europe, that are encouraging lower carbon emissions and the expansion of renewable energy sources (European SmartGrids Technology Platform 2006, p. 11).

The push for a consumer-interactive network within the electric industry is a cornerstone of smart grid technology. Advanced Metering Infrastructure (AMI) integrates electricity consumers with the grid to promote more efficient usage (U.S. DOE 2008, p. 12). AMI differs from traditional automatic meter readers in that its two-way digital communication provides an interactive grid for both power generation and consumption. Consumers can use the AMI's real-time information to adjust their behavior and operate more efficiently. Price signals can be used to help consumers shift their electricity use automatically to periods when power is cheaper. This information can be relayed to "smart" home devices, such as washers, dryers, and refrigerators.

The smart grid's transparency is provided by Phasor Measurement Units (PMUs), which are critical components that reduce peak electricity demand. PMUs record voltage and current multiple times per second at any given location. They are often referred to as the power system's "health meter" because they encourage consumers to modify their patterns of electricity usage, including timing and level of electricity demand, which helps to manage electricity demand at times of peak usage, with the goal of preventing blackouts (Figure 10.9 charts the peak hours of charging and

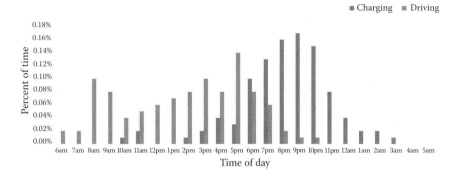

FIGURE 10.9

PHEV driving and charging patterns. (Reprinted with permission from Electrification Coalition. *Electrification Roadmap: Revolutionizing Transportation and Achieving Energy Security.* Washington, DC: Electrification Coalition, 2009. http://www.electrification coalition.org/sites/default/files/SAF_1213_EC-Roadmap_v12_Online.pdf.)

driving for plug-in hybrid vehicles). This measurement system works to ease congestion and bottlenecks by enhancing situational awareness and rapidly reconfiguring the transmission grid to prevent or limit blackouts. In a power system, it is very difficult to predict demand changes at any given point in time. Smart grids address this problem by offering visibility into the power system and making real-time grid response a reality. They reduce the high cost of meeting peak demand by enabling grid operators to control loads in a way that lowers peak demand. Because smart grids are capable of sensing system overloads and redirecting power to prevent potential outages, utility companies can be more confident that the introduction of EVs will not overload the grid.

Incorporating more renewable energies into the grid will also diminish doubts about the environmental effect and grid-straining potential of EVs. The smart grid is essential to the introduction of the newest wave of alternatively fueled vehicles—the plug-in EVs. As noted previously, the increase in electricity demand generated by EVs can be dealt with only through an integrated infrastructure that reroutes power overloads (or by a much more expensive expansion of generation capacity). The smart grid's two-way network can be harnessed to encourage renewable energy and facilitate the adoption of plug-in hybrid and all-electric vehicles (Lowe et al. 2011, p. 9). This is because, in part, renewable energy sources are intermittent, and full-scale adoption is stifled by their unpredictability and storage losses. The smart grid addresses these issues by continuously monitoring and adjusting all of its energy sources to ensure that a steady supply reaches its customers. Grid providers may use power generated from wind or solar farms and require utilities only to supplement these intermittent sources with gas-fired plants when supply is low. This will allow a smarter, more efficient grid to accommodate the 107,000 fully electric vehicles that are expected to hit the road in the United States by 2020 (J.D. Power and Associates 2010, pp. 4–6). Each smart grid will ensure that vehicle charging times are spread out to limit overloads during peak demand hours. More importantly, the two-way connection will also enable consumers to charge their vehicles at night, store power, and sell it back to the grid when needed—making it *less* difficult for utilities to supply energy during peak hours (Lowe et al. 2011, p. 12). This synergistic, mutually beneficial smart grid will thus help address energy storage issues arising from increased renewable usage.

10.5.2 Case Study of the U.S. DOE

Historically, the U.S. electric utility industry has devoted a small fraction of its revenues to research and development, with most utilities spending less than 1% on infrastructure investment. Underinvestment in the electricity grid has caused outages and power quality issues—resulting in five massive power blackouts over the past 40 years (MIT 2011). Since 1982, peak demand for electricity has exceeded transmission growth by nearly 25% annually. Smart grid technologies are expected to reduce power disturbance costs to the U.S. economy by US$49 billion per year. Theoretically, advances in smart grid deployment will reduce the need for massive infrastructure investments by US$117 billion over the next 20 years. As EVs gain in popularity in the United States, infrastructure investments in the smart grid must grow to prepare for the increased load on the electricity grid and to capitalize on the energy storage capacity of idle EV batteries. Making these investments even more necessary is that fact that EV charging has a greater effect on the distribution system due to the vehicles' high energy capacities. With electric cars targeted to replace internal combustion engines in the long term, significant investments need to be made in the electricity grid to support the increase in demand (MIT 2011).

Title XIII of the Energy Independence and Security Act of 2007 established the development of smart grids as a national priority, with the goal of modernizing the nation's electricity grid and maintaining reliable and secure electricity infrastructure that can meet future demand. Furthermore, the ARRA also made significant progress in smart grid deployment by funding the aforementioned US$2.4 billion program to establish 30 manufacturing facilities for EV batteries, funding the deployment of 877 PMUs, and funding US$812.6 million in federal grant awards for AMI deployments (U.S. DOE 2012).

The regional and local penetrations of EVs are critical factors that must be assessed when determining their effect on the electricity grid. For example, demand for plug-in hybrid electric vehicles (PHEVs) and battery electric vehicles (BEVs) in California, Oregon, and Washington are expected to be greater and will thus have a larger effect on the regional utility grid. To compensate for this and to simultaneously mitigate greenhouse gas emissions, the U.S. DOE has begun mandating that state governments shift to energy portfolios with higher shares of clean energy sources (up to 20%). Many states, including New York, Pennsylvania, and Colorado, have pledged to incorporate renewables into their energy supplies by 2020. Other states, such as California, had already agreed to add clean energy into their generation portfolios as early as 2010 (U.S. DOE 2012, p. 25).

In support of these new mandates, public–private partnerships have been created across the United States to supply utilities with renewable energy by utilizing smart grid technology. The Bonneville Power Administration and Mason County Public Utilities District in Washington State partnered with the technology company GridMobility to pilot an adaptive system that incorporates renewable energy sources into the power grid when they are available by synchronizing residential water heaters with wind generation. GridMobility harnesses smart grid technologies to increase the use of renewable energy and decreases power flow to a home's hot water heater when no renewable energy is available and if the family is not expected to need hot water. The heater will switch back on when real-time wind power is available, allowing the utility to harness clean and cheap renewable energy through the interactive electricity grid (National Science and Technology Council 2011).

The DOE has also been investing directly in the smart grid, supplementing the efforts of private–public partnerships. Through the ARRA's US$4.5 billion investment, the DOE aided utilities and state regulators to transition to a smarter grid. The municipal utility of Tallahassee, Florida, received US$8.8 million in smart grid Investment Grant Recovery Act funds to pay for a comprehensive demand–response program. Using the DOE funds, the Tallahassee municipal utility deployed smart grid technologies, including 220,000 smart meters, which improved efficiency and allowed consumers to control their thermostats remotely and pay lower rates during off-peak hours. The city is expected to save US$1.5 million over 15 years by utilizing smart grid technologies to operate the first combined "smart electric, water, and gas system in the United States" (National Science and Technology Council 2011).

10.6 CHARGING INFRASTRUCTURE CHALLENGE

As noted in Section 10.3, a pervasive charging infrastructure, in addition to a smarter grid, has become a precondition for the successful adoption of EVs. Although most battery-charging needs could be satisfied by charging units at home or at offices, public charging points will also be needed to achieve consumer acceptance (Ministry of Industry, Tourism and Commerce 2010). At current retail prices, it is not financially viable for electric utilities to bear the full investment in a public charging infrastructure. In addition, when faced with uncertainty about future demand for charging due to the risks inherent to the EV market, electric utilities may be hesitant to invest heavily in networks of charging stations. In this context, governments and local authorities may be required to play a major role in supporting the development of a suitable charging infrastructure, which would alleviate the need for EV drivers to plan their movements according to their vehicles' charging needs (Electrification Coalition 2009).

This section describes the charging process and the main characteristics and differences between private and public charging infrastructures. The case presented, Source London, is an example of effective local government support in the urban deployment of EVs. This plan sets out how the Office of the Mayor of London and other public sector local authorities plan to reach a short-term goal of having 100,000 EVs in use around London, paving the way for establishing an EV infrastructure network across the United Kingdom.

10.6.1 EV Charging Units

The EV charger connects the vehicle to the electricity grid. Charging stations are technically referred to as Electric Vehicle Supply Equipment (EVSE). Levels of charging vary according to the intensity of power available. In the United States, the Society of Automotive Engineers, the regulatory body responsible for developing automotive standards, defines three standardized charging levels: level I, level II, and level III (U.S. DOE: Energy Efficiency and Renewable Energy [U.S. DOE: EERE] 2008). For level I and level II charging, the actual charger is located within the vehicle itself. Level I charging units contain a cord set that integrates the EVSE and its required safety features into a box connected to the cord, which can plug into traditional 110-V outlets. Levels I and II have the same vehicle interface, but differ in how they are connected to the grid. Level I charging is considered relatively slow, as it can take up to 15 hours to charge the average battery of a pure EV, depending on its initial charge level. Level II EVSE requires a permanent wall mounting that needs to be attached and wired to an electrical panel with 220 V. Charging takes 4 to 8 hours. Most consumers opt for a level II EVSE because of its relatively short charging time. Level III charging, alternatively defined as "direct current," is designed for commercial applications. Through a power range of 30 to 250 kW, this level can achieve a complete charge in less than 10 minutes. The costs of these chargers range between US$25,000 and US$50,000, which is about equal to the cost of a traditional gasoline pump. Fast-charging rates are likely to be limited by the maximum grid infrastructure capability and by battery compound tolerance.

Based on the current technology for batteries, EV propulsion relies on frequent recharging. Therefore, pure EV drivers will require reliable access to charging units for their daily commutes and for all trips that exceed the basic range limit of the EV battery. Depending on the type of plug, the EV's charging time is expected to be in the range of hours rather than minutes. Although it is presumed that there will be a standard plug for all types of EVs, it is still unknown whether every car will be able to utilize all types of charging units. It is also unclear how ownership will be structured for large infrastructure investments, and what rates consumers will be assessed to recharge their EVs (U.S. DOE 2008). Overnight home charging is likely to reduce the need for public charging and to increase consumer confidence in EV deployment projects. Participants in the EV infrastructure market will face the challenge of determining how to provide overnight charging access to consumers who do not have dedicated private parking spaces.

Because most vehicles are idle at night, private charging at home is predicted to become the most common method, especially considering the convenience offered by level II (Mayor of London 2009a, p. 16). At the same time, however, many homes would require the installation of 220-V plugs in their garages or parking shelters, the cost of which can range from US$500 to US$2500, depending on whether electrical panels must be upgraded. Another potential shortcoming involves consumers who rely on street parking and do not have access to dedicated parking spaces. This complication, while not diminishing the importance of home charging, makes a reliable public-charging network a necessary component for achieving successful EV adoption.

This public-charging network will play an enormous role in alleviating consumers' worries. Experts believe it is in the best interests of all the market players to minimize the reticence of consumers concerned about EVs' driving ranges, a phenomenon often referred to as "range anxiety." However, setting up a ubiquitous infrastructure may raise concerns of utilization rates and electricity pricing, factors that may hinder the profitability of private owners operating such infrastructure (Deutsche Bank 2010, pp. 5–6). This doubt about private sector benefit notwithstanding, an extensive public-charging infrastructure will not only reduce range anxiety, but also provide public benefit by helping to avoid overloading the power grid. In the absence of widespread level II or level III chargers away from their homes, consumers are more likely to rely solely on home charging. If all home charging is heavily concentrated in a few hours, there may be a strain on the electric power sector, particularly at the local distribution level, where transformers are located.

Returning to private sector considerations, although EV advocates have encouraged private companies to invest in the public-charging infrastructure, the possibility of a profitable business model that can fully attract private sector investment is still being debated. The high cost of batteries is thought to limit consumers' willingness to pay for electricity to charge their vehicles. If the cost arbitrage with gasoline prices is eroded, it may become impossible for consumers to justify the purchase price of electric cars based on the total cost of ownership. Assuming that a level II EVSE owner can charge a 20% premium on the cost of electricity, achieve a charging unit utilization rate of 20%, and neglect installation costs, the period to repay the investment in public-charging infrastructure is estimated to exceed 25 years. Some critics have claimed that private firms might decide to install charging units as an incentive to attract customers to their businesses or as nonmonetary incentives for their employees—for example, retailers or large restaurant chains such as McDonald's. Experts believe, however, that relying on this type of deployment may result in an irregular network, which would not be sufficient to remove the range anxiety concern (Electrification Coalition 2009, p. 95).

The presence of economies of scale in deploying public-charging infrastructure, such as in parking lots or large areas, may provide opportunities to reduce the overall costs of setting up and operating charging stations. On the other hand, the greater the number of charging units installed within the same area, the lower the expected utilization rate of any one unit (Gartner and Wheelock 2010, p. 4). Most studies on EVs recommend the presence of a separate EVSE provider, such as a specialized firm or an electric utility (Electrification Coalition 2009, p. 98). In either case, firms will be subject to the concerns highlighted above regarding investment payback. The Network Operator model stands as a potential alternative. This business model entails the comprehensive provision of virtually all EV services. The best example of this model to date could be found in Better Place, as the company incorporated the cost of batteries and infrastructure into its own cost structure while charging consumers an all-inclusive fee based on mileage consumption. This business model is analyzed in greater detail in Section 10.7.2.

10.6.2 Case Study of Source London

In May 2009, the Office of the Mayor of London, headed by Boris Johnson, published the *Electric Delivery Plan for London*, which aimed to make London the EV capital of Europe. The plan laid out an integrated strategy to be delivered through partnerships among the central government, the boroughs, and the private sector. It was estimated to require funds on the order of GBP60 million, and the funding burden was to be split equally among the Greater London Authority, the central government, and the private sector (Mayor of London 2009a). The Greater London Authority expected to share a contribution of GBP20 million with the Department for Transport (ibid.). In early 2011, London had a network of 100 charging stations on the streets or in public parking lots. The standard used for these EVSEs is 240 V and 13 amps. The plan aimed to increase the number of charging points to 25,000 by 2015, to be achieved through cooperation with private sector partners (ibid.). In sum, the plan foresaw the installation of 500 on-street public charging units in residential areas, where off-street parking was limited, and 2000 units in public car parks. In addition, the plan aimed to engage with businesses and retailers to build 22,500 charging units in private car parks and customer car parks (ibid.). The charging network will comprise a mix of slow- and fast-charging points. The slow-charging network, similar to level I charging in the United States, is designed for vehicles that spend a considerable amount of time parked at the same location. Fast charging is designed to assist vehicle drivers who may stop for a short period (e.g., 30 minutes) and require a quick charge. The plan also examines the potential for rapid charging, which, like level III charging, allows vehicles to be fully recharged in a matter of minutes (ibid.).

As of late 2010, there were separate membership plans for each of the boroughs that operate existing charging points in London. As the size of the infrastructure expands across the city, the plan is to consolidate these disparate offerings into a single network that can be accessed by all users, irrespective of their borough (Mayor of London 2009b). In December 2009, the mayor's office published *London's Electric Vehicle Infrastructure Strategy*, emphasizing the need for an extensive infrastructure to ensure the successful adoption of EVs. This document detailed the proposed approach to deploying the charging infrastructure envisioned by the *Electric Vehicle Delivery Plan* (Mayor of London 2009b). In addition to reiterating the rationale for the rollout of charging points across London, the mayor's strategy sought to ensure a basic level of charging-point coverage to guarantee proximity to infrastructure and reduce range anxiety (New York City Global Partners 2010). Moreover, this strategy aimed to provide easier access to more convenient charging locations. It also addressed potential local concerns about health, safety, and controlled parking zones (Mayor of London 2009b).

London's EV charging network was expected to go live in spring 2011 under the name "Source London," which would serve as a common, recognizable brand for all the scheme's charging stations. Source London confirmed Siemens as its IT and services partner and had the ambitious target of creating more charging points than there are gasoline stations in the city (Mayor of London 2009a). It promises to deliver 1300 public charging points across the city by 2013, according to the mayor's plan. The network aims to create a visual identity for EV driving and will allow members to charge their vehicles at any of the public charging points, in exchange for an annual membership fee of £100. Additionally, members will be given the opportunity to sign up online and obtain a 100% exemption from the congestion charge that is applied to conventional vehicles that enter the center of London (New York City Global Partners 2010). The development of the Source London program is led primarily by the Transport for London agency, which cooperates with the London boroughs and a number of private sector partners. These partners play an important role in providing funding and locations for charging points (Mayor of London 2009b). The local authorities in London are now in discussions with other

U.K. cities to establish a national network of charging points. London has even offered the use of the trademarked Source brand to these cities, which would facilitate the creation of a unified national network under a commonly recognized identity (New York City Global Partners 2010).

Unfortunately, because of budget cuts resulting from the economic downturn, Boris Johnson announced in late 2010 that he had authorized a cut in the proposed investments for the London EV scheme. The proposed £20 million to be invested by the Greater London Authority was cut to £7 million, leaving a gap that the mayor hoped to fill through the use of corporate sponsorships. The mayor has received criticism for the lack of development of EV infrastructure throughout the city. He has already scaled back the goal of installing 25,000 charging points by 2015 and hopes, instead, to have 1300 charging points across London by the end of the 2013 (BBC 2011).

As of early 2012, there were only 2313 EVs in London, despite the mayor's announcement that there would be 100,000 such vehicles "as soon as possible" (Vaughn 2012). A report released by the Environment Committee in February 2012 analyzed the mayor's progress in making London the "electric car capital of Europe" (Environment Committee 2012). Clearly, EV adoption has been slower than anticipated, as reduced funding and the need for an extensive public-charging infrastructure persist. Source London was launched to improve air quality and reduce carbon emissions throughout London, but the latest report released by the London committee is now questioning the environmental benefits of the program if EV usage is not scaled up and the national grid cannot be powered by renewable energy. In addition to expressing uncertainty about environmental benefits, the committee has suggested addressing public concerns about charging facilities, the up-front costs of EVs, and information about EVs, such as their range capabilities and resale value (Environment Committee 2012).

10.7 BUSINESS MODELS

10.7.1 Overview

Despite increased interest in EVs, industry players need to develop business models that go beyond the ordinary. Demand growth in the initial stages of EV adoption is likely to introduce new players to the market, such as intermediaries who aim to create value at different points in the industry's value chain. In addition, a need for flexible partnerships is expected to appear among players from diverse and cross-functional industries such as media and telecommunications, energy providers, manufacturers, suppliers, and retailers.

As part of this reshaping of the competitive landscape in EV-related sectors, public entities—such as federal, regional, and local governments—are expected to play key roles by providing sizable financial resources in support of the technology. Research has pointed out that concerns about the successful deployment of EVs will increase if governments scale back financial support. Government support will prove especially crucial if oil prices decrease in the near term. Some business models rely on specific oil price assumptions while making comparisons with combustion engine cars. In the case of Nissan, for example, the Leaf business model is based on a US$70 per barrel oil price assumption, a reasonable level that provides confidence when comparing across different propulsion technologies (Electrification Coalition 2009, p. 162). Other companies' assumptions about oil prices may not be so conservative, however.

The development of adequate business models was still receiving little attention from the industry in 2010 (Little 2010). As vehicle manufacturers, utilities, infrastructure and leasing companies, and other service providers compete for consumers' mobility budgets, viable business models for this new landscape still need to be fleshed out. In its study, *Winning the E-Mobility Field,* Arthur D. Little identified four predominant business models. First, assuming the model of mega-manufacturers, on the back of the race for technology leadership and economies of scale, only a few large producers would be likely to survive the large investments required by EV production. Second, with the adaptation of the "Intel Inside" strategy, emerging specialized players would provide more established manufacturers with complete modular packages of EV technology. Third, under the "city-mobility-shop" model, local authorities would be "clean" alternative concepts of local mobility, that is, local authorities and private sector players would cooperate in the provision of EV services. Fourth, in the e-mobility provider model, traditional mobility providers in the rail, aviation, and car rental sectors would seek opportunities in the electro-mobility business by leveraging their established presences to develop the required infrastructure and to cater to affluent travelers who arrive from long-distance trips and seek mobility in their destination cities (ibid.).

The Fraunhofer Research Institute has stated that novel business models are required to transform the technological advantages of electric cars into value-added for end-users. According to the Institute, four main directions have been identified among successful business models: (1) better utilization of vehicle capacity—using innovative mobility concepts such as car-sharing and corporate fleets to extend their user bases at the lower operating costs of electric cars, thus spreading capital costs over a greater number of commuters; (2) extended utilization concepts—using batteries that can be charged cheaply with energy during off-peak times to balance the grid load; (3) secondary usage—using components that are no longer in use in the vehicles, such as the battery, for secondary applications to increase their residual value (e.g., use batteries as energy storage); and (4) increasing acceptance—constraints and restrictions in the driving range can be removed in part by offering so-called "mobility guarantees" (e.g., when purchasing an electric car, occasional use of a combustion engine car for longer journeys) (Fraunhofer Institute for Systems and Innovation Research [Fraunhofer ISI] 2010).

An example of an inadequate business model for the mass deployment of EVs is evident in the Chevy Volt, which was voted European Car of the Year in 2012, just 3 days after its manufacturer, GM, decided to halt production due to a lack of consumer demand (Adner 2012). Although the Volt's success was propelled by government investment and improvements in charging infrastructure, sales still lagged. New and improved models of EVs hitting the market will not likely improve their mass-market appeal for two neglected reasons—battery depreciation and power grid management. Electric car batteries account for a third of the EVs' cost, and current power infrastructure is incapable of handling the mass charging of these vehicles. These factors must be addressed to see a full-scale adoption of EVs.

In the next subsection, Better Place serves as a case study of a business model that addressed the high cost of EV batteries by separating battery ownership from vehicle ownership while building a network of quick battery-switching stations. This model relied on the provision of an integrated EV solution that is conceptually comparable to the service provided by a cellular phone operator, as customers are charged a tariff per mile driven, rather than being charged high costs up-front. Better Place set up a network of charging stations in which batteries could be swapped through an automated mechanical process to alleviate range anxiety (Adner 2012).

10.7.2 Case Study of Better Place

In October 2007, Shai Agassi, a former software executive at SAP, announced the formation of a business entity focused on developing a sustainable electric mobility solution. This business model, called Project Better Place, attracted US$750 million of first-round venture capital funding from Israel Corp., Morgan Stanley, VantagePoint Venture Partners, and a group of individual private investors. The company aimed to deploy infrastructure to support EVs on a country-by-country basis, leveraging battery exchange stations. This infrastructure, which also included electric charging spots at parking locations, was expected to provide consumers with the energy they needed to keep their vehicles charged while eliminating the need to wait for electricity (Deutsche Bank 2010).

To establish this unique infrastructure, Better Place formed a partnership with Renault to produce the only car model—the Fluence ZE—that ran on Better Place's switchable batteries. A Better Place membership provided unlimited access to these fully charged batteries at Battery Switch Stations. The fee also covered the cost for installing private Charge Spots at members' homes and offices. Thus, electricity was billed to Better Place for any private charging. Fixed-battery EVs could also use Better Place charging stations, representing a potential additional revenue stream for the firm (Deutsche Bank 2010).

Moving to the technology that truly characterized Better Place—the battery switch, which was first implemented in Yokohama, Japan—it was fairly revolutionary, as it provided a process for removing and replacing a battery in less than 60 seconds. This was an efficient method of eliminating range anxiety. The swap stations stored and charged multiple types of batteries. The battery swap lanes, where the process was performed, could accommodate various types of current and future battery and EV designs. The battery switch technology utilized an automated track system that run two robotic battery paths. Once parked on one of these tracks, a lift first lowered the car's current battery from its chassis. This battery was then stored away for charging, while a new battery was lifted into the car's now-empty battery compartment. Although seemingly mundane, this process was revolutionary because of the speed at which it was carried out (a typical swap lasted 5 minutes), liberating users from long charge times and their attendant range anxiety (Deutsche Bank 2010).

According to Agassi, his venture was inspired by a question posed in Davos at the 2005 World Economic Forum: "How do you make the world a better place by 2020?" Agassi's answer entailed a pragmatic solution that freed cars from oil, while reducing emissions and transitioning to an era of sustainable mobility. In recognition of his visionary leadership, Agassi appeared on *Time* magazine's list of the world's 100 most influential people in 2009. Among other awards, *Scientific American* named Agassi to the 2009 Scientific American 10, and *Foreign Policy* placed him on its Top 100 Global Thinkers list. In an interview with Wharton School Professor John Paul MacDuffie, Agassi explained how Israel's president, Shimon Peres, helped him turn his Project Better Place vision into a business. At the Young Global Leaders Forum, Agassi was inspired to figure out how to run a country without oil. He presented this idea to President Peres, who grasped the challenge and introduced Agassi to several Israeli government officials and a number of large industrial companies (Knowledge@ Wharton 2009).

These connections proved essential, as Agassi's idea would manifest itself as an unproven business model with many risks. The model Agassi envisioned was comparable to those employed by mobile phone operators, in which operators invest in the deployment of a network of towers to provide coverage in predetermined areas. Customers then pay fees according to their use of the network. Better Place partnered with car manufacturers and sourced batteries with the objective of providing consumers with a cost-effective alternative to their fuel-powered vehicles. According to Better Place's plans, consumers would own their cars, but they would be able to purchase them at subsidized prices and choose from a variety of models (Better Place 2009).

President Peres and Agassi met with Carlos Ghosn, the CEO of Renault–Nissan, at Davos in 2007 and were able to convince him to join their endeavor. It did not take long to convince Ghosn, who had already decided that Renault–Nissan's future was electric. Ghosn was considered somewhat of a maverick because he believed that hybrid vehicles did not make sense in the long term because of their dualized drive train, which resulted in a counterintuitive cost structure. He was quoted as saying, "If we go more electric, let's go all electric" (Knowledge@Wharton 2009). After an agreement with the Renault–Nissan Alliance on the deployment of a charging network for EVs, Better Place began its operations in 2008 in Israel, where the firm unveiled its first plug-in parking lots and installed hundreds of charging spots in Tel Aviv and Jerusalem (Better Place 2012). Agassi stressed the uniqueness of the Better Place business model and the firm's role in galvanizing public opinion about electric cars. The firm conveyed a powerful message in support of energy security to the public, as its mission was to eventually wean countries off of their oil dependence.

Unfortunately, Better Place will never achieve its goals due to its insolvency. To have actually achieved success, Better Place needed to increase its scale, which would have led to high volumes and lower per unit costs. Most of the reasons contributing to the company's bankruptcy revolve around this failure to increase scale, as the company's sales forecasts proved far too optimistic, which resulted in it being unable to cover its infrastructure expenditures. In fact, the company reported a cumulative loss of nearly US$560 million at the time of its liquidation filing (Pearson and Stub 2013).

As stated, a confluence of factors caused Better Place's bankruptcy. Its target was to offer an electric car costing between US$10,000 and US$15,000 (Knowledge@Wharton 2009). However, this would only have been possible at a large scale of production. The company unfortunately only sold approximately 2500 vehicles in total, far short of the number suggested by its initial contract with Renault, through which Better Place had contracted for up to 100,000 EVs (Pearson and Stub 2013). This lack of sales naturally led to major diseconomies of scale, leading to the aforementioned losses. Despite this huge failure, one of Better Place's former CEOs, Evan Thornley, stated that "I continue to believe that the Better Place vision is both accurate and commercially sound, and trust that whatever shortfalls we suffer are correctly seen as errors of execution not of strategy" (Gunther 2013). These "errors of execution" included both external and internal factors, such as more-difficult-than-expected regulatory approval processes for infrastructure in Israel; slow consumer adoption (and greater-than-expected consumer risk aversion in the face of Better Place's new business model); a lack of variety in EV options compatible with the company's charging stations; a lack of faith in the firm because of multiple changes of management; and no government support from Israel, despite EVs' positive externalities (ibid.).

We will now summarize Better Place's actual business model to support the idea that the firm's failure is in no way a death sentence for successor firms. The company owned the batteries and provided a service, which entailed unlimited driving, priced at a per mile basis. To scale up, therefore, the company needed to obtain new customers and to increase the miles driven by existing customers. To satisfy its customers' demands, the company bought kilowatt-hours and batteries to provide those kilowatt-hours through a comprehensive infrastructure. The consumer merely bought "kiloliters" (Knowledge@Wharton 2009).

To distribute these "kiloliters" in a new market, Better Place signed an agreement with DONG Energy to deploy a network infrastructure in Denmark aimed at the mass adoption of EVs powered by renewable energy. As of January 27, 2009, Better Place and DONG Energy had invested €103 million to begin installing EV charging infrastructure. Better Place tapped DONG Energy to supply renewable energy for the network while commissioning Renault–Nissan to provide the EVs. Before abandoning its Australian operations, Better Place partnered with AGL Energy and the financial advisor Macquarie Capital group to raise AUD one billion for deploying an infrastructure for EVs in that country. Although Macquarie Capital's role was to help raise the capital needed for network deployment, AGL Energy's role was to provide electricity from wind and other renewable sources to power the EVs. In 2009, Better Place announced its involvement, along with major Japanese automobile manufacturers, in the first EV project sponsored by the Japanese Ministry of Environment. In addition, Better Place partnered with the government of Ontario to establish a model for EV adoption in Canada. In 2010, as a first step for entering the Chinese market, Better Place signed a memorandum of understanding with Chery Automobile, the largest independent auto producer in China, to form a cooperation regarding the joint development of switchable battery EV prototypes. Furthermore, in October 2010, it announced its involvement in a program supported by the U.S. Department of Transportation to implement a switchable battery electric taxi fleet program in the Bay Area in partnership with the cities of San Francisco and San Jose (Better Place 2012).

Some people questioned whether Better Place's battery-swapping process would be too cumbersome and whether its strategy was financially viable, given the high investments required for building the infrastructure. At the same time, observers acknowledged that Agassi's innovation attracted support from savvy investors. It seems that despite Agassi's impressive backers, the skeptics were indeed correct regarding Better Place's viability. Even Agassi knew that his company required a favorable environment, as when asked about potential scaling into the U.S. market, he responded that, unless there was a willingness to tax gasoline at 3 dollars per gallon, headway into the United States was not possible. Countries with more favorable renewable energy policies, such as Israel and Denmark, were credited with adopting EVs and Better Place's business model more readily. With the presence of financially motivated private companies such as Better Place, the deployment of the EV infrastructure was expected to be accelerated.

In fact, in countries where similar companies are present, such as Israel, experts expect to see a steeper EV adoption curve (Deutsche Bank 2010, pp. 5–6).

Better Place seemed to provide market share growth opportunities for car producers focused on EVs. As part of their alliance, Better Place placed a 100,000 vehicle order with Renault—as an early deployment of EV models could become a driver for significant market share growth (ibid., pp. 5–6). Agassi's venture was considered the most advanced in an emerging group of infrastructure companies that believed EVs could be cheaper than combustion engine vehicles. Their belief was based primarily on the arbitrage opportunity between gasoline and electric driving. To illustrate this point, an analysis conducted by Deutsche Bank estimated that, for average vehicles, the cost of electric driving is US$.025 per mile vis-à-vis US$.089 per mile for gasoline driving. However, conveying this arbitrage to consumers would require that the battery be sold separately from the vehicle. The proponents of this theory argued that the battery was a consumable, much like gasoline, and, as such, it should not be packaged together with the sale of the vehicle (ibid., pp. 5–6).

Although charging stations were established in Israel, Denmark, and Hawaii during 2012, it was then still unclear whether Better Place would be able to take EVs to the next level (Better Place 2012). Its agreement to buy 100,000 swappable battery EVs from Renault by 2016 was already falling significantly short of expectations, with only 500 EVs sold by then. Financial reports released in mid-2012 announced that the company's first half loss had doubled to US$131 million, and losses were expected to continue in the ensuing quarters. Since its founding in 2007, the company had lost US$490 million as of 2012 (Mitnick 2012). In addition, as the company faced mounting losses, Agassi stepped down in 2012, with the CEO of Better Place's Australian headquarters, Evan Thornley, named to replace him. Unfortunately, he stepped down 2 months later. The increasing pressure on the company to raise more capital and generate revenue prompted both switches in leadership. Although Agassi had been credited with achieving the initial successes needed to take the company beyond its start-up phase, investors believed the CEO replacements would scale Better Place's business model throughout Europe (Woody 2012).

Despite these efforts, on May 26, 2013, news agencies worldwide announced that Better Place had filed for liquidation in an Israeli court, rendering the worst fears of its supporters a painful reality (Pearson and Stub 2013; Rabinovitch 2013). With fewer than 1500 cars on the road, with more than US$500 million in accumulated losses, and after exiting once-promising markets such as Australia, the company faced insurmountable scale issues and shareholders decided to liquidate its assets. Unfortunately, Better Place had found itself unable to institute the "paradigm shift" it had hoped for.

10.8 CONCLUSION

The governments of several developed countries have expressed their interest in reducing their dependence on oil because of its highly volatile price, which can create economic and political instability. Renewed global interest in EVs, a potential solution to this problem, is fueled by an increasing awareness of climate change issues and by a variety of factors. New modes of sustainable transportation are sought by governments and embraced by industry players, who are rushing to gain leadership positions in the EV market. Regardless of the uncertainty about adoption rates, car manufacturers are ramping up their production lines and devoting substantial R&D resources to these new vehicles. Governments around the world have introduced tax incentives and other types of subsidies to increase market interest.

Despite the considerable momentum in this sector, some issues pose a threat to the widespread adoption of EVs. The most pressing are the technical challenges associated with battery technology (which creates range anxiety). Other issues include criticism about how electricity is generated for the vehicles, the need to build an extensive charging infrastructure, and finally the choice of a sustainable business model.

The battery is key to widespread adoption of EVs. Its high cost—due to the materials employed in its production—and relatively short range may deter consumers who are considering an EV. Heightened demand for electricity resulting from an increase in EVs on the road poses implementation risks in the market. Nevertheless, the Johnson Controls case study in the United States demonstrates the willingness of governments to support growth in the domestic manufacture of advanced EV batteries and to encourage EV adoption among consumers.

The electricity flowing into these expensive batteries presents its own series of compounding problems. First, the environmental advantages of EVs will be insignificant if electricity is not generated from renewable energy sources. Thus, expanding renewable energy capacity is paramount for deriving environmental benefits from EVs. Second, greater numbers of EVs will require additional electricity, possibly adding to peak demand. Smart grids are increasingly considered a viable solution to ensure that the energy grid does not become overloaded. They allow more transparency and interaction within the power grid, allowing consumers to make smarter and more informed decisions about their energy usage. The U.S. DOE has increased its funding of smart grid technologies and has encouraged state legislators to start integrating them within the next decade.

The installation of an extensive infrastructure has become a necessary precondition for successful EV adoption. At current energy prices, it is considered risky and unprofitable for utility companies to invest in public recharging units. Support from local authorities may prove to be the key to overcoming this obstacle. The Source London case study sets a precedent for a successful initiative that places the capital of England at the forefront of innovation in the urban deployment of EVs.

Finally, despite receiving limited attention to date, the development of viable and sustainable business models is a necessary step for successful EV adoption in an unexplored market. Better Place's innovative model relied on an automated battery-switching process and on revenues derived from a charge per mile driven. The company designed this model so that its per mile charges covered the costs of both the vehicle's batteries and the charging infrastructure.

The bankruptcy suffered by Better Place, however, indicates that its model, although innovative, may not be ideal. Therefore, there is room in the emerging EV market for another business model. As stated above, some of the main issues that cut into Better Place–style, EV-integrated-service firms' profit margins were related to the utilization rate of their costly investments. Increasing utilization would, therefore, facilitate reductions in fixed costs per mile driven, leading to higher profits, because increased utilization would lead to a greater number of miles driven. Revenues increase linearly with this volume of miles driven, resulting in the aforementioned higher profit. This process of decreasing per unit costs is one of the fundamental concepts underlying economies of scale. Achieving such scale is, thus, key to generating profit from lines of business requiring large capital investments.

Consequently, a potential way forward may be to combine Better Place's infrastructure-owning model with a car-sharing business model, with the resultant firm itself owning the charging stations, battery-swapping stations, and EVs. This, at first, might seem counterintuitive, as this company would have even greater capital costs than Better Place, which does not own EVs. Total fixed costs, therefore, would increase! However, providing consumers—who would otherwise not purchase EVs—with ready-to-drive, conveniently located, plug-in EVs could encourage consumer adoption of EV-based mobility because consumers would not have to lay out cash up-front to purchase their own vehicles. Additionally, they could effectively "try-before-they-buy," as they could purchase a short-term membership with the company to see if the offered EV services mesh favorably with their lifestyles before making a long-term commitment. Thus, a new company that would hold title for the EVs, as car-sharing companies do, could encourage greater usage of EVs, in turn increasing utilization of the company's other infrastructure investments. This would simultaneously drive the adoption of EV use and demand for the company's own complementary offerings. The new firm could increase the number of EV miles driven, allowing it to spread its overall larger fixed costs out over an ideally much greater number of billable miles. This indicates that the larger capital investments due to EV ownership might pay off in the mid-to-long run because they could drive greater consumer demand for the firm's products.

That incorporating this car-sharing aspect into the Better Place model would be beneficial seems, at first glance, realistic, considering the recent success of other car-sharing firms. Zipcar, a leader in the industry, for example, expected to realize its first profit in 2012. This, in turn, led Avis, a much larger conventional car rental company, to agree to acquire Zipcar at a per share price premium of 49%. The addition of this fairly successful business model to the Better Place format could, therefore, unlock considerable value potential. Alternatively, it could prove difficult for a start-up to raise sufficient capital by itself to pay for EVs and infrastructure. A joint venture with an established car rental or car sharing player could help remedy this, as a new firm could obtain capital from the venture capital sector and from an established company looking for new growth opportunities. This would also allow the established company to mitigate its risk in the expansion to a new market by sharing the potential for losses with private investors. The established firm would also put itself in a position to easily exploit the success of the joint venture, as it could negotiate terms that give it the right to buy out the other investors.

Business structure notwithstanding, to minimize capital expenditures while maximizing utilization (i.e., maximizing driving time per vehicle), this modification of Better Place's business model would first need to be tested in an urban setting where high population densities would allow each EV to service many consumers. In addition, because of the limited geographic range serviced in an urban setting, fewer swapping stations and charging installations would be needed, making the magnitude of the investment more manageable. An urban setting is also the ideal market for this proposed company because there would be higher demand for relatively short trips than in rural areas, reducing the problems posed by range anxiety. On top of all this, the new firm could study the example of Source London, to learn from its experiences in establishing an urban EV infrastructure. This would mitigate the risks of the venture.

Despite the benefits posed by urban areas, the new firm would need to make its debut in a market in which it possesses a unique competitive advantage, as established car-sharing firms have already begun exploiting the benefits of urban regions. Considering that the new firm would probably not be able to undercut other car-sharers in terms of price, at least in the short-term, indicates that the perfect debut market would have a large population of affluent, environmentally conscious consumers. The new company could capitalize on these consumers' interest in protecting the environment, thereby gaining an advantage over its competitors.

Theoretically, the firm would also be aided in its market entrance by local, state, and federal government agencies, as the company's services would have the positive externality of reducing auto emissions in cities. In the United States, for example, this could lead to the firm securing sizable grants. Governments may be willing to distribute such funds because many—even most—cities, including New York, Philadelphia, Los Angeles, and San Francisco, consistently exceed the U.S. Environmental Protection Agency (U.S. EPA) National Ambient Air Quality Standards (NAAQS) for carbon monoxide (among many other harmful pollutants; U.S. EPA 2012). Conventional automobiles are among the primary emitters of carbon monoxide and also give off numerous other pollutants such as particulate matter, so the public sector should be willing to provide subsidies to reduce these emissions to improve public health.

Considering this potential for public sector aid, the proposed firm could obtain substantial market share in numerous urban markets by displacing conventional automobiles. This significant scale would then give the firm the power to decide which EVs are best suited for this displacement. Consequently, the firm could gain significant negotiating power with a major supplier of EVs, much like the relationship between

Better Place and Renault–Nissan. The firm could leverage this to drive down costs per vehicle, as the currently stagnant sales of EVs have left manufacturers with considerable idle capacity. The firm's business model would also allow it to plan its new car purchases effectively to minimize overcapacity, as it could demand delivery of more EVs only when its membership applications have already increased. In addition, although the firm may need to spend large sums of cash initially to purchase these EVs, these outlays would not necessarily reduce its net income, as these assets would be capitalized and depreciated over multiple years. Thus, the initial outlay would be expensed only gradually in the company's income statement. The firm could also benefit potentially from favorable tax treatment if the public sector grants accelerated depreciation for these assets.

Considering the many strengths of this new model—which in no way guarantees its success—we conclude that unconsidered strategies for accelerating the adoption of electric mobility certainly exist. Furthermore, these models may prove more effective than existing strategies. Nevertheless, as governments continue to press for greater energy security, better battery technology, and lower greenhouse gas emissions, EVs will play a role in their strategies. The finite nature of fossil fuels will supplement these policy goals, thereby creating a powerful case and bright future for EVs.

REFERENCES

Adner, R. Opinion: Can electric cars crack the mass market? *The Wall Street Journal*, May 30, 2012. http://blogs.wsj.com/drivers-seat/2012/05/30/opinion-can-electric-cars-crack-the-mass-market/ (accessed June 8, 2012).

BBC. Electric Car Plan for London "Half-Hearted." Last modified February 14, 2011 (accessed December 29, 2012). http://www.bbc.co.uk/news/uk-england-london-12453270.

Belgium Council of Ministers. New Eco-Tax Measures. Press Release, November 20, 2009 (accessed December 9, 2010).

Better Place. Global Progress. Last modified 2012 (accessed December 22, 2012) http://www.betterplace.com/global/progress.

Boston Consulting Group (BCG). *The Comeback of the Electric Car?* by Book, M., M. Groll, X. Mosquet, D. Rizoulis, and G. Sticher. The Boston Consulting Group, 2009. http://www.bcg.com/documents/file15404.pdf (accessed December 10, 2010).

Deutsche Bank. *Vehicle Electrification* by R. Lache, D. Galves, and P. Nolan. Deutsche Bank Securities, Inc., 2010. http://gm-volt.com/files/DB_EV_Growth.pdf (accessed December 22, 2010).

The Economist. The sound of silence. *The Economist*, May 7, 2009. http://www.economist.com/node/13606446 (accessed January 6, 2011).

The Economist. A sparky new motor. *The Economist,* October 7, 2010. http://www.economist. com/node/17202405 (accessed December 2, 2010).

Ehsani, M. *Modern Electric, Hybrid Electric, and Fuel Cell Vehicles.* Boca Raton, FL: CRC Press, 2004.

Electrification Coalition. *Electrification Roadmap: Revolutionizing Transportation and Achieving Energy Security* by Electrification Coalition. Washington, DC: Electrification Coalition, 2009. http://www.electrificationcoalition.org/sites/default/files/SAF_1213_ EC-Roadmap_v12_Online.pdf (accessed December 28, 2012).

Energy Watch Group. Ludwig-Bölkow-Stiftung. *Crude Oil: The Supply Outlook* by W. Zittel, and J. Schindler. Ottobrunn, Germany: Ludwig-Bölkow-Stiftung, 2007. http:// www.energywatchgroup.org/fileadmin/global/pdf/EWG_Oilreport_10-2007.pdf (accessed December 2, 2010).

Environment Committee. London Assembly. *Charging Ahead? An Overview of Progress in Implementing the Mayor's Electric Vehicle Delivery Plan.* London: Greater London Authority, 2012. http://www.london.gov.uk/moderngov/documents/s8512/ Appendix%201%20-%20Charging%20Ahead.pdf (accessed June 6, 2012).

EurActiv. EU Car Industry Gets "Green" Rescue Plan. Last modified November 27, 2008 (accessed December 3, 2010). http://www.euractiv.com/transport/eu-car-industry-gets-green-rescu-news-220831.

European Association for Battery Electric Vehicles. *Energy Consumption, CO2 Emissions and Other Considerations Related to Battery Electric Vehicles.* Brussels, Belgium: The European Association for Battery, Hybrid and Fuel Cell Electric Vehicles, 2009. http://ec.europa.eu/transport/themes/strategies/consultations/doc/2009_03_27_ future_of_transport/20090408_eabev_(scientific_study).pdf (accessed January 15, 2010).

European Conference of Ministers of Transport. *Assessment and Decision Making for Sustainable Transport.* Paris, France: OECD, 2004.

European Federation for Transport and Environment. *How To Avoid an Electric Shock— Electric Cars: From Hype to Reality.* Brussels, Belgium: Transport and Environment, 2009.

European SmartGrids Technology Platform. *Vision and Strategy for Europe's Electricity Networks of the Future.* Brussels, Belgium: European Commission, 2010.

Fraunhofer Institute for Systems and Innovation Research (Fraunhofer ISI). Working Paper Sustainability and Innovation by D. Dallinger, F. Kley, and C. Lerch. Karlsruhe, Germany: Fraunhofer ISI, 2010. http://www.isi.fraunhofer.de/isi-media/docs/e-x/working-papers-sustainability-and-innovation/WP5-2010_businessmodels_electric-mobility.pdf (accessed March 21, 2011).

Gartner, J., and C. Wheelock. Executive summary. In *Electric Vehicle Charging Equipment.* Washington, DC: Pike Research, 2010. http://urbact.eu/fileadmin/Projects/EVUE/ documents_media/EVCEU-12-Executive-Summary1.pdf (accessed January 23, 2011).

General Electric (GE). GE Announced Largest Single Electric Vehicle Commitment, Commits to Convert Half of Global Fleet by 2012. Last modified November 11, 2010 (accessed December 21, 2012). http://media.chevrolet.com/media/us/en/chevrolet/ news.detail.html/content/Pages/news/us/en/2010/Nov/1111_ge.html.

Greenpeace, European Renewable Energy Council (EREC), and Global Wind Energy Council (GWEC). Cost projections for efficient fossil fuel generation and carbon capture and storage (CCS). Last modified 2012 (accessed May 7, 2013). http://www. energyblueprint.info/1445.0.html.

Gunther, M. Yale Environment360. Why a Highly Promising Electric Car Start-up is Failing? Last modified March 5, 2013 (accessed May 29, 2013). http://e360.yale.edu/feature/gunther_why_israel_electric_car_startup_better_place_failed/2624/.

Harder, A. What's holding back electric cars? *National Journal*, October 22, 2012. http://energy.nationaljournal.com/2012/10/whats-holding-back-electric-ca.php (accessed December 22, 2012).

Hoffman, L. A123 bankruptcy, Clooney and Hubris. *Forbes*, October 17, 2012. http://www.forbes.com/sites/larahoffmans/2012/10/17/a123-bankruptcy-clooney-and-hubris/ (accessed January 28, 2012).

Hughes, J. FedEx, PG&E seek electric-car subsidy for corporate fleets. *Bloomberg*, November 15, 2010. http://www.bloomberg.com/news/2010-11-15/fedex-pg-e-seek-electric-car-credit-to-boost-u-s-fleet-to-200-000-by-15.html (accessed December 14, 2010).

IHS Global Insight. *Battery Electric and Plug-in Hybrid Vehicles: The Definitive Assessment of the Business Case.* Englewood, Colorado: IHS Global Insight, 2010.

Intergovernmental Panel on Climate Change (IPCC). Summary for Policymakers. In *Climate Change 2007: Mitigation. Contribution of Working Group III to the Fourth Assessment Report of the Intergovernmental Panel on Climate Change* by T. Barker, I. Bashmakov, L. Bernstein, J. Bogner, P. Bosch, R. Dave, O. Davidson, B. Fisher, M. Grubb, S. Gupta, K. Halsnaes, B. Heij, S.K. Ribeiro, S. Kobayashi, M. Levine, D. Martino, O.M. Cerutti, B. Metz, L. Meyer, G-J. Nabuurs, A. Najam, N. Nakicenovic, H.H. Rogner, J. Roy, J. Sathaye, R. Schock, P. Shukla, R. Sims, P. Smith, R. Swart, D. Tirpak, D. Urge-Vorsatz, and Z. Dadi. Cambridge, U.K.: Cambridge University Press, 2007. http://www.ipcc.ch/publications_and_data/ar4/wg3/en/spm.html (accessed April 5, 2011).

International Energy Agency (IEA). *Energy Technology Perspectives 2010: Scenarios and Strategies to 2050.* Paris, France: International Energy Agency, 2010. http://iea.org/w/bookshop/b.aspx (accessed December 14, 2010).

IRS. Qualified Vehicles Acquired After 12-31-2009. Last modified 2009 (accessed April 5, 2011).http://www.irs.gov/Businesses/Qualified-Vehicles-Acquired-after-12-31-2009.

Japan Automobile Manufacturers Association (JAMA). Issues and challenges of electric vehicles—JAMA speaks at the 4th Environmentally Friendly Vehicle Conference. News from *Jama Asia 37*, January, 2010. http://www.jama-english.jp/asia/news/2010/vol37/article2.html (accessed December 5, 2010).

J.D. Power and Associates. *Drive Green 2020: More Hope than Reality?* Columbus, Ohio: McGraw Hill Companies, November 2010.

Kampman, B. *Green Power for Electric Cars: Development of Policy Recommendations to Harvest the Potential of Electric Vehicles.* Delft, Netherlands: CE Delft, 2010.

Knowledge@Wharton. Shai Agassi, Israel's Homegrown Electric Car Pioneer: On the Road to Independence. Last modified August 13, 2009 (accessed February 12, 2011). http://knowledge.wharton.upenn.edu/article.cfm?articleid=2315.

LaMonica, M. Obama opens spigot on electric car grants. *CNET*, March 19, 2009. http://news.cnet.com/8301-11128_3-10200328-54.html (accessed November 4, 2010).

Little, A.D. *Winning on the E-Mobility Playing Field.* Arthur D. Little, 2010. http://www.e-connected.at/userfiles/AMG_2010_Winning_on_the_e-mobility_playing_field_final.pdf (accessed December 15, 2009).

Lloyd, J. *E2 (Energy + Environment) Study.* Livonia, Michigan: Market Strategies International, 2010.

Lowe, M., H. Fan, and G. Gereffi. *U.S Smart Grid: Finding New Ways to Cut Carbon and Create Jobs.* Durham, NC: Duke University Center on Globalization, Governance & Competitiveness, 2011.

Massachusetts Institute of Technology (MIT). Inventor of the Week Archive: Charles F. Kettering (1876–1958). Last modified January, 2000 (accessed May 6, 2013). http://web.mit.edu/invent/iow/kettering.html.

Massachusetts Institute of Technology (MIT). *The Future of the Electricity Grid.* Cambridge, MA: MIT, 2011. http://web.mit.edu/mitei/research/studies/the-electric-grid-2011.shtml (accessed December 28, 2011).

Massey-Beresford, H. Carmakers to struggle to beat electric car doubt. *Reuters,* May 12, 2010. http://www.reuters.com/article/2010/05/12/electricvehicles-idUKLNE64B03V20100512 (accessed November 20, 2010).

Mayor of London. *An Electric Vehicle Delivery Plan for London.* May, 2009a. https://www.london.gov.uk/sites/default/files/uploads/electric-vehicles-plan.pdf (accessed December 22, 2010).

Mayor of London. *London's Electric Vehicle Infrastructure Strategy.* London: Greater London Authority, 2009b. https://www.sourcelondon.net/sites/default/files/draft%20Electric%20Vehicle%20Infrastructure%20Strategy.pdf (accessed December 22, 2010).

Ministry of Industry, Tourism and Commerce. *Electric Vehicles Discussion Paper.* Madrid: Ministry of Industry, Tourism and Commerce, 2010.

Mitnick, J. CEO of electric-car network Better Place steps down. *The Wall Street Journal,* October 2, 2012. http://online.wsj.com/article/SB10000872396390444004704578032793162512254.html (accessed December 18, 2012).

Moffet, S. More governments power electric-car development. *New York Times,* October 21, 2009. http://online.wsj.com/article/SB125606654494097035.html (accessed March 11, 2011).

Mulder, D. PowerGrid International. Potential PHEV, EV issues. Last modified August 1, 2010 (accessed November 5, 2010). http://www.elp.com/articles/powergrid_international/print/volume-15/issue-8/departments/perspectives/potential-phev-ev-issues.html.

National Institute of Environmental Health Sciences. *Electric and Magnetic Fields Associated with the Use of Electric Power.* Research Triangle Park, NC: National Institute of Environmental Health Sciences, 2002. http://www.niehs.nih.gov/health/materials/electric_and_magnetic_fields_associated_with_the_use_of_electric_power_questions_and_answers_english_508.pdf#search=electric%20and%20magnetic%20fields%20associated%20with%20the%20use%20of%20electric%20cars (accessed December 5, 2010).

National Science and Technology Council. *A Policy Framework for the 21st Century Grid: Enabling Our Secure Energy Future.* Washington, DC: National Science and Technology Council, 2011.

New York City Global Partners. *Best Practice: Electrical Vehicle Development.* London, U.K.: New York City Global Partners, 2010. http://www.nyc.gov/html/unccp/gprb/downloads/pdf/London_ElectricVehicles.pdf (accessed December 21, 2010).

Pacific Northwest National Laboratory. U.S. Department of Energy. *The Smart Grid: An Estimation of the Energy and CO$_2$ Benefits.* Richland, Washington: Pacific Northwest National Laboratory, 2010.

Pearson, D. and S.T. Stub. Better Place's failure is blow to Renault. *The Wall Street Journal,* May 29, 2013. http://online.wsj.com/article/SB100014241278873238558045785072633247107312.html (accessed May 29, 2013).

Rabinovitch, A. Electric car venture Better Place files to liquidate. *Reuters,* May 26, 2013. http://www.reuters.com/article/2013/05/26/us-betterplace-idUSBRE94P0FU20130526 (accessed May 29, 2013).

Radowitz, B. RechargeNews. Solar Growth Leaves Renewables Set for 23% German Share. Last modified December 18, 2012 (accessed May 6, 2013). http://www.rechargenews.com/news/policy_market/article1301574.ece.

Rahim, S., and J. Leber. Obama admin issues "down payment" on electric cars, batteries. *New York Times*, August 6, 2009. http://www.nytimes.com/cwire/2009/08/06/06climatewire-obama-admin-issues-down-payment-on-electric-26697.html?pagewanted = all (accessed January 5, 2011).

Seetharaman, D., and A. Rascoe. NBC News. Government-Funded Battery Maker Files for Bankruptcy. Last modified October 16, 2012 (accessed December 24, 2012). http://www.nbcnews.com/business/government-funded-battery-maker-files-bankruptcy-1C6500980.

United Nations Economic Commission for Europe (UNECE). UNECE Vehicle Regulations—1958 Agreement (accessed December 15, 2010). http://www.unece.org/fileadmin/DAM/trans/main/wp29/wp29regs/r100a1e.pdf.

United States Council for Automotive Research, LLC. U.S. ABC awards $5.48 million battery technology development contract to Johnson Controls Inc. Last modified April 9, 2012 (accessed December 28, 2012). http://www.uscar.org/guest/news/593/Press-Release-U.S.ABC-AWARDS-5-48-MILLION-BATTERY-TECHNOLOGY-DEVELOPMENT-CONTRACT-TO-JOHNSON-CONTROLS-INC.

Universidad Carlos III de Madrid. 2010. The future of the electric car, analyzed from within the university. Last modified 2010 (accessed December 21, 2010). http://www.uc3m.es/portal/page/portal/actualidad_cientifica/actualidad/reportajes/electric_car.

U.S. Department of Energy: Energy Efficiency and Renewable Energy (U.S. DOE: EERE). *Plug-in Hybrid Electric Vehicle Charging Infrastructure Review* by K. Morrow, D. Karner, and J. Francfort. Idaho Falls, ID: Battelle Energy Alliance, 2008. http://avt.inl.gov/pdf/phev/phevInfrastructureReport08.pdf (accessed December 28, 2012).

U.S. Department of Energy: Energy Efficiency and Renewable Energy (U.S. DOE: EERE). Analysis Shows Significant Advances in Electric Vehicle Deployment. Last modified February 8, 2011 (accessed May 29, 2013). http://apps1.eere.energy.gov/news/daily.cfm/hp_news_id = 290.

U.S. Department of Energy (U.S. DOE). *Just the Basics: Electric Vehicles*. Washington, DC: U.S. Department of Energy, 2003.

U.S. Department of Energy (U.S. DOE). *Smart Grid: An Introduction*. Washington, DC: U.S. Department of Energy, 2008.

U.S. Department of Energy (U.S. DOE). *The Recovery Act: Transforming America's Transportation Sector*. Washington, DC: U.S. Department of Energy, 2010.

U.S. Department of Energy (U.S. DOE). *2010 Smart Grid System Report*. Report to Congress. Washington, DC: U.S. Department of Energy, 2012.

U.S. Energy Information Administration (U.S. EIA). Voluntary reporting of greenhouse gases program: Fuel emission coefficients. Last modified January 31, 2011 (accessed May 7, 2013). http://www.eia.gov/oiaf/1605/coefficients.html.

U.S. Environmental Protection Agency (U.S. EPA). Carbon Monoxide Information. Last modified December 14, 2012 (accessed May 6, 2013). http://www.epa.gov/airquality/greenbook/cindex.html.

Vaughn, A. London slow to become the "electric car capital of Europe." *The Guardian*, February 1, 2012. http://www.guardian.co.uk/environment/2012/feb/01/electric-cars-carbon-emissions (accessed December 29, 2012).

Woody, T. What's behind better place's ouster of Shai Agassi? *Forbes*, March 2, 2012. http://www.forbes.com/sites/toddwoody/2012/10/02/whats-behind-better-places-ouster-of-shai-agassi/ (accessed December 21, 2012).

World Energy Council. *Transport Technologies and Policy Scenarios to 2050*. London: World Energy Council, 2007. http://www.worldenergy.org/documents/transportation_study_final_online.pdf (accessed November 29, 2010).

Zacks Equity Research. Johnson Controls to Acquire A123. *Yahoo Finance*, October 19, 2012. http://finance.yahoo.com/news/johnson-controls-acquire-a123-192919774.html (accessed December 19, 2012).

11

Going Forward: Moving from Perspectives to Action

João Neiva de Figueiredo and Mauro F. Guillén

CONTENTS

The prospects for green energy production and consumption are key to efforts for attaining sustainability. It has become increasingly apparent that we need to adjust many of our individual practices and learn how to live in a way that does not jeopardize future generations' ability to obtain necessary resources for their own livelihood because (among many reasons) any behavior that enhances our present well-being while permanently decreasing the living standards of future generations, unless accompanied by some indemnifying offset, represents a societal contract violation of intergenerational justice. Furthermore, our species' very survival could well depend on our facing and overcoming extensively chronicled threats such as population growth, social inequality, and global warming, among others. These challenges have been analyzed thoroughly from diverse angles in the hope that sustainable solutions to our waste disposal, food, water, and energy problems would emerge. Such solutions have materialized on various scales—from that of the individual to that of society as a whole. Despite these successes, we cannot lessen our efforts until solutions to all these problems have been realized on a *global* scale. This book has addressed the electricity component of this vast, multidimensional equation, focusing specifically on the sustainable generation, transmission, distribution, storage, and consumption options achieved and envisioned by select countries.

The cases presented herein exemplify recent advances in renewable energy systems across a broad sample of geographies, cultures, and technologies. They have examined characteristics and conditions for the success of sustainable electricity generation in several specific instances in the hope that best practices become apparent and perhaps disseminated. It is clear, however, that best practices vary by setting. Indeed, to investigate the factors behind successful advances in green power, it is necessary to understand, on a country-by-country basis, the alignment among natural endowment (a function of climate and geographic location), institutional characteristics (a function of cultural and historic background), and technology. The cases of green power generation and usage depicted here illustrate the need for this three-pronged approach when examining instances of both successful and unsuccessful sustainable electricity deployment.

This was intentional, as the cases were selected with specific objectives in mind. First, they included countries with very different geographic, historical, cultural, and institutional idiosyncrasies to exemplify the many different ways in which green power policies can be implemented successfully. Chapters were dedicated not only to certain countries (namely, Germany, China, Spain, French Polynesia, and Brazil), but also to transnational themes that highlighted the absolute necessity of worldwide cooperation in transitioning to sustainable power generation. Chapters 2 and 10 tackled these global concerns with the aid of the experiences of other countries such as France, Saudi Arabia, Iceland, Canada, the United Kingdom, and Israel. This served the purpose of emphasizing that, although a country may need to tailor a solution to our common energy problems to its unique circumstances, the entire planet does indeed face common, fundamental problems. Second, the book deliberately offered perspectives on a variety of renewable electricity generation technologies such as wind, biomass, solar, and hydroelectricity, detailing their advantages and disadvantages. The goal here was to illustrate the fact that all technologies have pros and cons, and that the key to attaining a truly sustainable energy matrix is to match technologies with a given region's unique natural endowment. Third, the book also included cases that endorse open-minded investigations into as-yet unproven technologies that, if implemented successfully, could provide long-term sustainable solutions for global energy.

We identified the two main problems with electricity as being the fact that it cannot be stored easily and the fact that the negative effects of its generation are felt within the biosphere because it is endogenously generated. The latter challenge was addressed in the forward-looking Chapter 9, which argued both the business and ecological cases for space electricity generation for terrestrial use, that is, the feasibility of generating green power outside the biosphere and therefore exogenously to the system that will use it. The former challenge— that of storing electricity—is linked intimately with the fragmented storage possibilities associated with widespread adoption of electric vehicles and the large-scale implementation of smart grids, as analyzed in Chapter 10.

Despite the optimistic tone struck by some of the cases in this book, the reality is that much work needs to be done to achieve cost-efficient, wide-spread, sustainable electricity generation across the globe. Because the problems are extremely complex and situation-specific, there is no one-size-fits-all solution. On the contrary, the cases point to the need for solutions that are custom-fitted for each specific cultural and physical environment given its demographic, institutional, and geographic make-up. As a result, policy-makers must be willing to experiment with trial-and-error approaches, all the while carefully monitoring results and fostering self-adaptive ongoing adjustments. Top-down regulatory measures, bottom-up grassroots values, and market incentive mechanisms are all integral to positive results; and actions at and across all three levels must be encouraged and coordinated. Each of the many different economic agents involved has a unique set of objectives, and it is necessary to establish a framework for collaboration among those agents. Finally, a very long-term time horizon is needed, especially in the absence of a constituency representing future generations.

In an effort to synthesize actionable messages, we will explicitly enumerate four common themes. First, because of the large number of stakeholders and the risk of unintended consequences, these cases evince the necessity of a truly *systemic approach* when implementing sustainable energy practices. Within this approach, it is necessary to recognize and estimate hidden costs. It is also important, when considering this issue, to take an all-stakeholder–encompassing perspective to recognize that *no costs are external* to the issue of sustainability, that is, when thinking at this holistic, long-term systematic level, every cost, whether apparently internal or external, inevitably inflicts its full share of harm. This approach, therefore, must base itself on a *full cost* accounting framework. Traditional pricing mechanisms are incompatible with such a framework, as market prices may fail to account for unseen and unwanted medium-term and long-term effects that are often difficult to measure. They are thus inadequate. Second, and at a more tactical level, *incentive alignment* among stakeholders over time needs to be achieved through cooperation among regulatory initiatives, market mechanisms, and grassroots values formation. Third, because of the complexity of the issues involved, the cases point to the advisability of *simultaneously instituting top-down and bottom-up approaches.* Fourth, although it is desirable for countries to *tailor solutions* to leverage their natural endowments and cultural characteristics, *cross-border cooperation* will be increasingly important due to the sheer magnitude of some projects and the technological advisability of large-scale electricity grid linkages.

11.1 SYSTEMIC SOLUTIONS

The scope of the sustainability problem demands a systemic approach. Stakeholders with conflicting objectives, the long life cycles of electricity infrastructure, the occasionally narrow focus of traditional pricing mechanisms, and the quintessence of the aforementioned truly *full cost* accounting all demand a comprehensive, systematic approach when addressing sustainable energy options. Because any solution implies a combination of many components, it is not enough for each of them to be the ideal solution for the particular segment of the problem; the aggregate combination of all the components also needs to be considered and optimized. In China, for example, it is necessary to build high-capacity electricity transmission lines to the industrial centers from the northern deserts if vast wind farms are to be built there. Likewise, as European nations learn to take advantage of their respective natural energy endowments, as is the case of solar energy in Spain, integration with the rest of the continent should be at the forefront of the analysis.

Furthermore, because populations, economies, and ecological environments are complex adaptive systems, traditional systems theory may be insufficient to account for all potential outcomes. It will prove necessary to explore nonlinear (and unexpected or low probability) consequences of events when investigating implications in those systems. One uncomplicated, preliminary example of such an analysis was outlined in Chapter 8, which investigated the promise of biomass. The chapter found that, because the logistics cost component of converting biomass to electricity is high, there are limitations to its overall effect on the global energy matrix. Nevertheless, for certain regions, such as Brazil, or for certain applications, such as waste disposal, biomass-based electricity generation prospects are encouraging.

Using complex adaptive systems theory is particularly important when policymakers address bottom-up perspectives. Such grassroots movements have outcomes that are less predictable than top-down initiatives such as regulatory mandates and market incentives. For example, the uncontrollable speeds of dissemination of electric vehicles and of the implementation of smart grids will have mutually reinforcing effects, as more electric cars will create greater incentives for smart grid implementation and vice versa. Both of these developments will also be inputs into the equation of overall electricity system stability. Feedback loops such as this are plentiful in complex adaptive systems, as several examples in the chapters herein demonstrate, and need to be understood from a broad perspective to ensure long-term sustainable electricity generation.

11.2 INCENTIVE ALIGNMENT

The long-term sustainable electricity equation includes lowering consumption, increasing production efficiency, and accounting for hidden costs and externalities in a way that minimizes negative effects on the biosphere today and in the future. Although not the main focus of this book, mechanisms for lowering energy consumption were described in several chapters—most notably Chapter 3—and accounting for hidden costs is implicit in several policy efforts related to feed-in tariffs, as described for Germany, Spain, and Brazil. For these objectives to be reached, together with an increase in green electricity production efficiency, the right formula of incentives targeting all the important stakeholders must be implemented. The process of reaching such an optimal balance depends on many factors: the local geographic and climactic characteristics, the cultural make-up of the population, the main stakeholders' objectives, and the maturity of the relevant country's institutions, among others.

In addition to achieving incentive alignment among the various stakeholders to reach the desired sustainability objectives, it is necessary that policies remain in place over time to provide continuity in the face of an ever-changing environment. Proof of this is the example of Germany, which, over the past 40 years, has succeeded in overcoming administration changes to maintain progress toward a basic set of incentives that has helped turn the German Energy Revolution into reality. The consistency of regulations and of pricing policies over time has led to Germany surpassing its aggressive sustainable electricity generation targets. French Polynesia, on the other hand, may be seen as a counterexample as, during recent decades, there has been no continuity in policymaking. As a result, a country with ample renewable energy resources (e.g., wind, solar, tides, and marine) still depends on expensive imported oil for electricity generation. One common thread among all the successful cases investigated here is the presence of institutional maturity, which ensured both incentive alignment benefiting the greater good and policy continuity over time despite ruling party changes.

11.3 SIMULTANEOUS TOP-DOWN AND BOTTOM-UP APPROACHES

Another common thread among the successful cases of sustainable electricity generation is the simultaneous and synchronized deployment of mutually reinforcing top-down and bottom-up initiatives. This has been seen most clearly in Germany's Energy Revolution, as grassroots actions sprouting from the Germanic culture's link with and admiration for nature, as described in Chapter 3, were coupled with top-down institutional support for renewable energy sources that included legislation and market incentives. Bottom-up approaches are often based on values and on ethical principles, both of which depend on cultural traits and established traditions. However, bottom-up actions can also be encouraged by the right market mechanisms, as exemplified by certain smart grids that can purchase electricity generated by individuals' solar photovoltaic cells and stored in individuals' electric vehicle batteries. Another example of a convergence between a market-driven, top-down approach and a bottom-up desire is the centralized logistics processing of biomass for electricity generation in heavily populated small farm areas. In such settings, animal waste disposal is a major problem for individuals, spurring them to unite in cooperatives for anaerobic biodigestion and electricity generation to exploit top-down economic incentives such as feed-in tariffs, thus creating dichotomous benefits: waste processing and energy production.

Conceptually, there are at least three levers to encourage desired behavior for the greater good: institutional regulations (top-down), market incentives (usually top-down), and collective values (usually bottom-up). Because of the myriad stakeholders, the vast complexity of their interrelationships, and the idiosyncratic natures of each country's situation, careful joint examination of possible regulations, plausible incentives, and existing values is necessary to avoid unintended consequences. Furthermore, because of the dynamic nature of technology evolution and demographic flows, those factors are evolving over time, making the simultaneous consideration of top-down and bottom-up approaches essential. Although this coordination has been observed mostly in small countries such as Denmark, lessons seem to be transferable to larger contexts, as exemplified by the German example alluded to previously.

11.4 COUNTRY-SPECIFIC SOLUTIONS COUPLED WITH CROSS-BORDER COOPERATION

As mentioned frequently, solutions will necessarily be country-specific because of the unique geographic and climactic characteristics, the cultural and demographic make-up of the population, and the maturity and relative importance of institutions. Because China's electricity generation capacity is so heavily dependent on coal, it is unrealistic to presume or expect that coal will be replaced quickly by other sources. Therefore, a significant portion of China's efforts toward sustainability is devoted to cleaner coal technologies, which, although less desirable in other contexts such as those in Canada and Brazil, makes complete sense when more than 70% of electricity generation comes from that fossil fuel. As described in Chapter 5, China is also a leader in other renewable energy technologies such as solar and wind. Spain is another example of the search for country-specific solutions, in that most of the renewable energy efforts in that country have been devoted to solar and wind (available in huge quantities in that country), as described in Chapter 6. It is hoped that French Polynesia will follow this vision eventually, as noted in Chapter 7. However, the fact that solutions need to take local considerations into account and therefore need to be country-specific does not mean that lessons are not transferable. Not only have success stories been repeated elsewhere, such as the implementation of feed-in tariffs in several countries, but also countries have learned by observing lackluster outcomes abroad, such as the negative ecological effects of massive hydroelectric projects.

It seems straightforward that solutions need to be tailored to individual countries' needs, as exemplified by Iceland's primary dependence on geothermal energy. However, in reality, very few countries are so isolated (French Polynesia being one such instance): most share electricity grids to some extent with their neighbors. As a result, although solutions need to be tailored to each country's reality, cross-border energy policy cooperation will also need to develop globally. If this is true for efficiently tapping electricity sources endogenous to the biosphere, it will become a necessity when the scenarios outlined in Chapter 9 materialize, that is, when capital and technology hurdles to electricity generation in space for terrestrial use are overcome. Indeed, whereas the implementation of that technology is expected to solve mankind's energy challenge forever because it will tap a source of electricity exogenous to the biosphere, the large scale of the endeavor and the prerogative of sharing its benefits equally imply a degree of cooperation among nations without historical precedent.

11.5 LOOKING AHEAD

The search for sustainable sources of electricity—which are renewable in the sense that they are used at a pace at most equal to their restoration rate and are green in the sense that they do not generate waste that pollutes the biosphere—encompasses many challenges. These include technological hurdles in generation, adoption that may require collective behavioral changes, and appropriately accounting for hidden costs and externalities, among many others. The cases presented in this book seek to contribute to this quest by offering examples of successful implementation; by sounding cautionary notes regarding failed efforts; and by discussing conceptual alternatives that, although beyond today's possibilities, seem to be within our future collective grasp. Not all societies or civilizations, however, have developed in sustainable ways, and many have declined and disappeared throughout history, as Jared Diamond (2005) has argued. The authors of the case studies in this book believe that although it is possible to find solutions to the present challenges of energy sustainability, only concerted action and political will can bring them to fruition once they become available.

REFERENCE

Diamond, J. *Collapse: How Societies Choose to Fail or Succeed.* New York City: Penguin Group, 2005.

Index

Page numbers f and t indicate figures and tables, respectively.